Into the Unknown Together

The DOD, NASA, and Early Spaceflight

MARK ERICKSON
Lieutenant Colonel, USAF

Air University Press
Maxwell Air Force Base, Alabama

September 2005

Air University Library Cataloging Data

Erickson, Mark, 1962-
     Into the unknown together : the DOD, NASA and early spaceflight / Mark Erickson.
     p. ; cm.
     Includes bibliographical references and index.
     ISBN 1-58566-140-6
     1. Manned space flight—Government policy—United States—History. 2. National Aeronautics and Space Administration—History. 3. Astronautics, Military—Government policy—United States. 4. United States. Air Force—History. 5. United States. Dept. of Defense—History. I. Title.

     629.45'009'73—dc22

Disclaimer

Air University Press
131 West Shumacher Avenue
Maxwell AFB AL 36112-6615
http://aupress.maxwell.af.mil

To Becky, Anna, and Jessica
You make it all worthwhile.

Contents

About the Author

Lt Col Mark Erickson was born and raised on a small farm south of Ogden, Iowa. After graduating from Ogden Community High School, he earned his bachelor's degree from Augustana College, Rock Island, Illinois, in 1985. Commissioned as a second lieutenant in the United States Air Force in 1987, his first assignment as a space and missile officer was to Comiso Air Station (AS), Italy, to serve as a launch control officer with the ground launched cruise missile system. He then served in the same capacity with the Minuteman Intercontinental Ballistic Missiles at Whiteman Air Force Base (AFB) from 1988 to 1992 while simultaneously earning his master of arts in history from Central Missouri State University. After his first tour as an instructor of history at the United States Air Force Academy (USAFA) in Colorado Springs, Colorado, Colonel Erickson completed a doctorate in history at George Washington University before serving at Woomera AS, Australia, as a space warning officer. He returned to USAFA as an assistant professor of history in 1999 and then attended Air Command and Staff College at Maxwell AFB, Alabama, starting in 2001. He was a staff officer and executive officer at US Strategic Command, Offutt AFB, Omaha, Nebraska, from 2002 through 2004 and the chief of safety at the 91st Space Wing, Minot AFB, North Dakota, from 2004 to 2005. In the summer of 2005, he took command of the 326th Training Squadron (Basic Military Training), Lackland AFB, Texas. Colonel Erickson is married to Rebecca (Kessler) Erickson and has twin daughters Anna and Jessica.

Chapter 1

Necessary Preconditions

Between the 4 October 1957 launching by the Soviet Union of the first artificial earth satellite, *Sputnik I*, and the successful American landing and return from the moon in July 1969, the United States sponsored five human-spaceflight programs. The National Aeronautics and Space Administration (NASA) started and completed Projects Mercury and Gemini while its Project Apollo would land on the moon five times before December 1972.[1] Meanwhile, different administrations cancelled the Department of Defense's (DOD) Project Dynasoar in December 1963 and its successor the Manned Orbiting Laboratory (MOL) in June 1969.[2] Therefore, the US Air Force (USAF), as the agency directly responsible for both programs, failed in its attempts to evaluate and use humans in space for military purposes. This book examines the NASA-DOD relationship, with a special focus on these human-spaceflight projects, and the larger context in which this relationship existed. By examining the geopolitical, domestic political, and bureaucratic environments in which decisions concerning these projects were made, the relationships between America's first five human-spaceflight projects will become clear.

The author examines the NASA-DOD relationship in human-spaceflight programs by looking at three issues. First, what was the attitude of presidents Dwight D. Eisenhower, John F. Kennedy, and Lyndon B. Johnson toward the use of space exploration as a tool to secure international prestige and national pride as part of the Cold War struggle with the Soviet Union? While a complete examination of each president's Cold War policies and general beliefs is outside the scope of this work, it is necessary to touch upon the highlights of Eisenhower's, Kennedy's, and Johnson's fundamental perspectives on the Soviet Union and the Cold War. More important, however, is to examine what each man specifically believed concerning the role space exploration was to play in the geopolitical struggle with the Union of Soviet Socialist Republics

1

(USSR), by analyzing each president's pronouncements on such topics as space for peaceful pursuits, human spaceflight, and space for prestige purposes. Each president's specific actions in the field of space policy, human-spaceflight projects, and cooperation with the USSR in space will also be a key piece of the puzzle. In essence, Eisenhower was not at all keen on such a construct; he did not believe the United States should race to the moon in search of prestige. Kennedy believed and reoriented American space policy toward the moon. Johnson continued this lunar landing goal but refused to expand American space policy beyond it as he grappled with the demands of Vietnam and the Great Society.

Second, what institutional relationship existed between NASA and the DOD—the level of support, coordination, and rivalry during each president's term(s)? What specific instances and programs illustrate these dynamics? What role did individual personalities play in this interaction? How did NASA achieve greater independence by lessening its reliance on the DOD over these 12 years? Equally important to the NASA-DOD relationship was the relationship within the DOD between the Office of the Secretary of Defense (OSD) and the Air Force. A level of tension far in excess of any that may have existed between NASA and the DOD resulted from the conflict between Kennedy's and Johnson's secretary of defense (SECDEF) Robert S. McNamara's reluctance to authorize and fund DOD human-spaceflight projects and the Air Force's conclusion that these very programs were necessary to guarantee national security.

The third examination will focus on the actual projects themselves: Mercury, Gemini, Apollo, Dynasoar, and MOL. What was each designed to accomplish and why? Under Eisenhower, Dynasoar and Mercury achieved their initial momentum. The DOD offered critical support for Mercury, but Mercury's capabilities did not seriously endanger the existence of Dynasoar. In Kennedy's administration, the Gemini program was born and matured to the point where McNamara came to view Dynasoar as largely redundant and canceled it in December 1963, a few days after Kennedy's death, and simultaneously initiated MOL. Under Johnson, MOL and Apollo matured, and while MOL maintained a tenuous hold on life as a reconnaissance platform, newly inaugurated president Richard M. Nixon cancelled it shortly after taking office. Therefore, before the decade ended and before the actual lunar landing in July 1969, the Air Force saw both its human-spaceflight projects canceled. Neither project failed due to NASA's urging, rather, there was a complex mixture of financial, political, international, and institutional factors that eventually led to their demise.

The October 1957 launch of the *Sputnik I* "officially" opened the space age. After World War II, the military services of the United States had begun thinking about, and had even taken tentative steps toward, military operations in space; however, contextual factors limited concrete developments. The United States did initiate separate civilian and military reconnaissance satellite programs, which then proceeded at a relatively leisurely pace before October 1957; both were limited by appropriations far below the level of the ballistic missile program. Nevertheless, they had laid the groundwork to support a dramatically expanded post-Sputnik space program for both the military and civilian arenas. In the end it became clear that the quest for reliable reconnaissance of the Soviet Union was the fundamental driving force behind Eisenhower's space programs and policy. This chapter discusses the salient developments in space policy during that leisurely period before October 1957.

Ambling toward Sputnik

The idea that international prestige could be enhanced by space exploration did not appear until the Cold War was in full bloom. Before that, the three significant pioneers in the field of astronautics thought space travel necessary to satisfy the human urge to explore and glean scientific knowledge. Russian Konstantin E. Tsiolkovsky (1857–1935) is credited with the idea of liquid-fueled rockets and the design of reaction-rocket engines. In a 1929 essay, "Cosmic Rocket Trains," Tsiolkovsky proposed the idea of linking rockets together and then sequentially firing them—a concept known today as rocket stages.[3] Robert H. Goddard (1882–1945) was on the American vanguard of astronautical thinking and went beyond Tsiolkovsky and actually fab-

3

ricated, experimented with, and launched rockets. His first successful rocket flight on 16 March 1926 flew for 2.5 seconds, rose to 41 feet, and traveled 64 miles per hour (mph). By the time of his death, near the end of World War II and while working for the US Navy, his rockets had reached speeds of 700 mph and an altitude of 15 miles.[4] Hermann J. Oberth (1894–1989), a German, was the final pioneering thinker in space exploration and foresaw a complex mix of scientific knowledge, commercial potential, and military applications.[5] Yet none of these gentlemen foresaw space exploration as a tool for enhancing national prestige in the Cold War's very competitive geopolitical environment.

World War II had cemented the incipient link between space technology and military applications. The atomic bomb and the ballistic missile were the two most important technological innovations of that war and were soon integrated to form a weapons system, which made possible the emergence of the space age— the intercontinental ballistic missile (ICBM). It was a rocket from Russia's first-generation R-7 ICBM that carried *Sputnik I* to orbit on 4 October 1957. *Explorer I*, America's first satellite, was likewise carried to orbit in January 1958 by a modified Army Jupiter—an intermediate range ballistic missile (IRBM).[6] As the Cold War hardened near the time of these initial satellite launches, prestige-oriented competition made its entry. This followed the post–World War II decade that was not enthusiastic about the development of missiles, satellites, and space technology. The Air Force's reconnaissance satellite was not approved until March 1955, and its budget was limited to $3 million in 1956. Research and development (R&D) funds for ballistic missiles are shown in table 1 and were the necessary precursor for any space program.[7]

Table 1. Research and development funding for ballistic missiles

Year	Prior to 1953	1953	1954	1955	1956	1957	1958
Funding ($ million)	<1	3	14	161	515	1,380	1,349

Adapted from Robert A. Divine, *The Sputnik Challenge*, (Oxford: Oxford University Press, 1993), 29.

Before Eisenhower's first inauguration in 1953, resources had not existed to develop the boosters necessary to put anything into space—mainly because the scientists advising the government did not believe it was possible to create an ICBM. Vannevar Bush, head of the R&D board in World War II and dean of the scientific community advising the federal government, testified to Congress in December 1945 concerning long-range ballistic missiles saying, "I don't think anybody in the world knows how to do such a thing, and I feel confident it will not be done for a very long period of time to come."[8] The early 1950s studies of satellites by organizations such as the Navy's Bureau of Aeronautics (BuAer) and the Air Force's newly created think tank, the Research and Development Corporation, or Project RAND could most accurately be classified as theoretical yearnings of institutions with very little likelihood of being implemented in the near term.[9] More than anything, these studies revealed the degree to which interservice rivalries were emerging and characterizing the formative period of the space age. For example, a 1945 Navy investigation concluded that "in view of the recent progress in the field of rocket missiles it may prove advantageous to review the possibility of establishing a space ship in an orbit above the surface of the earth. . . . This orbit may prove more desirable for communications or for scientific observations."[10] However, when in March 1946, the Navy requested the Army Air Forces (AAF) join its satellite studies, the AAF concluded that a "joint program of evaluation, justification, and, if warranted, construction and operation . . . was not agreeable," and as a result the services "would conduct separate investigational programs."[11] While the Navy's BuAer was conducting its studies, the AAF tasked RAND to study the issue of satellite feasibility. In 1946 RAND conducted technical and engineering analysis and reported that an artificial earth satellite was entirely feasible. RAND believed a satellite would cost $150 million and require five years of R&D but concluded it had neither military nor scientific utility "commensurate with the presently expected cost. . . . No satellite should be built until utility commensurate with the cost is clearly established."[12] As budgets became increasingly stringent, the Navy dropped its satellite studies on 22 June 1948 after the USAF refused to join. The best the Air Force could do was a policy statement in January 1948 stating, "The USAF, as the Service dealing primarily with air weapons—especially strategic—has logical responsibility for the Satellite."[13] James Forrestal, the nation's first SECDEF, had the final word on the early satellite studies, as documented in his first annual report. That 1948 report stated, "The Earth Satellite Vehicle Program, which was being carried out independently by each military service, was assigned to the Committee on Guided Missiles for Coordination. . . . The committee recommended that current efforts in this field be limited to studies and component designs."[14] This first public mention of the military satellite program caused bemused journalists to query, "Will America possess moons of war?"[15] A cloak of silence descended on the subject, and "Satellites were not publicly mentioned again until November 1954.[16] When pressed about its participation in the International Geophysical Year (IGY), the DOD tersely admitted that unspecified satellite studies were continuing."[17] In February 1956, former president Harry S. Truman characterized Eisenhower's civilian Vanguard satellite as a lot of "hooey."[18] His administration had canceled all research into ICBMs in 1947.[19]

The Air Force continued its RAND study efforts throughout the early 1950s, but with a very low level of funding. Increasingly, these studies and the numerous classified conferences that discussed them focused on the use of satellites for overhead photoreconnaissance. "All of them could agree by the early 1950s that the most valuable, first-priority use of a satellite vehicle involved one strategic application: a platform from which to observe and record activity on the Earth."[20] Collectively, these study efforts foretold the two objectives on which Eisenhower's space policy would focus. First, diminish the likelihood of a surprise attack on America by gathering photographic intelligence information on the Soviet Union. Second, establish *freedom of space* as a legal regime in which these reconnaissance satellites could operate. Kennedy endorsed these principles during his administration and added an emphasis on human spaceflight as a prestige-gathering instrument. As Walter McDougall pointed out, "In these few pages the RAND Corporation spelled out the central political problem attending the birth of the Space Age."[21]

The Air Force's ability to act on RAND's recommendations for a reconnaissance satellite was made possible by the June 1950 Communist North Korean invasion of South Korea, which led to a tripling of defense budgets. In late 1951, the Air Force authorized RAND to solicit specific reconnaissance satellite designs in an effort titled Project Feed Back.[22] That produced responses from various defense firms that were documented in a final report in March 1954. In addition, the Eisenhower administration in 1953 had an increased interest in ICBMs. Apparently, there was still enough Air Force and RAND activity to catch the attention of even President Truman. He tasked A. V. Grosse, a physicist from Temple University, to examine the question of satellites. While Grosse's report was not completed during Truman's term, it was presented to Eisenhower in August 1953. Grosse discussed a satellite's scientific research value, its military utility as "a valuable observation post," and its psychological/propaganda value as "a highly effective sky messenger of the free world . . . [that would create a] psychological effect" that must be "considered of utmost value by members of the Soviet Politburo. . . . [Finally] if the Soviet Union should accomplish this ahead of us it would be a serious blow to the technical and engineering prestige of America the world over. It would be used by Soviet propaganda for all it is worth."[23] "The expectation that development of the ICBM was a practical option gave a new impetus to studies on space missions and space vehicles."[24] Gen Bernard A. Schriever recalled that RAND's final Project Feed Back report identified all the support missions (navigation, communications, meteorological reconnaissance, and photoreconnaissance) that satellites could perform.[25] It recommended the Air Force "undertake the earliest possible completion and use of an efficient satellite reconnaissance vehicle" as a matter of "vital strategic interest to the United States."[26] On 16 March 1955, Headquarters Air Force issued General Operational Requirement No. 80, which officially ordered the development of an advanced reconnaissance satellite to provide continuous surveillance of "preselected areas of the earth" in order "to determine the status of a potential enemy's war-making capability."[27] That order officially put the Air Force in the space business, and the reason was reconnaissance.

Nevertheless, the president's budget for fiscal year (FY) 1957 that was passed to Congress included only $3 million for continued R&D of an advanced reconnaissance satellite.[28] That amount was "a major disappointment to all involved, since it was less than ten percent of the amount needed to go to full-scale development."[29] Indeed, before Sputnik, the funding was lean for the military space program, which was principally the [Weapons System] WS-117L—the Air Force's reconnaissance satellite effort. The WS-117L ran into two difficulties: first, the economic policy cutting R&D funds had crippled the project badly, and second, the top officials within the office of the SECDEF and the Air Force showed academic interest but warned that insistence on more funding would create unfavorable repercussions at high political levels.[30] Actual pre-Sputnik funding for WS-117L was limited to $4.7 million in FY 56 and was then decreased to $3 million in the president's budget for FY 57 but increased to $13.9 million through the efforts of General Schriever "pounding the halls" of the Pentagon to secure another $10 million. He finally got it but "with the instructions that we could not use that money in any other way except for component development. No systems work whatsoever. Ten million dollars!"[31] The funding, while still small, grew slightly in FY 58 to $15.5 million—before Sputnik's dramatic launch. Additional "reactive funding increased that amount to $65.8 million."[32] "When Sputnik came along in October, the floodgates opened [and the limit increased even more]."[33] Most observers who described WS-117L funding as insufficient were probably unaware of Eisenhower's entire space policy or his priorities. He wanted a civilian/scientific program to be first into orbit to help establish the legal right of overflight for later military reconnaissance satellites. He also funded a significant secret space effort under the Central Intelligence Agency (CIA) under the recently declassified code name Discoverer.

While Eisenhower's unclassified budgets may not have funded reconnaissance satellites (or even the civilian IGY satellite Vanguard) as robustly as some would have preferred, he did dramatically increase funding for the ballistic missiles, whose dual-use rockets were necessary to launch satellites. When Eisenhower became president, he reportedly had "looked around and said, 'Where are the rockets?'"[34] During his administration, all three services concurrently developed six separate ICBM systems. One DOD official explained such duplication: "We charge it off to insurance—expensive but necessary. . . . But the intense race between the Army and Air Force goes on—and each regards it essentially as a matter of survival."[35] The ICBM fit in well with Eisenhower's *New Look* defense concept. From American soil and using American nuclear-weapon superiority, it could provide more defense for less money—when compared to establishing and maintaining masses of conventional forces. As an added bonus, General Schriever noted that "90 percent of the developments in the ballistic missile program can be applied to advancing in space, satellites, and other vehicles" because it is a "normal transition to step from these ballistic missiles into satellites, moon rockets, going to planets."[36]

The funding for the Atlas, America's first ICBM, increased rapidly from $3 million in FY 53, the first year in which the DOD spent over $1 million, to $161 million in FY 55. By 1957 the overall ballistic missile program had grown to $1.3 billion and included the Air Force Atlas, Titan, Thor, and Minuteman; the Army Jupiter; and the Navy Polaris programs.[37] This growth represented a 433-fold increase during President Eisenhower's administration. National Security Council (NSC) Action No. 1433, 13 September 1955, declared, "There would be the gravest repercussions on the national security and on the cohesion of the Free World, should the USSR achieve an operational capability with the ICBM substantially in advance of the United States. In view of the known Soviet progress in this field . . . the secretary of Defense will prosecute the program with maximum urgency, and all other executive departments and agencies will assist the Department of Defense as required."[38] The ICBM was given the highest priority of all DOD programs. Interservice rivalry relating to missile development and operations led Eisenhower to lament the competitive publicity among the services because it was "highly harmful to the Nation, and thought it should be stopped."[39] The situation led Charles E. Wilson, SECDEF, to declare that the Navy would be responsible for missiles launched at sea (Polaris) and that IRBMs (Thor and Jupiter) and ICBMs would be the sole responsibility of the Air Force.[40] The services continued bickering and after Sputnik's launch, it quickly metastasized into the space roles and missions field, helping convince Eisenhower to create a civilian space agency and assign it the responsibility for the human-spaceflight mission.[41]

Three final crucial events remain from the pre-Sputnik era: the report of the Technological Capabilities Panel (TCP), America's first space policy NSC 5520, *U.S. Scientific Satellite Program*, and the establishment of a civilian scientific satellite program (Vanguard) as a concrete expression of the nascent dual civilian-military space policy. The creation of the TCP rested in Eisenhower's desire to avoid another Pearl Harbor–like surprise attack. Eisenhower's biographer concluded that for Eisenhower's generation, "Pearl Harbor burned into their souls in a way that younger men, the leaders in the later decades of the Cold War, had not."[42] Somehow obtaining the information necessary to detect preparations for such an attack was linked to the need for better intelligence on the USSR's strategic capabilities and intentions. Developments in the late 1940s and early 1950s "included a failure to accurately predict: when the Soviets would develop an atomic weapon, the pace and nature of the Soviet nuclear weapons program and its progress on an hydrogen bomb, North Korea's surprise attack, and a possible 'bomber gap.'" [43] Eisenhower turned to a group of academic and industrial scientists for help with this challenge. On 27 March 1954, Eisenhower tasked his Science Advisory Committee (SAC) to undertake a "searching review of the whole status of our weapons development programs."[44] James R. Killian Jr. was the president of the Massachusetts Institute of Technology and a member of the SAC. Eisenhower asked for a special emphasis on "the present vulnerability of the United States to surprise attack and ways whereby science and technology can strengthen our offense and defense to reduce this hazard."[45] The responsible group became known as the TCP, a brain trust on which Eisenhower would depend for invaluable space- and intelligence-policy advice during the remainder of his administration.[46]

Although relatively unknown, the TCP's February 1955 final report is one of the seminal documents of the Cold War and American space policy. Including its classified annexes, it contained recommendations that led to the Thor, Jupiter, and Polaris IRBMs; the supersecret U-2 reconnaissance aircraft; and support for the development of reconnaissance satellites. In addition, its reasoned analysis of the threat of surprise attack divided the immediate future into four phases and recommended specific actions for each to minimize the risk. It correctly foretold how by phase four, possibly within a decade, both the United States and the USSR would be able to destroy each other, and neither could achieve an advantage in a nuclear exchange assuming one side did not develop ballistic missiles before the other.[47] Its general section on intelligence gathering concluded, "We must find ways to increase the number of hard facts upon which our intelligence estimates are based, to provide better strategic warning, to minimize surprise in the kind of attack, and to reduce the danger of gross overestimation or gross underestimation of the threat. To this end, we recommend adoption of a vigorous program for the extensive use, in many procedures, of the most advanced knowledge in science and technology."[48] Quite simply, "The TCP report of 1955 set the pace and direction of American strategic policy for years to come."[49]

Recently declassified documents illuminate the central role the TCP report played in codifying the civil-military bifurcation in American space policy. The TCP's space-related recommendations for general policy stated, "Freedom of Space. The present possibility of launching a small artificial satellite into an orbit about the earth presents an early opportunity to establish a precedent for distinguishing between 'national air' and 'international space,' a distinction which could be to our advantage at some future date when we might employ larger satellites for intelligence purposes."[50] The TCP endorsed the idea that the primary purpose for satellites was reconnaissance and intelligence gathering. However, it also stressed that a small civilian or scientific satellite should pave the way by serving as a "stalking horse" to establish the precedent of the legal right of overflight before the military reconnaissance satellites that would come later.

The second important pre-Sputnik legacy was the creation of America's first space policy. Donald A. Quarles was a key figure in developing and documenting that policy. Beginning in September 1953, he served as the assistant secretary of Defense for R&D; then as secretary of the Air Force (SAF) from August 1955 to April 1957; and finally, until his death in 1959, he filled the position of deputy secretary of Defense. He carefully examined the TCP's report and its recommendations and was privy to more closely held U-2 information. He endorsed those recommendations and drafted what would become NSC 5520, "Draft Statement of Policy on U.S. Scientific Satellite Program," and submitted it to the NSC.[51] It emphasized the necessity of establishing the legal right of overflight and became the official American policy concerning the use of space for reconnaissance when President Eisenhower endorsed it on 27 May 1955.

The policy recognized that "a small scientific satellite will provide a test of the principle of 'Freedom of Space,'" but cautioned that the small scientific satellite must not hinder the development of a larger surveillance satellite. NSC 5520 also touched on the prestige factor that would become associated with space during the Cold War. It recognized that "prestige and psychological benefits will accrue to the nation which first is successful in launching a satellite." Therefore, the DOD sought to develop a small scientific satellite by 1958 "with the understanding that this program will not prejudice continued research directed toward large instrumented satellites for additional research and intelligence purposes, or materially delay other major Defense programs. . . . [and] does not involve actions which imply a requirement for prior consent by any nation over which the satellite might pass in orbit, and thereby does not jeopardize the concept of 'Freedom of Space.'"[52]

The outlines of the new American space program were clear: a civilian scientific satellite program would be initiated under the IGY to gather scientific information about outer space and to establish a legalized regime for satellite overflight. However,

12

this civilian effort would not be allowed to impede the military's reconnaissance satellite effort or other high-priority DOD programs, that is, the ballistic missiles.[53] It should be noted that when Eisenhower approved NSC 5520, he referred it to the SECDEF for implementation "in consultation with the Secretary of State and the Director of Central Intelligence."[54] A month later, the CIA reported "a proposal to undertake a small satellite program in connection with the International Geophysical Year and for propaganda and scientific purposes has been presented to the NSC Planning Board by Department of Defense, Central Intelligence Agency and Department of State representatives." The document explained the CIA's interest in terms of "the psychological warfare value of launching the first earth satellite makes its prompt development of great interest to the intelligence community and may make it a crucial event in sustaining the international prestige of the US. There is an increasing amount of evidence that the Soviet Union is placing more and more emphasis on the successful launching of the satellite."[55] Given the primacy of reconnaissance and intelligence concerns in the new space policy, CIA involvement from day one was not surprising. By mid-1955 not only had the principles of long-term US space policy been established, but a threefold organizational structure of the IGY program-DOD-CIA emerged and foreshadowed the structure into which it eventually evolved: National Aeronautics and Space Administration (NASA)-DOD-National Reconnaissance Office (NRO).

Vanguard, the IGY satellite program and the final legacy of pre-Sputnik developments, began with the DOD playing a central role and the CIA maintaining a shadowy and at-the-fringes type of presence. In fact, the beginnings of the Vanguard project took place during the same time frame in which the TCP report was released and Eisenhower approved NSC 5520, making these months some of the most momentous in the American space program.

The origins of the IGY go back at least to 5 April 1950, when geophysicist James Van Allen gave a small dinner party for some of his colleagues that included, among others, Sydney Chapman, Lloyd Berkner, and S. Fred Singer. They concluded that there was a scientific need for numerous, simultaneous observations at many points around the earth to enable conclusions to be drawn about the earth as a whole. Over the next several years, this core group of American scientists gradually incorporated the IGY concept into numerous and diverse scientific conferences and succeeded in achieving near-unanimous support. A period of maximum solar activity from July 1957 to December 1958 was selected to be the duration of the IGY.[56] Soon scientific objectives were incorporated into geopolitics, the TCP report, and NSC 5520. At every step of the way, this cadre of American geophysicists, with DOD support—particularly Quarles—ensured the IGY satellite was an agenda item that received prompt domestic and international attention.[57] The NASA historian noted that "the fingerprints of these core leaders are all over every decision relative to [the] IGY satellite program and the US decision by Eisenhower to sponsor a satellite."[58] The government was able to trust these leaders to establish the right of overflight—its real concern. That strategy allowed it to focus its attention on operations behind the scenes and still enjoy the luxury of defining Vanguard as civilian and scientific in nature. By March 1955, the scientific concern started to merge with the military. The two most important scientific officials in this process, Alan Waterman (director of the National Science Foundation [NSF]) and Detlev Bronk (president of the National Academy of Sciences [NAS]), regularly attended NSC meetings and were privy to discussions concerning the TCP. When Joseph Kaplan, the chairman of the United States National Committee for the IGY, wrote Waterman on 14 March 1955 to explain that the US IGY representatives felt "a small, approximately fifty-pound, earth-circling satellite . . . would yield new geophysical data of considerable interest" and recommended the US government include such vehicles in its rocket program, Waterman passed this suggestion on to Deputy Undersecretary of State Robert Murphy. He explained that the US IGY committee had been considering "at the suggestion of the assistant secretary of defense for research and development [Quarles] the feasibility and scientific importance of inclusion in the United States program of the launching of a small satellite. . . . Accordingly, in consideration of the interests of the Department of Defense and other agencies in this subject, and because of its importance from a public and international relations standpoint, Dr. Bronk and I wish to discuss . . . the initiation of such steps as may be necessary in arriving at the position of the Government with respect to this matter."[59]

An initial key step in the IGY program was selecting the person and organization that would be responsible for managing the production, assembly, and launching of its satellite and launch vehicle. Not surprisingly, Quarles was given this responsibility not long before he became SAF. The selection of an organization was fairly constrained, given the NSC admonitions to not let the civilian-scientific satellite interfere with the military reconnaissance satellite, ICBM and IRBM development, or any other high-priority defense projects and that the IGY satellite effort should appear as civilian and scientific as possible. The Air Force entry was rejected because it was based on the Atlas rocket, and they could not guarantee that it would not interfere with IRBM and ICBM efforts. The Army's proposal was likewise rejected because it was based on using the rocket from the Redstone and Jupiter IRBM programs and also carried the additional baggage of being a modification of the Nazi V-2 rocket. The Navy's entry, however, was based on using Viking sounding rockets that were operated by its Naval Research Laboratory (NRL) for atmospheric research. That proposal was a better fit with the NSC's concerns, and in August 1955, the NRL was selected to develop the Vanguard satellite for the IGY.[60] A comment by one member of the selection committee provided insight into the future of the space age: "We finally decided that breaking the space barrier would be an easier task than breaking the interservice barrier."[61]

The IGY satellite was officially announced on 29 July 1955 and emphasized its civilian pedigree: "This program will for the first time in history enable scientists throughout the world to make sustained observations in the regions beyond the earth's atmosphere."[62] Press Secretary James Hagerty explained, "The only connection the Department of Defense will have with this project is actually getting these satellites up in the air."[63]

Vanguard's budget, under Eisenhower, mushroomed between August 1955 and the October 1957 launch of *Sputnik I*—but was still not large enough to please its proponents. In the summer of 1955, Vanguard's original budget estimate was

$20 million. Eisenhower permitted numerous supplemental appropriations that increased its total cost to over $110 million by the time of its completion but complained about the "very costly instrumentation" on Vanguard. He emphasized that "the element of national prestige, so strongly emphasized in NSC 5520, depended on getting a satellite into its orbit, and not on the instrumentation of the scientific satellite."[64] Even with this growth, Vanguard's backers still could not conduct an all-out, competitive race with the Soviets to capture international prestige because of Eisenhower's insistence that it not interfere with the priority military projects and not receive unlimited funds. Vanguard's funding came mostly from DOD's emergency funds, with supplements from the NSF, and additional contributions from the CIA—cementing its role as a participant in the early American space program.[65] Clearly the question of Eisenhower and space for prestige is not a simple black and white matter but rather deals with shades of gray. At a seminal May 1957 NSC meeting, CIA director Allen Dulles emphasized that "if the Soviets succeeded in orbiting a scientific satellite and the United States did not even try to, the USSR would have achieved a propaganda weapon which they could use to boast about the superiority of Soviet scientists." Secretary of State Christian Herter concurred with the sentiment to continue with Vanguard "because of the prestige it would confer on the United States." Eisenhower said he "did not see how the United States could back out of the earth satellite program at this time," but he "was much annoyed by this tendency to 'gold-plate' the satellite in terms of instrumentation." Wilson summarized, "The satellite program had too many promoters and no bankers." In the end, Eisenhower requested a supplemental appropriation specifically for Vanguard. In August 1957, Congress provided $34.2 million, and Vanguard had its own source of funds until NASA took over the project in October 1958.[66]

During this period, the prestige factor of being first to launch an earth satellite was not considered important enough by Eisenhower to merit granting Vanguard an open-ended budget or permitting it to interfere with either the top-priority missile programs or the military reconnaissance satellite. However, in July 1955, the NSC established an ad hoc working group on information aspects of NSC 5520 as a subdivision of the NSC called the Operations Coordinating Board (OCB). This group regularly defined one of Vanguard's purposes as deriving "the maximum psychological advantage obtainable for the United States through domestic/foreign information output as generated by the US decision to launch earth satellites." Despite numerous staff meetings and reports highlighting the potential competitive aspects of the Soviet and American IGY satellite programs, Eisenhower did not conceive of the Vanguard program as engaging in any kind of a *race* with the USSR.[67] The charge that Eisenhower ignored the space-prestige angle before Sputnik is not accurate. Indeed, over the course of 1957 when Eisenhower accepted increases in Vanguard's cost, Congress was in the midst of a drive to lower the national budget. Far from frugal, his defense budget was "the largest peacetime request in the history of the United States," and Congress threatened to cut $2 billion from Eisenhower's request for missiles and aircraft.[68] With this budgetary backdrop, Eisenhower did accept the original IGY satellite proposal: a fivefold increase in its budget and a top priority for Vanguard—just below ballistic missiles—a status no other research project enjoyed. He did not, however, write a blank check for Vanguard, which is what would have been required for an all-out prestige race with the Soviet effort. While the machinery of the executive branch repeatedly emphasized the program's competitive and prestige-related aspects, Eisenhower did not wholeheartedly subscribe to it and made only limited concessions.[69] This single reservation was sufficient to permit the Soviets to launch a satellite first. However, after the national and international reaction to the October 1957 Sputnik launch, Eisenhower was forced to reassess and accept the importance of the aspect of national prestige, which had been tied to space operations.

While the Eisenhower administration was cognizant of the prestige value of being first into space, there were other considerations that might help explain why the race was not joined. As explained by McDougall, either of two sets of circumstances could prepare the way for military reconnaissance satellites:

"One was if the United States got away with an initial small satellite orbiting above the nations of the earth 'for the advancement of science'—and had no one object to it. The other way was if the Soviet Union launched first. The second solution was less desirable, but it was not worth taking every measure to prevent."[70] By way of example, few dispute that the Army's Jupiter, the first US rocket with orbital-launch capability, could have placed a satellite into orbit in 1956. During its development testing, it had, in fact, carried a sand ballast to represent future payloads. But as Eisenhower explained, the DOD and the NSF "showed little inclination either to drop Vanguard, already well under way, or to divert the Redstone group from missiles to satellite work. Since no obvious requirement for a crash satellite program was apparent, there was no reason for interfering with the scientists and their projected time schedule."[71]

The collective "space-for-peace" policy was an effort to gather scientific information about space and the upper atmosphere with a civilian-scientific satellite; a satellite that would, if it were also the world's first satellite, establish a right of passage for later military reconnaissance satellites. This policy "constituted the intellectual medium in which the program took shape during its early years" and clearly dominated the pre-Sputnik space policy of the Eisenhower administration.[72] While the idea that superpowers might compete for prestige was not unknown in the Eisenhower pre-Sputnik space policy, it was not a prime mover. More important was the expectation and "hope that international agreements would recognize some specific distance above the earth as analogous to the three-mile limit [at sea], beyond which there would be freedom of space comparable to freedom of the seas."[73]

NASA's Predecessor Organization and the DOD

Between 1908 and 1913, the United States spent only $435,000 on aviation development, less than nations such as Japan, China, Bulgaria, Greece, and Brazil. As a result, when

World War I began in 1914, the United States had only 23 military aircraft, all technologically obsolete, when compared to France with 1,400, Germany with 1,000, Russia with 800, and England with 400.[74] In the wartime environment, most European governments encouraged their scientists, engineers, and governments to further aeronautical R&D, but the United States lagged where airplane development was left to "a host of amateur inventors."[75] Some prominent Americans began to see this backwardness as "not only a national disgrace, but [also] a possible danger to our security."[76] Backers of an American national aeronautical laboratory included Smithsonian Institution secretary Charles Walcott and Alexander Graham Bell. Their efforts were stymied until the crisis environment of World War I served as a catalyst.

The National Advisory Committee for Aeronautics (NACA) was the nucleus from which NASA was formed. Its history is intimately tied to the military, forms an important foundation stone of the civil-military story of the space age, and is important to the understanding of pre-Sputnik developments. Its founding legislation was attached as a rider to a 1915 naval appropriations bill, "a piece of legislation assured of passage, what with the war in Europe and the bipartisan support then abounding for a strong Navy."[77] From this time and until it was transformed into NASA in October 1958, the history of the NACA and its R&D was closely tied to national security and the fortunes of the military services. From a first-year budget of $5,000, NACA's appropriations swelled to $85,000 by the end of World War I.[78] In the period before World War II, the NACA become mainly a research organization working on questions raised by its primary clients: the growing American aircraft industry but especially the War and Navy Departments. For instance, one of the NACA's premier accomplishments in the interwar era was the invention of a cowling that provided superior cooling for radial aircraft engines. But, Alex Roland points out, "What is less well known is that the military services had been the first to ask the NACA to investigate cowling of radial engines." This 1926 request, "Like all requests from the military . . . was assigned a research authorization and work began on a prototype."[79]

This close relationship with the military greatly assisted the NACA in justifying its existence and securing funding during the Great Depression—during which time its budget fell by one-third. The aircraft industry was still relatively small during the interwar period; in 1929 its expenditures were less than the sales of straight pins.[80] Overall, the NACA remained "obscure, humble, and poor," with peak peacetime funding of only $3.1 million in 1940.[81] It was not an operating agency in the sense of conducting missions or actual flights. The NACA did research on aircraft loaned by the military or industry and owned no aircraft. It had no contracting authority, "received its meager funds through military appropriations, and most of its facilities were co-located at military air bases." Its total budget during the period from 1915 to 1940 was $31 million.[82] NACA facilities, such as wind tunnels, and NACA research on topics such as laminar flow, retractable landing gear, and all-metallic aircraft structures were indispensable in the development of the World War II military aircraft. George Lewis, NACA's director for aeronautical research, often remarked, "If the NACA ever sets itself aside from the Army and Navy, it is a 'dead duck.'"[83]

During World War II, the NACA "worked for the military essentially on a support basis" and exchanged personnel, facilities, and equipment almost casually. Between 1938 and 1945, its staffing grew from 480 to 5,453, and funding jumped from $1.28 million to $40.9 million.[84] NACA engineers and scientists used their wind tunnels and other research equipment to create new aerodynamic theories and solve specific problems with particular aircraft. "The military services and aviation industry took the job from there and designed and produced the airplanes."[85] The NACA's basic wartime task was "testing, cleanup, and refinement of military prototypes [for] immediate use in the war."[86] NACA executive secretary John Victory declared in 1943, "All of the research activities of the National Advisory Committee for Aeronautics are connected with immediate and vital problems of the Army and Navy air organizations."[87]

Former NASA historian Roger Launius explains that World War II transformed the NACA "from a sleepy R&D organization

20

created to experiment and solve the problems of flight for the military, the civil aviation industry, and the airlines . . . to a much larger institution that, after 1939, was more firmly wedded to military aviation." He continues, "Relations between NACA and the military had always been amicable, but they became especially so after wholesale changes on the committee reoriented it toward acquiescence in military prerogatives."[88] NACA chairman Jerome Hunsaker said that by Pearl Harbor, 71 percent of NACA work was on specified military projects.[89] Therefore, "Without NACA, American aerial supremacy, won and held at least by the first part of 1944, would have been less complete. Every airplane that fought in the war was tested and improved in NACA laboratories."[90]

After World War II, the NACA evolved along a slightly more independent line while continuing to work closely with the DOD in propulsion research (the famous rocket-powered X-series aircraft in which Chuck Yeager and his successors broke, and then flew well beyond, the sound barrier, as well as the more down-to-earth jet-engine research), perfecting aircraft and ballistic missile designs.[91] By 1949 the Bureau of the Budget (BOB), the predecessor to the Office of Management and Budget (OMB), shifted NACA's budgetary classification from "Transportation and Communications" to "National Defense." The BOB had concluded that all of NACA's growth in the previous decade "had been based entirely on military considerations" and "all NACA officials agree that the primary mission of the agency for the foreseeable future was military in nature."[92]

For instance, the NACA's H. Julian Allen in 1951–52 discovered a solution to a serious problem associated with ICBMs: how to deal with the high temperatures generated by aerodynamic heating during reentry. In place of a sleek rifle shell configured with a sharply pointed nose, he proved the efficacy of a blunt-body shape designed to build up a powerful, bow-shaped shock wave that deflected the heat safely outward and away from the reentry vehicle's main structure. This slightly curved, blunt-body design was incorporated into America's first-generation ICBMs (until ablative reentry materials were perfected) and into NASA's later Mercury, Gemini, and Apollo space capsules. After World War II, the NACA did enough research into missiles and rockets to merit the establishment of the Pilotless Aircraft Research Division at Wallops Island, Virginia. NASA officials later characterized the NACA's missile work as consisting of "studies of basic problems in aerodynamics, structures, and propulsion . . . undertaken often on request of the sponsoring military services. . . . Its [NACA's] function is to provide fundamental scientific information that will be useful to the military services and to manufacturers in the design and development of missiles of superior performance." By January 1950, and using a very liberal definition of applicability, the NACA reported that approximately 30 percent of its research effort was applicable to missiles.[93]

Before Sputnik, the NACA was not enthusiastic about doing space-related R&D. While it continued to make excellent progress in aeronautics, "Space flight, however, was something else."[94] When NACA chairman Hunsaker was informed in 1940 by the Guggenheim Aeronautical Laboratory, California Institute of Technology, of the military's interest in rockets, he replied with a distinct lack of interest, "You can have the Buck Rogers jobs."[95] Christopher Kraft was a long-time NACA employee who would become famous as director of NASA's Flight Operations. He recalled that before Sputnik, the word *space* was considered a dirty word in the NACA, and it "wasn't even allowed in the NACA library. The prevailing attitude in the NACA at that time was that if it was anything that had to do with space that didn't have anything to do with airplanes, then why were we working on it?"[96] Robert Seamans would serve as associate and then deputy administrator of NASA, as well as SAF, but was on an NACA subcommittee in 1948 that openly asked what the NACA was doing to prepare America for possible space activity. He reported, "We had our wrists slapped. We were told that the NACA was for aeronautics, period. Forget space."[97]

Gen James H. Doolittle was appointed in 1956 to replace Jerome Hunsaker as the chairman of the NACA. Hunsaker had concluded that he was personally ill equipped by temperament and training to cope with the new technologies and challenges of the space age.[98] Doolittle had an earned doctorate in aeronautics from the Massachusetts Institute of Technology, was a recipient of the Medal of Honor for leading the raid on Tokyo, rose to the rank of lieutenant general in the AAF during World War II, and was a successful executive with the Shell Oil Company. He explained that the NACA leaders had "sort of dragged our feet" about astronautics. "We knew the rocket was coming [and] while we knew that space must be explored, we were hesitant to turn over to the missile people and their supporters all of the funds that we had been receiving for the development of the airplane and associated equipment."[99] He admitted in hindsight that the NACA was wrong on this count but emphasized,

> We in the old NACA were I think mentally circumscribed, to the extent that we never could have realized the potential of growing not six times but sixty times bigger in a short period of time, because we had fought very hard each year in order to get the little increases that we needed in order to build up over a period of a great many years to $100 million a year. . . . It was only that we began to take quantum steps when we began to get quantum bucks. . . . NACA, like every other governmental agency, had to fight every year for its appropriations. It never got what it wanted to do its job, and frequently it got appropriations on the basis of "You will use it for this and nothing else."[100]

The combination of directed appropriations and a constrained fiscal environment meant the NACA was only too happy to leave space exploration to the Air Force with its reconnaissance satellite and to the NRL's Vanguard program. In turn, this facilitated a continued smooth relationship between the two organizations because the DOD enjoyed NACA's responsiveness to its research requests with missiles, and the DOD did not feel that the NACA had any desire to poach on the new and potentially glamorous field of space R&D. After Sputnik, the NACA realized that its institutional existence depended on being named the organization responsible for America's civilian space exploration program. The NACA then employed the most liberal definitions of space-related R&D to claim that 50 percent of its activities were "space related."[101] McDougall's assessment was more objective and stated, "By the mid-1950s, the venerable NACA was slumping."[102]

The DOD and Air Force would probably have been happy to see the status quo of their relationship with the NACA continue after Sputnik. Schriever stated that the NACA "worked extremely well with the military and commercial sides. There were no sandboxes, no jealousies among the organizations. It was a happy family."[103] Others closely familiar with the relationships between the NACA-DOD and the NACA-Air Force concur. An admiral who later headed NASA's Office of Defense Affairs stated that during the 43 years before NASA, the NACA and the DOD enjoyed "a very harmonious and productive relationship. . . . The relationship was a simple and direct one, generally devoid of any contest in roles and missions."[104] Additional praises, such as "the long history of NACA's relationship with the military has been the relationship of a trusted supplier to an active orderer," could occupy many pages.[105] With Sputnik's repeated beeping, all this would soon change.

Notes

(All notes appear in shortened form. For full details, see the appropriate entry in the bibliography.)

1. Apollo hardware and its Saturn family of launch vehicles were also used for the three Skylab missions in 1973 and the Apollo Soyuz Test Project in 1975. *NASA Pocket Statistics*, 1996.

2. The word "Dynasoar" is alternately rendered as "Dyna-Soar," "Dyna Soar," and capitalized versions of all three. It is a contraction neologism of "dynamic soaring" created by the Air Force.

3. Tsiolkovsky, *Works on Rocket Technology*, 95, 250.

4. Goddard, "A Method of Reaching Extreme Altitudes," 5; and Oberth, *Ways to Spaceflight*, 93–96, 363–71, 515–16.

5. Oberth, *Ways to Spaceflight*, 93–96.

6. IRBM and ICBM boosters launched most civilian payloads. Redstone rockets were used for the unmanned Mercury flights while manned Mercury flights were flown on the Air Force's Atlas ICBM. All Gemini flights were launched using the Air Force's Titan ICBM rocket. The Saturn family was developed for the Apollo program and became the first NASA-manned space booster not to be taken directly from the Air Force's stable of ICBMs. Even the Saturn program had some ballistic missile heritage since it was started by the Army Ballistic Missile Agency, which had previously developed the Redstone. The Saturn was officially transferred to NASA in 1960.

7. Divine, *The Sputnik Challenge*, 29.

8. Van Dyke, *Pride and Power*, 10. There was little support for developing state-of-the-art technologies to meet a requirement already being met by existing weapons systems—American heavy bombers—in the post–World War II environment. Maier, "Science, Politics, and Defense in the Eisenhower Era," xxxiii.

9. Hall, "Early U.S. Satellite Proposals," 410–34.

10. US Navy, "Investigation of the Possibility of Establishing a Space Ship in an Orbit above the Surface of the Earth," 1–2.

11. Joint Research and Development Board, "Minutes of 4th Meeting," 6 March 1947, 2–3. That meeting summarized the events of 1946, as well as recounting progress in early 1947.

12. Douglas Aircraft Corp., *Preliminary Design of a World-Circling Spaceship*, 236–39. This is the report of the Technical Evaluation Group of the Committee on Guided Missiles of the Research and Development Board, 29 March 1948. The National Security Act of 1947 established the Air Force as a separate service on 18 September 1947. Bowen, *An Air Force History of Space Activities, 1945–1959*, 36.

13. Logsdon et al., *Exploring the Unknown*, vol. 2, 272. In addition to once again pointing to incipient interservice rivalry in this new field, the practical impact of this policy declaration was that RAND received a low level of funding to continue its investigation into the feasibility of various types of satellites and the missions they could perform.

14. Forrestal, *First Report of the Secretary of Defense, 1948*, 129.

15. Futrell, *Ideas, Concepts, Doctrine*, vol. 1, 541.

16. Hungerford, *Organization for Military Space*, 8.

17. Bulkeley, *The Sputniks Crisis and Early United States Space Policy*, 80.

18. Ibid., 83.

19. Three sources provide an excellent overview of the American ICBM program's roots and development. See Schwiebert, *A History of U.S. Air Force Ballistic Missiles*, for an overview; Neufeld, *Ballistic Missiles in the United States Air Force, 1945–1960*, for emphasis on the administrative, technical, and logistical aspects of the first generation of ICBMs; and Beard, *Developing the ICBM: A Study in Bureaucratic Politics*, for details about the bureaucratic, cultural, and political environment in which the ICBM matured.

20. Hall, "Origins of U.S. Space Policy," 215. RAND had anticipated the use of civilian satellites in their studies, and in 1958 Eisenhower used *Vanguard I* (the first solar-powered satellite and the second artificial satellite successfully placed in orbit by the United States) as part of the US exhibit to the 1957–58 IGY. The satellite was six inches in diameter, weighed only three pounds (lbs.), and carried scientific experiments that for seven years measured the size and shape of the earth, air density, temperatures, and impacts of micrometeorites. That was when its solar cells failed and its transmissions stopped—it continues to orbit and 2005 marked its 47th year in orbit. Its most notable accomplishment was to help establish the overflight precedent and establish freedom of space as a legal regime. Kecskemeti, *The Satellite Rocket Vehicle*, 1–5, 14, 17, 20–22.

21. McDougall, *The Heavens and the Earth*, 109. Similar conclusions were reached about collecting weather data from a higher orbit of approximately 350 miles: "The development of all the suggested methods mentioned in this report appears to be feasible" and so "the analysis of synoptic weather from satellite observations is also feasible." Greenfield and Kellogg, *Inquiry into the Feasibility of Weather Reconnaissance from a Satellite Vehicle*, 31.

22. Klass, *Secret Sentries in Space*, 76. Klass claims that Project Feed Back was "indirectly sponsored by the CIA." The author of this book was unable to discover any primary source evidence or documentation to support this assertion.

23. Grosse, *Report on the Present State of the Satellite Problem*, 266–69.

24. Davies and Harris, *RAND's Role in the Evolution of Balloon and Satellite Observation Systems and Related U.S. Space Technology*, 47.

25. Schriever, oral history interview with the author, 2 July 1996.

26. Davies and Harris, *An Analysis of the Potential of an Unconventional Reconnaissance Method*, 53. This final report stated that developing such a satellite would require approximately seven years and $165 million; it would have a resolution capability of approximately 144 feet from 300 miles while scanning a strip of land 375 miles wide.

27. Ibid., 61. The Air Force opened a design competition code named Pied Piper between the RCA Corporation, Glenn L. Martin Company, and Lockheed Aircraft. On 29 October 1956, Lockheed received the development contract. From this point, the program was generally referred to as WS-117L, and "The military satellite program was now committed to development and testing of actual satellites." Augenstein, *Evolution of the U.S. Military Space Program, 1945–1960*, 275.

28. Richelson, *America's Secret Eyes in Space*, 13.

29. Day, "CORONA: America's First Spy Satellite Program," *Quest: The History of Spaceflight Magazine*, 9.

30. Bowen, *An Air Force History of Space Activities, 1945–1959*, 48.

31. Neufeld, ed., *Research and Development in the United States Air Force*, 88.

32. Davies and Harris, *RAND's Role in the Evolution of Balloon and Satellite Observation Systems*, 95.

33. Schriever, "Comments," in Needell, ed., *The First 25 Years in Space*, 28.

34. Harlow, oral history interview, 11 June 1974, 6.

35. Gale, "Post-Sputnik Washington from an Inside Office," 228.

36. Futrell, *Ideas, Concepts, Doctrine*, vol. 1, 545.

37. Eisenhower, *Waging Peace: 1956–1960*, 208.

38. NSC 6108, "Certain Aspects of Missile and Space Programs," 1.

39. Goodpaster, Memorandum of Conference (Memcon) with the President, 30 March 1956, 2.

40. Wilson, SECDEF Memorandum (Memo), 307–11.

41. Launius, *A History of the U.S. Civil Space Program*, 16.

42. Ambrose, *Eisenhower*, vol. 2, 257.

43. Hayes, "Struggling Towards Space Doctrine," 53.

44. Killian, oral history interview, 9 November 1969–16 July 1970, 14.

45. Technological Capabilities Panel (TCP), *Meeting the Threat of Surprise Attack*, vol. 1, 14 February 1955, v. Document is now declassified.

46. Killian, *Sputnik, Scientists, and Eisenhower*, 68.

47. TCP, "Meeting the Threat of Surprise Attack," final report (U), 1–38. (Top Secret) Information extracted is unclassified. TCP suggested dispersal of the American bomber force, extension of the distant early warning line, and numerous research and development projects.

48. Killian, *Sputnik, Scientists, and Eisenhower*, 79. Unfortunately, the entire part 5 of the TCP report, which dealt with intelligence gathering and includes the entire space- and satellite-related sections, remains classified. Other excerpts can be found cited in documents, created by associated governmental agencies, such as the NSC. Eisenhower was pleased that not one leak was associated with the TCP and grew to trust the scientists associated with it, especially Killian. That trust enabled Eisenhower, after Sputnik's launch, to task Killian with creating an organizational structure for a space program that would remain under civilian control (ibid., 67, 86).

49. Roland, *SP-4103 Model Research*, 280. Roland discusses the TCP in the context of NACA's history and explains why NACA's budget stabilized in the mid-1950s after falling for several years. Roland credits that stabilization to Eisenhower's increasing concern with American scientific and technological progress.

50. NSC 5522, *Comments on the Report to the President by the Technological Capability Panel of the Science Advisory Committee*, 8 June 1955; and Eisenhower, *Waging Peace*, 209.

51. Watson, *The Office of the Secretary of the Air Force, 1947–1965*, 162.

52. NSC 5520, "Draft Statement of Policy on U.S. Scientific Satellite Program," 2–4, 6, 11. Its principles, along with Kennedy's space policy modification to include landing on the moon, are applicable to the entire period covered by this book. The SPI file also includes recent declassified material.

53. Eisenhower, *Waging Peace*, 209. Eisenhower recalled, "We were careful to keep the earth satellite program separated from the Defense Department's work on long-range ballistic missiles. . . . It was not to interfere with our top priority work on missiles" (ibid).

54. NSC Memo of Discussion, 250th Meeting of the NSC, 26 May 1955, 2. These memoranda of discussions often had subsequent actions appended to them.

55. NSC 5522, *Comments on the Report to the President by the Technological Capabilities Panel*, 8 June 1955.

56. Holmes, *America on the Moon*, 46–47. Previously in 1882 and 1932, International Polar Years had taken place, and Berkner recalled making the spontaneous suggestion for a third.

57. Bulkeley, *Sputniks Crisis*, 89–131.

58. Launius, "Eisenhower, Sputnik, and the Creation of NASA," 131.

59. Emme, "Presidents and Space," 17. Document is now declassified.

60. Green and Lomask, *Vanguard*, 30–55. They describe the complicated process by which the Advisory Group on Special Capabilities, eight civilian scientists appointed by Quarles, selected a specific project and contractor for the IGY satellite.

61. Ibid., 51. Clifford Furnas, chancellor of the University of Buffalo, is quoted. He would take Quarles's place as assistant SECDEF for R&D when Quarles became SAF in August 1955.

62. Hagerty, "Statement by White House Press Secretary," 29 July 1955.

63. White House, "Question and Answer Session," reprinted in *New York Times*, 30 July 1955, 22.

64. Glennon, *FRUS, 1955–1957*, vol. 11, 748–49.

65. Day, "Invitation to Struggle," 245.

66. Glennon, *FRUS*, vol. 11, 750–53.

67. NSC, "Public Information Program with Respect to the Implementation of NSC 5520," 1.

68. Eisenhower, *Waging Peace*, 209.

69. NSC, "Pentagon Briefing Memo on Earth Satellite Program," 1.

70. McDougall, *The Heavens and the Earth*, 123–24.

71. Eisenhower, *Waging Peace*, 209. A Jupiter-C launched on 20 September 1956 had the propellant in its fourth-stage engine replaced with sand by direct order of the OSD. York and Greb, "Strategic Reconnaissance," 39. This missile flew to an altitude of 682 miles at a velocity of 13,000 mph while traveling 3,355 miles down range, all well within acceptable parameters for inserting a satellite into orbit. McDougall, *The Heavens and the Earth*, 130.

72. Bowen, *An Air Force History of Space Activities, 1945–1959*, 58.

73. Ibid., 60.

74. Hunsaker, "Forty Years of Aeronautical Research," 4.

75. Ibid., 3.

76. Ibid., 4.

77. Roland, *SP-4103 Model Research*, 5.

78. Hunsaker, "Forty Years of Aeronautical Research," 5.

79. Roland, *SP-4103 Model Research*, 115.

80. Neufeld, *Research and Development in the United States Air Force*, 19.

81. McDougall, *The Heavens and the Earth*, 75.

82. Wilson, "Lyndon Johnson and the Legislative Origins of NASA," 363.

83. Roland, *SP-4103 Model Research*, 141.

84. Ibid., 471–72, 489.

85. House, Committee on Government Operations, *Government Operations in Space*, 2.

86. Roland, *SP-4103 Model Research*, 167.

87. Ibid., 179.

88. Launius, "Never Was Life More Interesting," 361, 366.

89. Ibid., 367. Reliable figures for the percentage of all NACA research authorizations which were military requests: 1920–25, 27 percent; 1926–30, 55 percent; 1931–35, 50 percent; 1936–40, 63 percent; 1940–41, 88 percent. This clearly indicates the close ties between the NACA and the military both before and during WWII. Hansen, *Engineer in Charge*, 161.

90. Launius, "Never Was Life More Interesting," 371.

91. Dean, "Mounting a National Space Program," 221. Other scholarly examinations of the X-15 can be found in Houston, Hallion, and Boston's "Transiting from Air to Space: The North American X-15"; and Hallion, ed., *The Hypersonic Revolution: Eight Case Studies in the History of Hypersonic Technology.*

92. Roland, *SP-4103 Model Research*, 261.

93. NACA, "NACA Research on Missiles," 1.

94. Swenson, Grimwood, and Alexander, *This New Ocean*, 11.

95. Hirsch and Trento, *The National Aeronautics and Space Administration*, 7.

96. Hansen, *Spaceflight Revolution*, 17–18.

97. Seamans, *Aiming at Targets*, 61.

98. Roland, *SP-4103 Model Research*, 283.

99. Doolittle, oral history interview, 21 April 1969, 6, 30, 32.

100. Ibid., 30, 32.

101. Rosholt, *An Administrative History of NASA: 1958–1963*, 6.

102. McDougall, *The Heavens and the Earth*, 164.

103. Schriever, oral history interview, 2 July 1996.

104. Boone, *NASA Office of Defense Affairs: The First Five Years*, 6. A number of official agreements documented the official NACA-DOD relationship. Typical agreements follow: "On Assignment to the National Advisory Committee for Aeronautics Certain Officers of the United States Army for Reserve for Extended Active Duty," 27 July 1956, and similar agreements with the same date for the Air Force and Navy, see NACA, "Wartime Role of NACA in Support of Department of Defense," 21 March 1957; and NACA, "Air Force Support for NACA Research Activities," 8 July 1957.

105. House, *The National Space Program*, 13.

Chapter 2

Eisenhower Act I: Reaction to Sputnik and the Birth of NASA

The initial reaction of many in the Eisenhower administration was to deprecate the Soviet's accomplishment of launching *Sputnik I*, the first artificial satellite to orbit the earth.[1] Rear Adm Rawson Bennett, director of the Office of Naval Research (ONR)—the organization ultimately responsible for the Vanguard effort—declared that it was "a hunk of iron anybody could launch." Sherman Adams, Eisenhower's chief of staff, quipped, "the serving of science, not high score in an outer space basketball game, has been and still is our country's goal."[2] Bryce Harlow, deputy assistant to the president for congressional affairs, later concluded he did a great disservice to Adams by documenting the "basketball" remark but stated that it reflected "the context of that time inside the thinking of the White House."[3]

As national alarm appeared to grow, more perceptive thinking prevailed. Vice President Nixon noted, "We could make no greater mistake than to brush off this event as a scientific stunt. We have a grim and timely reminder . . . that the Soviet Union has developed a scientific and industrial capacity of great magnitude."[4] Over the long term, the administration's tone reflected President Eisenhower's philosophy. The president believed that Sputnik was not a military threat that necessitated a crash response. He also thought that America should remain calm and take a reasoned, rational approach to determining the proper pace and structure of a civilian organization for space activities. Meanwhile, the military's research and development efforts associated with reconnaissance satellites continued and in February 1958 were placed under a new organization called the Advanced Research Projects Agency (ARPA) whose objectives included tempering interservice rivalries. Sputnik's free passage was concluded; however, it established the international principle of a legal right of overflight, and therefore, Vanguard's significance in this matter receded.

Eisenhower Attempts to Calm the Nation

In a conference with his advisors on 8 October 1957, President Eisenhower set the tone saying, "His intent was not to belittle the Russian accomplishment. He would like, however, to allay histeria [sic] and alarm, and to bring out that the Russian action is simply proof of a thrust mechanism of a certain power, accuracy and reliability."[5] NSC's OCB issued guidance that same day instructing agencies of the government to "play down competitive aspects and implication of a 'race.'. . . Keep the accomplishment within a peaceful context, stressing the usefulness of the experiment towards increasing knowledge. . . . Avoid any material indicating that this demonstrates Soviet superiority in science and material indicating that this strengthens the Soviet hand in dealing with the West."[6] During his press conference on the following day, Eisenhower struck the same chord: "I think I have time and again emphasized my concern about the nation's security. . . . Now, as far as the satellite is concerned, that does not raise my apprehensions, not one iota. I see nothing at this moment, at this stage of development, that is significant in that development as far as security is concerned." Eisenhower, however, inadvertently let slip the real motivation of his overall space policy when he said that the Russians, with their fine scientists and even in spite of their dictatorial society "have put one small ball in the air. I wouldn't believe that at this moment you have to fear the intelligence aspects of this."[7]

The Eisenhower administration was perfectly willing to admit that Sputnik's launch indicated an advanced level of Soviet competence in ICBMs—while unexpected, it was nothing to panic about. Eisenhower responded to congressional Democrats such as Stuart Symington by stating, "In total military strength, the US, in our judgment, is still distinctly ahead of the USSR."[8] He later stated, "The possibility of the Russians having intercontinental missiles before we do was not catastrophic since that by no means removed the power of our bombers."[9] Secretary Quarles flatly stated that the "Soviets possess a competence in long-range rocketry and in auxiliary fields which is even more advanced than the competence with which we had credited them; although, of course, we had always given them the capability of orbiting an earth satellite."[10] James Killian, soon to be named Eisenhower's first special assistant for science and technology, summarized Eisenhower's general demeanor,

> I think that Eisenhower was in no way upset about the Russian achievement, that I think he knew enough about our military strength to have no doubts that we were still in a position of superiority at that time. I think too that he felt the public had overreacted to the event, and that his problem was more a political problem than it was one of dealing actually with a major weakness in our government or in its policies. I think a number of us also took the view that it was silly to conclude from the Russian's launch of *Sputnik I* that all of our scientific programs both within and without government had been brought into serious question, or that it meant any really significant weakness.[11]

President Eisenhower was concerned, however, by the confirmation of Soviet ICBM abilities, which topped administration worries. The Soviet space accomplishment caused the president little concern because it—as one high administration official said—"was regarded as a stunt more than a gigantic event of worldwide crucial significance. . . . I think the 'sophisticates' regarded it more as a stunt for worldwide publicity purposes by the Soviet Union rather than as a matter of grave significance. The gravity was regarded as what they would do with their weaponry, not what they were doing with Sputnik."[12] Eisenhower's point man on space up to this point, Secretary Quarles, concurred and wrote to Eisenhower three days after Sputnik that the facts "appear to be that the satellite success does indicate competence in long-range ballistic missiles and does tend to corroborate their ICBM claim of August 27."[13] The Democrats also shared these observations. "You know, it's not the satellite that is so significant today. It's what put it there," remarked an aide to Johnson, the senate majority leader.[14] Another Johnson staffer explained "the simple fact is that we can no longer consider the Russians to be behind us in technology. It took them four years to catch up to our atomic bomb and nine months to catch up to our hydrogen bomb. Now we are trying to catch up to their satellite."[15]

The Soviets wasted no time feeding the American public's growing concern. Three days after the *Sputnik I* launch, Nikita

Khrushchev stated that "we now have all the rockets we need: long-range rockets, intermediate-range rockets and short-range rockets." They launched *Sputnik II* on 3 November 1957, which carried a live dog—a precursor to human spaceflight. Afterwards he declared, "I think that it is no secret that there now exists a range of missiles with the aid of which it is possible to fulfill any assignment of operational and strategic importance. . . . The Soviet Union has intercontinental ballistic rockets with hydrogen warheads [which] now make it possible to hit a target in any area of the globe."[16] Khrushchev even challenged the United States to a rocket "shooting match" to prove his assertions that the Soviets were ahead.[17]

Americans were concerned, justified or not, about the synergism of Soviet capabilities in missiles, nuclear warheads, space capabilities, science and technology, and manufacturing. The fear that they were outpacing American technological growth led to a sense of panic that eventually impelled Eisenhower to create NASA. Although he did not believe that America was threatened by the new issues Sputnik raised, the call for action was severe enough to require something be done, and NASA's creation was one of the steps Eisenhower approved.[18] Outside the group of top Eisenhower administration officials, numerous individuals intimately involved with the American side of the Sputnik equation testified to the sense of alarm—even panic—that pervaded Washington in the fall of 1957 and spring of 1958. Senator Lyndon Johnson remembered a "profound shock of realizing that it might be possible for another nation to achieve technological superiority over this great country of ours. Most Americans shared my sense of shock that October night. . . . [Sputnik] plunged the America of 1957 into spiritual depression [and] depreciated our prestige. Russia's image as a technological leader suddenly increased to alarming proportions and our own image diminished, especially among the people of the developing nations."[19] One congressman, also a historian, summarized, "The prairie fire of demands for action swept across the Nation. The clamor rose to a roar."[20]

This was not merely partisan posturing as Killian also sensed a "climate of near hysteria" among many people, "some of whom should have known better." He concluded that Sputnik had created "a crisis of confidence that swept the country like a windblown forest fire. Overnight there developed a widespread fear that the country lay at the mercy of the Russian military machine and that our own government and its military arm had abruptly lost the power to defend the homeland itself."[21] The tone shown in a *Washington Post* article represented that taken by most major media outlets: "Not even the most dim-witted State Department official needed more than a second glance at those news bulletins on Sputnik to realize that the United States had suffered the worst psychological licking in the history of its relations and struggle with the Soviet Union and the Communist World. The United States could no longer proclaim the supremacy of its industrial machine or of the capitalist free system of economics."[22]

T. Keith Glennan, NASA's first and only administrator during the Eisenhower administration, commented on the catalyst for NASA's creation: "I think you ought to realize first that NASA was born out of a state of hysteria; that, indeed, if Sputnik number one had not been put into orbit, it is highly improbable that there would be a NASA."[23] Eisenhower himself concurred, later saying NASA's "whole program was based on psychological values. . . . The furor produced by Sputnik was really the reason for the creation of NASA."[24] Eisenhower did not like being forced to react to the nation's unmerited panic. His son recalled, "I think the public became hysterical, and he couldn't figure out why they were," which caused his father to question, "What the hell are they [the public] worried about?"[25] Eisenhower believed his challenge was "to find [a] way of affording perspective to our people and so relieve the wave of near-hysteria."[26]

Eisenhower's attempt to calm the nation's anxiety also highlighted the two major objectives of his space policy. First, space exploration must not endanger—in any way—the process of gathering intelligence on the Soviet Union through the use of reconnaissance satellites, and second, space exploration was not to be regarded as a prestige-oriented race with the Soviet Union. NASA's creation supported both of these objectives.

The Right of Overflight

Quarles noted that none of the countries—the United States included—had protested Sputnik's transit of the sky above their homelands. He therefore concluded that the Soviets had established the legal principle of *freedom of overflight* for reconnaissance satellites.[27] He explained during an 8 October 1957 conference with the president that, "the Russians have in fact done us a good turn, unintentionally, in establishing the concept of freedom of international space—this seems to be generally accepted as orbital space, in which the missile [Sputnik] is making an inoffensive passage."[28] A more attentive observer of this meeting recounted that "Quarles made the important point that the Russians having been the first with their Satellite to overfly all countries, they have thereby established the international characteristic of orbital space. We believe we can get a great deal more information out of free use of orbital space than they can."[29] Quarles explained to a full NSC meeting two days later that one of the United States' objectives in the Vanguard program "was to establish the principle of the freedom of outer space—that is, the international rather than the national character of outer space. In this respect the Soviets have now proved very helpful. Their earth satellite has over flown [*sic*] practically every nation on earth, and there have thus far been no protests. . . . The outer space implications of the launching of this satellite were of very great significance, especially in relation to the development of reconnaissance satellites."[30]

The apparent international consensus in approving of, rather than objecting to, satellite overflight resulted in the tentative establishment of the right of satellite overflight. That principle was very important to the administration and had to be protected; the "space for peace" policy that had already been widely publicized continued to receive heavy emphasis. Eisenhower's civilian space and defense officials wanted the American space program to appear as peaceful, scientific, and civilian as possible to avoid provoking the Soviets and possibly endangering the tentative right of overflight. This policy was a primary cause for the tension that developed between space-oriented Air Force officers and civilian executive-branch leaders, the OSD included, and persisted until the mid-1960s. Those Air Force officers wanted to explore the possibilities of fully using space for national defense, to include basing offensive weapons in space. Both the Eisenhower and Kennedy administrations, however, disapproved of that approach and largely quashed those efforts to protect the vital reconnaissance satellites (and later, during Kennedy's and Johnson's tenure, to avoid tainting the lunar landing's prestige with the stigma of an offensive military label).[31]

No Race for Prestige

The other major objective of Eisenhower's space policy, during the period after *Sputnik I* and leading up to the creation of NASA, was his desire to avoid a crash speculative program in a race for prestige. Hagerty, Eisenhower's press secretary, emphasized this point when he briefed the press the day after Sputnik, saying that "I would also like to make it quite clear that the Soviet launching did not come as any surprise and that we have never thought of our program as one which was in a race with the Soviet program."[32] Likewise, during an 8 October 1957 meeting discussing Sputnik, Eisenhower interjected, saying "timing was never given too much importance in our own program, which is tied to the IGY."[33] He emphasized that "no pressure or priority was exerted by the U.S. on timing, so long as the Satellite would be orbited during the IGY 1957–1958."[34]

Such declarations led historians to conclude that Eisenhower ignored the clear warning in NSC 5520 of a potentially negative psychological impact on the nation if the Soviets were to launch a satellite first. Nor was he swayed by similar NSC and OCB's entreaties to mitigate those likely prestige ramifications when determining Vanguard's schedule. Although the words were present in the pre-Sputnik-policy documents, his presidential commitment was lacking. Eisenhower acknowledged this during a press conference on 9 October 1957 as he recounted some of the Vanguard program deliberations: "More than once we would say, well, there is going to be a great psychological advantage in world politics to putting this thing up. But that didn't seem to be a reason, in view of the scientific character of our development, there didn't seem to be a reason for just trying to grow hysterical about it." The written statement distributed to the press stated, "Our satellite program has never been conducted as a race with other nations."[35] The same pattern continued—even after NASA's creation. The Eisenhower administration understood that great prestige would accrue to the nation that first flew a human in space, but its efforts continued to be conducted as though it was not a race.

Eisenhower expressed his displeasure with those trying to create a race dynamic as he swore in Neil H. McElroy as the new SECDEF, also on 9 October 1957. Quarles, the civilian service secretaries, and the joint chiefs were in the audience and heard him say, "When military people begin to talk about this matter, and to assert that other missiles could have been used to launch a satellite sooner, they tend to make the matter look like a 'race,' which is exactly the wrong impression."[36] He repeated this position again in the summer of 1958 after signing the bill to establish NASA. He sent Dr. Hugh L. Dryden, its deputy administrator, to Congress to explain NASA's first-year budget and explain that "it most decidedly is not a crash program to catch up with anybody." Following Dryden's presentation, Congressman Overton Brooks asked if that meant NASA's program was not—in any way—competitive with the Soviet program. Dryden replied, "I would say that this program is not at a level at which we could guarantee to do that."[37] Following his term as president, Eisenhower explained to a historian that "under no circumstances did we want to make the thing a competition, because a race always implies urgency and special progress regardless of cost or need. . . . Neither then nor since have I ever agreed that it was wise to base any of these projects on an openly and announced competition with any country. This kind of thing is unnecessary, wasteful and violates the basic tenets of common sense."[38]

Closely related, of course, was Eisenhower's immediate post-Sputnik lack of enthusiasm for prestige-oriented space spectaculars, or "stunts." Quarles testified to Congress on 18 November 1957: "We must not be panicked or pushed into any sudden dispersion of effort. . . . We must not be talked into 'hitting the moon with a rocket,' for example, just to be first, unless by doing so we stand to gain something of real scientific or military significance."[39] Nevertheless, before endorsing the creation of a civilian organization to conduct the civilian space program, Eisenhower had to accept—to a small degree—the legitimacy of the prestige factor to help justify NASA. On 4 February 1958, "The President stressed the importance of picking out the phases of activity in which we should undertake to compete with the Soviets, and to beat them. We should not try to excel in everything. He added that psychological as well as technical considerations are important—at times appearances are as significant as the reality, if not more so."[40] This indicated a shift in Eisenhower's thinking—away from no prestige-oriented projects to an attitude in which some carefully selected projects could be designed to compete with the Soviets. Although human spaceflight would not be one of them, he did approve the orbit of a 100-foot-wide balloon, a passive (reflective) communications satellite.

According to the President's Science Advisory Committee (PSAC), Eisenhower approved this project because the balloon's "psychological value from the standpoint of free use for every nation," and of the options available, it "appears to be the best psychological-scientific experiment."[41] Similarly, in December 1958, he authorized Project SCORE (Signal Communication Orbit Relay Experiment) that launched a payload made up of a stripped-down Atlas booster, weighing 9,000 lbs.—100 lbs. of which was communications equipment. That payload played, for eight days, a tape-recorded message from Eisenhower stating, "I convey to you and to all mankind America's wish for peace on earth and good will toward men everywhere."[42] It also allowed the United States to boast that it had orbited a "satellite" of over four tons—even though most of the weight was an expended booster. Said one historian, "Technically, it was all a stunt."[43]

Although Eisenhower's antipathy toward competing with the Soviets was occasionally interrupted by his endorsement of small stunts, his general "approach was if we're doing the right thing in about the right way we'll let the prestige work itself out."[44] For example, in February 1958 (shortly after America orbited its first satellite), Eisenhower resisted calls for a crash-lunar-probe program because he would rather have a good IRBM "than be able to hit the moon, for we didn't have any enemies on the moon!"[45] Herbert York, director of defense research and engineering (DDR&E), explained that "Eisenhower, Killian, and Kistiakowsky [George B. Kistiakowsky, Killian's successor as Eisenhower's science advisor] were not the kind of people who would accept prestige as the sole reason for doing something. . . . With them, I think it's fair to say, prestige could be a fine dividend but there had to be a better reason than simply prestige alone."[46]

Eisenhower expressed his administration's impression of Sputnik's significance to the American people during his first "chins up" speech on 7 November 1957. It was designed to calm a growing national alarm, based on the fear that national security and the American way of life were imperiled. That fear grew out of the erroneous perception that the USSR had suddenly surpassed the United States in science and technology and was now vastly superior. Eisenhower had access to U-2 reconnaissance photographs, which made it clear to him that claims of a large Soviet ICBM force menacing America were highly unlikely. Unfortunately, he could not share that U-2 data with the nation to help justify his confidence—to do so would risk compromising it and the vital information only it could obtain.[47]

In that first "chins-up" address, Eisenhower emphasized that "our nation has today, and has had for some years, enough power in its strategic retaliatory forces to bring near annihilation to the war-making capabilities of any other country." That power can be delivered through its hundreds of bombers and diversified missiles, "adapted to every kind of distance, launching and use." He explained, "Our scientists assure me that we are well ahead of the Soviets in the nuclear field, both in quantity and in quality. We intend to stay ahead." Although the majority of his speech focused on the adequate nature of America's national defense structure, he addressed space with only one comment: "Earth satellites, in themselves, have no direct present effect upon the nation's security." They do, however, imply the Soviets have powerful rockets. He closed by saying, "What the world needs today even more than a giant leap into outer space, is a giant leap toward peace."[48] Eisenhower elaborated in his second chins-up speech a week later, 13 November 1957: "The sputniks have inspired a wide variety of suggestions. These range from acceleration of missile programs, to shooting a rocket around the moon, to an indiscriminate increase in every kind of military and scientific expenditure. Now, my friends, common sense demands that we put first things first. The first of all firsts is our nation's security!" He explained that if a satellite were solely for scientific purposes, then its size and cost must be tailored to the scientific job it was going to do. If it were for defense purposes, then "its urgency for this purpose is to be judged in comparison with the probable value of competing defense projects."[49] The first chins-up speech reflected Eisenhower's perception that space as an issue primarily related to national defense and not to international prestige or propaganda—a perception he would hold until the spring of 1958. He also authorized the first federal funding for colleges and universities, to increase the production of scientists and engineers, and finally increased defense spending for three years—primarily for strategic bombers and ballistic missiles.

He added a $1.3 billion supplemental in FY 58, which brought the total defense budget to $44.5 billion, and a further increase in FY 59 brought that fiscal year's defense budget to $46.6 billion. Once the furor over Sputnik faded, Eisenhower decreased the defense budget in FY 60 to $45.9 billion, the lowest authorization since FY 54—the year he assumed office. Eisenhower's post-Sputnik actions were neither rash nor damaging, and his additions to the defense budget, when viewed as a percentage of the gross national product (GNP), actually decreased from 9.9 to 9.1 percent during the three years following Sputnik's launch.[50] Eisenhower noted that in responding to Sputnik, "Somehow the United States had to put on hair shirt and sackcloth yet avoid scaring people."[51]

President's Scientific Advisory Committee and Civil Space

The central role Killian and PSAC played in creating NASA was crucial to the civil-military relationship in space that developed in the period immediately following Sputnik. These scientists firmly believed a civilian organization should direct all space exploration programs.[52] Once Eisenhower had tasked Killian with determining how America should structure its space program, it was no great surprise that his recommendation greatly expanded the NACA into NASA, while preserving the DOD's right to weapons-systems-related space activities. Quite simply, Killian "exerted enormous influence on the manner in which the American space program was structured and conducted."[53] Killian recounted, "I was greatly helped in achieving admission to the inner sanctum of the Eisenhower White House by several earlier appointments" such as the TCP and serving as chairman of the President's Board of Consultants on Foreign Intelligence Activities.[54] Killian's authority from Eisenhower was significant. He was to "have the active responsibility of helping me [Eisenhower] follow through on the program of scientific improvement of our defenses [so that the] entire program is carried forward in closely integrated fashion, and that such things as interservice competition or insufficient use of overtime shall not be allowed to create even the suspicion of harm to our scientific and development program. . . . [He was to] see to it that those projects which experts judge have the highest potential shall advance with the utmost speed. . . . It is my full desire that you have full access to all plans, programs, and activities involving science and technology in the Government, including the DOD, AEC [Atomic Energy Commission], and CIA."[55] When Killian left full-time government service in July 1959, Eisenhower praised his work and credited it with helping the United States avoid the urge "to plunge headfirst and almost blindly into the space age. . . . No one did more than you, in those early days, to bring reason, fact, and logic into our plans for space research and adventure."[56] Jerome Wiesner would become Kennedy's science advisor and was a PSAC member in the Eisenhower administration. He corroborated the highly influential role Killian and other scientists occupied, saying Killian was always "extremely careful about what he does and says." As a result, "I think the president understood that and appreciated it so that on the whole he trusted him completely; and really, I had the impression he was very supported by having Killian around."[57] Killian's appointment was the first time that a scientist had key access to the White House, and one scholar called it "the most important step that Eisenhower took following *Sputnik II*."[58]

The esteem in which Eisenhower held Killian seems also to have been true for the PSAC and its members as a whole. In fact, key PSAC members were one of the first groups Eisenhower convened after Sputnik. After their 15 October 1957 meeting, "The President concluded by saying that he was delighted with this conversation. . . . He found no solace in crying over spilled milk. He was not concerned about the Soviets beating us in the Satellite field."[59]

Eisenhower's trust of what he later termed "my scientists" grew throughout the remainder of his administration. Shortly before Eisenhower died, Killian visited him and Eisenhower volunteered, "You know, Jim, this bunch of scientists was one of the few groups that I encountered in Washington who seemed to be there to help the country and not help themselves."[60] Killian was, in fact, uncomfortable with Eisenhower's confidence and reliance on the elite scientists' input.

> One of the qualities of Eisenhower that troubled me during the course of my service to him was his almost exaggerated confidence in the judgment of the scientists that he had called upon to help him. He sometimes came to have a feeling that this group of scientists were endowed with an objectivity that he couldn't expect to find in other contacts that he had in government. And I think he over-estimated the capacity for objectivity that any kind of professional people . . . could demonstrate in regard to controversial problems. . . . [Nevertheless] he used the President's Science Advisory Committee and its panels constantly to appraise programs where there were interservice rivalries involved.[61]

Thomas Gates, Eisenhower's final SECDEF concurred, "All of a sudden the scientists became very important. . . . They had great veto power."[62] York, DDR&E's first director, explained that in the post-Sputnik Eisenhower administration the PSAC "reviewed virtually every program of the Department of Defense, and many of those of the AEC [Atomic Energy Commission] and CIA as well. Few programs or ideas that did not meet their approval got very far."[63] In sum, one effect of Sputnik was that "scientists were rushed to the most important single center of power, the Office of the President."[64]

The scientists' influence on the creation of a civilian space organization could be seen by the end of 1957. Groups of civilian scientists not affiliated with the government, such as the American Rocket Society (ARS), were submitting plans within a week of Sputnik's launch for a civilian space organization.[65] From then and until Killian's formal recommendation to create NASA in March 1958, numerous other scientific organizations either submitted similar plans or endorsed the general concept.[66]

Three days after his appointment, Killian said that Eisenhower had tasked him to determine the organizational structure of the US space program. Killian recalled, "It was perfectly obvious that the military was terribly anxious—at least, below the level of the civilian top command—to have responsibility for the space program. . . . There were strong indications from the DOD that the space program ought to be lodged in the DOD."[67] However, there was no chance of that happening. The predominant attitude of the government-affiliated civilian scientists was revealed at a PSAC meeting on 10 December 1957. Bronk, NAS president, summarized for that group, "There are many aspects of the NACA worth looking into." Berkner, NAS member, concurred and added, "We want the controlling agency outside the DOD. Inevitably this type of space activity will be a powerful binding force." Killian agreed and said the key enabler would be "if we could say NACA should have increased funds. . . . NACA is used to getting hardware from DOD. Its relationship to the military has enabled NACA to have experimental hardware built."[68]

Killian wrote a memo summarizing his initial thoughts on 30 December. In that memo he assumed that DOD would soon form a special organization to manage defense-related space R&D, even if a separate civilian agency were established and that "the DOD must play a major role in space research and development if we are to use the nation's manpower and facilities in this area to the greatest advantage." DOD would be "primarily concerned with those aspects of space research and development which will have military value" although it is hard to separate the civilian from the military elements in space. While it would be "entirely feasible" for the DOD to handle all civilian and military space research and development, "such an arrangement might improperly limit the program to narrowly concerned military objectives. In the second place, it would tag our basic space research as military and place the U.S. in the unfortunate position before the world of apparently tailoring all space research to military ends." Killian therefore viewed his basic challenge as "devising the means for non-military basic space research while at the same time taking advantage of the immense resources of the military missile and reconnaissance satellite programs." Killian had, in that sentence, identified the next decade's central challenge to the NASA-DOD relationship for the next 10 years.[69]

Killian foresaw that in two months he would officially recommend that DOD "might confine itself to its military mission, and some other agency or agencies external to the DOD might engage in basic research. One obvious way of doing this would be to encourage NACA to expand its space research and to provide it with the necessary funds to do so." Killian understood that "it would be necessary to carefully work out a cooperative arrangement with the DOD, for the DOD would have to be an active partner with these agencies." Killian's concept emphasized the necessity for fundamental scientific research in the space program—not prestige-related stunts—saying that "we must have far more than a program which appeals to the 'space cadets.' . . . If we do not achieve this, then other nations will continue to hold the leadership."[70] After developing his concept, Killian's task during the first few weeks of 1958 was to convince Eisenhower of its wisdom.

The DOD's civilian leadership had no problems with Killian's basic concept. Richard E. Horner was assistant SAF for R&D and would become NASA's associate administrator in June 1959. He recalled that he, Quarles, and James Douglas, SAF, had discussed the organizational structure for space and had concluded that, "the best thing for the nation was to put this in the NACA. The rationalization as far as Donald Quarles was concerned, was that the Air Force had too many important eggs in the ballistic missile basket to divert its attention to doing

45

other things. But there obviously was going to be a national response to Sputnik. Of course, Don was considerably troubled by the fact that he had made the Vanguard decision over the broken and bleeding body of the Air Force. . . . The Air Force was very acquisitive in those days, and they wanted to do everything, but they wanted more money than anybody else." Therefore, Horner said, the Air Force and DOD leadership decided to support the general idea of a civilian organization.[71] The enabling bill was drafted in March 1958 and submitted to Congress on 2 April 1958.

Johnson and the Preparedness Investigating Subcommittee

To understand the context of the Eisenhower administration's draft of the National Aeronautics and Space Act of 1958 (Space Act), it is necessary to discuss the hearings before Johnson's Preparedness Investigating Subcommittee and the NACA campaign to be given responsibility for civilian space exploration. Johnson's hearings added congressional fuel to the fires of many who were calling for action and, when combined with the NACA campaign, set the stage for the Space Act to make official the civil-military split in the US space program. Eilene M. Galloway was an acknowledged expert in science and technology and national defense in the Legislative Reference Service of the Library of Congress. She was frequently detailed to congressional committees and became one of the key behind-the-scene players in the congressional response to Sputnik. Galloway authored many reports and briefed senators and representatives on the political and technological implications of the space age.[72] She recalls that shortly after Sputnik, Lyndon Johnson telephoned her and said, "Eilene, I want to make me a record in outer space, and I want you to help me."[73] George Reedy was on the staff of the Senate Armed Services Committee's (SASC) Preparedness Investigating Subcommittee, which Johnson chaired. Reedy wrote Johnson on 17 October 1957 that Sputnik was an issue that "would blast the Republicans out of the water, unify the Democratic Party and elect you president. . . . Eye [sic] think you should plan to plunge heavily into this one." Reedy explained that the racial integration issue was "a potent weapon which chews the Democratic Party to pieces" and is not going to go away. Therefore, the only possible response "is to find another issue which is even more potent. Otherwise, the Democratic future is bleak."[74] A close reading of the 2,475 pages of the three-volume transcript that documents the testimonies of 73 witnesses during the subcommittee's 110 days of hearings makes clear that the hearings were designed to meet partisan objectives and not, as is often incorrectly stated, to provide an objective look into the state of America's satellite and missile programs.

Johnson admits in his memoirs that even before the hearings started in November 1957, "I was already convinced that our country was in trouble," and the hearings would have to "determine what steps can be taken to strengthen our position and restore the leadership we should have in technology. . . . I knew one thing beyond doubt—we had to catch up."[75] Johnson entered the hearings with certain presumptions: there was a crisis that merited a dramatic response, and there was a loss of American leadership in technology that had to be reclaimed. No witnesses were called from the Truman era to explain the relative lack of ICBM and satellite R&D between 1945 and Eisenhower's inauguration. Johnson opened the hearings by declaring all witnesses had to give "a clear definition of the present threat to our security, perhaps the greatest that our country has ever known" and then offer specific recommended responses because "Our goal is to find out what is to be done. The facts that I learned so far give me no cause for comfort." Johnson emphasized, "It is not necessary to hold these hearings to determine that we have lost an important battle in technology. That has been demonstrated by the satellites that are whistling above our heads."[76] Not surprisingly, all the witnesses selected and called by Johnson shared his presuppositions.[77]

Secretary of Defense McElroy, and other administration officials, tried to bring some balance by explaining that assessing Soviet and American defense strengths was a matter of "toting up" the assets and capabilities of each and then comparing them in total, one to the other. That approach is necessary because "they have certain strengths in excess of ours and we have certain strengths in excess of theirs."[78] Quarles added that by "taking the missile program as a whole, and comparing their own program with our own, I estimate that as of today our program is ahead of theirs." He also supported the United States decision to conduct Vanguard at a pace so that it would "not interfere with the top priority of the ballistic missiles program. . . . I believe there is no question that our near-term position is sound."[79] Nevertheless, the headlines regularly went to military officers such as Lt Gen James M. Gavin, USA, and others, who claimed, "From the straight estimate of the balance of military power, our position is exceedingly difficult." When asked if this meant the United States was behind the USSR, General Gavin replied, "Yes, I would say we are."[80]

The question of the appropriate balance in the civil-military management of the response to Sputnik received relatively little attention in the subcommittee's hearings; the vast majority of time and witnesses focused directly on military issues such as missiles, bombers, and the nuclear balance. The Army Ballistic Missile Agency's (ABMA) Wernher von Braun did muse, "Suppose a National Space Agency were set up, either under the Secretary of Defense or as an independent agency, and this agency were given its own budget." Such an agency could conduct the American space program for $1.5 billion per year. Von Braun said that scientists "would prefer having it an independent agency. . . . But I am convinced it would work either way."[81] Not surprisingly, representatives of the defense industry supported an independent agency.

General Schriever openly discussed the Air Force's reconnaissance satellite, saying there has been a great deal of interest in it within the government, "But we got no approval for proceeding with this on a systems basis either on the Air Force secretarial level or at the Department of Defense secretarial level until just recently." He said that with adequate funding, the Air Force could launch a reconnaissance satellite "with a recoverable [film] capsule" by the spring of 1959. In addition, Schriever emphasized, "at least 90 percent of what we are doing in the Air Force ballistic missile program, 90 percent of all this work can be directly applied to an astronautics or space program."[82] Neither Schriever nor the other senior military officers were keen on the idea of an independent agency. Gen John Medaris, ABMA commander, said, "I cannot in conscience endorse an independent agency. . . . There is no need for creating a separate agency with operating characteristics outside the Defense Department for doing this job." Creating another bureaucracy "will create a confusion that will set our program back a year."[83]

Most of the preparedness subcommittee's 17 recommendations, issued on 23 January 1958, focused on military actions such as dispersing Strategic Air Command assets; building more bombers, missiles, and submarines; and improving the early warning system.[84] Number 15, however, touched on the organizational issue in its recommendation to "accelerate and expand research and development programs, providing funding on a long-term basis, and improve control and administration within the Department of Defense or through the establishment of an independent agency."[85] The tenor of the subcommittee's conclusions was that "we are engaging in a race for survival and we intend to win that race."[86]

Although the preparedness subcommittee did not specifically recommend the creation of NASA, it did offer that option. More importantly, it fed the crisis atmosphere that existed in early 1958 and, through the extensive media coverage of its hearings, created an expectation that the Eisenhower administration should create some kind of a program. In a private meeting with the Democratic conference on 7 January 1958 and following the completion of the preparedness subcommittee's hearings, Johnson summarized his thoughts and said, "The peril of the hour is obvious." Sputnik had opened up the realm of space and Johnson believed, "The exploitation of these capabilities by men of selfish purposes holds the awful threat of a world in subjugation. The mastery of such capabilities by men wholly dedicated to freedom presents, instead, the prospect of a world at last liberated from tyranny, liberated in fact from the fear of war."[87] These remarks to his colleagues are important because they better indicate his true perceptions and conclusions on the importance of space, sincerely expressed to his congressional associates and not "spun" for media consumption.

In this candid context, Johnson declared that the American evaluation of the proper role for space had, so far, not been made by the "men most qualified to make such an appraisal. Our decisions, more often than not, have been made within the framework of the government's annual budget. This control has, again and again, appeared and reappeared as the prime limitation upon our scientific achievement." This must change, Johnson concluded, because:

> Control of space means control of the world, far more certainly, far more totally, than any control that has ever or could ever be achieved by weapons, or by troops of occupation. From space, the masters of infinity would have the power to control the earth's weather, to cause drouth [sic] and flood, to change the tides and raise the levels of the sea, to divert the gulf stream and change temperate climates to frigid. . . . The urgent race we are now in—or which we must enter—is not the race to perfect long-range ballistic missiles. There is something more important than any ultimate weapon. This is the ultimate position—the position of total control over earth that lies somewhere out in space. . . . Whoever gains that ultimate position gains control, total control, over the earth, for purposes of tyranny or for the service of freedom. . . . Our national goal and the goal of all free men must be to win and hold that position. Total security perhaps is possible now, for the first time in man's history. Total security—and, with it—total peace.[88]

These unfiltered statements of Lyndon Johnson on the importance of space, particularly those oriented toward national security, represented those basic beliefs that motivated his actions in Congress and beyond. That motivation was evident during his term as vice president when President Kennedy tasked him to determine how to beat the Russians in space. It was still evident early in his presidency, before the realities of the overall federal government budget's large and diverse requirements settled in and modified his conviction on the preeminence of space.

The preparedness subcommittee was far from being a nonpartisan fact-finding group and seemed to be part of the movement that was pushing the Eisenhower administration into creating a formal organizational structure for space exploration. One Space Act analyst said that the subcommittee had concluded, "the country lacked leadership and that the Democrats would provide it, whether or not the administration went along."[89] Walter McDougall added, "Day by day the witnesses rose to confirm the committee's suspicions and provide quotes for the next day's front pages" as well as "general and specific accounts of American humiliation [which] flowed through the press and public mind together, weakening faith in the administration and its values."[90] In short, the subcommittee, "markedly refrained from anything like a thorough and objective review of the development and implementation of the policies of the Eisenhower administration on missiles and satellites."[91] The subcommittee's hearings can only be understood in light of the fact that they "were aimed primarily at achieving a transfer of the policy-making initiative."[92]

The NACA Enters the Fray

The pace of policy making within the Eisenhower administration picked up after the first of the year. Killian's general recommendations for a civilian space agency in late December 1957 led directly to the Space Act's official submission to Congress by early April 1958. In this swirl of events, the National Advisory Committee for Aeronautics made the case that it was the most logical choice to head the civilian space exploration program. It was significant that Hugh Dryden, NACA's director, and James Doolittle, NACA's chairman, were both members of the PSAC during the 1957–58 time frame when that group, under Killian's leadership, forged the organizational structure for space.[93] The NACA's position was well-represented within the very group charged with making the decision.

On 21 November 1957, the NACA had established the Special Committee on Space Technology to consider how to best use human capabilities in space exploration and outline how the NACA could develop its resources for space exploration. Although this committee did not issue its formal report until after NASA was created, it did show the NACA's early concern for evaluating and establishing its own role in US space efforts.[94] On 14 January 1958, Dryden released a formalized space R&D plan that he had directed his staff to develop. That plan, titled "A National Research Program for Space Technology," stated:

> It is of great urgency and importance to our country both from consideration of our prestige as a nation as well as military necessity that this

51

challenge [Sputnik] be met by an energetic program of research and development for the conquest of space. . . . It is accordingly proposed that the scientific research be the responsibility of a national civilian agency working in close cooperation with the applied research and development groups required for weapon systems development by the military. The pattern to be followed is that already developed by the NACA and the military services. . . . The NACA is capable, by rapid extension and expansion of its effort, of providing leadership in space technology.[95]

Dryden elaborated on the plan's proposal for a civilian space agency in a 27 January 1958 speech. He recognized that many scientists feared that "the extremely important nonmilitary aspects of space technology would be submerged or perhaps even lost if included as a mere adjunct to a military program." His proposed alternative, the NACA, was "old and well-tested," and its proposal for a space exploration program "can be most rapidly, effectively and efficiently implemented by the cooperative effort" of the NACA, DOD, NSF, NAS, civilian universities, research institutions, and industry.[96] The NACA concept called for a multi-institutional space-exploration program in which it would take the lead role, work closely with other interested parties, and not infringe on the DOD's prerogatives.

A document that originated in either the White House or the BOB states the NACA's attempts in January 1958 to claim a leading role in space exploration "had been cleared with Dr. Killian and possibly also with the White House. They received a favorable reaction among staff of the Bureau of the Budget who had already been thinking of NACA as the logical nucleus of a new aeronautics and space agency."[97] The NACA's specific plan, dated 10 February 1958, was titled "A Program for Expansion of NACA Research in Space Flight Technology with Estimates of the Staff and Facilities Required." In that plan the NACA went into extreme detail to outline the research program that would be conducted and the additional facilities, staff, and budgetary augmentations that would be necessary for implementation.[98] "It seems to me that the NACA is on the 'horns of a dilemma,'" Doolittle wrote in a personal and confidential letter to Dryden. He continued, "Unless it is given at least some part of the space program, it will decline with the airplane."[99] Dryden verified the

NACA's concern for its own existence in those early days, recalling,

> We've either got to be in space or run out of business. We did decide that we wanted to stake out a role for NACA in whatever happened right from the beginning. And the minimum role . . . was to bear the same relation to whatever agency was set up to carry out the actual operations in space as NACA had had with the Defense Department. We also felt that rather than take an aggressive position in the matter that the best attitude was to play it down a bit . . . to express this minimum claim. . . . This paid off in the long run. . . . We never took the offensive.[100]

The PSAC and "Introduction to Outer Space"

Eisenhower held his first important meeting dealing with space organization with Republican congressional leaders on 4 February 1958. When asked, he stated that he favored keeping everything within the DOD:

> The president's feeling was essentially a desire to avoid duplication, and priority for the present would seem to rest with Defense because of paramountcy of defense aspects. However, the president thought that in regard to non-military aspects, Defense would be the operational agent. . . . The president was firmly of the opinion that a rule of reason had to be applied to these Space projects—that we couldn't pour unlimited funds into these costly projects where there was nothing of early value to the Nation's security. . . . he didn't want to just rush into an all-out effort on each one of these possible glamor [sic] performances without a full appreciation of their great cost.[101]

However, Eisenhower did not wholly rule out the possibility of a separate civilian organization focused on space exploration, pending the recommendations of his President's Science Advisory Committee. At the end of the space-related meetings on 4 February, Eisenhower tasked Killian to work out, once and for all, a concrete organizational structure for space exploration to include the civilian and military division of labor.

At the following day's press conference, Eisenhower explained, "I have gotten a group of fine scientists under the chairmanship of Dr. Killian. . . . He is getting the scientists to give for the United States a program of outer space achievement."[102] During the next month, the president expunged his predilection for DOD to handle all aspects of space R&D because: "Killian persuaded Eisenhower that a civilian agency was the better choice."[103] Congressional opposition to DOD primacy and interservice rivalry in space may have helped sour Eisenhower on the DOD being responsible for the primarily scientific arena of civilian space exploration.[104]

Killian's memorandum to the president, "Organization for Civil Space Programs," dated 5 March, was the key document leading to Eisenhower's sanctioning the creation of NASA. It responded to the president's 4 February tasking and concluded, "An aggressive space program will produce important civilian gains in the form of advances in general scientific knowledge and the protection of the international prestige of the United States. These benefits will be in addition to such military uses of outer space as may prove feasible." It said civilian direction of the space program was supported by overwhelming civilian interests inherent in it and also by the "public and foreign relations considerations. However, civilian control does not envisage taking out from military control projects relating to missiles, anti-missile defense, reconnaissance satellites, military communications, and other technology relating to weapons systems or direct military requirements."[105]

Killian's memo listed several reasons why leadership of the civil space program should be lodged in a "strengthened and redesignated" NACA. First, it was a "going Federal research agency" with 7,500 employees and $300 million worth of laboratories and test facilities, which could expand its research program "with a minimum of delay." Second, its aeronautical research "has been progressively involving it in technical problems associated with space flight" such as rocket engines, materials, and designs, and its future would be in doubt without responsibility for space. Finally, "NACA has a long history of close and cordial cooperation with the military departments" and so "the tradition of comity and civil-military accommodation which has been built up over the years will be a great asset in minimizing friction between the civilian space agency and the Department of Defense."[106] Killian recommended an all-out effort to draft a new law and submit it to the current session of Congress, which could establish an organizational structure and begin a civilian space-exploration program before the end of 1958.

During the following day's 6 March NSC meeting, Eisenhower approved Killian's memo and said, "Let's get a bill prepared at the earliest possible opportunity."[107] Killian briefed his conclusions, and Eisenhower vigorously nodded his approval, pleased to have confirmation of the viewpoint he had reached in the period since the first week of February.[108] Eisenhower was so impressed, he had Killian assemble a PSAC team to brief the rest of the government on the pending space program. These briefings were in turn so successful that Eisenhower directed Killian and the PSAC to design a small booklet for nationwide distribution. That booklet, *Introduction to Outer Space*, was released on 26 March 1958.

Introduction to Outer Space and the Space Act are the two seminal policy documents that succinctly state the fundamental space-policy principles that guided the remainder of Eisenhower's term.[109] The president called the former "the most interesting and fascinating thing in this field that I have seen, and I want to make it available to the entire public."[110] It subsequently appeared in major newspapers and other periodicals including the *New York Times* and *Reader's Digest*. He also wrote its introduction and stated, "This is not science fiction. This is a sober, realistic presentation prepared by leading scientists. . . . It clarifies many aspects of space and space technology in a way which can be helpful to all people as the United States proceeds with its peaceful program in space science and exploration." Eisenhower sounded the space-for-peace clarion call: "We and other nations have a great responsibility to promote the peaceful use of space and to utilize the new knowledge obtainable from space science and technology for the benefit of all mankind."[111]

The booklet's introduction offered four answers to the question: why explore space? First, "the compelling urge of man to explore and to discover, the thrust of curiosity." Second, the "defense objective" in which "we wish to be sure that space is not used to endanger our security. If space is to be used for military purposes, we must be prepared to use space to defend ourselves." (Note the use of the conditional "if" in the PSAC's formulation.) Third, "To be strong and bold in space technology will enhance the prestige of the United States among the peoples of the world and create added confidence in our scientific, technological, industrial and military strength." Finally, space offered "new opportunities for scientific observation and experiments which will add to our knowledge and understanding of the earth, the solar system, and the universe."[112]

The booklet's third answer reinforces Eisenhower's willingness to accept some prestige-oriented projects. While the second answer addressed defense, the use of the conditional "if" reflected the PSAC's lack of enthusiasm for any military use of space beyond communication and reconnaissance.[113] "Much has been written about space as a future theater of war, raising such suggestions as satellite bombers, military bases on the moon, and so on. For the most part, even the more sober proposals do not hold up well on close examination or appear to be achievable at an early date. Granted that they will become technologically possible, most of these schemes, nevertheless, appear to be clumsy and ineffective ways of doing a job. . . . In short, the earth would appear to be, after all, the best weapons carrier."[114]

The PSAC's conclusion effectively retarded the development of expensive, forward-looking military spaceflight projects throughout the balance of the Eisenhower administration and also into Kennedy's. The Air Force struggled year after year to convince civilian policy makers that there was a legitimate reason for military officers to operate in space. The Air Force's quest was dealt a serious blow with Dynasoar's cancellation in 1963 and ultimately failed with the MOL's cancellation in 1969.

Finally, the PSAC timetable for space accomplishments was categorized within the chronological labels of *early, later, still later,* and *much later.* The *early* objectives included physics, meteorology, and experimental communications. Astronomy and human flight in orbit would come *later,* while human lunar exploration and return would be *still later.* Human planetary exploration was planned for *much later still.* The PSAC closed the booklet by saying the United States must be "cautious and modest in our predictions and pronouncements about future space activities—and quietly bold in our execution."[115] The race mentality was clearly not present. As Eisenhower said in his memoirs, information from purely scientific exploration should "be made available to all the world. But military research would naturally demand secrecy. The highest priority should go of course to space research with a military application, but because national morale, and to some extent national prestige, could be affected by the results of peaceful space research, this should likewise be pushed, but through a separate agency."[116]

Although Eisenhower was not overly enthusiastic to create a civilian space agency, he saw it as "a preemptive strike to prevent something less wise from being done. . . . Eisenhower had to act authoritatively or take a political beating from his rivals in Washington. . . . Left to his own devices, President Eisenhower would have been quite pleased to undertake a modest space program that was oriented toward practical applications. His type of space program was motivated by a realistic desire to invest limited funds in space systems with military and other applications rather than to engage in what he characterized as space stunts." However, lacking the luxury of a perfect world, Eisenhower endorsed the creation of NASA "because it was the least bureaucracy he could get away with in the post-Sputnik crisis atmosphere."[117]

Balancing Civilian and Military Responsibilities in Space

The Space Act was drafted during the period between the 6 March NSC meeting and its submission to Congress on 2 April 1958. Representatives from the PSAC and the BOB crafted the bill; one historian explains, "The Department of Defense was not brought into the picture until the end of March when the draft bill was sent to various agencies for comment."[118] Nonetheless, during the March–April period, the DOD had "no strong objections" to the idea of expanding the NACA into a space agency. This was because, as Killian recalled, Quarles acted as DOD's point man for space and "was extremely sensible and understanding about the whole program, and very tough about his views of the Department of Defense taking on more than it needed for defense purposes."[119]

During deliberations on the Space Act, Congress modified the civil-military balance of power, as defined in Eisenhower's proposal, before the act became law. Concurrent service and interservice activity in the military space field, after Sputnik, also helped convince Eisenhower to endorse the idea of a civilian agency for scientific space exploration.

Interservice Rivalry and the Creation of the ARPA

In the military's post-Sputnik program, the Army Ballistic Missile Agency launched *Explorer I*—America's first satellite— on 31 January 1958. As the post-Sputnik clamor grew for an American satellite and it was clear that the Naval Research Laboratory's Vanguard program could not easily be accelerated, Eisenhower told the DOD on 8 October to "do what is necessary to have the Redstone ready as a backup."[120] In response the secretary of defense authorized the Army on 8 November 1957 to configure a Jupiter C (a Redstone modified into an IRBM) to launch a satellite. On 6 December 1957, Vanguard was given the first chance—it rose a few inches from its launch pad and exploded. Lyndon Johnson called it "one of the best publicized and most humiliating failures in our history." Pundits quipped that it should be called "Dudnik," "Flopnik," "Stayputnik," or "Kaputnik."[121] At the United Nations, the Soviets took advantage of the American failure and offered the United States the same kind of technical assistance it made available to underdeveloped nations.[122] The Army proceeded expeditiously and within eight weeks launched America into space, which "invoked an all but audible sigh of relief across the country."[123]

The relatively independent and uncoordinated efforts of the services are one of the reasons Eisenhower established a dual program in the Space Act. Both before and after Explorer, the "Army sought a major role in military space technology."[124] The Army had launched the first satellite, and the ABMA was doing most of the work in very large rocket engines, over one million pounds of thrust (the Air Force's rocket engines for the Atlas

ICBM developed approximately 300,000 lbs. of thrust).[125] The Air Force's WS-117L concept had been intensely studied since 1950, got officially underway in March 1955, and, at that time, was drawing up detailed plans for a man-in-space program. The Navy NRL had responsibility for the Vanguard-satellite effort. Although Eisenhower had played a responsible role by allowing this proliferation, he was not impressed by its duplication and took steps to control it: first by creating NASA and second by creating the ARPA.

The Air Force was perhaps most vocal in its post-Sputnik drive for increased space responsibilities and programs. It created a special panel under Edward Teller soon after Sputnik "to examine possible Air Force contributions to a United States technical demonstration which would counter world reaction to the USSR earth Satellite."[126] Teller recommended putting "the ballistic missile and space flight programs on a maximum effort basis in all its aspects, without reservation as to time, dollars or people used." He maintained that the DOD's R&D budget should be "inviolate against financial restrictions" because "if we continue to lag behind the USSR in the conquest of space, we risk losing our deterrent ability." The Air Force, he thought, should spearhead the efforts that we would "now undertake to equal and surpass the Russian achievement. Existing Air Force programs will, if vigorously supported and pushed forward, give our nation the needed capability."[127]

Air Force leaders soon began to voice the idea that space was a natural extension of their responsibility and incorporated those concepts into its official doctrine. Gen Thomas D. White, Air Force chief of staff (CSAF), declared on 29 November 1957: "I feel that in the future whoever has the capability to control space will possess the capability to exert control of the surface of the earth. . . . We airmen who have fought to assure that the United States has the capability to control the air are determined that the United States must win the capability to control space. . . . I wish to stress that there is no division, per se, between air and space. Air and space are an indivisible field of operations."[128] By December 1959, the Air Force had coined the term *aerospace* and incorporated it into its official doctrine manuals: "The aerospace is an operationally indivisible medium consisting of the total expanse beyond the earth's surface. The forces of the Air Force comprise a family of operating systems—air systems, ballistic missiles, and space vehicle systems. These are the fundamental aerospace forces of the nation."[129] These concepts seem moderate when compared to those of others elsewhere in the Air Force.

In a 28 January 1958 speech, Brig Gen Homer A. Boushey posited, "the moon provides a retaliation base of unequaled advantage. . . . It has been said that 'He who controls the moon, controls the earth.' Our planners must carefully evaluate this statement, for, if true (and I for one think it is), then the United States must control the moon."[130] The Air Force's deputy chief of staff for development, Lt Gen Donald L. Putt, supported a military base on the moon while testifying to Congress in March 1958 and declared this was "only a first step toward stations on planets far more distant from which control over the moon might be exercised."[131]

Generals Boushey and Putt were not unrepresentative crackpots. On 10 December 1957 the Air Force attempted to create a new directorate of astronautics and named Boushey as its commander. However, one OSD official "reacted unfavorably," stating the Air Force "wanted to grab the limelight and establish a position." The SECDEF also opposed the Air Force creating a space organization and felt it was an Air Force bid for popular support.[132] Eisenhower reportedly "hit the roof" and phoned Secretary McElroy from a NATO meeting in Paris to express his displeasure.[133] Bending to that pressure, the Air Force disbanded its new astronautics directorate on 13 December, becoming the Air Staff's shortest-lived directorate. Eisenhower's space-for-peace policy was designed to overtly emphasize civilian scientific exploration and divert attention away from covert reconnaissance satellites. This type of bold Air Force rhetoric worked against that strategy and was, most certainly, not welcomed at the presidential level.[134]

In response to *Sputnik I*, the Air Force drafted bold plans for a large space program. Schriever's Ballistic Missile Division (BMD) submitted a $1.5 billion plan to the OSD in January 1958 for FY 59 that included the following programs: R&D test vehicles, satellite reconnaissance systems, lunar-based intelligence-gathering system, orbital-defense systems, logistics requirements for lunar transport, and strategic communications.[135] Malcolm A. MacIntyre, the undersecretary of the Air Force explained, "A space warfare capability on the part of the United States is vital to the survival of the free world. . . . We must seek out every possible means of acquiring a military capability to control space—or to deny that capability to an enemy."[136] The gulf in space strategy that existed between senior leaders in the OSD and the Air Force was revealed when the OSD limited the Air Force to only $177 million of its FY 59 $1.5 billion request.[137] To help counter the effect of statements emphasizing military control of space and to limit the role of the services, Eisenhower turned to his scientists: Killian and the members of the PSAC.

Killian has written that he and the other civilian scientists affiliated with the government "felt compelled to ridicule the occasional wild-blue-yonder proposals by a few air force officers for the exploitation of space for military purposes. . . . These officers, often more romantic than scientific, made proposals that indicated an extraordinary ignorance of Newtonian mechanics, and the PSAC made clear to the president the inappropriateness of these proposals."[138] Dr. Lee A. DuBridge, a member of the PSAC and president of the California Institute of Technology, told Congress that "in many cases it will be found that a man contributes nothing or very little to what could be done with instruments alone" and added that a military lunar base was not necessary because "it is clearly easier, cheaper, faster, more certain, more accurate to transport a warhead from a base in the United States to an enemy target on the other side of the earth than to take the same warhead . . . and then shoot it back from the moon."[139] DuBridge continued, warning against "wild programs of Buck Rogers stunts and insane pseudomilitary expeditions."[140] These comments were perhaps harsh, but indicated the Eisenhower administration's lack of tolerance for open speculation about the military uses of space.

Killian reported, "President Eisenhower was disturbed by the numerous space proposals by the military services which did not contribute to national security. The services were fighting for 'weapons systems in space,' which neither PSAC nor BOB regarded as consistent with the president's view."[141] The military's tendency for aggressive rhetoric and its interservice rivalry led Eisenhower to eventually support the creation of NASA. The bitter interservice rivalry that bedeviled Eisenhower had begun when he accelerated the ballistic missile program in 1954 and now appeared likely to spread to the space arena.[142] In response to both fact and perception, Eisenhower not only created NASA but authorized the creation of the ARPA as a separate OSD-level space agency to manage the military space projects and hopefully curtail the interservice conflict.

A week after Sputnik, Eisenhower volunteered, "he sometimes wondered whether there should not be a fourth service established to handle the whole missiles activity. . . . The president suggested that Mr. McElroy let people know . . . that he will deal with a very heavy hand in putting his own ideas into effect."[143] Yet several days later someone leaked to the trade press specific information concerning the WS-117L program.[144] One scholar describes that leak as deliberate and part of "a stream of sensitive information [which] began to flow from individuals within the Air Force directly to congressmen considered sympathetic to Air Force views on military research and procurement."[145]

General Medaris's memoirs and other primary sources amply support these assessments. He described several vitriolic attacks on the Air Force and its managerial competence.[146] On 19 November 1957, the Army submitted to the OSD a proposal for a satellite reconnaissance system that would largely duplicate the Air Force's WS117L because "the Army can satisfy the Nation's and its allies' urgent requirement for accurate and timely intelligence from within the USSR in less time, for less cost, and with a greater assurance of success than any other agency." The Army stated that with its ABMA and von Braun, "Nowhere else in this nation does there exist a comparable reservoir of proven experience and competence."[147] The Army even proposed Project Adam, its own human-spaceflight program, which it justified as research into "large scale transport by troop-carrier missiles." That program, the Army continued, would also "enhance the technological prestige of the United States in the eyes of its friends, allies and citizens."[148]

The Air Force worried about the precedent if America's first satellite was launched by the Army as evidenced by a comment made by the Air Force liaison officer for Project Vanguard: "The Army can certainly be expected to beat the drums for the assignment of satellite and space projects exclusively to the Army. . . . This will place the Army in a most favorable position in regard to future space problems."[149] The Air Force deputy chief of staff for development was also concerned about the consequences of a successful Vanguard launch by the NRL: "If the Vanguard program continues and has a lucky success, two things must be considered: a. The Navy will have a basis for claims on space roles. b. The civilian scientists will be able to claim success."[150] Such dissension distressed Eisenhower. He often expressed, as he did on 4 February 1958 that "he has come to regret deeply that the missile program was not set up in OSD rather than in any of the services." When York, DDR&E's first director, mentioned to Eisenhower that the ABMA was highly competent and interested in a permanent role in the space program, "The president quickly interjected a caution not to put the satellite job in any of the services." Eisenhower emphasized he wanted to see the space program "kept out of service politics."[151] Under these circumstances, an expanded military role in the space program was highly unlikely. Each service, York summarized, could justify building any rocket or satellite it desired and could state why it could accomplish that task better than any of the other services: "There just was confusion, chaos, [and] unnecessary duplication at the highest level."[152]

Eisenhower's response not only placed the civilian space-exploration program under NASA but also was to create the ARPA. McElroy first discussed the administration's intention to create a special weapons laboratory on 1 November 1957. That institution "would limit its operations to research and exploratory development, it would not affect military department roles and missions."[153] Then on 20 November he explained to Congress his plans for a "special projects" agency whose duty would be to unify the various space projects then scattered

63

among the three services. OSD could then control "all our effort in the satellite and space research field" and mitigate the interservice rivalry.[154] McElroy told Johnson's preparedness subcommittee that the ARPA meant, "There is not going to be any satellite program in the services except as directed by the Advanced Research Projects Agency . . . the entire program will be directed and controlled by a single agency."[155]

The Joint Chiefs of Staff (JCS) opposed the ARPA's creation, and that was one of the few things the Army, the Air Force, and the Navy could agree on. Commenting on earth satellite and other space vehicles, the Air Force deputy chief of staff for development said, "In my opinion the national interest would best be served by a firm decision assigning these two mission responsibilities to *one* military service. I believe the Air Force should receive confirmation that both areas are within its purview of mission assignment" (emphasis in original).[156] But Secretary of Defense McElroy overruled them all. Eisenhower acknowledged that and included the pending ARPA creation in his 9 January 1958 State of the Union address saying, "In recognition of the need for single control in some of our most advanced development projects, the Secretary of Defense has already decided to concentrate into one organization all antimissile and satellite technology undertaken within the Department of Defense."[157]

Congress passed the legislation funding the ARPA on 12 February 1958 and authorized the secretary of defense to "engage in such advanced projects essential to the Defense Department's responsibilities in the field of basic and applied research and development which pertain to weapons systems and military requirements . . . and for a period of one year from the effective date of this Act, the Secretary of Defense or his designate is further authorized to engage in such advanced space projects as may be designated by the president."[158] The ARPA was, in fact, America's first space agency. It started operations in February 1958, whereas NASA did not begin functioning until October 1958. Congress clearly limited the ARPA's funding and authority over civilian space projects to one year because, as a later congressional report explained, the ARPA was an "emergency measure to provide coordination and leadership . . . for space projects already underway or en-visioned within the Defense Department. ARPA was our only attempt at giving immediate direction to the space effort on a fairly high level. It came also as a response to a feeling that the far-ranging space exploration projects were hard to reconcile with individual services' missions. At the time of its founding, plans for a civilian space agency were already developing."[159] Eisenhower transferred all military and scientific space proj-ects to the ARPA on 2 March 1958. Three weeks later the ARPA's conditional nature was emphasized in a memo to the SECDEF when he wrote the transfer was "with the under-standing that when and if a civilian agency is created, these projects will be subject to review to determine which would be under the cognizance of the Department of Defense and which under the cognizance of the new agency."[160] Eisenhower's staff secretary summarized the ARPA's purpose: "We simply had to get above this very difficult situation involving the services. You see, if you were dependent on the services that meant you were going to be affected and afflicted by this rivalry rather than having an agency which would go at the problems from a national point of view."[161]

The ARPA's period of importance in space lasted through September 1959. During that time, the agency had managerial responsibility for all of America's space programs. In most cases the ARPA immediately contracted the projects back to the original organization for execution. The services continued their space efforts: the Air Force developed the WS-117L, the NRL administered the Vanguard program, and the ABMA worked on heavy boosters. Nevertheless, the services resented losing the final say in their space projects and felt the ARPA posed the danger of evolving into a fourth service.

The ARPA's space-project responsibilities were transferred back to the services by September 1959. Although it has con-tinued as an R&D organization investigating cutting-edge technologies, it faded from importance in the space field. The ARPA had been another layer of the OSD bureaucracy that the Air Force had to contend with in addition to NASA's increas-ing power and bureaucratic competition. As the ARPA space influence faded, the new director of DDR&E began and continued to exercise tight control over Air Force space projects throughout the 1960s.[162] General Schriever, and other Air Force officers, asserted repeatedly that "the services were under proper supervision at the OSD level, we didn't need an ARPA. We were doing that kind of work."[163] Even before the ARPA was official, Schriever had said, "Any program to establish a separate astronautics management agency would result in duplication of capabilities already existing in the Air Force ballistic missile programs at a cost in funds and time."[164] Strict OSD-level supervision, through the ARPA and later the DDR&E, was a reality that the Air Force had to learn to live with, just as it would have to share the playing field with NASA in a few months. One of the consequences of the creation of the ARPA and the pending creation of NASA was that by the summer of 1958, "The identity of the well-thought-out Air Force space program had been lost."[165]

Dividing the Indivisible

Eisenhower's space-for-peace policy had been emerging since early 1955 and culminated in the submission of his administration's version of the Space Act on 2 April 1958. One important component of the space-for-peace policy was the separation of the US space program into distinct civilian and military components. Before continuing with the discussion of the Space Act, one must consider the question, Was it possible to separate civilian and military space programs? Maurice Stans, Eisenhower's budget director, told the attendees during the congressional hearings on the Space Act that the president would simply have to assign many projects that exist in the gray area between NASA and the DOD to one organization or the other. Because, he continued, the BOB "has found it almost impossible in legislation to establish precise division between agencies with closely related programs. I don't think it can be done here."[166]

Space policy analyst Galloway provided the most cogent analysis of this question to the congressional committees considering Eisenhower's Space Act in her reports. In one dated 11 May 1958, she explained, "The fact that the satellite as an instrument is practically indivisible as between military and civilian use has not been stressed, with the result that some people are trying to divide things which cannot be divided without increasing the cost beyond necessity." She also warned against the tendency to characterize the DOD as *military* and the soon-to-be NASA as *civilian* because the DOD was in fact under civilian control and NASA's predecessor, the NACA, spent much of its time on military matters:

> The fact to emphasize is that <u>both</u> ARPA and NASA are scientific. . . . The fact that one scientist wears a uniform while his co-worker wears a civilian suit does not mean that the uniformed scientist is an incipient Napoleon who threatens popular government. . . . Control by a group of scientific specialists is just as dangerous to democratic government as control by a group of military specialists. [The important point is the] concept of control of policy by the elected representatives of the people over the various professional specialists who lack the breadth of vision required for guarding the common welfare and the public interest. . . . The main reason we must have a civilian agency in the outer space field is because of the necessity of negotiating with other nations and the United Nations from some non-military posture. . . . If all we wanted to achieve was maximum efficiency at minimum cost in a satellite program, we could leave it all in ARPA as presently constituted.[167] (emphasis in original)

The committees reprinted Galloway's sentiments in their final reports. House members were convinced that "it is extremely difficult to separate scientific discoveries directly applicable to the military from those most important to peaceful uses. Discovery is impartial and impersonal. It can be controlled by no blueprint. It can be contained by no laws. . . . The job of a space agency is to turn a sword into something of a cosmic plowshare."[168]

Galloway's comments are very helpful in attempting to sort out the decisions and events that took place between October 1957 and October 1958, especially the role of Congress. Although it might not be possible to categorize the nature of all space projects as either civilian or military, the international nature of space during the Cold War made the attempt necessary. Further, the real concern was not reining in a corps of out-of-control "Colonel Blimps" but rather ensuring that America's elected representatives, including the president, maintained firm control over all parties who might want to use the American space program to advance their own agenda.

The Congressional Role in Balancing
Civil-Military Responsibility

There were, in fact, four documents referred to as the Space Act: Eisenhower's submission to Congress on 2 April 1958, the separate versions passed by the House and Senate, and the final version crafted by a conference committee which Eisenhower signed on 29 July 1958. Fortunately, they agreed on all but two of the main points relevant to this discussion. Those exceptions were the exact wording used to divide the responsibility for space projects between NASA and the DOD and how to effect the coordination of subsequent efforts on space projects. The relevant points of agreement require little discussion except to point them out. First, there was consensus that US space activities "should be devoted to peaceful purposes for the benefit of all mankind." Second, there was agreement on the fact that "such activities shall be the responsibility of, and shall be directed by, a civilian agency exercising control over aeronautical and space activities sponsored by the United States, except . . . [the "exception clause," the language that followed the word *except*, was where the differences occurred]." Third, there was general agreement on the US space objectives that follow: expanding human knowledge, studying the potential benefits of space vehicles, preserving "the role of the United States as a leader in aeronautical and space science and technology [and] making available to agencies directly concerned with national defenses . . . [those] discoveries that have military value or significance," and cooperating with other nations in space.[169]

Eisenhower's version of the exception clause read, "Except insofar as such activities may be peculiar to or primarily associated with weapons systems or military operations, in which case the agency may act in cooperation with, or on behalf of, the Department of Defense."[170] This language, and the discussions in the Eisenhower administration preceding its version of the Space Act, led many officials within the DOD to conclude the proposed NASA would simply be an extension of and expansion upon the old NACA. As noted in chapter 1, the DOD and the Air Force had a very cozy and comfortable relationship with the NACA. In April 1958, Quarles, deputy secretary of defense, concluded that NASA would be a "logical extension" of the NACA and stated, "It is assumed the operation of the new agency would bear the same relationship to the Department of Defense . . . as the NACA now does in the aeronautics field, and specifically, that NASA would continue to perform aeronautical research that is basic to military aeronautics."[171] When it became clear during the subsequent congressional hearings that NASA would become an operating agency, not just a research agency, the DOD became much more concerned with the president's exception clause.

Some DOD officials expressed these assumptions during the congressional hearings on the Space Act that occurred in both houses between 15 April and 15 May. ARPA director Roy Johnson stated the DOD felt its relationship with NASA would be "basically an extension of the relationship with NACA as it existed in the past and there was not much concern about the language or the change in relationship."[172] The top USAF R&D officer believed NASA "should perform almost the same role across the board" as the NACA had "with all the agencies of the Government in essentially the same manner and the same method that has been practiced in the past. . . . So I view their role and relationships as just remaining practically the same except extending in scope from conventional aeronautics into space. It would seem to me that NASA would still function in an advisory capacity in the same way that they have in the past." Even Doolittle, the NACA chairman, said, "I see no change in relationship between the military services and the NACA as a result of the establishment of the NASA."[173] The bulk of the congressional hearings on Eisenhower's Space Act are the story of how this DOD perception changed and the resulting modification of the exception language.

Since the preparedness subcommittee's inquiry termination in January 1958, Congress had been waiting for the White House to submit legislation so it could resume an active role. Prior to the president's proposals arriving on 2 April, both Houses had established new standing committees (for the first time since 1946) to deal with the issue of space.[174] Both Houses felt America's organizational response to Sputnik was so important that both selected their majority leaders to chair the committees: Senator Lyndon Johnson (D-Tex.) and Rep. John McCormack (D-Mass.). Likewise, very senior congressmen were selected to serve on those new committees.

The Senate created its Special Committee on Space and Astronautics on 6 February 1958 and the House followed on 5 March with its Select Committee on Astronautics and Space Exploration. Rhetoric dominated both occasions, with Johnson declaring, "The exploration of outer space will dominate the affairs of mankind, just as the exploration of the Western Hemisphere dominated the affairs of mankind in the 16th and 17th centuries."[175] The Senate had previously become familiar with much of the technical information during its preparedness hearings. Its new special committee therefore confined its hearings narrowly to Eisenhower's bill and met for just six days (6–15 May) and called only 20 witnesses. Since the House had lesser institutional experience with space issues, its select committee met for 17 days between 15 April and 12 May and called 48 witnesses. Although both committees were reacting to the same bill, they had different concerns. The House seemed concerned that the DOD would have too much power in the space arena while the Senate questioned whether they would have too *little* say in space R&D and operations. In both bodies, however, the key issue was the same: the proper balance between civilian and military control in America's space effort.

None of the witnesses who testified before the committees questioned the fundamental wisdom of civilian control. Many questions focused on how the balance of power between NASA and the DOD would be influenced by the various permutations of the exception clause. There was also much concern that the two organizations would properly coordinate their activities to ensure America had a rational program with as little duplication as possible. On this second point the solutions differed. The House favored the creation of a relatively large military liaison committee that would meet at the organizational level to ensure programs and projects were properly coordinated. The Senate preferred an approach by the National Space Council similar to the National Security Council that would involve fewer but higher-ranking members such as the secretary of defense, secretary of state, and the NASA administrator, to provide guidance and overall policy direction to America's space program. In a typical bureaucratic compromise, the final legislation ultimately created both bodies.

The House's select committee's concern with the bill's tasking language was that it would enable the DOD to control almost all of the space program by simply declaring that virtually everything was related to military weapons or operations. Chairman McCormack repeatedly objected: "You create a civilian situation but then you accept everything that is peculiar to or primarily associated with weapons systems or military operations. That covers everything. . . . Through the word 'except' you take all the powers away from it [NASA] practically unless the Defense Department says it is all right. . . . The Defense Department might hold that the sending up of satellites is primarily military. Then you realize under the terms of the bill the military makes the decisions, does it not, unless we change the language?"[176]

Typical responses to such queries stated that unresolved issues would be solved in a fashion similar to disputes between any two executive branch departments and agencies. Although most, given proper coordination, could probably be worked out at lower levels, unresolved issues would ultimately go to the president for resolution.

DOD personnel, uniformed and civilian, repeatedly urged the House committee members not to accept language that would prohibit the DOD from engaging in R&D or space operations that they considered part of their national security responsibility. A parade of witnesses emphasized, "The bill . . . should not have language in it which says we can only work on things for which there is a well-defined requirement." Dr. York, the ARPA's chief scientist, said the DOD had to have the freedom to engage in very basic exploratory R&D that might or might not lead to militarily useful hardware.[177] When they realized that NASA would not be an advisory R&D organization similar to the old NACA but, rather, another agency operating in space on a day-to-day basis, Rear Adm John Hayward, assistant chief of naval operations for R&D, emphasized that it was "of great importance that the delineation between military and civilian interests be made clearly and justly to avoid jurisdictional disputes."[178] The importance of modifying the exception clause became clear to DOD leaders when civilian scientists, such as the well-known physics professor Dr. James Van Allen, testified before Congress and said that he had "the strong feeling that the Department of Agriculture . . . might have more cognizance and basic interest in the research of outer space than Defense."[179] The developing interpretation of the Eisenhower exception language indicated that the DOD would have to clear its projects through NASA and that NASA would handle a large portion of military space R&D, which "aroused the DOD to legislative counterattack."[180]

Unfortunately, this short-term task of drawing a clear delineation between civilian and military concerns in space was probably impossible. General Medaris, already experienced in space as the director of the Army's Ballistic Missile Agency, explained, "Neither this bill nor succeeding events can completely define in all cases where the division point is. . . . I find it very difficult in my own mind, with assurance, to divide out the scientific, the peaceful, and the military."[181] DOD witnesses urged Congress to adopt some mechanism for close cooperation and granting the DOD flexibility to pursue a wide variety of R&D that could lead to national security hardware at some point in the future. General Schriever, director of the Air Force's ICBM development team, expressed a sentiment common among military witnesses when he said that "I think any civilian agency that is established should not have an inhibiting influence on the military's being able to carry out its requirements."[182] Most DOD officials would have agreed with Deputy Defense Secretary Quarles in urging "administrative latitude" in working out the NASA-DOD relationship.[183] Roy Johnson, the combative ARPA director, bluntly expressed the DOD's concern: "The legislation setting up a civilian group should not be so worded that it may be construed to mean that the military uses of space are to be limited by a civilian agency. This could be disastrous. It behooves the writers of this legislation to state positively this freedom clearly and without equivocation. . . . If the DOD decides it to be militarily desirable to program for putting man into space, it should not have to justify this activity to the civilian agency."[184] ARPA's chief scientist added, "If the Department of Defense wants to put up reconnaissance satellites I don't see why the civilian agency should have anything to say about it."[185] By the end of the hearings, even McCormack was convinced of the necessity for such language to guarantee the DOD freedom to conduct R&D efforts: "I realize the difficulty of divorcing what is civilian from military, and I think any doubt should be resolved . . . on the side of safety. . . on the side of the military."[186]

Accordingly, the House amended Eisenhower's original language. They gave the DOD freedom to conduct R&D and *directed* NASA to cooperate with the DOD, as opposed to the Eisenhower language, which said NASA *may* do so. Section 102 of the House's bill said NASA, "shall act in cooperation with (A) the Department of Defense insofar as such activities are peculiar to or primarily associated with weapons systems, military operations, or the defense of the United States (including the research and development necessary to make effective provision for the defense of the United States)."[187] Although the House vested more responsibility in the DOD and mandated cooperation, the House still did not grant the DOD sole responsibility for that part of the space program *peculiar to* or *primarily associated* with defense—that would have to wait for the Senate. The House had clearly felt that authority was unnecessary. One of its staff reports stated the House language "makes clear the Space Agency is civilian and free from military domination, yet organized so that neither civilian nor military activities will be slighted or obstructed."[188] In the House members' minds, the provision for the Military Liaison Committee ensured, through its agency-coordinating function, that no slighting or obstructing would take place.

The Senate Special Committee took the final step and established the DOD's authority to completely control those aspects of the space program *peculiar to* or *primarily associated* with defense. DOD officials had expressed to the Senate almost exactly the same concerns they had presented to the House. The Senate was initially much more concerned that the civilian agency could inhibit the DOD's role in space, which led the DOD's testimony to have more impact in the Senate's proceedings than it had in the House. Quarles reemphasized: the *peculiar to* language must not "define by exclusion or otherwise the proper activities of the Department of Defense. I would construe this language as not limiting the clear responsibility of the Department of Defense for programs that are important to the defense mission, including the support of research that is closely related to the defense mission."[189] He recommended clear language tasking the DOD to be the responsible agent for such activities. The ARPA's Johnson explained that the DOD was "certain that a high order of cooperation must exist if the national program is to be accomplished. . . . I believe what is really important here is that the Department of Defense not be precluded from going into a scientific exploration for defense reasons."[190] In other words, the DOD should be able to pursue programs "it believes have a reasonable chance of fulfilling military ends without having a civilian agency say yes or no."[191] A number of uniformed officers made the same points.

Senate Special Committee members on both sides of the aisle seemed more than amenable to this train of thought. Senator Bourke Hickenlooper (R-Iowa) believed the military aspects might be placed at risk of being "deteriorated under perhaps certain imagined or possible civilian attitudes" if applied under the language Eisenhower had put forth.[192] Senator Styles Bridges (R-N.H.), the ranking minority member, suggested the language "must be tied closer to the military than is now proposed in the bill. . . . I am for this space exploration, but the primary purpose of it . . . is the defense of the country."[193] Johnson's final committee report, which was forwarded to the whole Senate, pointed out, "Your committee believes great mischief could be wrought by delegating to the civilian Space Agency authority over military weapons systems and military operations." Therefore, the committee had rewritten the bill's language based upon its "universal recognition that the proposed legislation should not restrict or hamper the Department of Defense . . . [because] the military aspects of the problem are grave, involving as they do the very survival of the nation."[194]

The Senate's modifications were twofold. They ensured that coordination would be accomplished through a small but high-ranking space policy board. More importantly, they tightened up the tasking language and designated America's space program to be the responsibility of, and directed by, a civilian agency "except that activities peculiar to or primarily associated with the development of weapons systems or military operations shall be the responsibility of, and shall be directed by, the Department of Defense."[195] Both the DOD and BOB expressed their approval of this wording.[196]

The Eisenhower administration seemed amenable to the language contained in the House, Senate, and various compromise versions. The White House only insisted that the space policy board not usurp presidential power. Initially it appeared that reconciling the House and Senate versions would be difficult.[197] Despite that initial pessimism, a compromise was reached by early July 1958. One participant cites the willingness of the White House to take part in the discussions key to resolving the differences.[198] The only remaining difficulty was reconciling Johnson's insistence on a strong space-policy board with the White House's concern that such a board would diminish presidential authority. That impasse was broken on 7 July when Johnson visited Eisenhower in the White House. Johnson suggested that the president be designated the policy board's chairman and that it function similarly to the National Security Council. Eisenhower said he felt that would work and agreed to what would become known as the Space Council.[199]

In the end a House-Senate conference committee required only a single day, 15 July, to draft a final version of the National Aeronautics and Space Act of 1958. McCormack agreed to Johnson's Space Council and the Senate's tasking language, which was more friendly to the DOD. In return, Johnson agreed to the House's Military Liaison Committee and backed away from his insistence on a joint Senate-House standing space committee (both Houses would, however, establish their own standing space committees). The final version of the bill passed both houses of Congress by unanimous voice votes on 16 July 1958 with no debate and no amendments.

The Final Product

Eisenhower signed the National Aeronautics and Space Act of 1958, Public Law (PL) 85-568, on 29 July 1958. Its tasking language still contained the House's R&D proviso but was, overall, a victory for the Senate's interpretation. NASA then and now exercises control over, has responsibility for, and directs US aeronautical and space activities, "except that activities peculiar to or primarily associated with the development of weapons systems, military operations, or the defense of the United States (including the research and development necessary to make effective provision for the defense of the United States) shall be the responsibility of, and shall be directed by, the Department of Defense; and that determination as to which such agency has responsibility for and direction of any such activity shall be made by the president."[200]

The House-Senate conference report explained the common ground. Both organizations were afforded the necessary freedom to fully develop their respective peaceful and defense uses to avoid delay and to "exclude the possibility that one agency would be able to preempt a field of activity so as to preclude the other agency from moving along related lines of development. . . . However, because there is a gray area between civilian and military interests, and unavoidable overlapping, it is necessary that machinery be provided at the highest level of Government to make determinations of responsibility and jurisdiction."[201] NASA thus came into existence on 1 October 1958 with only a very general framework to describe its role, mission, and particular responsibilities. The specific division of projects and programs would take place over the next few years by means of bureaucratic give-and-take. Sometimes the process of division would be mutually agreed upon, but on other occasions, the decisions created a measure of hostility.

The Space Act created two organizations designed to facilitate NASA-DOD coordination. The National Aeronautics and Space Council's (NASC) charter was "to advise the president with respect to the performance of the duties" prescribed in the Space Act.[202] It could, therefore, become involved in DOD-NASA disputes. The second organization was the Civilian-Military Liaison Committee (CMLC). The CMLC's tasking was to provide a forum for NASA and the DOD to "advise and consult with each other on all matters within their respective jurisdictions relating to aeronautical and space activities and shall keep each other fully and currently informed with respect to such activities."[203] While the NASC was occasionally used as a forum in which high-ranking administration officials discussed overarching space policy, the CMLC failed to achieve any measure of effectiveness because the appointed members had no authority in either NASA or the DOD and so were bypassed with impunity.

Throughout its deliberations, Congress demonstrated its inclination to regard space as a competitive tool in the Cold War struggle. The House report stated, "The United States must leapfrog these Soviet accomplishments. This will take some years, and will require a genuine mobilization, on a national scale, of the vast scientific and technical capabilities of this country."[204] Senate reports contained similar sentiments. One pair of scholars concluded "there can be no question that for the moment the overriding concern within and outside the Congress was to get the United States in a position to *compete* effectively with the USSR" (emphasis in original).[205]

The Space Act enshrined the concepts of a dual civil-military space program, overall civilian control, and adequate leeway for the DOD to conduct space technology R&D that is related to American national defense. A Senate report could simply note, "The essentiality of civilian control is so clear as to be no longer a point of discussion." At the same time, "There is universal recognition that the . . . legislation should not restrict or hamper the Department of Defense in conducting its aeronautical and space activities which are vital to national security. . . . Each will maintain its own sphere of primary interest, but necessarily there will be areas within which those separate interests overlap."[206] Not all members of the military were entirely happy with the outcome. General Schriever boldly stated that he "was very much opposed to the organizational arrangements right from the very beginning. NACA should never have been disturbed. Creating NASA was an unnecessary creation of an organization." Schriever continued, saying the government simply "took the military, put them over in NASA and started the manned spaceflight program. They would've done much better had they allowed the military to carry out the operational type of flying. We proved that we could do it. We had our people running the programs. Eisenhower was sold a bill of goods by Jim Killian."[207]

However, the die was cast. NASA was a reality and the DOD, and the Air Force, would have to forge some kind of working relationship with it. NASA had powerful congressional allies who were proud of their creation and would serve as effective checks on any perceived DOD-USAF hegemony. President Johnson looked back on his entire political career and the "dozens and dozens" of laws he had sponsored and concluded, "There is not a single one that gives me more pride than the Space Act."[208] Certainly one could quarrel with portions of the civil-military balance struck by the Space Act, maintaining it "sewed as many snarls as stitches in the fabric of American government"[209] or that it "would mark only the beginning of the fight to ensure full civilian control over the nation's space program."[210] There is a degree of legitimacy in both charges. But Eisenhower and the 85th Congress did remarkably well in creating an organizational structure that produced a proper civil-military split in the American space program without unduly restricting either organization's freedom of action. Perhaps the Space Act achieved a more important, though largely unspoken objective. It created an aura of space for peaceful purposes and maintained that dominant impression of the US space program while still ensuring that the quest for operational reconnaissance satellites could continue unimpeded.

"The golfer [Eisenhower] actually knew a great deal more than he was letting on," noted one very perceptive analyst.[211] Although one can say that he failed to appreciate the psychological vulnerability of the American people or anticipate their panicked reaction to Sputnik, he succeeded in resisting the calls for dramatic increases in all sorts of federal expenditures—most particularly defense. Likewise, Eisenhower remained poised when others used Sputnik to repeatedly accuse him of permitting a dangerous missile gap to develop. He knew from U-2 provided intelligence that this was not the case and "took the heat, grinned, and kept his mouth shut."[212] While he did permit the creation of a new civilian space agency that, perhaps, he had not originally supported, he ensured the Space Act protected his fundamental space policy. Beyond NASA's creation, Eisenhower refused to sanction major increases in federal expenditures. As a result, his biographer concluded that "Eisenhower's calm, common-sense, deliberate response to Sputnik may have been his finest gift to the nation, if only because he was the only man who could have given it."[213]

The next two chapters detail how Eisenhower, during his terms as president, continued to rein in the impulses for a prestige-oriented space race with the Soviets and calls for a massive human-spaceflight program. His efforts strongly influenced developing the NASA-DOD relationship but only delayed human spaceflight—an effort that, under President Kennedy, would soon become the primary component of a space program that satisfied the earlier calls for a prestige-oriented space race.

Notes

1. Johnson, *The Vantage Point: Perspectives on the Presidency*, 273.
2. Witkin, *The Challenge of the Sputniks*, 6.
3. Harlow, transcript of oral history, 11 June 1974, 46.
4. Witkin, *The Challenge of the Sputniks*, 6, citing Nixon's 15 October 1957 speech.
5. Goodpaster, memcon, 5:00 p.m., 8 October 1957, 1.
6. NSC, "Working Group on Certain Aspects of NSC 5520," 2.
7. Eisenhower, *Public Papers of the Presidents*, vol. 3, no. 210, 724, 730.
8. Eisenhower, letter to Symington, 29 October 1957, 1.
9. Minutes of Cabinet Meeting, 3 January 1958.
10. NSC, Memorandum of Discussion at the 339th meeting of the NSC, 11 October 1957, 4.
11. Killian, oral history interview, 9 November 1969–17 July 1970, 42, 44. Killian has written about Eisenhower's general approach to the panic following *Sputnik I* saying Eisenhower called him one morning out of the blue. "He wanted me to know, he said, that his own judgment led him to the conclusion that we would not be involved in any hostilities with the Soviets during the oncoming five years and that the Soviets were not as strong as many claimed." Killian, *Sputnik, Scientists, and Eisenhower*, 222.
12. Harlow, oral history interview, 11 June 1974, 18.
13. Quarles, memorandum to president, 7 October 1957.
14. Siegel, oral history interview, 8 June 1976, 10.
15. Reedy, cited in Burrows, *Deep Black: Space Espionage and National Security*, 89.
16. Horelick and Rush, *Strategic Power and Soviet Foreign Policy*, 43, 45.
17. United Press, "Khrushchev Invites U.S. to Missile Shooting Match," 16 November 1957, 1.

18. Some of Eisenhower's other responses to *Sputnik I* were the Defense Reorganization Act strengthening the powers of the SECDEF and the National Defense Education Act, which, for the first time, put the federal government in the business of rendering financial assistance to colleges and universities. See Divine, *The Sputnik Challenge*, for an overview and individual monographs and Clowse, *Brainpower for the Cold War*, for the individual responses.

19. Johnson, *Vantage Point*, 271, 273.

20. Hechler, *The Endless Space Frontier*, 2.

21. Killian, *Sputnik, Scientists, and Eisenhower*, xv, 7. Noted physicist and political conservative Edward Teller, known as "the father of the atomic bomb," declared the United States had lost "a battle more important and greater than Pearl Harbor" (ibid., 8).

22. Roberts, "Sputnik Healthily Destroyed Some Illusions," 20 October 1957, E1. With conflicting public-opinion surveys and other divergent evidence, the question of cause and effect remains: Did the overreactions of the media and Congress cause the people to panic or did the panic spring from the grass-root level and spread to the leadership, which, in turn, was simply reported by the media? Whatever the cause, the important conclusion for this study is that a growing sense of alarm soon reached crescendo and caused Eisenhower to make a response.

23. Glennan, speech to the Industrial College of the Armed Forces, 20 November 1959, 1.

24. NSC, memorandum of discussion, 415th meeting of the NSC, 30 July 1959, 7–8.

25. John S. D. Eisenhower, oral history interview, 28 February 1967, 94–95.

26. Eisenhower, *Waging Peace*, 211, 226.

27. This assumption would turn out to be incorrect, as demonstrated by the continuing Soviet diplomatic protests against reconnaissance satellites throughout the Kennedy administration.

28. Goodpaster, memcon, 5:00 p.m., 8 October 1957, 2.

29. Glennon, *Foreign Relations of the United States, 1955–1957*, vol. 11, 755–56.

30. NSC, memorandum of discussion at the 339th meeting of the NSC, 10 October 1957, 4–6. In response to a question from Nixon on whether the United States still planned to make information from Vanguard available to all, Quarles responded in the affirmative, leading Nixon to agree that this "would be a great propaganda advantage for the United States to give out such information." Eisenhower concluded the meeting by stating, "We should answer inquiries by stating that we have a plan—a good plan—and that we are going to stick to it" (ibid., 4–6). To the full cabinet on 18 October 1957, Quarles explained the US IGY satellite program "had been separated from the military programs so as to keep it purely scientific and thus perhaps obviate or weaken Soviet protests on over-flights. Ironically, the Russians themselves . . . had now established the acceptability of over-flights." Minutes of Cabinet Meeting, 18 October 1957, 2–3.

31. Quarles's response on 16 October 1957 to an Air Force briefing that provided update on the progress of its WS-117L reconnaissance satellite and included other military uses of satellites is a good example of the administration's policy: "Mr. Quarles took very strong and specific exception to the inclusion in the presentation of any thoughts on the use of a satellite as a (nuclear) weapons carrier and stated that the Air Force was out of line in advancing this as a possible application of the satellite. He verbally directed that any such applications not be considered further in Air Force planning." Air Force leaders objected but "Mr. Quarles remained adamant." Col Frederic C. E. Oder, the director of WS-117L program, documented Quarles response in a memorandum for record that was quoted in Hayes, *Struggling Towards Space Doctrine*, 114. The Air Force was repeatedly chastised in a similar fashion over the next several years.

32. Hagerty, transcript of press conference, 5 October 1957, 2.

33. Goodpaster, memcon, 8:30 p.m., 8 October 1957. This was after Quarles stated, "There is no doubt that the Redstone, had it been used, could have orbited a satellite a year or more ago" (ibid.).

34. NSC, "Conference in the President's Office," 8 October 1957, 1.

35. Eisenhower, *Public Papers of the Presidents*, no. 210, 728, 735. The paper is actually dated 9 October 1957.

36. Goodpaster, memcon, 8:30 a.m., 8 October 1957, 1. By the first week of 1958, Eisenhower was almost philosophical: "It seemed ironic . . . that we should undertake something in good faith only to get behind the eight-ball in a contest which we never considered a contest." He added, a bit disingenuously given the prestige-related sections of NSC 5520 and the OCB's pre-Sputnik meetings, "Only very recently has this psychological factor of beating the Russians to it been introduced." Minnich, "Legislative Leadership Meeting, Supplementary Notes," 7 January 1958, 1–2.

37. Hechler, *The Endless Space Frontier*, 11.

38. Eisenhower, letter to Professor Loyd Swenson, 4. Swenson was primary author of *This New Ocean: A History of Project Mercury*.

39. Senate, *Inquiry into Satellite and Missile Programs*, 302.

40. Goodpaster, memcon, 4 February 1958, 2.

41. Piland, memorandum to James Killian, 15 June 1958, 1.

42. Eisenhower, *Public Papers of the Presidents*, no. 322, 865.

43. Barber Associates, *The Advanced Research Projects Agency*, 24. Herbert York concurred, "It was propaganda from the very beginning, and I was opposed to a propagandistic approach. I felt it's hollow and people are going to know it's hollow." York, oral history interview, 12 June 1973, 94.

44. Goodpaster, oral history interview, 22 July 1974, 56.

45. Ambrose, *Eisenhower*, vol. 2, 457.

46. York, transcript of oral history interview, 24 January 1989, 46–47.

47. Beschloss, *Mayday: Eisenhower, Khrushchev and the U-2 Affair*. The U-2 had been tracked by Soviet radar since its maiden journey in mid-1956. One would eventually be shot down in May 1960, causing immense embarrassment to the administration and forcing the United States to stop the

manned overflights. Discontinuing those flights resulted in the loss of virtually all hard data on the USSR until the first successful reconnaissance satellite was orbited in August 1960.

48. Eisenhower, *Public Papers of the Presidents*, no. 230, 789–98. He also announced, at that time, the appointment of James Killian to serve as the first special assistant to the president for Science and Technology. The president had been impressed with Killian's management of the TCP, and this action would be his first concrete response to Sputnik. The president tasked Killian with coordinating federal policy concerning scientific R&D and technology, including space policy. From this point forward, Killian and the scientists who collectively formed the PSAC would play the central role in creating NASA. Other actions Eisenhower took at that time included the creation of the position of DOD guided-missile director, to tackle the rampant interservice rivalry in ballistic missiles.

49. Eisenhower, *Public Papers of the Presidents*, no. 234, 811–12.

50. Gaddis, *Strategies of Containment*, 185. See also Kinnard, *President Eisenhower and Strategy Management*, 69.

51. Minutes of Cabinet Meeting, 3 January 1958, 3.

52. The PSAC files in the Eisenhower Library make clear the prevalence of this attitude among PSAC members. PSAC membership included scientific luminaries from both academia and industry.

53. Hall, "The Eisenhower Administration and the Cold War," 65.

54. Killian, *The Education of a College President*, 326–27.

55. Killian, *Sputnik, Scientists, and Eisenhower*, 28, 36.

56. Eisenhower, letter to Killian, 16 July 1959.

57. Wiesner, oral history interview, 24 July 1974, 23.

58. Divine, *The Sputnik Challenge*, 47.

59. Science Advisory Committee, Notes on Meeting of the Office of Defense Mobilization (ODM), 15 October 1957, 5.

60. Killian, *Sputnik, Scientists, and Eisenhower*, 240.

61. Killian, oral history interview, 53–54, xxix.

62. Maier, "Science, Politics, and Defense in the Eisenhower Era."

63. York and Greb, "Military Research and Development: A Postwar History," 24.

64. Schilling, "Scientists, Foreign Policy, and Politics," 156.

65. The ARS plan on 10 October 1957 suggested creating an astronautical R&D agency and recommended "that a national space flight program be initiated; and second, that an agency having independent status similar to that of the Atomic Energy Commission or the National Advisory Committee for Aeronautics, be created to manage the program" consisting of all space-related R&D except "strictly military applications of space-flight techniques." American Rocket Society, "Space Flight Program Report," 10 October 1957, 2.

66. For instance, the National Academy of Sciences lobbied for a National Space Establishment and the Society of Professional Engineers for a Federal Space Exploration Commission. Rotunda, *The Legislative History of the National Aeronautics and Space Act of 1958*, 21.

67. Killian, oral history interview, 23 July 1974, 14.

68. Roland, minutes of PSAC meeting, 10 December 1957, 9, 12. If nothing else, George Kistiakowsky (PSAC member and Eisenhower's second science advisor) recalled, "PSAC held that NACA had to be included simply to avoid creating two competitive bureaucracies." Kistiakowsky, oral history interview, 22 May 1974, 21.

69. Killian, memorandum for record, 30 December 1957, 628–31.

70. Ibid., 630–31.

71. Horner, oral history interview, 13 March 1974, 46–47.

72. For a verification of Galloway's role, see Griffith, *The National Aeronautics and Space Act*, 72.

73. Extracted from interviews documented in Space Policy Institute, *The Legislative Origins of the Space Act: Proceedings of a Videotaped Workshop*, 40; and confirmed by Eilene M. Galloway, interview by author, 2 June 1995.

74. George Reedy, memorandum to Lyndon Johnson, reprinted in Glen P. Wilson, "Lyndon Johnson and the Legislative Origins of NASA," 365.

75. Johnson, *Vantage Point*, 273.

76. Senate, *Inquiry into Satellite and Missile Programs*, 2–3.

77. For example, Edward Teller recommended accelerating and expanding the ballistic missile and submarine programs and building more civil defense shelters. Vannevar Bush concurred and added science education should be strengthened. In the middle of the hearings Johnson offered a satirical poem expressing his opinion of Eisenhower's defense policy: "I'd rather be bombed than be bankrupt, I'd rather be dead than be broke. 'Tis better by far to remain as we are, And I'm a solvent if moribund bloke" (ibid., 122).

78. Ibid., 244.

79. Ibid., 265, 284, 301.

80. Ibid., 511.

81. Ibid., 603–4.

82. Ibid., 1635, 1649.

83. Ibid., 1710.

84. A DOD official explained how the Preparedness Subcommittee obtained its recommendations. "They ask Defense to submit a week or so in advance a written report on what we are doing to catch up with the Soviets. Then, following the hearing, they issue to the press a report in which they urge Defense to do the very things we have said we are doing. The picture is clear: they are directing Defense, leading the nation in its frantic rush to reduce the state of peril, and we are gratefully—or perhaps even reluctantly—doing as we are told. We go along partly because we have no choice, and partly because these are the same individuals who have to approve our military budget and we can do nothing but lose if we fight them." Gale, "Post-Sputnik Washington from an Inside Office," 232.

85. Senate, *Inquiry into Satellite and Missile Programs*, 2428.

86. Ibid., 2429.

87. Johnson, statement of Democratic leader, 7 January 1958, 1.

88. Ibid., 2–3, 5–6.

89. Schoettle, "The Establishment of NASA," 220.

90. McDougall, *The Heavens and the Earth*, 153–54.

91. Bulkeley, *The Sputniks Crisis and Early United States Space Policy*, 11.

92. Ibid., 196.

93. Killian, *Sputnik, Scientists, and Eisenhower*, appendix 2.

94. Dryden hosted a dinner on 18 December to determine the sentiments of some of the NACA's younger employees concerning its role in space R&D. Reportedly, the "sentiment was overwhelmingly in favor of NACA moving into the space field." Rosholt, *An Administrative History of NASA: 1958–1963*, 34.

95. NACA, "A National Research Program for Space Technology," 14 January 1958, 1–3. That plan went on to propose the areas in which to vastly expand R&D activities to meet the space challenge. Those included the following specific fields: propulsion, vehicle configuration and structures, navigation and guidance, launch and rendezvous, and bioastronautics (ibid., 1–3).

96. Dryden, (address to the Institute for Aeronautical Sciences, 27 January 1958), 1–3. Dryden explained that "the development and operation of military missiles, military satellites, and military space vehicles is clearly the function of the Department of Defense" but that additional vehicles for scientific research and exploration should be operated by the NACA "when within its capabilities or jointly by the appropriate agencies of the Department of Defense and the National Advisory Committee for Aeronautics." The NAS and NSF would cooperate with the NACA to select and plan scientific experiments, assign priorities for research, and render financial support (ibid., 1–3).

97. White House, memorandum, 8 April 1958, 1.

98. Roland, *Model Research: The National Advisory Committee for Aeronautics, 1915–1958*, 730.

99. Doolittle to Dryden, letter, 28 March 1958, 1. At the end of this letter Doolittle wrote, "There are no other copies of this letter except yours and mine."

100. Dryden, oral history interview, 1 September 1965, 2–3, 6.

101. Minnich, Legislative Leadership Meeting—Supplementary Notes, 4 February 1958, 1–2. Nixon pressed the point that the United States should set up a separate agency for "peaceful research projects because the military would not undertake projects without potential military value." But Eisenhower "thought Defense would inevitably be involved since it presently had all the hardware, and he did not want further duplication. He did not preclude having eventually a great Department of Space" (ibid). At a conference after this meeting with Killian, "The President said that space objectives relating to Defense are those to which the highest priority attaches, because they bear on our immediate safety. He recognized that the psychological factor is of importance to our security. . . . He did not think that large operating activities should be put in another organization, because of duplication, and did not feel that we should put talent etc. into crash programs outside the Defense establishment." Eisenhower's only proviso was that "Defense get its own organization correct, i.e., that there is a central organization to handle this in defense." Goodpaster, memcon, 4 February 1958, 3–4.

102. Eisenhower, *Public Papers of the Presidents*, no. 28, 142.

103. Hall, "The Eisenhower Administration and the Cold War," 65.

104. Robert Divine notes Eisenhower's evolving thoughts on a proper organizational structure for US space efforts saying that by the first week of March, "Whatever early attraction he had to a purely military agency now had given way to strong support for NASA as a body that would appeal not only to American scientists but to world opinion in general." Divine, *Sputnik Challenge*, 104.

105. Killian, Brundage, and Rockefeller, to President Eisenhower, memorandum, 5 March 1958, 2.

106. Ibid., 3–5. Killian recognized the proposal's liabilities: (1) the NACA had little experience on projects beyond aircraft and missiles or with large-scale developmental contracts; (2) the DOD already employed most of the scientists working on rocket engines and space vehicles and had contracts with the industrial firms in those areas; (3) the NACA "is not in a position to push ahead with the immediate demonstration projects which may be necessary to protect the nation's world prestige" and so the military may have to handle such projects for a period of time; and (4) the NACA only spends approximately $100 million while the space program will be "substantially in excess" of that amount (ibid., 6–7).

107. Killian, oral history interview, 23 July 1974, 36.

108. Ibid., 24. See also Divine, *The Sputnik Challenge*, 105.

109. There were two official NSC space policies for internal administration use that were successors to NSC 5520. NSC 5814/1, 18 August 1958, and NSC 5918, 26 January 1960 (more correctly referred to as National Aeronautics and Space Council [NASC], *U.S. Policy on Outer Space* because Eisenhower directed it be circulated within the government not as an NSC document but as an NASC document). These two policy statements did not go much beyond the principles elucidated in *Introduction to Outer Space* with one exception. The *Introduction to Outer Space* had nationwide distribution and only mentioned the general idea of reconnaissance from space. By contrast, 5814/1 and 5918 were internal, classified government documents and discussed the reconnaissance satellites in detail.

110. Eisenhower, *Public Papers of the Presidents, 1958*, no. 56, 233.

111. PSAC, *Introduction to Outer Space*, 26 March 1958, in Logsdon et al., *Exploring the Unknown*, vol. 1, 332. It also should be noted that Killian contacted Lyndon Johnson's staff with an offer to brief the administration's conclusions. "And we got back the response that the committee [Senate Special Committee on Space and Astronautics, formed 6 February 1958] did not need any advice or material from the White House with regard to space." Killian, oral history interview, 23 July 1974, 27.

112. PSAC, *Introduction to Outer Space*, 26 March 1958, 1–2, 5.

113. PSAC's *Introduction to Outer Space* also laid out the Eisenhower administration's policy on the military use of space: "There are important, foreseeable, military uses for space vehicles. These lie, broadly speaking, in the fields of *communication* and *reconnaissance*. To this we could add meteorology."

PSAC noted that while reconnaissance from 200 or more miles up would be a challenge, it said that telescopic cameras meant "it is certainly feasible to obtain reconnaissance with a fairly elaborate instrument, information which could be relayed back to earth by radio" (emphasis in original) (ibid., 12).

114. Ibid.

115. Ibid., 14–15.

116. Eisenhower, *Waging Peace*, 257.

117. Launius, "Eisenhower, Sputnik, and the Creation of NASA," 128–29, 138.

118. Rosholt, *An Administrative History of NASA*, 10.

119. Killian, oral history interview, 23 July 1974, 25–26.

120. Goodpaster, memcon, 5:00 p.m., 8 October 1957, 3.

121. Cited in Elliott, *Finding an Appropriate Commitment*, 62.

122. Green and Lomask, *Vanguard: A History*, 202.

123. Ibid., 210.

124. Swenson, Grimwood, and Alexander, *This New Ocean*, 79.

125. Polmar and Laur, *Strategic Air Command*, 299.

126. Teller, *Report of the Teller Ad Hoc Committee*, 28 October 1957, 1.

127. Ibid., 9–11.

128. Cited in Futrell, *Basic Thinking in the United States Air Force, 1907–1960*, vol. 1, 550.

129. Cited in Bowen, *Threshold of Space*, 18. There are many other examples of the developing Air Force institutional mind-set toward space. For instance, in a memorandum from its R&D branch: "The Air Force, with greater justification than any other service, should be primarily responsible for the Astronautics (space) mission. . . . With any stretch of the imagination, the Air Force is the Service legitimately having the greatest responsibility for extending its present three-dimensional mobility out further into space." The only element lacking was "the administrative intestinal fortitude to take appropriate actions within its own sphere of prerogative to begin such work, and at the proper time so notify the DOD." Weitzen to Richard Horner, memorandum: Astronautics Planning, 1. Document is now declassified.

130. Cited in Emme, *The Impact of Air Power: National Security and World Politics*, 872. He elaborated, "the moon represents the age-old advantage of 'high ground.'. . . Lunar outposts and even launch sites could be located on the far side of the moon—never to be seen from earth—yet earth locations could be viewed by telescope from the moon." Boushey, "Who Controls the Moon Controls the Earth," 54.

131. Cited in Divine, *The Sputnik Challenge*, 98.

132. Bowen, *Threshold of Space*, 109–11.

133. Putt, oral history interview, 30 April 1974, 36.

134. In July 1958, the Air Force was finally permitted to establish an innocuously titled position, directorate of advanced technology, under Boushey to supervise space projects.

135. Air Force, Air Force Systems Command (AFSC), *Chronology of Early Air Force Man-in-Space Activity*, 13.

136. Senate hearings, *National Aeronautics and Space Act*, 192.

137. Bowen, *An Air Force History of Space Activities*, 4. This type of rhetoric was by no means limited to the Air Force, nor was the striving for institutional advantage. The ABMA's Wernher von Braun told Congress, "I have not the slightest doubt that the question of whether we or another nation has control of the spaces around the earth will have a very great impact on our military position on the earth itself. In other words, space superiority, control of the spaces around the earth, will soon be just as important as air superiority is today." House, *Astronautics and Space Exploration*, 37.

138. Killian, *Sputnik, Scientists, and Eisenhower*, 112.

139. House, *Astronautics and Space Exploration*, 80–81.

140. Cited in Divine, *The Sputnik Challenge*, 98. A Killian staffer simply pointed out, "We can discount at this point most of the 'Buck Rogers' type of thinking which anticipates hordes of little men in space helmets firing disintegrators into each other from flying saucers." See Johnson to Killian, memorandum, 21 February 1958, 633.

141. Killian, oral history interview, 23 July 1974, 5.

142. For a full account of interservice problems in the early IRBM programs, see Armacost, *The Politics of Weapons Innovation*, 1969.

143. Goodpaster, memcon, 11 October 1957, 2.

144. *Aviation Week*, "USAF Pushes Pied Piper Space Vehicle," 26.

145. Prados, *The Soviet Estimate*, 59. One anonymous OSD official quipped, "I have not heard it suggested that any of the services has employed poisonous drugs or physical violence in its struggles against the others, but few other weapons are neglected." Van Dyke, *Pride and Power*, 200.

146. See Medaris, *Countdown for Decision*.

147. US Army, briefing, 19 November 1957, 1–2, 11.

148. Army, *Development Proposal for Project Adam*, 1–2. Even the Navy had a human-spaceflight proposal called "Manned Earth Reconnaissance," which called for a cylindrical spacecraft with spherical ends that could become a delta-winged inflatable glider once in orbit. Ezell, *NASA Historical Data Book*, vol. 2, 94. Also, see Swenson, et al., *This New Ocean*, 100.

149. Gibbs to Putt, memorandum, 17 January 1958, 1–2.

150. Office of the Deputy Chief of Staff for Development to Horner, memorandum, 6 February 1958, 2.

151. Goodpaster, memcon, 4 February 1958, 2.

152. York, oral history interview, 24 January 1989, 41.

153. McElroy, transcript of remarks, MUS document 270, 3.

154. Bowen, *Air Force History of Space Activities*, 103. McElroy added that the new agency would have "single control in some of our most advanced development projects." House, *Government Operations in Space*, 31.

155. Cited in Hungerford, *Organization for Military Space—A Historical Perspective*, 26.

156. Putt to chief of staff Air Force, memorandum, 22 November 1957.

157. Cited in Griffith, *The National Aeronautics and Space Act*, 11.

158. House, *Advanced Research Projects Agency*, 12 February 1958, 4.

159. House, *The National Space Program*, 9.

160. Eisenhower to the secretary of defense, memorandum, 24 March 1958, 1.

161. Goodpaster, oral history interview, 22 July 1974, 31.

162. The muddled organizational situation of late 1958 and 1959 was a result of the Defense Reorganization Act, PL 85-599, 6 August 1958, which created yet *another* OSD bureaucracy, the DDR&E, to supervise all DOD research, development, technical, and engineering activities. To view this bill, see House, *Defense Space Interests: Hearings before the Committee on Science and Astronautics Hearings*, March 1961, 219. There was an initial period of confusion about whether Roy Johnson, as the ARPA director, was subordinate to or superior to Herbert York, the first DDR&E and whose previous job had been the ARPA chief scientist, serving under Roy Johnson. Finally, Deputy Secretary of Defense Quarles declared in 1959 that Roy Johnson and the ARPA "will be subject to the supervision and coordination of Dr. York's office just as are those of the military departments. House, *Department of Defense Appropriations for 1960*, pt. 6, 1959. DOD Directive 5105.15, 17 March 1959, officially made the ARPA and its projects "subject to the supervision and coordination of the Director of Defense Research and Engineering." Bowen, *An Air Force History of Space Activities*, 176. From this point forward, DDR&E replaced the ARPA director as the voice of the SECDEF in R&D matters.

163. Gen Bernard A. Schriever, oral history interview with the author, 2 July 1996.

164. Futrell, *Basic Thinking in the United States Air Force, 1907–1960*, vol. 1, 590.

165. Bowen, *An Air Force History of Space Activities*, 142. Bowen notes that when Boushey's Directorate for Advanced Technology began operations on 15 July 1958, the nature of the Air Force space program it directly supervised consisted of seven studies, which included a manned reconnaissance system, a lunar observatory, a satellite interceptor, and a 24-hour reconnaissance satellite. Although each of these studies had the possibility of becoming an active program, the USAF did not orbit an actual space vehicle until Project Score was launched in December 1958.

166. Senate, *National Aeronautics and Space Act of 1958*, hearings, 282.

167. Galloway, "Reasons for Confusion about Space Law," 1–4. Her March report stated much the same thing: "The line between the peaceful and military uses of outer space is much more difficult to draw than is the case with atomic energy. . . . Practically every peaceful use of outer space appears to have a military application," such as weather, communication, reconnaissance, and even biomedical research. "We can establish civilian control within the United States, but if it turns out that peaceful uses cannot be scientifically separated from military implications, then how are we to regulate the international civilian-military situation?" She asked, "Upon the basis of what scientific facts can a line be drawn between military and non-

military outer space activities?" She concluded, "It will be a difficult legislative task to devise a law for the effective organization and administration of these far-flung operations in which the military and non-military are so closely associated." Galloway, "Problems of Congress in Formulating Outer Space Legislation," 8–9.

168. House, *The National Space Program*, H. R. 1758, 3, 5.

169. Senate, *National Aeronautics and Space Act of 1958*, reprinted in Logsdon et al., 335.

170. House, *Legislative History of the Space Act of 1958*, 18.

171. Quarles to Stans, letter, 1 April 1958, 1.

172. Senate, *National Aeronautics and Space Act*, 168.

173. House, *Astronautics and Space Exploration*, 131–32, 767, 930.

174. McDougall, *The Heavens and the Earth*, 169.

175. Cited in House, *Legislative History of the Space Act*, 3.

176. House, *Astronautics and Space Exploration*, 837, 862, 981.

177. Ibid., 40.

178. Ibid., 274.

179. Ibid., 864. Dr. James Van Allen, a long-time leader in astrophysics research at the University of Iowa, discovered the radiation belts that surround the earth and which now bear his name.

180. Schoettle, "The Establishment of NASA," 242.

181. House, *Astronautics and Space Exploration*, 144–45.

182. Ibid., 627.

183. Ibid., 1105.

184. Ibid., 1165.

185. Ibid., 1533.

186. Ibid., 1172.

187. House, *Legislative History of the Space Act*, 828.

188. House, *Comparison of H. R. 12575 as Passed by the House and as Passed the Senate*, 18 June 1958, Committee Print.

189. Senate, *National Aeronautics and Space Act of 1958*, hearings, 67.

190. Ibid., 147–48.

191. Ibid., 178.

192. Ibid., 24.

193. Ibid., 73–74.

194. Senate, *National Aeronautics and Space Act*, S. Rep. 1701. Johnson emphasized in the Congressional Record that his intention was to say to the DOD: "You shall have complete responsibility for those aeronautical and space activities primarily associated with research into and development of our weapons systems and with military operations, both in peacetime and wartime." At the same time, "There is no dispute here as to whether we shall have civilian or military control over our aeronautical and space activities. That control will clearly be civilian." Congressional Record, vol. 104, no. 98 for 16 June 1958, reprinted in *Legislative History of the Space Act*, 1107.

195. Senate, *A Bill Establishing a National Aeronautics and Space Agency, with a Section by Section Analysis and Staff Explanation and Com-*

ments, 1958, S. 3609, 18. This bill drafted by the Senate to include President Eisenhower's proposal of 2 April 1958 on space science and exploration was introduced by Johnson as S. 3609 with widespread bipartisan support. The final version, enacted as the National Aeronautics and Space Act of 1958 (PL 85-568), established the National Aeronautics and Space Administration (NASA) and defined the relationship between NASA and the Defense Department.

196. Ibid., 2.

197. The House was worried that under the Senate's language, the DOD would have full responsibility over military space. This "could prevent effective planning of the national space program and critically hamper its coordination. . . . The Department of Defense would have the controlling voice in determining what were military and what were civilian space activities." House, *Comparison of H.R. 12575*, 1–2. The House and its Military Liaison Committee preferred its own language because they felt there would be "a continuous two-way street of information and decision-making." House, *National Space Program*, H. R. 1758, 4. Therefore, the full House did not modify its committee's bill and unanimously passed it by voice vote on 2 June 1958 after only two hours of debate dealing with issues such as patents and salary levels. Nevertheless, the Senate remained committed to its version as the best balance of guaranteeing civilian control while taking care "to insure that the Department of Defense and the military services have the necessary authority and responsibility to carry out those programs and projects which are needed to maintain military security." Senate, Special Committee on Space and Astronautics, *Final Report*, S. Rep. 100, 6. It remained committed to a high-level space-policy board that would go beyond the advisory function envisioned by the House. The Senate's SPB would actually craft America's space program and, through that process, ensure that proper civil-military cooperation and coordination would occur. The Senate version of the bill unanimously passed that body by voice vote on 16 June.

198. Hechler, *Endless Frontier*, 21. Hechler explains that the liaison between Bryce Harlow, Eisenhower's deputy assistant for congressional affairs, and Edward A. McCabe, the House Select Committee's administrative assistant, greatly facilitated the compromising process.

199. Memorandum for record, "Off-the-record meeting between President Eisenhower and Sen. Lyndon Johnson," 7 July 1958, 1.

200. Public Law 85-568, *National Aeronautics and Space Act of 1958*, sec. 102.b.

201. House, *National Aeronautics and Space Act of 1958*, H. R. 2166, 16.

202. Public Law 85-568, *National Aeronautics and Space Act of 1958*, sec. 201. The NASC's members were the president, secretary of state, secretary of defense, NASA administrator, AEC chairman, one other federal-government member appointed by the president, and three other members from the civilian community appointed by the president. Although Eisenhower acceded to the NASC's creation to facilitate passage of the overall Space Act, he refused to hire any staff for it during the remainder of his term, and it met irregularly.

203. Ibid., sec. 204. CMLC's members included a chairman appointed by the president, one or more representatives from the DOD, one or more representatives from the military services selected by the SECDEF, and an equal number of total representatives from NASA appointed by the NASA administrator. The numbers were not fixed.

204. House, *Establishment of the National Space Program*, H. R. 1770, 4. Another House document said the "direct connection between science and the world power struggle will be appreciably extended by the race into space. It is a race, no matter how sincerely we long for some form of viable international cooperation. And it will be viewed as such by the eyes of the world." House, *The National Space Program*, 4.

205. Harvey and Ciccoritti, *U.S.-Soviet Cooperation in Space*, 36.

206. Senate, *National Aeronautics and Space Act of 1958*, S. Rep. 1701, 4.

207. Schriever, oral history interview with the author. Gen Curtis LeMay, vice-chief of staff of the Air Force at the time of NASA's creation, agreed. "We made a costly error when we formed NASA. The Air Force had a good relationship with NACA. NACA did basic research and knew what we wanted. When NASA was formed, it expanded from basic research to an operating organization. It had no management talent." LeMay, oral history interview, January 1965.

208. Johnson, *Public Papers of the Presidents*, no. 616, 9 December 1968, 1174.

209. McDougall, *The Heavens and the Earth*, 176.

210. Divine, *The Sputnik Challenge*, 112.

211. Burrows, *Deep Black*, 90.

212. Senator Barry Goldwater, cited in Divine, *The Heavens and the Earth*, 41.

213. Ambrose, *Eisenhower*, vol. 2, 435.

Chapter 3

Eisenhower Act II:
Forging a NASA-DOD Relationship

This chapter examines the institutional climate that developed between NASA and the DOD in the final 27 months of Eisenhower's term—after NASA began operations on 1 October 1958. To understand this organizational relationship, however, it is necessary to revisit the larger issues of Eisenhower's philosophy and the Cold War environment. In addition, the particular beliefs of the Air Force concerning the necessary level of effort in space and the resulting tension with its civilian supervisors in the OSD are integral parts of the NASA-DOD relationship in the Eisenhower administration and later.

The first "big picture" factor is Eisenhower's beliefs concerning participation in a competitive race with the Soviet Union for prestige using space exploration as a tool. The previous chapter explained that before NASA's creation Eisenhower was generally against this concept, but did not totally rule out certain competitive projects. This principle held true during the balance of his term, when he authorized development of the large Saturn rocket for what can only be surmised as prestige-related reasons. Chapter 4, the final "Eisenhower" chapter and the one focusing on human spaceflight, makes clear that human spaceflight was *not* an area he regarded as legitimate for prestige-related competition.

To Compete with the Soviet Union

The quest for prestige, the search for balance, and the open issue of actually cooperating instead of competing with the Soviets in space comprise three important topics of the US-USSR space race. The collective understanding of these topics sheds light on the overarching issue of the relationship of Eisenhower's space program to that of the Soviet Union.

Prestige

The policy document issued in the summer of 1958 to bring some sense of order to the rapidly changing field of space exploration was NSC 5814/1, *Preliminary U.S. Policy on Outer Space*. The report concluded, "The USSR, if it maintains its present superiority in the exploitation of outer space, will be able to use the superiority as a means of undermining the prestige and leadership of the United States and of threatening U.S. security." Space exploration had "an appeal to deep insights within man which transcend his earthbound concerns" and result in a tendency "to equate achievement in outer space with leadership in science, military capability, industrial technology, and with leadership in general."[1] This did not mean the United States must launch numerous crash-space projects designed to foster prestige. On the contrary, the United States should "judiciously select projects for implementation which, while having scientific or military value, are designed to achieve a favorable worldwide-psychological impact" and also develop information programs "to counter the psychological impact of Soviet outer space activities and to present U.S. outer space progress in the most favorable comparative light."[2]

Eisenhower had a significant challenge in resisting congressional calls for a project-by-project race with the Soviets. The Soviets launched the first satellite to escape earth's orbit with *Luna I* on 2 January 1959, the first lunar impact with *Luna II* on 12 September 1959, and the first photographs of the moon's far side with *Luna III* on 4 October 1959.[3] These types of "spectacular firsts" led Representative McCormack, House majority leader, to declare the United States faced "national extinction." In March 1959 the Senate voted 91 to 0 in favor of authorizing $27.6 million to expand space research and $20.7 million to accelerate Project Mercury.[4]

NASA administrator Glennan expressed the Eisenhower administration's position: "To get into a race with Russia and operate our space program solely because we think they are going to do this or that, and then try to beat them at it, would guarantee their always being in command of the situation. We're in a race all right, but we must run it the way we want and towards goals we set for ourselves."[5] Privately, Eisenhower also set the tone. In a 17 February 1959 conference, his budget director informed Eisenhower that despite planned FY 60 expenditures of $830 million for NASA and the ARPA, Senator Johnson had said, "He will add substantially to the Administration's program, whatever it is." Eisenhower explained,

> He could stand the pressures himself, but he was sure that the Congress would break loose under the pressure. He stated that world psychology on this matter has proven to be tremendously important—even if not too well informed. He thought it was indisputable that we must show considerable performance in this field. The pressures are great, and people are demanding miracles. . . . The President said this is a stern chase only in one field—that of propulsive capability; by concentrating on this field the Soviets are ahead of us. He said he did not minimize the importance of ourselves attaining the propulsive capabilities that we need. . . . He would like to see NASA reprogram its operations in order to put maximum effort behind the achievement of boosters of greater thrust—which is the visible element in affecting world psychology. . . . In the present circumstances, he felt we must lay more stress on not going into debt by spending beyond our receipts. At the same time, the relationship of the program to the Soviet rate of advance must be clearly recognized.[6]

All the important points are clear: a reluctance to race in general but an acceptance of its necessity in particular instances; an acceptance that a large rocket booster would be necessary for those cases in which prestige-related competition was necessary—this meant Eisenhower would support the Saturn program that gave President Kennedy a solid technological foundation to approve Project Apollo to go to the moon; and a continuing concern with the avoidance of deficit spending. Eisenhower's statements over the balance of 1959 and throughout 1960 support this general approach to space policy.[7]

NASA did not take over management of America's heavy-lift space booster, the Saturn program, from the DOD's ABMA until October 1959; before then NASA's ability to influence Saturn's developmental priority and funding was extremely limited. It was only in association with this transfer of most of the ABMA to NASA that the conditions existed for NASA to develop the Saturn into a vehicle available for whatever prestige-related uses Eisenhower wished. Eisenhower explained on 21 October 1959 that this transfer of von Braun's team from the DOD to NASA "will force us to focus on the development of a super-booster, which to him is the key to a leading position in space activities. . . . He thought the super-booster is the key to successful competition and we should concentrate on that. He recalled his principle of attacking one enemy or one principal objective at a time."[8] Eisenhower's concept of a program of priority needs—DOD's reconnaissance satellites, NASA's prestige-related projects, and NASA's scientific R&D—is key to understanding the space program during his administration.[9] The development of the Saturn rocket, which was much larger than the then-current ICBM-based space boosters, remained far enough into the future that Eisenhower never had to specifically define exactly what prestige-oriented projects he would authorize, except to make clear that human spaceflight was not likely to be one of them.

By January 1960, "The President thought that the big booster is the only thing that will have major psychological effect, and if we are going to build it we should build it fast."[10] Glennan recounted a 12 January 1960 meeting with Eisenhower. Concerning its cost, "He said he was quite certain that we were going to have to spend an extra $100 million on Saturn during the course of the spring, and he thought it ought to be settled at once."[11] Eisenhower tasked Glennan to prepare an official request for the extra funds for Saturn and added, "Consistent with my decision to assign a high priority to the Saturn development, you are directed, as an immediate measure, to use such additional overtime as you may deem necessary."[12] On 1 February 1960, Eisenhower approved adding $113 million to FY 61 appropriations to accelerate Saturn and other elements of the US superbooster program such as studies for an even larger rocket, the Nova that would be Saturn's successor. In addition, on 18 January 1960, Eisenhower placed Saturn in the "DX" category of the budget, signifying that it had the highest priority when scarce resources were allocated or when labor shortages emerged.[13] Eisenhower, therefore, was willing to spend significant sums on one project, a next generation space launcher, that would give the United States the capacity at some point in the future to launch unspecified prestige-related payloads in a competitive context with the USSR.

Eisenhower repeatedly tried to assure the public that the US-space program was in fine shape. At a 26 January 1960 press conference, when asked if the United States should not move with a greater sense of urgency in competing with the USSR in space, he replied, "Not particularly, no." He explained that the United States had achieved in five years what the Soviets had been working on since 1945. Therefore, "I don't think that we should begin to bow our heads in shame. . . . I think that once in a while we ought just to remember that our country is not asleep, and it is not incapable of doing these things; indeed, we are doing them."[14] Therefore, he had decided to spend the extra $100 million on Saturn.[15]

Eisenhower felt that NASA's program of scientific R&D had as much potential for winning prestige for America, as did the Soviet pattern of lifting huge payloads into space. In August 1960, he commented on assorted American space accomplishments such as the *Pioneer V* solar satellite, the *Tiros I* meteorological satellite, the *Transit I* navigation satellite, and the *Echo I* passive-communications satellite and emphasized, "All these are the result of a well planned and determined attack on this new field—an attack that promises very real and useful results for all mankind. . . . The United States leads the world in the activities that promise real benefits to mankind."[16] One tally showed that by the end of Eisenhower's term, the United States had launched 31 earth satellites and two deep-space probes—the Soviets seven and one.[17] Glennan said that by 1960 he considered the United States to be behind the Russians in total thrust available and in thrust from first-stage boosters, but "In all other areas, it is my considered opinion that we are not behind the Russians, that we are equal or the better of the Russians."[18] Eisenhower asserted, "The significance of the space program is that it affects the morale of our people. In the field of space there are a certain number of things that affect defense directly. Basically, however, the program is scientific."[19]

Balance

Therefore, the key word for analyzing Eisenhower and space-related prestige is "balance." He did not totally discount the quest for prestige, as evidenced by his strong backing of

97

the Saturn program and his transfer of it from the DOD to NASA. But prestige could only be part of a balanced program in which the DOD's interests were paramount and in which NASA's scientific programs played an important role. The American space program could only achieve stability if it refused to lurch from one priority to another, perpetually reacting to whatever spectacular feat the Soviets accomplished. The PSAC's "Introduction to Outer Space" in March 1958 clearly stated the four reasons for exploring space (national defense, urge to discover, prestige, and scientific knowledge), and the balanced-program principle continued throughout NASA's history under Eisenhower. His commitment to balance also meant refusing to be pushed into deficit spending by panicking over supposed Soviet accomplishments or superiority in space or defense. On 10 March 1959, he reminded congressional Republicans, "Once you spend a single dollar beyond adequacy, you are weakening yourself. . . . Anyone who has read even a little bit on Communism . . . all the way back to Lenin, knows that the Communist objective is to make us spend ourselves into bankruptcy. This is a continuous crisis."[20] Eisenhower's commitment, enshrined in the Space Act, was to ensure the United States was *a* leader in space, not *the* leader in space.

Eisenhower concluded his funding of NASA was entirely adequate for a well-balanced program: "The program, of course, that is already set up is, to my mind, a rather—well, indeed it *is* quite generous. . . . Now remember, Glennan and his crowd are supposed to have the peaceful uses; this, therefore, is not involved except you might say psychologically, in our defending the United States. This seems to me to be a quite splendid program; I mean, a very well supported one."[21] Glennan explained that Eisenhower's request for $230 million for Saturn in FY 61 would lead to an expected first-operational launch before the end of 1964: "I doubt that the Soviet Union will exceed us in thrust capability after that time."[22]

Some budget figures help illustrate the principles of balance, priority-of-defense needs (reconnaissance satellites), and limiting expenditures devoted solely to prestige in space. NASA's first budget was for the period from October 1958 to the start of the next fiscal year in June 1959. NASA received $242 million of which $58 million was transferred from the USAF and $59 million from the ARPA. By comparison, the military space program (assigned to the ARPA at the time) totaled $294 million, of which the USAF's reconnaissance satellite Sentry received $186 million. Therefore, the entire space budget was $536 million.[23] An interesting prefatory note for the next chapter is that only $87 million of NASA's budget and $10 million of ARPA's was devoted to human-spaceflight technology.[24]

By the end of the Eisenhower administration, Glennan and others were actually fighting congressional attempts to cut the space budget below what Eisenhower requested. For instance, Glennan pleaded with the Senate Appropriations Committee on 19 May 1960 for a restoration of the House's $39 million cut from the $915 million presidential request.[25] In FY 60 Congress appropriated $23 million less than Eisenhower asked and in FY 61 $1 million less.[26] The budgeted dollars for years 1959–61 are shown in table 2. Eisenhower's NASA request for 1962 was $965 million. His hope was to level off NASA's budget at approximately the $1 billion level. Therefore, he permitted a five-fold increase in civilian-space spending in his final term.[27] It is true that Eisenhower did not authorize space expenditures on the scale that presidents Kennedy and Johnson would. On the other hand, he was building from ground zero and did in fact permit a several-fold increase in space spending; in addition, Congress reduced his requests at the end of his administration.

Table 2. General-space budgetary trends for the Eisenhower administration (in millions of real-year dollars)

YEAR	NASA	DOD
1959	261	490
1960	462	561
1961	926	814

Reprinted from NASA, *Aeronautics and Space Report of the President, Fiscal Year 1995 Activities*, A-30. (Both FY 61 sums augmented by Kennedy.)

Cooperation

A subsidiary factor to mention in the space for prestige and competitive race discussion for Eisenhower, Kennedy, and Johnson is the question of cooperating in space with the Soviet Union. All three presidents explored this area, and all three failed to achieve major breakthroughs. The reality of the Cold War competitive dynamic in space consistently overshadowed the rhetoric from both sides concerning the desirability of cooperation. Even though Eisenhower was not enthusiastic about competing in space, only reluctantly accepting the need to do so with the Saturn project, his efforts at space cooperation nevertheless came to naught. Much more was this the case for Kennedy who featured prestige-based competition via human spaceflight as the centerpiece of his space policy (even while apparently offering to make the lunar-landing project a joint one with the Soviets).

For Eisenhower, international cooperation meshed nicely with his space-for-peace policy. He saw no reason the United States and the USSR could not jointly pursue scientific projects in space, thereby emphasizing that weapons systems had no place in space while simultaneously paving the way for reconnaissance satellites. If both nations were working together on scientific satellites that overflew each other's territory and neither nation protested, the legal regime of overflight would be established for subsequent reconnaissance satellites. Important in this scheme is the fact that the Eisenhower administration did not see reconnaissance satellites as "weapons systems" and ensured the Air Force changed its nomenclature from WS-117L to the more innocuous Sentry.[28] Reconnaissance satellites were viewed as wholly peaceful because they conducted only the defensive operations of gathering information and did not have any capability to deliver bombs or offensive power of any kind. Endorsing and pursuing "space for peace" was wholly consistent with endorsing and pursuing reconnaissance via satellites at the earliest possible moment because the satellites were seen as an effective deterrent to war. The reconnaissance satellites would lessen the danger of surprise attack through an "unrelenting and increasingly sophisticated effort to peel away the mask that concealed the enemy's most important military and industrial secrets."[29]

On 12 January 1958—before NASA's creation—Eisenhower wrote Soviet Premier Nikolai Bulganin to propose that "we agree that outer space should be used only for peaceful purposes. We face a decisive moment in history in relation to this matter. . . . The time to stop is now. Should not outer space be dedicated to peaceful uses of mankind and denied to the purposes of war?"[30] The Soviets' response set their pattern of intransigence as they accused the United States, which had yet to launch a satellite, of wanting "to prohibit that which they do not possess."[31]

This Cold War confrontational tone tended to characterize the attempts at US-USSR space cooperation until the mid-1960s.[32] NSC 5814/1, official United States space policy, clearly stated the United States should pursue international cooperation in space "as a means of maintaining the U.S. position as the leading advocate of the use of outer space for peaceful purposes," and therefore, the United States should "be prepared to join with other nations, including the USSR, in cooperative efforts." Why? Because, the United States should "seek to achieve common agreement to relate such negotiations to the traversing or operating of man-made objects in outer space, rather than to define regions of outer space." The legalized right of overflight would thus be facilitated.[33] Scholars of this time period extending to the mid-1960s correctly conclude, "The simple but historic fact was that it had become fully evident that there was no prospect of the United States and the USSR getting together in a way that would have forestalled the extension of their existing differences and rivalries into the new domain of space. . . . The frame of reference would henceforth be one of an ongoing competitive race for national advantage in space."[34] As the ambassador to the Soviet Union during the Kennedy administration, Foy D. Kohler, explained:

> There stood for more than a decade a single compelling fact: it proved impossible in practice to effect anything more than token cooperation between the two great space powers of the world. . . . After some ten years of effort at direct cooperation between the two countries, nothing to speak of had actually happened. How could this be? The answer is

simple and straightforward: despite our hopes and expectations, the Soviet leadership has repeatedly and consistently refused to approach any relationship in the space area outside the context of the overall relationship between the two countries.[35]

The Soviets' space program provided them a valuable worldwide image of a progressive, technologically advanced nation. The Soviets saw no reason to cooperate in any substantive manner when they could continue to enjoy the geopolitical benefits of this perception.

The remainder of the history of international cooperation and bilateral US-USSR cooperation in space during the Eisenhower administration consists of the December 1958 passage of a United Nations resolution establishing a Committee on the Peaceful Uses of Outer Space (COPUOS), the USSR's and its allies' boycotting it for two and one-half years, and the United States' diplomatic attempts to jump-start the COPUOS. Eisenhower would plead to the United Nations and, indirectly, the USSR shortly before he left office, "Will outer space be preserved for peaceful use and developed for the benefit of mankind? Or will it become another focus for the arms race—and thus an area of dangerous and sterile competition? The choice is urgent. And it is ours to make."[36] Nevertheless, the dreary story of UN diplomatic wrangling continued into the Kennedy administration with little progress.[37] The salient point is that US-USSR cooperation, or lack thereof, is another illustration of the Cold War dynamic permeating space policy during this era. It also illustrates how the Eisenhower administration tended to filter many space-related possibilities through the lens of how they would affect the space-for-peace policy and the concern for reconnaissance satellites underlying it.

Space for Peace?

The interrelated complex of reconnaissance satellites, freedom of space, and space for peace set the tenor not only for international cooperation in space as well as the overall Eisenhower space policy but also set the stage for the NASA-DOD institutional relationship. Historians must be clear as to the central importance of reconnaissance satellites and the associated idea of freedom of space which, when combined with space for scientific research, formed the space-for-peace policy outlined in previous chapters. The space-for-peace policy was as important after NASA's creation as it was before. First, the policy provided the environment within which NASA-DOD relations would develop. Second, the policy limited the degree to which presidents and civilian OSD leaders would permit independent USAF projects and action in space because they feared the USAF might endanger the delicate principle of freedom of space by somehow "militarizing" space through either words or deeds.

NSC 5814/1 was the space policy document approved in August 1958 as NASA was being created. It declared that the United States had not and would not recognize "any upper limit to sovereignty" nor would the United States take any "public position on the definition" to maintain both "flexibility in international negotiations with respect to all uses of 'space'" and "freedom of action with respect to the military uses of 'space.'" The basic United States position would continue to be that "the right of passage through outer space of any orbiting object that is so designed that it cannot physically interfere with the legitimate activities of other nations is completely acceptable."[38] Therefore, administration officials did not appreciate General Gavin writing, "It is inconceivable to me that we would indefinitely tolerate Soviet reconnaissance of the United States without protest. . . . It is necessary, therefore, and I believe urgently necessary, that we acquire at least a capability of denying Soviet overflight—that we develop a satellite interceptor."[39] Clearly administration officials had a legitimate concern about the space-for-peace principle being endangered by certain military pronouncements. Eisenhower's final science advisor Kistiakowsky recalled how he had to make it clear to officials still active within the administration that Eisenhower discouraged such "dangerous statement[s] about destruction of enemy satellites if they overfly the United States. My point was that later this would prejudice the use of our own reconnaissance satellites."[40] Eisenhower permitted only low-level studies of offensive space weapons systems, such as antiballistic missile systems, satellite interceptors, and manned orbital bombers, because they could threaten the free overflight precedent.[41]

Quarles continued to emphasize shortly before he died the original point he made immediately after Sputnik's launch: "The USSR has already established an international practice with respect to orbital space vehicles and objects by orbiting Sputniks over the U.S. and other territories and sending out other space objects without seeking prior permission to do so." Therefore, the United States should avoid making any policy statements defining exactly where space began or ended because this "might conceivably limit or hamper its own freedom of action. Thus, it is to the advantage of the U.S. that no legal restrictions on the use of outer space be established" because the freedom of the United States and the free world "may depend upon our freedom to make use of outer space. Thus, it would be dangerous to impose limitations upon the types of activity we may find necessary to conduct there."[42]

Eisenhower's final space policy document was issued under the auspices of the NASC in January 1960 and similarly declared, "It should be noted that definitions of 'peaceful' or 'non-interfering' uses of outer space have not been advanced by the United States" because the United States considered as already established "the right of transit through outer space for orbital space vehicles or objects not equipped to inflict injury or damage."[43] The extremely delicate international sensibilities surrounding the issue of aircraft and satellite overflight were apparent at the brief Paris summit meeting of the United States and the USSR in May 1960. The summit was quickly aborted due to lingering hostility generated by the Soviet's downing of a supersecret U-2 reconnaissance aircraft in May 1960 and Soviet resentment at having been overflown since 1956. As Khrushchev "read a long diatribe denouncing the U-2 flights," he screamed, "I have been overflown." To this Pres. Charles de Gaulle of France countered that France, too, had been overflown, but by Soviet satellites. Khrushchev appeared "startled" and replied the USSR was innocent. De Gaulle then asked how the Soviets got photographs of the far side of the moon from its Lunik satellites. Khrushchev replied, "In that satellite we had cameras." De Gaulle sarcastically countered,

"Ah, in *that* one you had cameras! Pray continue." Khrushchev demanded Eisenhower apologize; Eisenhower refused; the summit ended (emphasis in original).[44]

Apparently, the assumption by some officials within the administration that the Soviet Sputniks had de facto established the right of satellite overflight was in no way a de jure reality in the international diplomatic arena. Khrushchev declared on 16 May 1960, "As long as arms exist our skies will remain closed and we will shoot down everything that is there without consent."[45] The United States therefore had to proceed with extreme caution in the reconnaissance-satellite overflight area. It would not define exactly where space began or ended. It would support the concept of peaceful uses of space and the prohibition of the deployment of weapons of mass destruction in space as part of this peaceful-uses doctrine, while considering reconnaissance satellites not to be such weapons. Finally, the executive branch would ensure the military services did not exacerbate the delicate international environment regarding overflight by discussing anything that could be construed as the militarization of space or the consideration of placing offensive weapons there. As the State Department lamented near the end of Eisenhower's term, "A Soviet political and propaganda attack on our launching a spy satellite at this time seems inevitable."[46] Unfortunately, as one noted space historian concludes, "Despite these and subsequent messages that canceled offensive space-based, weapon-research programs, Air Force military leaders at that time seemed unable to grasp—or unwilling to accept—the meaning of President Eisenhower's 'peaceful uses of outer space,' or the rationale behind it."[47]

The USAF and Space for Peace

The Air Force perspective was slightly different. It believed national security demanded an investigation of the defensive and offensive potential of space. The USAF considered its presence in space to be no different from the Navy's on the high seas—ensuring the medium's peaceful use and availability for transit to all parties.[48] As one space historian explains, the USAF viewpoint was that "restrictions on the military did not match the obligations of the military to ensure the security of the nation." Until all nations subscribed to the space-for-peace ideal, "They believed the United States needed the capability to control space to ensure the liberty of free people everywhere."[49] For instance, when Gen Nathan Twining, chairman of the JCS and Air Force, provided his input to NSC 5814/1, he said the United States should "place primary emphasis on activities related to outer space necessary to maintain the overall deterrent capability of the United States and the Free World."[50] The Air Force's operative mantra was and is *faster, farther, higher.*

The input to NSC 5814/1 of Gen Thomas Power, Strategic Air Command's commander, included the kind of statements that top officials of the Eisenhower administration felt might endanger the space-for-peace policy's goals. Power said prestige comes through leadership in the clash with communism, while admitting reconnaissance was probably the most important immediate military space possibility. He maintained that "close behind lies a true potential for unique and effective weapons system development. . . . We must not, in the fashion of decadent nations, permit our gross potential to be bled off into purely defensive weapons. As we enter the space era the primacy of the offensive has never been more clearly defined. . . . Because space offers the ultimate in mobility and dispersal for weapons which can be addressed at the enemy heartland, the ultimate in deterrence may well be in this direction." He believed the Air Force must "emphasize constantly the positive contribution of offensive weapons systems. The logic of this fact must be identified for scientific and national leaders."[51] In January 1959, the Air Force concluded, "We <u>must</u> investigate the possibility of military utilization of the moon. If we do not develop the capability to more than match each Soviet space move, we may find ourselves outflanked in the new dimension of space" (emphasis in original).[52]

Indeed, it was not unknown for even these calls to arms to be overshadowed by declarations such as, "In twenty years, I believe both the moon and Mars will have permanent, manned outposts. . . . Another use [of satellites] will be purely military—bombardment—and accomplished by space vehicles. I use the term vehicles rather than satellites because I believe these systems will be manned. . . . It appears logical to assume we will have antisatellite weapons and space fighters." This general opined that the only thing that would cost more than such systems "would be the failure to be first on the moon. We cannot afford to come out second in a territorial race of this magnitude. . . . This outpost, under our control, would be the best possible guarantee that all of space will indeed be preserved for the peaceful purposes of man."[53]

In addition to highlighting that part of the USAF's space philosophy that desired to explore the possible offensive potential of space for national security purposes, Power and Boushey also displayed another important component of the Air Force's space thinking—the central role that humans would play in space systems. Power declared, "For the long term, the critical requirement is to establish man in the space environment. In the early-unmanned exploratory stages of the conquest of space, unmanned vehicles can be used for many scientific purposes, and certain specific military applications. However, to fully exploit the space medium, man is the essential ingredient." The Air Force must, therefore, "Identify the mandatory presence of man in the space environment before significant fulfillment of either military or economic potentials can be enjoyed."[54] Therefore, "In reaching the objective of extraterrestrial 'high ground,' there must be a progressive development and employment of Air Force experience in manned flight."[55]

The third part (in addition to exploring the offensive potential of space and asserting the vital role of humans in space) of the Air Force's space philosophy was introduced in the last chapter—the belief that the USAF was the proper organization to conduct the nation's military operations in space. This illustrates that continuing interservice rivalry even after NASA's creation was one reason Eisenhower administration officials concluded they had made the correct choice then and that NASA must become a strong and independent organization. An important USAF meeting took place in late January 1959 as it tried to determine exactly what its position was in the post-NASA space structure. At this meeting the services' top generals briefed the services' top civilian officials such as the secretary of the Air Force and its chief scientist. The officers emphasized that "Air Force responsibility extends outward into space, and that there can be no line of distinction between air and space as far as operational responsibilities of the Air Force are concerned." Further, "The operational means for the overall control and direction of space activities does not and cannot exist outside a military service." Which is to say, not in NASA. In addition, "The control of space activities and operations for military purposes is but a normal extension of the control of air activities by the Air Force." This is to say, not part of the Army, Navy, or the ARPA. Therefore, "The Air Force has no quarrel with NASA and ARPA but the basic responsibility for the overall space defense of the United States, and the military position of the United States in space, cannot be abandoned. No organization other than the Air Force exists or is contemplated which would carry out such a mission."[56]

Part of the USAF's concern for securing the space mission was the belief that its future may very well have depended on it. In the late 1950s, the Air Force was operationally deploying its ICBMs, and there was some institutional concern that the Air Force officer corps would be transformed from dashing and courageous pilots into the "silent silo-sitters of the seventies."[57] Eisenhower told General White, CSAF, that the USAF's success in rocketry "has made possible and necessary reductions in aircraft programs. It is a change in our thinking." White replied that this raised "the question of what is the future of the Air Force and of flying. This shift has a great impingement on morale. There is no follow on to the fighter, and no new opportunity for Air Force personnel. A natural extension of Air Force activity would be into space as flying drops off." He wanted the predominant role in space for the Air Force.[58]

The three-fold Air Force space philosophy—of using space if necessary to guarantee American security; designating the Air Force as the primary institution to carry out this security role; and assuring humans would play a central role in space systems—made little headway with the civilian policy makers of the Eisenhower administration (nor later in the Kennedy administration). A historian points out the fundamental problem when he explains that the tendency over history has been for airpower theorists to promise more than their chosen technological instrument could deliver. However, concerning space, exactly the opposite had been true: "The technology has far outpaced any coherent doctrine on how to employ space systems effectively."[59] The Air Force's inability to articulate convincingly and precisely what humans would do in space, to the satisfaction of its civilian overseers in the OSD and higher in the executive branch, meant it could not establish an independent, long-term, human presence in space.

Air Force Philosophy Made Little Headway

The primary reason for the administration's reluctance to endorse this Air Force space philosophy was the simple fact that it directly contravened the intent of Eisenhower's space-for-peace policy and risked casting a military aura onto the American space program, exactly what Eisenhower wanted to avoid. Several secondary reasons also contributed to the policy makers' aversion toward the Air Force's space philosophy, of which the financial and the interservice rivalry factors are paramount. When the Air Force discussed its "aerospace" with the inherent idea that only the Air Force had a legitimate military mission in space, Rep. Daniel J. Flood sarcastically responded, "This is a beauty. . . . That means everybody is out of space and the air except the Air Force, in case you didn't know it. Has the Air Force, without consulting anybody, taken the Navy out of air and space? . . . They have to have something to stay in business. You had better get there, or you won't be around."[60] Meanwhile, the Army continued to strive for an active role in space and would continue to do so until Eisenhower authorized transfer of the ABMA to NASA in the fall of 1960. The culmination of the Army's effort was Project Horizon of June 1959—a four-volume Army study for a lunar base and all the associated supporting systems (launch vehicles, space capsules, etc.).[61] The ABMA commander, General Medaris, later commented that Horizon was "shot down in flames by the assignment of all space vehicles to the Air Force."[62]

It should come as no surprise that in April 1960 the Air Force released its own lunar-base study. It claimed the USAF could send a man to the moon and return him in 1967, have a fully operational lunar base by June 1969, and perform earth surveillance at a total cost of $7.7 billion. The Air Force posited a lunar base was necessary because it provided "a site where future military deterrent forces could be located. . . . A military lunar system has the potential to increase our deterrent capability by insuring positive retaliation."[63] As R. Cargill Hall summarizes, "Besides flying in the face of stated administration commitments to explore and use outer space for peaceful and defensive purposes only, these proposals gained few adherents other than those who already viewed the Soviet sputniks with unalloyed hysteria."[64]

As Eisenhower's second science advisor, Kistiakowsky, recalled this and many other self-aggrandizing service proposals for grandiose military space projects "were quite partisan, to put it mildly. . . . Rather awful! . . . I still recall becoming indignant on discovering that the cost of exclusively paper studies in industrial establishments on 'Strategic Defense of Cis-Lunar Space' and similar topics amounted to more dollars than all the funds available to the NSF for the support of research in chemistry. I tried to raise hell about this with [DDR&E] York."[65] NASA administrator Glennan watched the interservice maneuverings with some bemusement. He called one USAF-Navy dispute concerning space responsibilities on the west coast "an argument that has bordered on the ridiculous. . . . The situation reminded me of two little boys arguing over which of their fathers could lick the other."[66]

The other secondary reason the USAF space philosophy made little headway during the Eisenhower, or subsequent, administrations was the financial issue—duplication, wasteful expenditures, overlap, and so forth. This meant that unless performing a particular task in space offered identifiable functional efficiencies (such as reconnaissance, meteorology, communications, or navigation) or financial savings (this, arguably, never materialized in the military arena because of the continuing high cost of launching payloads to orbit), then that task would not be performed in space, and little exploratory

110

R&D for it would be authorized. As early as April 1959, the civilian undersecretary of the Air Force said, "Future military needs will be satisfied by the use of whatever future weapons and techniques will provide improved capabilities or effectiveness. If so-called 'space systems or techniques' can improve the military potential, they undoubtedly will be used. However, space is not a function, it is a location, and as such it may or may not permit the traditional military missions to be performed more effectively."[67]

Try as they might, however, to limit military space spending to only those subjects likely to enhance current capabilities, administration officials such as Kistiakowsky could still listen to Air Force briefings on the proposed USAF space program and be "shocked by the incredible wastage of taxpayers' money. For instance, $8 million spent in paper studies such as lunar defense systems."[68] Two months before the end of his administration, Eisenhower reacted to a briefing on the proposed military space program: "The President said that he did not know where the money for such programs was going to come from. It seemed to him that we should finally reach the point where these programs were not constantly going up until they absorbed nine-tenths of our research money. We should determine some sort of level of effort and set a dollar ceiling which would be changed only if there were some sort of startling development."[69] An important point to highlight in the overall NASA-DOD framework is that "much of the struggle over the military uses of space was as much between elements within DOD as between DOD and NASA."[70]

The NASA-DOD Relationship Phase I

The foregoing discussion sets the stage for the specific NASA-DOD relationship that emerged. The president wanted to protect his space-for-peace initiatives, and so Air Force space proposals were kept under control, and NASA was nurtured. The first task relevant to the NASA-DOD relationship was the division of projects and facilities when NASA began operations in October 1958. The most important decision, the assignment of the human-spaceflight mission to NASA and the program's subsequent development, is covered in the next chapter. Other project and facility assignments occupy an important supporting role in the human-spaceflight story.

Division of Labor

The division of labor process started on 2 April 1958, the same day Eisenhower submitted his version of the Space Act to Congress. He wrote the SECDEF and the NACA chairman to explain his philosophy concerning which organization would do what under the new legislation: "It is appropriate that a civilian agency of the Government take the lead in those activities related to space which extend beyond the responsibilities customarily considered to be those of a military organization." Eisenhower said it was "especially felicitous" that the NACA and the DOD have such a close and harmonious relationship because, "this relationship will ease the period of transition that lies ahead and will provide a basis for the close cooperation that will be needed to solve the difficult problems that will be encountered." The NACA and the DOD should therefore "formulate such detailed plans as may be required to reorient present programs, internal organization, and management structures" in accordance with the pending Space Act and form recommendations concerning which programs would be transferred to NASA.[71]

Later that month, the NACA and the DOD responded with a general guide as to the appropriate division of labor. They had decided that the military unquestionably should be responsible for these missions: reconnaissance and surveillance, countermeasures against space vehicles, weapons in space, and navigational aids.[72] Missions going to NASA without dispute would be unmanned spaceflights for scientific data such as vertical probes, lunar and interplanetary probes, and scientific satellites. However, a gray area termed *common interest programs* included human spaceflight; large rocket engines; and communications satellites and meteorological satellites.[73] A neat and orderly division of effort to include projects and facilities was clearly not going to be an easy task. The BOB stated its opinion: "From our review, it appears to us that the only major project proposed for FY 1959 that is 'peculiar to or primarily associated with weapons systems or military operations' is the so-called 'Advanced Reconnaissance Satellite' project."[74] This was the technical name for Sentry, the renamed WS-117L.[75] At a minimum, an ambitious military space program was going to be a difficult row for the Air Force to hoe.

Eisenhower's assessment was that anything "not yet proved as to [its] technical feasibility should be the concern of this agency [NASA], and that non-military applications should also be the concern of this agency."[76] He told the NSC on 14 August 1958, "We should put, as far as possible, all space projects under the space agency [NASA, which] must prove the military practicability or feasibility of a given space project or activity before the Defense Department takes over such a project or activity. . . . Not every activity in outer space is going to turn out to have military use."[77] Also in August 1958, Eisenhower awarded NASA, and not the DOD, the human-spaceflight mission. One week before NASA began operations, "The President reaffirmed that NASA should [be] the heart of the whole activity; unless a project is a very definite application to a specific military purpose, it should be in NASA. . . . The President said that, unless definite military purpose can be shown, the responsibility and the funds should be in NASA."[78] Again, the general situation was not a fertile one for the development of a robust and diverse military space program.

Accordingly, Eisenhower's Executive Order 10783 on 1 October 1958 transferred from the DOD to NASA: Project Vanguard, lunar probes, scientific satellites, passive communication satellites, and most rocket engine research (but not Saturn or its management agency, the ABMA).[79] One primary source recounts these transfers "had left some feeling in DOD that the Services had been deprived of something which was theirs by right of initiation and, in some cases, ultimate user status. This, in turn, had caused some reluctance to enter into a fully cooperative partnership of mutual support in aerospace activities."[80] There was enough grumbling within the military space agency over the scope of the transfers that ARPA director Roy Johnson felt it necessary to inform his staff it was "ARPA's policy to provide the fullest kind of support and assistance to the National Aeronautics and Space Agency [*sic*] in all areas. . . . It is, moreover, ARPA policy to support fully the transfer of functions prescribed by the statutes establishing NASA."[81]

One source explains this "major change" in the old NACA-DOD relationship: "Whereas in prior years NACA had been a valuable support agency fulfilling military research requirements, now NASA, elevated into the big league of government departments and agencies, with major budgetary demands of its own . . . loomed as a competitor for funds as well as for Presidential and public attention."[82] NASA *did* become an operating agency with its own contracting and management centers and was no longer simply an R&D organization supporting the DOD in a client-server relationship: "NASA became the biggest single rival and competitor of the mammoth Defense Establishment. It would not be, as NACA was, a research activity working mainly for the military. It would initiate its own programs, build its own facilities, develop its own procurement and management organizations."[83]

The main cause of disagreement was, according to DDR&E York, a "conflict between NASA and ARPA about roles and missions at the high end, that is to say large rockets and man." Roy Johnson's view was that these were essential military activities, and Keith Glennan's view was that the Space Act of 1958 gave him a set of responsibilities to explore space and so forth, that it ought to be carried out with large rockets and men. York added that he and Killian also believed, "It was NASA who needed men in space and who needed large rockets in order to carry out its mission, not ARPA."[84] As one scholar concluded, "ARPA and the services were fighting a lost battle. The president's policy of space for peace made him reluctant to grant any space activity to the military that could be considered of scientific interest."[85]

The ABMA as the Central Issue

Battles ensued nonetheless. The most important one centered on control of the Army Ballistic Missile Agency. The ABMA was one of two military organizations skilled in the design and construction (or managing the construction) of large rockets. The other was the USAF's BMD. Clearly, the administration would

not permit NASA to take over the BMD because it was responsible for developing the bulk of the US ballistic missile deterrent
force. The ABMA was a different matter. Its main project was the
huge Saturn rocket, an order of magnitude larger than any
single IRBM or ICBM. The DOD was unsure in late 1958 if there
was any military requirement for such a large missile; in 1959
it would conclude there was not, and the Saturn project along
with most of the ABMA was transferred to NASA in 1960. For
NASA the ABMA's capabilities were absolutely essential to the
process of NASA becoming a viable space exploration agency.
Obviously, without the ability to construct the large rockets
needed to launch heavy scientific payloads into earth orbit and
into deep space, NASA's institutional capabilities would be extremely circumscribed.

Therefore, one of Glennan's first orders of business in the
fall of 1958 was to petition for the transfer of the ABMA to
NASA. The ultimate outcome of the complex-bureaucratic maneuvering associated with the ABMA transfer throughout late
1958 and most of 1959 depended on Eisenhower. As early as
March 1958, he stated that he "thought the Huntsville force
[the ABMA was located at the Army's Redstone Arsenal in
Huntsville, Alabama] should be promoted to space and similar
activities. He thought consideration should be given to taking
them out of their present assignment and assigning them to
ARPA, or even to NASA."[86] This was less than a week after approving Killian's memo recommending creation of a NASA, and
he telegraphed the ultimate outcome of the ABMA's transfer to
NASA which occurred less than two years later.

Glennan was on duty throughout September 1958 before
NASA's official standing up on 1 October, and he toured DOD
installations to determine their potential value to NASA. Concerning the ABMA, Glennan observed, "I became convinced
that the talents of this group—so dedicated to space exploration and so hemmed-in by the fact that the Air Force had
been given control of air and was intent on extending that control to space—would be a useful part of NASA." The obstacle
would be the ABMA's commander, General Medaris, who had
treated Glennan "in a somewhat cavalier fashion." Glennan
characterized Medaris as "a martinet, addicted to 'spit and

polish,' never without a swagger stick, and determined to beat the Air Force. He simply did not have the cards."[87] Glennan felt he had the support of McElroy, Quarles, Roy Johnson, and York. However, "I was not prepared for what then transpired."[88] Glennan proposed to SECDEF McElroy on 15 October 1958 that the DOD relinquish control of the ABMA and the Jet Propulsion Laboratory (JPL)[89] because they are "vitally important to accomplishment of the NASA mission," and since current trends indicate that "it may be expected soon that the major effort of the ABMA will be in support of NASA programs." Therefore, Glennan added, "We believe that the transfer of the space capability of these organizations to NASA is in the national interest."[90]

Secretary of the Army Wilber Brucker called Glennan into his office and "became irate" at Glennan's transfer proposal and "said he could not countenance such a move." Glennan regretted that he "hadn't realized how much of a pet of the Army's von Braun and his operation had become. He was its one avenue to fame in the space business. . . . I finally left with my tail between my legs and called a session of our people to determine strategy."[91] Brucker believed, "Currently, 85 percent of the existing capabilities at the Army Ballistic Missile Agency and the Jet Propulsion Laboratory are required for—and committed to—the Army's missile programs. . . . The damage done by disrupting the existing organization at this time would be irreparable. . . . The proposal to absorb at this time part of ABMA and to take over JPL is not in the national interest."[92] Brucker and Medaris leaked the situation to the press, and soon the entire situation became public knowledge.[93] By the end of October, "The President said that he is completely nonplused at the spirit of bureaucracy which seems to become predominant in such affairs—the lack of any spirit of give and take to try to work out the best national interest."[94]

Glennan quickly enlisted the assistance of Killian and Quarles. They soon hammered out a compromise solution, whereby Glennan agreed to drop his request for the ABMA and in return the Army ceded the JPL to NASA and promised that the ABMA would be completely responsive to NASA work orders.[95] However, as Glennan wrote to McElroy, "We must recognize that as time passes important changes will undoubtedly occur in the nature of the requirements of both the Department of Defense and NASA." Therefore, the agreement called for a review and a report in one year "on the success of these arrangements."[96] Eisenhower told Glennan that he felt the partial transfer was a mistake because "he would prefer to make the ABMA shift right away" but was unwilling to intervene in the compromise solution his subordinates had crafted.[97] Glennan told Congress in January 1959 he was keeping open all options: "I shall certainly avail myself of the opportunity, if I think I need it, to ask again for the transfer of this agency, if it seems important."[98]

Two developments helped secure the ABMA's transfer to NASA one year later in 1959. First and most important was the OSD's conclusion that the Saturn rocket had no immediate military utility and was becoming too expensive; therefore, they would not oppose its transfer along with that of the von Braun team developing it. Second, Glennan changed his tactics by waiting for the DOD to offer von Braun and the Saturn project to NASA, refusing to deal with Brucker and dealing directly and only with OSD officials such as DDR&E York. Numerous sources indicate that in April 1959 York declared, "I have decided to cancel the Saturn program on the grounds there is no military justification."[99] Kistiakowsky observed that Glennan and NASA were "in constant jurisdictional conflict with the United States Army which, using Wernher von Braun and his rockets, was feverishly trying to carve a bigger role in space for itself."[100] In August 1959, an internal NASA document explained, "Recently, the Department of Defense has stated that due to budgetary limitations they would like to reopen the question of transferring the ABMA to NASA. . . . Army opposition can be expected to vary inversely as the amount of pressure applied by the Department of Defense."[101]

Glennan made sure the DOD took the initiative for the transfer, however, the memories of his 1958 experiences with Brucker were still fresh in his mind. He recounted a discussion with McElroy in which McElroy "was trying to find out whether or not we were sufficiently interested to make it worth his while to move forward with his plan to carry out York's recommendations" to transfer Saturn and the ABMA to NASA instead of canceling it outright. Glennan said, "The impression is left that this is a move on the part of Defense Department— not NASA. Naturally we are insisting that this is the posture."[102] Glennan informed York on 23 September 1959 that NASA would be ready to reexamine the ABMA transfer question, "but on the basis that the initiative is being taken by Defense (recalling the very bad experience with the Army of last year.)"[103]

By September 1959, SECDEF McElroy reported that the DOD was negotiating to "turn ABMA over to some agency other than the Army—probably NASA—since it was getting too expensive to support."[104] York explained, "We believe that we need the bigger boosters, but we do not at this time have firm requirements. For this reason, we would be satisfied to have NASA build the big boosters."[105] By 29 September, any doubt as to the ABMA's future evaporated when Eisenhower stated that he "didn't want the NASA budget to go much over half a billion dollars a year; that we weren't in a race with the Soviets, but were engaged in a scholarly exploration of space." He added that "ABMA should be put under NASA and on my warning conceded that he will have to defend Glennan publicly."[106] The only remaining task was drafting an official plan and submitting it to Congress. Brucker and Medaris realized they faced a fait accompli and so raised no serious objection. Eisenhower announced his intention to transfer most of the ABMA, essentially von Braun's team, called the Development Operations Division, and the Saturn project to NASA on 21 October 1959. The detailed NASA-DOD agreement was ready by December, and the transfer became effective on 14 March 1960.[107]

The entire ABMA transfer episode, which one historian termed the "most significant event in NASA's history" between its establishment and Kennedy's May 1961 lunar-landing decision, indicates the key role that top OSD officials played in the overall NASA-DOD relationship.[108] NASA might desire an organizational realignment such as the ABMA's transfer in 1958, but lacking top-level OSD backing, it did not occur. The next year the ABMA was smoothly transferred to NASA because the DDR&E and SECDEF concluded it was an organizational and financial liability to the DOD. The principle of the OSD's input acting as a crucial determinant in the NASA-DOD relationship continued into the 1960s in that the Air Force's drive for a human-spaceflight mission would be largely circumscribed by OSD-level officials. Another legacy of the ABMA affair was that from the fall of 1959 on, the Army no longer played any significant role in space and was largely absent from the remaining discussion of the NASA-DOD relationship. The dean of space historians John Logsdon explained that with the transfer of the von Braun team to NASA, "Army plans for manned space flight came to an end."[109]

ARPA's Space Role Fades

The autumn of 1959 witnessed a second organizational change relevant to the NASA-DOD relationship: the ARPA faded from importance in the space-organizational scheme. While the DDR&E had already become the OSD's point man on space issues, ARPA's receding from the scene did mean the military space projects under its active management, such as the Sentry reconnaissance satellite, were returned to the control of the individual military services. From this point forward, NASA would interact directly with either the OSD or the Air Force, not the ARPA, in forging agreements or arranging project support. This reduction of ARPA's role apparently came at the initiative of DDR&E York and was motivated by the desire for even more centralized OSD-level control of military space projects. Kistiakowsky explained,

> It is rather clear that York intends to reduce the role of ARPA and restrict it to the field which is defined by its name. He wants to put all space activities directly into the Air Force except for specific missions to be assigned to the Army and Navy, but even those are to use booster vehicles of the Air Force. He feels that making the program part of the Air Force budget will automatically restrain the wildest boys, whereas at present they simply write fantastic requirements and expect ARPA to take care of them.[110]

Kistiakowsky agreed with York's initiative, saying, "We simply do not have the means to support all-out development efforts in all 'important' areas."[111]

119

Therefore, on 15 September 1959, Eisenhower approved DDR&E York's memo transferring the various military space projects from the ARPA back to the military services. Eisenhower seems to have been persuaded by Kistiakowsky's argument that farming the projects back to the services would create a more clear-cut assignment of authority along reasonably functional lines, thereby reducing duplication. In addition, "Since the projects will be carried out on Service rather than ARPA budgets, a more effective restraint against indefinite multiplication and elaboration of projects will be established."[112] The Air Force received management for the military reconnaissance satellite program (now called Samos) as well as the early warning against the ballistic missile attack satellite called Midas, as well as another program called Discoverer to be discussed in the next chapter. The reorientation also granted the Air Force responsibility for developing all military boosters, integrating the satellite payloads with boosters, and launching the complete package. The Navy received developmental responsibility for a navigation satellite called Transit, and the Army received a family of communication satellites.[113] Clearly, the Air Force was consolidating its hold on the vast majority of military space responsibilities but would still be under close OSD-level scrutiny.[114] In February 1960, DDR&E York explained that because of these actions the ARPA "ceased to exist as an independent agency reporting to the Secretary of Defense. It no longer does play a role in the space program. . . . We have taken ARPA out of the programs which are virtually near the operational stage."[115]

Amending the Space Act

Given the rationalization of organizational structure taking place in late 1959 within the military space context, it is reasonable to ask if a similar process had been taking place between the military and civilian space fields (beyond the ABMA transfer discussed above). Glennan had no serious complaints about the situation, expressing in a confidential setting, "I don't mean to imply that the relationships between NASA and the Department of Defense have been anything but amicable. We have worked out our immediate problems in a cooperative spirit and, with the help of other agencies, have made reasonable progress."[116] Nevertheless, Glennan felt improvements to the Space Act's division-of-labor language were in order. Starting in mid-1959, he was the driving force behind an effort to amend the Space Act to reflect more rationally the actual relationship between NASA and the DOD. Ultimately, this revised Space Act did pass the House in 1960, but Lyndon Johnson refused to permit its consideration in the Senate and the legislation died. The incident does reveal important clues concerning Eisenhower's space policy.

The Civilian-Military Liaison Committee and the National Aeronautics and Space Council

One prefatory note to the discussion of the effort to revise the Space Act is to mention that the two bodies Congress created to facilitate NASA-DOD coordination, the NASC and the CMLC, were not important policy-making bodies during the Eisenhower administration. They were sufficiently superfluous so that Eisenhower recommended their abolition in his proposed Space Act amendments. In the case of the CMLC, the central problem was its members' lack of authority; they could neither make nor enforce decisions because they did not hold positions of responsibility in either NASA or the DOD. Neither NASA nor the DOD ever delegated any authority to them.

The CMLC's original charter of 22 October 1958 outlined its primary function: "Provide a channel for official advice, consultation and exchange of information and maintain a flow of this information adequate to keep . . . [NASA and DOD] fully and currently informed of each other's aeronautical and space plans, programs, and activities." Its authority was negligible: "When requested by the Administrator, National Aeronautics and Space Administration, or the Secretary of Defense, study and recommend courses of action where jurisdictional differences . . . have arisen, or might arise, unnecessary duplication of effort might develop, or coordination of jointly sponsored or related programs is required."[117] Thus, CMLC chairman William Holaday (its only full-time member) could not initiate any action unless requested by Glennan or McElroy (or Eisenhower's

121

last SECDEF, Gates). By the CMLC's January 1959 meeting, neither the OSD representative nor his alternate was present; two of four NASA members were absent and sent lower-ranking alternates. This pattern of either absence or sending subordinates was soon the norm.[118] Its March 1959 meeting agenda contained only one item, a NASA presentation on the national space-vehicle program.[119]

The next month Chairman Holaday was candid with Congress, testifying that the CMLC was not "contributing much to the space effort," though he, the SECDEF, and Glennan were trying to devise more useful functions. Holaday stated, "It is recognized that normal project activities can be conducted in a more expeditious manner if carried out at a project officer to project officer level."[120] In July 1959 Holaday complained he was being completely cut out of the information-exchange process between NASA and the DOD: "The Chairman is finding it impossible to carry out his responsibilities due to lack of complete information on discussions and decisions that are being made by the separate offices."[121] A Senate report the next month concluded, "The Civilian-Military Liaison Committee is not organized or authorized to perform effectively its coordinating functions between NASA and the Department of Defense. Coordination between NASA and the Department of Defense is being carried on by numerous and informal personal contacts. At times the Civilian-Military Liaison Committee is not even advised. . . . We have no authority." Holaday added, "If we do not get something more constructive to do than what the Committee is now doing, I can see no need for continuing the Committee" because its only current function was in the "'exchange of mail' area," a post office.[122]

The last of the CMLC's 13 meetings was in December 1959. Before he resigned in April 1960, Holaday told Congress, "The formal actions of the Committee are few in number. . . . The role of the Committee has been of relatively minor importance. . . . A Committee, because of its usual composition, that is, a membership made up of representatives who are subject to a higher internal authority, is incapable of making firm decisions. . . . The activities of the Civilian-Military Liaison Committee are limited to recommended courses of action to the heads of the two agencies for their consideration and decision."[123] The CMLC faded from the scene, an organization that was created for the express purpose of coordinating the NASA-DOD relationship but which never had an impact on the relationship because it completely lacked any authority to take action.[124]

Much the same story holds true for the role of the second organization expected to facilitate NASA-DOD interaction, the NASC. Eisenhower said that he "did not expect the Council to function too formally or elaborately." He also indicated he would not hire an executive director for it or any full-time staff, "indicating what he had in mind was someone to serve as a recording secretary rather than an Executive Secretary." Eisenhower added the NASC should function "very much as a Board of Directors" considering only those issues brought to it by the NASA administrator or the SECDEF.[125] In 1959 when asked about the NASC's ineffectiveness, Eisenhower simply replied that he "had not sought the creation of the Space Council but had been forced to accept it as a compromise with the Democratic leadership."[126] Not surprisingly, the NASC met only eight times between NASA's establishment and its final meeting in January 1960. As with the CMLC, it exercised no important policy-making role.[127] Eisenhower recommended that both the CMLC and the NASC be abolished in his Space Act amendments.[128]

Space Act Amendments Stymied

Glennan's drive for a revised Space Act containing a more realistic reflection of the NASA-DOD situation and eliminating the CMLC and NASC gathered momentum with a long memo he submitted to Eisenhower on 16 November 1959. Glennan began by discussing the CMLC and NASC, saying, "Neither of these activities has been particularly useful or effective. . . . It is doubtful that either of these agencies can usefully be employed in the management of the nation's space program." Second, he laid out what he felt DOD's position about space to be, namely, that "space is a place—not a program" and so "space projects in the DOD are undertaken only to meet military requirements," not scientific research or exploration. Therefore, military space projects must compete with more conventional means of accomplishing the same or similar military objectives. In addition, Glennan now believed the Space Act needed no specific mention of what the DOD or NASA would do in space because, "What the military needs to do in whatever medium . . . they can and should do under the statutory responsibilities for defending the nation" that already existed and needed no further addressing in the Space Act. Glennan's idea was to remove any specific tasking language for the DOD from the Space Act and simply allow the DOD to act in space in accordance with legislation already tasking it to defend America, primarily the National Security Act of 1947.[129] SECDEF Gates agreed with Glennan's changes, saying they "made a good deal of sense from a management standpoint" and "the law should have been written this way in the first place."[130]

When Glennan and Eisenhower met with Senate leaders to sell their proposed Space Act revisions, Glennan explained that "the difficulties between the Defense Department and NASA began to disappear approximately four months ago. . . . The President commented that the Defense Department is satisfied with the proposed agreement. Lyndon Johnson promised that if the President wanted to do away with the CMLC and NASC 'I'm certain it will be all right with me' and that he would begin hearings later in January 1960."[131] Neither of these Johnson statements was true. On 14 January 1960, Eisenhower publicly released and explained his proposed amendments to the Space Act designed to "clarify management responsibilities and to streamline organizational arrangements concerning the national program of space exploration." In addition to deleting the CMLC and NASC clauses, the new Space Act would also eliminate the exception clause. Eisenhower went on to say that the DOD had ample authority outside the Space Act to do R&D on space systems and to use space for defense purposes, "and nothing in the Act should derogate from that authority." But, "the statute should go no further than requiring that NASA and the Department of Defense advise, consult, and keep each other fully informed with respect to space activities . . . it should not prescribe the specific means of doing so."[132]

Eisenhower's proposed Space Act amendments highlight two important points about the Eisenhower administration's space policy. First, it regarded NASA and the DOD space programs as two separate entities, not as subcomponents of one overall program. The Kennedy-Johnson administrations would take exactly the opposite approach under McNamara's management philosophy of eliminating redundancy and duplication in pursuit of efficiency. Second, Eisenhower made clear that military space projects would be authorized only if there were a definitely identifiable and specific requirement for it. On this point, the Kennedy-Johnson policy would be the same.

The House considered and passed on 9 June 1960 Eisenhower's new Space Act with one major change: it added an Aeronautics and Astronautics Coordinating Board (AACB) to take the place of the CMLC. However, it was expected that the AACB's members would be high-ranking officials from NASA and the DOD, who would be able to speak with authority, make decisions, and return to their respective organizations and enforce the AACB's decisions. The House also inserted a phrase to protect the DOD's interests: "The Department of Defense shall undertake such activities in space, and such research and development connected therewith, as may be necessary for the defense of the United States."[133]

Lyndon Johnson, however, refused to let the Senate consider the legislation. Glennan recounts Johnson told him on 23 June 1960, "Look now, doctor, you haven't a chance to get that legislation. . . . I don't see any reason for giving you a new law at the present time. If I am elected president, you will get a changed law without delay."[134] Johnson's entry into the *Congressional Record* was more diplomatic: "Analysis of the key issues involved fails to uncover any persuasive reasons for pressing for Senate action on these amendments. . . . One fact is of overriding importance. A new President will take office on January 20, 1961. The next President could well have different views as to organization and functions of the military and civilian space programs. Any changes in the Space Act at this session . . . could restrict the freedom of action of the next President."[135]

The Aeronautics and Astronautics Coordinating Board

Therefore, the only result of consequence from the attempt to revise the Space Act throughout late 1959 and 1960 was the AACB. The AACB would function throughout the 1960s with a higher degree of importance than the CMLC. It and its six panels met regularly not to engage in the policy-making function but to ensure proper coordination between NASA and the DOD efforts in certain space technology fields.[136] Any decisions concerning improving coordination or reducing duplication were usually carried out because the AACB's cochairmen were NASA's deputy administrator (number two in the NASA hierarchy) and the DDR&E (responsible for all DOD R&D, engineering, and technical activities) who reported directly to the SECDEF. The AACB's charter explained that it was "essential" to coordinate space activities of NASA and the DOD. Therefore, "Where policy issues and management decisions are not involved, it is important that liaison be achieved in the most direct manner possible, and that it continues to be accomplished as in the past between project-level personnel on a day-to-day basis." The AACB existed simply to identify any problems in this area and ensure that exchange of information ". . . be facilitated between officials having the authority and responsibility for decisions within their respective offices."[137]

As NASA cochairman Dryden explained, "In the case of the AACB, the Co-Chairmen, being placed at a very high level in their respective organizations, can, indeed, arrive at decisions regarding a great many interagency problems and proceed to carry them out."[138] This change in the structure of the leaders and members of the AACB when compared to the CMLC would be the difference enabling the AACB to act as an effective mid-level coordinating entity throughout the 1960s. Policy, of course, continued to originate at higher levels. However, Logsdon's assessment is also relevant: "As the separate NASA and defense programs became more institutionalized in the 1960s and 1970s, there has been a tendency for coordination between the programs to be defensive in character, i.e., aimed at protecting each agency's own programs and 'turf.'"[139]

The NASA-DOD Relationship Phase II

This relationship raises at least two questions. First, what were some of the points of cooperation, support, and rivalry that existed between NASA and the DOD during Eisenhower's terms apart from the division of labor and the ABMA issues discussed earlier? Second, how did the multifaceted pattern of assistance and conflict emerge? These two questions tend to involve NASA and the Air Force because, as explained above, the Air Force became the agency responsible for conducting most of the DOD's space program after the ARPA receded from the scene and the ABMA was transferred to NASA.

One early illustrative example of USAF-NASA tension centered on the Agena upper-stage vehicle.[140] The Agena began as the upper stage for the Air Force's reconnaissance satellite and so was part of the 117L program when the original contract went to Lockheed in 1956.[141] When NASA and the DOD were supposed to be coordinating their overall launch-vehicle programs to avoid duplication, the Air Force failed to inform NASA of the existence of the Agena program in late 1958 for what one source terms "reasons of 'national security.'"[142] As a result, NASA began an entirely separate upper-stage project called Vega that had very similar performance characteristics when compared to Agena. Nowhere in the official NASA-DOD report on launch vehicles from January 1959 is the Agena's existence mentioned—however, the Vega is extensively discussed.[143] At some point between January and September 1959 (one source says between January and May), NASA became aware of Agena's existence, and pressure grew for NASA to cancel Vega.[144] A 30 September 1959 report said the United States' fleet of launch vehicles was basically sound, except that Agena should replace the Vega for NASA use.[145]

Glennan had to then inform Eisenhower that "there has apparently been a departure by the Department of Defense from the President's instruction that 'no substantial changes in the program presented early in the year are to be made without specific Presidential approval.' The Defense Department initiated a project named AGENA, which substantially duplicates NASA projects. They have gone so far with contracting and actual work under the project that to cancel it now would save very little money. . . . The President said he thought he had cleared up such duplications. Dr. Glennan said he thought so too."[146] After Eisenhower noted that "coordination in matters of this sort should occur before millions of dollars are committed," he "requested the Administrator, NASA, and the Secretary of Defense develop a scheme that would further coordination and, where possible, meld the NASA and Defense contributions to the National Space Vehicle program."[147] Kistiakowsky recorded that Eisenhower was "obviously very angry" about the Agena-Vega duplication "and made references to subordinates disobeying orders in connection with this duplication."[148] NASA canceled the Vega on 11 December 1959. The DOD and the Air Force appeared to have been sufficiently chastised because there are no other recorded incidents of such blatant duplication resulting from a failure by the DOD to inform NASA about the status of its programs. One source calculated the duplication cost as $16 million.[149] In the future, the AACB's Launch Vehicle Panel ensured NASA and the DOD had a forum wherein each could be promptly informed of the other's launch vehicle work. The AACB's other panels performed a similar role in their respective functional areas.

The Agena-Vega episode was not the only indication of rivalry in the NASA-DOD/USAF relationship. Certainly the Air Force was none too pleased to have lost some of its space responsibilities and projects to NASA. Vice-chief of staff of the Air Force (the service's second-highest ranking officer), Gen Curtis LeMay, shortly after NASA's creation, "complain[ed] forcefully about the lack of military input into the new NASA and to assert that in his opinion the manned satellite project would be delayed by a year or more" as a result of its transfer to NASA.[150] Some Air Force leaders resented the fact that large numbers of Air Force officers were transferred to NASA so that their management skills, acquired in USAF ballistic missile and space programs, could be used on NASA projects because NACA personnel lacked such experience. Several aspects of this Air Force personnel transfer to NASA gave rise to one of the most celebrated instances of supposed USAF-NASA tension of the Eisenhower administration.

Until early 1960, USAF officers were being transferred to NASA as individuals, and the Air Force honored virtually all NASA requests. But during the last year of the Eisenhower administration, NASA started requesting the transfer of entire project teams from USAF projects to NASA. For instance, NASA requested in April that the entire Project Centaur (another Air Force upper-stage vehicle) management team consisting of a colonel, a lieutenant colonel, and three majors be transferred to NASA. The deputy CSAF for development wrote to General White, the CSAF, that "the USAF just can't afford a continued dissipation of its in-service technical capability. . . . I recommend that the Air Force resist the reassignment of the officers in question."[151] In this context of increasing resistance to NASA personnel requests, General White decided the time was right for a "sermon from the Chief of Staff to his staff" because he believed the Air Force had to continue to support NASA to the absolute limits of the USAF's ability.[152] Therefore, White wrote his subordinates on 14 April 1960,

> I am convinced that one of the major long range elements of the Air Force future lies in space. It is also obvious that NASA will play a large part in the national effort in this direction and, moreover, inevitably will be closely associated, if not eventually combined with the military. It is perfectly clear to me that particularly in these formative years the Air Force must, for its own good as well as for the national interest, cooperate to the maximum extent with NASA, to include the furnishing of key personnel even at the expense of some Air Force dilution of technical talent. . . . I want to make it crystal clear that the policy has not changed and that to the very limit of our ability, and even beyond it to the extent of some risk to our own programs, the Air Force will cooperate and will supply all reasonable key personnel requests made on it by NASA.[153]

The "not eventually combined with" phrase was later taken out of context in an attempt to prove the Air Force was engaged in some type of a "campaign" to usurp NASA's authority. When this was investigated in the early Kennedy administration, White had to carefully explain this was not the case by emphasizing that the context of the letter was to ensure his subordinate generals were unequivocally clear that they would continue to honor NASA personnel requests. White told Congress, "The sole purpose of this memorandum—and I think I stated it very clearly—is that I want to make it crystal clear that the policy is we will cooperate with NASA—and to the very limit of our ability and even beyond, to the extent of some risk in our own programs." When asked if he had any thoughts of taking over any portion of NASA's mission, White responded, "Absolutely not. None then [April 1960], none now [March 1961], and I know of no one else who has contrary views in the Air Force. I would like to point out that this is not a statement of advocacy, but a statement of possible fact. . . . No planning whatsoever."[154] Indeed, when asked at the same March 1961 hearings if the military should take over any part of NASA, Kennedy's deputy SECDEF Roswell Gilpatric replied, "We have plenty of problems today. We don't need any more."[155]

The Air Force certainly should have been more careful with some of its public-relations/public-affairs type of activities. Chapter 6 examines in detail the supposed Air Force campaign during the Eisenhower-Kennedy interregnum. The fundamental point, however, is that while the Air Force was heavy-handed in attempting to create a greater awareness of its space capabilities, there was probably no orchestrated drive to shut down NASA and to take over its civilian space exploration and experimentation programs. Glennan wrote that he did not believe the Air Force wanted to take NASA over but that "the blatant nature of its propaganda is a little bit disturbing to me."[156] Glennan even met with the top USAF civilian and military leaders in December 1960

> to try to find out whether or not there was anything seriously wrong between NASA and the Air Force. The publication of stories of strife, vying for position, stealing each other's projects, etc. have been very frequent these past two or three weeks. It was a pleasant discussion with much agreement on both sides. Certainly at the top of our organizations there is no real difference or need for concern. I am sure, however, at the 'colonel' level, there is a good deal of envy and flexing of elbows.[157]

Accusations of Air Force poaching on NASA's territory clearly made good newspaper copy but were not supported by NASA's top official. Glennan later summarized that with the exception of the ABMA affair, "I had no real battles, very little trouble with the Pentagon actually. Sometimes members of the staff locked horns with somebody over there, but I'd go see Jim Douglas [secretary of the Air Force from May 1957 to December 1959] or Tommy White for five minutes and the problems would be solved."[158]

This is not to say there was not rivalry between NASA and the Air Force. Perhaps this was inevitable. One political scientist explains that any new and rapidly expanding bureaucracy will "soon engender hostility and antagonism from functionally competitive bureaus. Its attempt to grow by taking over their functions is a direct threat to their autonomy. Hence the total amount of bureaucratic opposition to the expansion of any one bureau rises the more it tries to take over the functions of existing bureaus."[159] For instance, Hall maintains that the loss of many space-related projects to NASA particularly galled the Air Force, "which still nursed a deep resentment over a civilian space agency's preempting a field it called its own."[160] Air Force frustration could flare over seemingly minor issues. NASA did not permit Air Force officers on duty at NASA to wear their uniforms, causing retired chairman of the JCS Nathan Twining to lament, "Yet these regular career men have to go around in semi-masquerade as civilians. In this regard I feel that as a nation we went overboard in our efforts to show peaceful intent."[161]

Similarly, the Air Force felt slighted because it believed NASA was not keeping it adequately informed on NASA's growing lunar studies, while the Air Force did regularly brief NASA on USAF lunar studies.[162] For instance, NASA formed a working group to prepare a lunar-exploration program that included the JPL, the ABMA, and the California Institute of Technology, but not the Air Force. On 17 April 1959, NASA announced plans for long-range scientific exploration of the moon, much to the USAF's surprise. That same month NASA responded to an Air Force briefing on the status of the USAF's strategic lunar-system studies by declaring that the lunar area was "exclusively NASA property." Some within the Air Force felt that NASA also took over the Air Force's nascent human-spaceflight program with no acknowledgment of indebtedness. One Air Force historian summarized, "NASA's uncooperative attitude in the lunar field became more noticeable. . . . The developing relations was [sic] discouraging." Another historian emphasized, "NASA was kept informed of progress but seemed less and less inclined to reciprocate. Gradually a background of unhappy incidents in NASA-USAF relations built up. . . . This far-from-cooperative attitude by NASA in the lunar field became more noticeable as weeks passed, and it came to cover much wider areas."[163]

Part of the problem was traceable to an issue from the previous chapter: the fundamental difficulty and perhaps impossibility of crafting a neat division between the civilian and military uses and responsibilities in space, given the similarity of technology used in each. Since there were large and unavoidable gray areas of overlap, "There were endless opportunities for disagreements and rivalries that at any time might delay projects of vital interest to the United States." From the Air Force's perspective, "To have the space program taken over by ARPA was a serious blow, and to have the program divided again with NASA was yet more disturbing. . . . The leaders of the civilian agency thought neither in terms nor interests of the military but pursued space flight and space exploration as ends in themselves. Yet national defense was at stake." Nevertheless, the Air Force was savvy enough to know that cooperation with NASA was in its best interests because in the long run the United States would develop the building blocks of space technology, albeit in NASA instead of the Air Force: "There were of course occasions of misunderstanding, but the Air Force kept its goal of cooperation."[164]

It becomes apparent that the concrete areas of DOD/USAF support of and coordination with NASA (see table 3) were more important than the areas of rivalry or tension described above. Many Air Force officers served in NASA, thereby giving NASA vital leadership experience in large project management which NACA personnel simply lacked and that NASA could obtain from no other source: "When NASA was established, the only persons with experience in the kinds of projects the agency was expected to implement were officers involved in weapons systems development."[165] This flow of needed individuals was codified in a 13 April 1959 NASA-DOD agreement that laid out the bureaucratic procedures for the three-year assignments, with a one-year extension possible.[166] When NASA lost Richard Horner as associate administrator, its number-three position, Glennan had difficulty securing another quality individual. He

Table 3. Long-term figures for military personnel assigned to NASA

1958	1959	1960	1961	1962	1963	1964	1965	1966	1967	1968	1969
66	67	77	117	161	239	249	280	323	318	317	268

Adapted from Jane Van Nimmen, Leonard C. Bruno, and Robert Rosholt, *NASA Historical Data Book*, vol. 1, *NASA Resources 1958–1968*, NASA SP-4012 (Washington, DC: USGPO, 1988), 80ff. The 1969 data extracted from Ihor Gawdiak and Helen Fodor, *NASA Historical Data Book*, vol. 4, *NASA Resources, 1969–1978*, NASA SP-4012 (Washington, DC: USGPO, 1994), 68.

went to CSAF White to inquire about the availability of an Air Force general for this position. White "promised that we would have our choice of three or four on very short notice although he fully agreed with us that we ought to bend every effort toward getting the civilian."[167] In the end, Glennan was able to hire a civilian, Robert C. Seamans.

Another type of support alluded to already was the family of ballistic missiles the DOD made available to NASA to use as space launchers. After all, "A launch vehicle is only a modified ballistic missile; and it cannot be overstated that for everything between sounding rockets and the Saturn I, NASA relied on vehicles successfully developed by the Air Force—between 1954 and 1959," particularly the Atlas, Thor, and Titan.[168] NASA's launch-vehicle dependence included relying on use of Air Force's launch facilities at Cape Canaveral on Florida's east coast and on the DOD's extensive worldwide network of tracking and data-acquisition stations. The Navy entered the support picture because its ships were used in the process of recovering astronauts returning from orbit. NASA's associate administrator Seamans quipped, "The Navy fell into this quite gladly. They didn't mind the visibility of having admirals greet astronauts when they arrived from the moon."[169] The Army's role in support came largely through its Corps of Engineers. NASA relied on the corps for designing and constructing the mammoth rocket stands and huge launch complexes, providing ship transportation, and so forth during NASA's period of rapid expansion in the late 1950s and early 1960s—"one of the wiser decisions in this hectic period."[170] In addition, given the NACA's experience with contracting and procurement,

NASA conducted its operations in these areas in accordance with the Armed Services Procurement Regulation.[171] The areas of support the DOD, and in particular the USAF, granted NASA in its early years were of undeniable importance in NASA becoming a viable organization and NASA's ability to conduct a robust civilian space program.

In addition to support, coordination was an important element of NASA-DOD interaction. The AACB's creation in the spring of 1960 provided a formal structure to the day-to-day coordination that had been taking place on myriad NASA-DOD topics since NASA's establishment. There were also numerous committees, ad hoc groups, and project-level consultations that greased the cogs of America's space program and the civilian-military interaction within it. In NASA's very first month, there were already 13 separate committees devoted to coordinating R&D topics between the two organizations.[172] As the AACB's six panels matured, they tended to form even more specialized subpanels to ensure that NASA and the DOD were reciprocally informed as to the other's activities in virtually every area of project development and facility construction. As Glennan told a House member in April 1960, "It seems clear to me that separate but closely related and properly coordinated management of military and non-military space activities is the sound procedure to be followed. . . . It is my conviction that we are well on the way to achieving a satisfactory management-level coordination that will work."[173]

It comes as no surprise, given such an extensive network of interagency committees, that interagency agreements and memoranda of understanding would proliferate. One government report from 1965 listed 88 separate "major" NASA-DOD agreements.[174] A comprehensive NASA accounting from 1967 listed 176 NASA-DOD accords.[175] One of the most important of these agreements from the perspective of later developments would be the one concerning how the two agencies would reimburse each other for services rendered. The November 1959 agreement on this subject basically stated that if the DOD received an order from NASA that the DOD had to then subcontract out, NASA would only have to reimburse the direct cost of the subcontract—there would be no overhead or administrative charges. If the DOD had the capability to fulfill the contract at one of its facilities, NASA's costs would be limited to the costs directly attributable to performance of the contract; there would be no charges for depreciation, rent, overhead, and so forth.[176] No attempt at coordinating two large programs such as the civilian and military space programs could have completely eliminated all traces of duplication and waste. Nevertheless, it appears NASA and the DOD made a good-faith effort (after the Agena-Vega fiasco) to reduce inefficiencies to a minimum.

The overall NASA-DOD situation in the Eisenhower administration was therefore a complex mixture of support, coordination, and rivalry in which no one facet predominated over the others. Glennan told the Senate six months into his tenure, "NASA and the military have functioned without undue friction or duplication of effort. . . . We are facing the same management problems confronting any large government or industrial complex. . . . Thus far, there have been no instances in which reasonable solutions to questions of jurisdictions have been impossible to reach."[177] In private sessions with Eisenhower, Glennan also "reported that he is finding his working relationships with Dr. York and Secretaries Gates and Douglas extremely fine."[178] There were no radical departures from this assessment during his remaining two years as NASA's administrator. Only a few days before departing NASA in January 1961, Glennan summarized, "People in both NASA and the Department of Defense are ambitious and imaginative. In such a situation there will always be pulling and hauling. But there has been less controversy and more cooperation in the last year than anyone had any right to expect." Schriever concurred in 1961 when asked about the NASA-USAF relationship: "It is completely satisfactory. I think that we had some growing pains at first when NASA was first created. During the past year our relationships—at least from my level—and I think this is true at the higher levels—has been extremely good."[179]

This chapter shows how the initial stages of the NASA-DOD relationship unfolded as well as how the relationship evolved as part of the overall Eisenhower space philosophy in action. On the one hand, Eisenhower did not want a full-speed crash program of space spectaculars. On the other hand, he recognized what he termed the *psychological* component of space exploration and did authorize the Saturn program as the vehicle that would eventually enable America to launch the large payloads that tended to be viewed as spectacular firsts by the world. Eisenhower also ensured that the space-for-peace philosophy continued to be America's primary statement on space affairs. He had little tolerance for "space cadets" in the military who wanted to discuss lunar bases or antisatellite weapons because such statements might endanger the fragile principle of freedom of overflight for reconnaissance satellites that lay behind the space-for-peace philosophy. As one perceptive historian wrote,

> The clear mandate from the Eisenhower administration . . . was that NASA's space efforts would be nonmilitary in character and highly visible to the public. This would serve two distinct but necessary purposes. First, NASA's projects were clearly cold war propaganda weapons that national leaders wanted to use to sway world opinion about the relative merits of democracy versus the communism of the Soviet Union. The rivalry was not friendly, and the stakes were potentially quite high, but at least this competition had the virtue of not being military in disposition. . . . Second, NASA's civilian effort served as an excellent smoke-screen [sic] for the DOD's military space activities, especially for reconnaissance missions. NASA's civilian mission, therefore, dovetailed nicely into cold war rivalries and priorities in national defense.[180]

The only remaining question—and the focus of the next chapter, concerning the Eisenhower administration—is how early human-spaceflight projects will fit into his philosophy and into the NASA-DOD relationship.

Notes

1. NSC 5814/1, *Preliminary U.S. Policy on Outer Space*, 2. Document is now declassified. The Eisenhower administration's final space policy document was written and coordinated as NSC 5918 but approved and issued by Eisenhower on 26 January 1960 as an NASC document, "US Policy on Outer Space." It does not modify in any significant way the prestige-related sections of NSC 5814/1. For "US Policy on Outer Space," see SPI document 92. NSC 5814/1 continued by stating that if the United States did not have some type of comparable advance in space, this condition "may dangerously impair the confidence of . . . peoples in U.S. over-all leadership . . . . To be strong and bold in space technology will enhance the prestige of the United

States among the peoples of the world and create added confidence in U.S. scientific, technological, industrial and military strength" (ibid., 3).

2. Ibid., 20.

3. *Aviation Week and Space Technology*, "Council [NASC] Compiles List of Space 'Firsts,' " 100. Another term often used to refer to the first generation series of Soviet lunar satellites is Lunik.

4. Cited in Elliott, *Finding an Appropriate Commitment*, 91–92. Senator John Stennis added, "We can expect to be spending billions of dollars a year on various types of space vehicles unless there is a drastic change in the world situation." Concerning Stennis's "billions" of dollars proposal, NASA's space-related budget for FY 59 was $231 million and the DOD's was $490 million. See NASA, *Aeronautics and Space Report of the President, Fiscal Year 1995 Activities*, A-30.

5. Glennan, "Our Plans for Outer Space," 99.

6. Goodpaster, memcon, 17 February 1959, 1–4.

7. For instance, Glennan recalled there was only one time in his tenure as NASA administrator (and he was Eisenhower's only NASA administrator) that Eisenhower ever directly told him to do anything. In the summer of 1959, "As I started to walk out the door, Ike called to me, 'Keith, there's just one thing that I'm very anxious that we get done. I want to see a booster rocket that will loft a house into orbit.'" Glennan responded this would take hundreds of millions of dollars, but Eisenhower simply responded, "Keith, go on back to your shop and get the figures put together. Let me see them." Glennan recalled that two months later, around September 1959, Eisenhower approved acceleration of the Saturn program and increased funding for it. Glennan, interview by Collins and Needell, 29 May 1987, 151. Eisenhower himself explained in September 1959, "If we must compete, we must focus the competition on some one or two key items where we have the best chance to do something that has great impact. We must also look for by-product contributions to our defense establishment." Goodpaster, memcon, 21 September 1959, 3.

8. Goodpaster, memcon, 21 October 1959, 2. Eisenhower recapitulated his space philosophy in three principles: "The first is that we must get what Defense really needs in space; this is mandatory. The second is that we should make a real advance in space so that the United States does not have to be ashamed no matter what other countries do; this is where the super-booster is needed. The third is that we should have an orderly, progressive scientific program" (ibid.).

9. The space historian must never forget that terms or phrases such as *military space* or *defense needs in space* are essentially veiled references to reconnaissance satellites. Eisenhower's final space-policy statement in January 1960 explained, "Space technology constitutes a foreseeable means of obtaining increasingly essential information regarding a potential enemy whose area and security preclude the effective and timely acquisition of these data by foreseeable non-space techniques." This is not to say other military functions such as meteorology, communications, navigation, and geodesy could not be *supported* by military satellites, but reconnaissance had to be wholly *conducted* by

space-borne platforms after the U-2 was shot down in May 1960. That final Eisenhower space policy of January 1960 defined the reconnaissance satellites as "satellite systems to provide optical, infrared and electronic intelligence and surveillance on a world-wide or preselected area basis" and emphasized it was the *only* satellite application currently assigned the highest national priority for both R&D and operational capability. NASC, "U.S. Policy on Outer Space," 3, 7. Document is now declassified.

10. Goodpaster, memcon, 11 January 1960, 2.

11. Hunley, *The Birth of NASA*, 44.

12. Eisenhower to Glennan, letter, 14 January 1960, 1.

13. Bilstein, *Stages to Saturn*, 50. This Bilstein book is the best source for detailed information on the extremely complex Saturn program and the many vehicles belonging to the Saturn family of boosters. For instance, the *Saturn V* that took Americans to the moon had many characteristics in common with those proposed for the Nova vehicle in the early NASA studies. Other programs sharing the top priority DX rating were the various ballistic missiles and the reconnaissance satellite program. See Boggs, "Missiles and Military Satellite Programs," 14 December 1960, 3. For an examination of America's fleet of space-launch vehicles in general, see Launius and Jenkins, *To Reach the High Frontier*, 2002.

14. Eisenhower, *Public Papers of the Presidents*, no. 21, 26 January 1960. 127.

15. Eisenhower, *Public Papers of the Presidents*, no. 24, 3 February 1960, 146.

16. Eisenhower, *Public Papers of the Presidents*, no. 264, 7 August 1960, 643–44.

17. Emme, *A History of Space Flight*, 161. Nevertheless, the total of Soviet payloads launched was 87,000 lbs., while the United States' was only 34,240. In addition, of the 33 US launches, 24 were conducted by the Air Force, five by the Army, three by NASA, and one by the Navy. Of the satellite payloads themselves, the Air Force had built 15, NASA 10, and the Army and Navy four each. Swenson, Grimwood, and Alexander, *This New Ocean*, 303.

18. Emme, *A History of Space Flight*, 159. Eisenhower concurred in his final State of the Union message, stating that US scientific achievements in space "unquestionably make us preeminent today in space exploration for the betterment of mankind." Eisenhower, *Public Papers of the Presidents*, no. 410, 12 January 1961, 913–30.

19. Goodpaster, memcon, 21 November 1959, 3.

20. Minnich, Notes on Legislative Leadership Meeting, 10 March 1959, 5. Eisenhower said in late 1959 that "if he had to approve another unbalanced budget he would be obliged to regard his Administration as discredited." Goodpaster, memcon, 17 November 1959, 2.

21. Eisenhower, *Public Papers of the Presidents*, no. 172, 29 July 1959, 556.

22. Glennan, statement to the NASA Authorization Subcommittee of the Senate Committee on Aeronautical and Space Sciences, 28 March 1960, 11.

23. Stans to Eisenhower, memorandum, 29 July 1958, 1. The USAF's reconnaissance satellite was now called Sentry and later Samos so as to have no connotations with weapons systems via the previous "WS"-style designation.

24. Killian, brief summary prepared for the first NASC meeting, 5 August 1958, 14. Document is now declassified.

25. Glennan said, "This reduction will materially restrict, if not substantially jeopardize, our progress toward the national objectives of scientific and technical leadership in the aeronautical and space fields. . . . On the one hand we are repeatedly urged to 'leapfrog the Russians' with our technological efforts and on the other, we are expected, apparently, to carry out space efforts with reductions made to a carefully crafted, conservative budget." Glennan, Statement to the Subcommittee on Independent Offices of the Senate Appropriations Committee, 19 May 1960, 1–2.

26. NASA, *Preliminary History of NASA*, II–11.

27. Eisenhower, *Public Papers of the Presidents*, no. 414, 16 January 1961, 970. By 1965 Kennedy and Johnson orchestrated *another* five-fold increase to over $5 billion.

28. ARPA director Roy Johnson on 20 October 1958 ordered the Air Force to cease using the WS designation "to minimize the aggressive international implications of overflight. . . . It is desired to emphasize the defensive, surprise-prevention aspects of the system. This change . . . should reduce the effectiveness of possible diplomatic protest against peacetime employment." Johnson to Schriever, letter, 20 October 1958, cited by Hall, "Origins of U.S. Space Policy," 229, n. 72.

29. Burrows, *Deep Black*, 134.

30. State Department to McGeorge Bundy, report, 10. Eisenhower continued to address letters to Bulganin because he was premier and chairman of the Council of Ministers, although Khrushchev, as first secretary of the Communist Party, exercised the real power.

31. Branyan and Larsen, *The Eisenhower Administration*, 650. On 15 February Eisenhower tried again: "If this peaceful purpose is not realized, and the worse than useless race of weapons goes on, the world will have only the Soviet Union to blame. . . . A terrible new menace can be seen to be in the making. That menace is to be found in the use of outer space for war purposes. The time to deal with that menace is now. It would be tragic if the Soviet leaders were blind or indifferent toward this menace" (ibid., 650–51).

32. It should be noted at this point that an issue raised concerning the pre-Sputnik era still held true after Sputnik. Eisenhower's attitude toward the Soviet Union and the overall Cold War had Janus-like, looking-in-both-directions quality. So would Kennedy's approach. Hope for arms control, conciliatory gestures, and a spirit of bipartisanship alternated with confrontational rhetoric and brinkmanship. For instance, NSC 5810/1, *Basic National Security Policy*, 5 May 1958, stated, "The United States should continue its readiness to negotiate with the USSR whenever it appears that U.S. interests will be served thereby. . . . Agreements with the USSR should be dependent upon a balance of advantages" and not implied goodwill or trust.

"Safeguarded arms control should be sought with particular urgency, in an effort to reduce the risk of war." Richelson, *Presidential Directives on National Security from Truman to Clinton*, no. 556, 19. Conversely, after the failure of disarmament and test-ban talks, Eisenhower wrote, "The Soviet Union, far from following a comparable [US] policy of restraint appears to have undertaken with deliberate intent a policy of increasing tension throughout the world and in particular of damaging relations with the US." The USSR "has threatened rocket retaliation against . . . the United States on the pretext of contrived and imaginary intentions. . . . The Soviets have unilaterally disrupted the ten-nation disarmament talks in Geneva" and, therefore, bears full responsibility "for the increased tension and the failure to make any progress in the solution of outstanding problems." Eisenhower, *Public Papers of the Presidents*, no. 313, 2 October 1960, 743. As Eisenhower summarized in his memoirs, "Of the various presidential tasks to which I early determined to devote my energies, none transcended in importance that of trying to devise practical and acceptable means to lighten the burdens of armaments and to lessen the likelihood of war. . . . In the end our accomplishments were meager, almost negligible. . . . That failure can be explained in one sentence: It was the adamant insistence of the Communists on maintaining a closed society." Eisenhower, *Waging Peace*, 467–68.

33. NSC 5814/1, *Preliminary U.S. Policy on Outer Space*, 20.

34. Harvey and Ciccoritti, *U.S.-Soviet Cooperation in Space*, 22.

35. Kohler, forward to *U.S.-Soviet Cooperation in Space* by Harvey and Ciccoritti, xxi. Arnold Frutkin, NASA's long-time director of International Programs, provided another primary source attestation to this assessment when he said shortly after the end of the Eisenhower administration that the USSR "has, so far at least, rejected or failed to follow through on every proposal for substantive cooperation in space science made by the United States or the scientists of other nations. . . . The fact is that the Soviet Union neither leads nor follows in international efforts in space research." Harvey and Ciccoritti, *U.S.-Soviet Cooperation in Space*, 47.

36. Eisenhower, *Public Papers of the Presidents*, no. 303, 22 September 1960, 720.

37. The historical minutiae are not relevant to this study. For the complete story see Harvey and Ciccoritti, *U.S.-Soviet Cooperation in Space*; Frutkin, *International Cooperation in Space*; and Kash, *The Politics of Space Cooperation*.

38. NSC 5814/1, *Preliminary U.S. Policy on Outer Space*, 4, 21.

39. Gavin, *War and Peace in the Space Age*, 224.

40. Kistiakowsky, *A Scientist at the White House*, 245.

41. Ibid., 229–30, 239–40, 245–46.

42. Quarles to the acting secretary of the NASC, memorandum, 15 April 1959, 1. Therefore, US space policy officials decided that international organizations such as the UN's COPUOS should be regarded "as bearing essentially on gaining acceptance for use of reconnaissance satellites as a legitimate outer space activity. It was suggested that discussion in the UN forum be oriented

toward establishing a 'freedom of outer space' concept." NASC, Minutes of Meeting, 27 April 1959, 6.

43. NASC, "U.S. Policy on Outer Space," 26 January 1960, 8, 12.

44. Prados, *The Soviet Estimate*, 101.

45. State Department, paper on *Samos* satellite, 1.

46. State Department, internal memorandum, 14 September 1960, 3.

47. Hall, "Origins of U.S. Space Policy," 229, n. 72.

48. Schriever, oral history interview by the author, 2 July 1996.

49. Houchin, *The Rise and Fall of Dyna-Soar*, 105. Page numbers from computer disc copy provided by Houchin to author.

50. Twining to Secretary of Defense, memorandum, 11 August 1958, in *Exploring the Unknown*, vol. 1, 360. A fundamental premise of Air Force doctrine was then, still is, and almost certainly will be "that a decisive margin of advantage goes to the nation whose delivery vehicles can attain the greatest speed, the greatest range, and the greatest altitude." House, Statement of Brig Gen Homer A. Boushey, 522.

51. Power to White, letter, 18 August 1958, 1–3. To the USAF, weapons in space were peaceful, just like "a watchman is peaceful but he must be armed. It is the intent, rather than the weapons, which determines what is and what isn't peaceful. Likewise, our weapons of space would be peaceful—since we would never use them for aggressive purposes." Air Force, "Air Force Policy on Space," briefings, 7.

52. Air Force, "Air Force Policy on Space," briefings, 19.

53. Boushey, "Blueprints for Space," 239, 241, 252–53.

54. Power to White, letter, 18 August 1958, 2–3. The Air Force simply assumed, "It is inconceivable that the ability of man to deal with new situations, his judgment or ability to take many unrelated facts and decide upon a course of action to accomplish his assigned mission" would not prove invaluable in space. Air Force Policy on Space, briefings, SAF, 28–29 January 1959, 14.

55. Ogle, cited in Gantz, *Man in Space*, 3.

56. Ibid., 1.

57. "Silent silo sitters" phrase from Dyke, *Pride and Power*, 171. It should be noted that while the author has attempted to maintain an objective approach toward all research questions and conclusions throughout this book, I am an active duty officer in the Air Force. In addition, my career classification is Space and Missile Operations and as a missile combat crew commander in the late 1980s and early 1990s I was, in fact, a "silent silo sitter" for several years.

58. Goodpaster, memcon, 18 November 1959, 8. In public forums this institutional concern often took the form of the Air Force emphasizing the defense aspects of space. As White wrote in 1959, "The United States must win and maintain the capability to control space in order to assure the progress and preeminence of the free nations. If liberty and freedom are to remain in the world, the United States and its allies must be in a position to control space. We cannot permit the dominance of space by those who have repeatedly stated

they intend to crush the free world. . . . Only through our military capability to control space will we be able to use space for peaceful purposes." White, "Space Control and National Security," 11, 13.

59. Meilinger, *10 Propositions Regarding Air Power*, 84.

60. Bowen, *An Air Force History of Space Activities*, 189. Flood was an ardent congressional supporter of the US Navy.

61. The Project Horizon report concluded, "Military, political and scientific considerations indicated that it is imperative for the United States to establish a lunar outpost at the earliest practicable date." It would not only facilitate communications and surveillance but also would "establish and protect U.S. interests on the moon." If the United States did not start a lunar outpost program quickly, it would forfeit "the chance of defeating the USSR in a military-technological race which is already recognized as such throughout the world." The Army anticipated 149 Saturn 1B launches to build and equip the base with the first-manned landing in April 1965 and a total cost of $6.01 billion. US Army, *Project Horizon*, vol. 1, 1, 5, 56. The Army also played the prestige card: "The primary implication of the feasibility of establishing a lunar outpost is the importance of being first." Failure to be first in space produces implications that the Army considered a matter of "public record" (ibid., 46–47).

62. Medaris, *Countdown for Decision*, 298.

63. Air Force, "Military Lunar Base Program," 1, 4–7.

64. Hall, "Origins of U.S. Space Policy," 226, n. 63.

65. Kistiakowsky, *Scientist at the White House*, 120, 141. Kistiakowsky refers to seven Air Force study programs active during the final years of the Eisenhower administration: SR 178, Global Surveillance System; SR 181, Strategic Orbital System; SR 182, Strategic Interplanetary System; SR 183, Lunar Observatory (the program discussed in the preceding paragraph); SR 184, 24-Hour Reconnaissance Satellite; SR 187, Satellite Interceptor System; and SR 192, Strategic Lunar System. See Air Force, "The Air Force Space Study Program," n.d., 1. The total budget for the studies as described in this document, probably for 1958, was $2.7 million.

66. Glennan, *The Birth of NASA*, 85.

67. MacIntyre to Roy Johnson, memorandum, 14 April 1959, 1. This sentiment echoed strongly throughout the remainder of the Eisenhower administration and into the Robert McNamara era at the DOD. Secretary of Defense Thomas Gates stated early in 1960 that the DOD was "not interested in space flight and exploration as ends in themselves. Our space efforts are an integral part of our over-all military program and will complement our other military capabilities." Gates address, in *Air Force Information Policy Letter, Supplement for Commanders, Special Issue*, 27 January 1960, 4. The president borrowed this language in his annual space report to Congress, declaring that the DOD's space programs "are a means toward achieving a more effective military posture for the United States and its allies, rather than space flight and exploration as ends in themselves. Therefore, the space efforts of the Department of Defense are an integral part of our overall military program and will complement

or supplement other military capabilities." House, *U.S. Aeronautics and Space Activities, January 1 to December 31, 1959*, H. Doc. 349, 24 February 1960, 22.

68. Kistiakowsky, reflecting on a 5 August 1960 briefing. Kistiakowsky, *Scientist at the White House*, 383.

69. NSC, Memorandum of Discussion at the 466th Meeting of the NSC, 7 November 1960, 2.

70. Levine, *Managing NASA in the Apollo Era*, 211.

71. Eisenhower to the secretary of defense and the NACA chairman, memorandum, 2 April 1958, 1–2.

72. Burrows differentiates between the two as follows: reconnaissance "has to do with the active pursuit of specific information, such as the performance characteristics of a ballistic missile. Surveillance entails the passive, systematic watching or listening for something to happen, such as a ballistic missile being fired." Burrows, *Deep Black*, xxv.

73. NACA chairman and SECDEF to the president, joint memorandum, April 1958, 1–2.

74. Military division of the BOB to the BOB director, memorandum, 2.

75. The final name for WS-117L, after being referred to as Sentry for a period of time, would be Samos (some said this referred to Satellite and Missile Observation System; some said it simply was the name of an island in the Aegean Sea, picked at random). America's first operational reconnaissance satellite, the Corona project, was outside of this strictly Air Force framework, as will be seen at the end of chap. 4.

76. Goodpaster, memcon, 17 July 1958, 2.

77. NSC, Memorandum of Discussion at the 376th Meeting of the NSC, 14 August 1958, 6.

78. Goodpaster, memcon, 23 September 1958, LOC, reel 18, 2.

79. Eisenhower, Executive Order 10783, 1–2.

80. Boone, *NASA Office of Defense Affairs*, 6.

81. Johnson admitted that some of these transfers "will initially appear to be contrary to the apparent requirements of the Department of Defense. I am satisfied that with good will and cooperation among all parties, a middle course will be developed. . . . I desire that all ARPA personnel adhere strictly to the policy of supporting the programs of the National Aeronautics and Space Agency [sic] and the letter and the spirit of the statute under which the relationships of the Department of Defense and that Agency are prescribed." Roy Johnson to all ARPA staff, memorandum, 14 October 1958, 1–3.

82. Hirsch and Trento, *The National Aeronautics and Space Administration*, 30.

83. House, Committee on Government Operations, H. Rep. 445, 4 June 1965, 12.

84. York, oral history interview, 24 January 1989, 43. York added, "There was very serious consideration at the top of the Air Force to changing their name to the United States Aerospace Force" (ibid., 44).

85. Cantwell, *The Air Force–NASA Relationship in Space*, 12.

86. Goodpaster, memcon, 11 March 1958, 2.

87. Glennan, *Birth of NASA*, 9. One should note Glennan's candor through-out his diary. He repeatedly offered his honest assessments of the individuals with whom he came into contact, even if, as in this case, his opinions were disparaging. In another context he commented on Medaris, "I hope he gets into heaven, now that he's a priest." Glennan, oral history interview, 29 May 1987, 147; Medaris became an Episcopal priest after his retirement from the Army. This is important because he never spoke of any Air Force leaders in such terms. In fact, he speaks fondly of CSAF general Thomas White throughout his diary and respectfully of General Schriever. This lends some credibility to the overall conclusion that while NASA and the Air Force had their points of difference during Glennan's tenure, the relationship was fundamentally sound.

88. Ibid., 10.

89. The JPL was basically responsible for building the scientific payloads that the large rockets would launch. It was an Army facility managed under contract by the California Institute of Technology. The JPL would ultimately become the NASA laboratory responsible for the construction and operation of most of NASA's robotic planetary and deep-space probes.

90. Glennan to McElroy, letter, 15 October 1958, 6.

91. Glennan, *Birth of NASA*, 10–11.

92. Brucker, Army Position Paper, 15 October 1958, 1–2.

93. Glennan, *Birth of NASA*, 10–11. In his memoirs, Medaris describes the intentional leak. Medaris, *Countdown for Decision*, 243.

94. Goodpaster, memcon, 31 October 1958, 1.

95. NASA/Department of the Army, "Cooperative Agreement on Army Ordnance Missile Command," 3 December 1958.

96. Glennan to McElroy, letter, 1 December 1958, 2.

97. Glennan, *Birth of NASA*, 12. Space Council meeting minutes reveal that most members felt that "although the solution that is being recommended does not wholly meet NASA's needs, it is considered the best arrangement which can be achieved at this time." NASC, Minutes of Meeting, 3 December 1958, 1.

98. Ambrose, *The National Space Program Phase*, 22.

99. Bilstein, *Stages to Saturn*, 39. See also Emme, "Historical Perspectives on Apollo," 372; and Benson and Faherty, *Moonport*, 13.

100. Kistiakowsky, *Scientist at the White House*, 10.

101. NASA, Considerations Preparatory to Establishing a NASA Position on ABMA, 20 August 1959, 1.

102. Glennan to Dryden, letter, 28 August 1959, 1.

103. Glennan added, "The only way he would consider the takeover would be for Defense to propose it and to deliver the Army and Von Braun in support of the transfer." Goodpaster, memcon, 21 September 1959, 2. Glennan stated in his memoirs, "I made it clear that I proposed to make a new deal, if any, only with the Office of the Secretary of Defense and that I expected Brucker to be told the results of the deal once it had been made." Glennan, *Birth of NASA*, 22.

104. Goodpaster, memcon, 16 September 1959, 1.

105. Glennan, Notes on Discussion, 23 September 1959, 2.

106. Kistiakowsky, *Scientist at the White House*, 100, 111.

107. House, *Transfer of the Development Operations Division*, hearings, February 1960, 30.

108. Rosholt, *An Administrative History of NASA*, 107.

109. Logsdon, *The Decision to Go to the Moon*, 53.

110. Kistiakowsky, *Scientist at the White House*, 57–58.

111. Ibid.

112. Kistiakowsky to Goodpaster, memorandum, 1.

113. McElroy to the chairman of the Joint Chiefs of Staff, memorandum, 18 September 1959.

114. It should be noted that this same memorandum from McElroy to the chairman, JCS, also rejected the official request the Army and Navy had made for creation of a joint, multiservice Defense Astronautical Agency to exercise control over all military space projects. Both services saw their input into and involvement with the military space environment slipping inexorably into the hands of the Air Force. This last-ditch attempt by the Army and Navy to maintain some active command role in military space projects was rebuffed by SECDEF McElroy in the September memo and again by new SECDEF Thomas Gates on 16 June 1960 when the Army and Navy renewed their request for a joint astronautical command. See, among others, Bowen, *Threshold of Space*, 31; and DOD, "General Proposal for Organization for Command and Control of Military Operations in Space," 1959.

115. Senate, Committee on Aeronautical and Space Sciences, H. J. Res. 567, 18 February 1960, 29–30.

116. NASC, Minutes of Meeting, 2 March 1959, 1.

117. CMLC, Terms of Reference, 22 October 1958, 2.

118. CMLC, Minutes of Meeting, 13 January 1959, 1.

119. CMLC, Minutes of Meeting, 10 March 1959, 1.

120. Holaday, testimony on 29 April 1959, Senate, Subcommittee on Government Organization for Space Activities, Hearings, 1959, 504–5.

121. Holaday, memorandum for record, 22 July 1959, 1.

122. Glennan concurred: "They do not have any authority. It is entirely a communications channel." Glennan said he could do his job without it. Senate, Subcommittee on Government Organization for Space Activities, S. Rep. 806, 25 August 1959, 4, 46–48.

123. Holaday, transcript of testimony, House Committee on Science and Astronautics, 10 March 1960, 5–6.

124. NASA, *Aeronautics and Astronautics, 1965*, 351. The CMLC was not legislatively and officially abolished until Reorganization Plan No. 4 of 1965.

125. Lester, *The Diaries of Dwight D. Eisenhower, 1953–1961*, microfilm, LOC, reel 18, 1.

126. Gray, memorandum for record, 3 August 1959, 1. After Killian left the administration, the task of chairing NASC meetings fell to his replacement Kistiakowsky, who termed it "another useless job" that "spoils my plans for a week's vacation." Kistiakowsky, *Scientist at the White House*, 46.

127. While neither the CMLC nor the NASC *made* policy during the Eisenhower administration, their records nevertheless are useful for the historian because they sometimes contain documents, reports, statements, and so forth from the agencies and persons who *were* making the important decisions and explaining in the context of an NASC or CMLC meeting why they had made a decision.

128. Eisenhower, *Public Papers of the Presidents*, no. 11, 4 January 1960, 34–36. The NASC enjoyed a brief resurgence when Vice Pres. Lyndon Johnson used it as the forum through which to conduct his investigation responding to Kennedy's April 1961 tasking asking how the United States could beat the Soviets into space. Even in this instance, however, Johnson was clearly in control of the process, and the NASC was largely a vehicle for his research. Subsequently, the NASC continued to meet throughout the Kennedy and Johnson administrations not as a policy-making body but as a forum for discussion and exchange of ideas. Its records do, therefore, contain some important documents, even though the NASC did not technically make policy or decisions. Richard Nixon abolished the NASC in Reorganization Plan No. 1 of April 1973. For a good sketch of the NASC's history, see Day, "Space Policy-Making in the White House," in Launius, *Organizing for the Use of Space*.

129. Glennan to Eisenhower, memorandum, 16 November 1959, reprinted in Hunley, *Birth of NASA*, 24–29. Glennan explained his thoughts to senior NASA staffers after the first of the year: "There is no need for the 'except' clause in the law. The military services have all the authority they need to make use of the space environment to satisfy military requirements." His changes would eliminate the "except" clause "with the statement that nothing in the Act prevents them from using the space environment to satisfy military requirements. This has been done in order that the responsibility for the nation's space 'exploration' program may be given to NASA. . . . It places responsibility squarely where it should be placed. It protects the right of the DOD to utilize the space environment for military purposes." Glennan to NASA senior leadership, memorandum, 2 January 1960, 1.

130. Glennan, *The Birth of NASA*, 33.

131. Ibid., 46. See also Harlow, memorandum for record, 13 January 1960, 2–4.

132. Eisenhower, *Public Papers of the Presidents*, no. 11, 4 January 1960, 34–37.

133. House, *Amending the National Aeronautics and Space Act of 1958*, H. Rep. 1633, 5–6.

134. Glennan, *The Birth of NASA*, 171.

135. Johnson, Memorandum on Proposed Amendments to the Space Act, H. R. 12049, 18508.

136. The six panels were: Manned Space Flight, Unmanned Spacecraft, Launch Vehicles, Space Flight Ground Environment, Supporting Space Research and Technology, and Aeronautics.

137. House, *The NASA-DOD Relationship*, 1964, 10–11.

138. NASA, Minutes from the Williamsburg Conference, 21 October 1960, 1.

139. Logsdon, "Opportunities for Policy Historians," 94.

140. Lower-stage vehicles are those designed to lift a rocket and its payload off the ground and through the dense lower portions of the earth's atmosphere. Upper-stage vehicles are designed to insert the payload into its final orbit and to maneuver it once it is there, or, alternatively, to boost the payload to such a velocity that it can escape the earth's atmosphere and begin its flight into interplanetary space.

141. York, *Race to Oblivion*, 105. See also Hall, "Origins of U.S. Space Policy," 223.

142. Hall, "Civil-Military Relations in America's Early Space Program," 12.

143. NASA, *A National Space Vehicle*, 27 January 1959.

144. Hall, *Lunar Impact*, 23.

145. Holaday to the SECDEF and NASA administrator, memorandum, 30 September 1959, 1–5.

146. Goodpaster, memcon, 26 October 1959, 1.

147. NASC, Minutes of 7th Meeting, 26 October 1959, 3.

148. Kistiakowsky, *Scientist at the White House*, 128.

149. Means, "Vega-Agena-B Mix-Up Cost Millions," 19.

150. Wood to Glennan, memorandum, 29 October 1958, 1.

151. Wilson to White, letter, 7 April 1960, 1. Bernard Schriever also explained that he deeply regretted the loss of several key individuals from his command such as Maj Gen Don Ostrander, whom Schriever "considered to be greatly needed for his own developmental programs" and who had become chief of NASA's launch-vehicle programs. Futrell, *Ideas, Concepts, Doctrine*, vol. 1, 605.

152. Futrell, *Ideas, Concepts, Doctrine*, vol. 1, 605.

153. White to Landon and Wilson, letter, 14 April 1960, 1.

154. House, *Defense Space Interests*, Hearings, March 1961, 92–93.

155. Ibid., 35.

156. Glennan, *Birth of NASA*, 224.

157. Ibid., 284. The "colonel" reference is an allusion to the fact that within the military the colonels tend to be the highest-ranking members of the teams who actually work on a particular project in the sense of supervising its day-to-day management and operation. Colonels, as a general rule, do not get involved in the higher-level policy-making decisions such as how NASA and the Air Force would interact as agencies. Such policy issues would be settled by the highest-ranking generals and civilian executive-branch presidential appointees.

158. Glennan, oral history interview, 5 April 1974, 9.

159. Anthony Downs, *Inside Bureaucracy*, 64.

160. Hall, *Lunar Impact*, 27. Another Air Force historian added, "As a new government agency, NASA had the normal human and institutional instinct to build an empire and reach out for control of all space vehicles." Cantwell, *The Air Force–NASA Relationship in Space*, 16.

161. Twining, *Neither Liberty Nor Safety*, 291.

162. Cantwell, *The Air Force–NASA Relationship in Space*, 10. See also Bowen, *An Air Force History of Space Activities*, 159–61.

163. Bowen, *An Air Force History of Space Activities*, 160.

164. Ibid., 123, 155.

165. Levine, *Managing NASA in the Apollo Era*, 122. For instance, Gen Samuel Phillips would become director of the Apollo Program and perform in such a superb manner that NASA called him back after the *Challenger* disaster of January 1986 to conduct an investigation of the accident; General Ostrander headed up NASA's launch-vehicles programs; Maj Gen James Humphreys would become NASA's director of Space Medicine; Brig Gen Edmund O'Connor was Marshall Space Flight Center's director of Industrial Operations; and Brig Gen C. H. Bolender was program director for Apollo's lunar module.

166. DOD, Agreement between the DOD, Army, Navy, Air Force and the NASA, 13 April 1959.

167. Glennan, *Birth of NASA*, 171.

168. Levine, *Managing NASA in the Apollo Era*, 212.

169. Seamans, oral history interview, September 1973–March 1974, 539. Document is now declassified.

170. Bilstein, *Orders of Magnitude*, 60.

171. NASA, Agreement between the Department of the Air Force and the NASA, 15 October 1959.

172. Glennan to all military service secretaries and Deputy Secretary of Defense Gates, letter, 31 October 1958, 1–2. These committees were Fluid Mechanics; Aircraft Aerodynamics; Missile and Spacecraft Aerodynamics; Control Guidance and Navigation; Chemical Energy Processes; Nuclear Energy Processes; Mechanical Power Plant Systems; Electrical Power Plant Systems; Structural Loads; Structural Design; and Structural Dynamics, Materials, and Aircraft Operating Problems.

173. Glennan to Rep. James M. Quigley, letter, 4 April 1960, 4.

174. House, *Government Operations in Space*, 123–32.

175. NASA, Inventory of NASA Interagency Relationships, 13 October 1967, folder: Copies of Agreements, DOD subseries, Federal Agencies series, NHDRC.

176. DOD, Agreement between the DOD and NASA Concerning the Reimbursement of Costs, 12 November 1959, 293–96.

177. Glennan, Statement before the Subcommittee on Governmental Organization for Space Activities of the Senate Committee on Aeronautical and Space Sciences, 24 March 1959, 15–16.

178. Goodpaster, memcon, 8 April 1960, 2.

179. Quoted in Grossbard, *The Civilian Space Program*, 168.

180. Launius, *A History of the U.S. Civil Space Program*, 34–35.

Chapter 4

Mercury, Dynasoar, and the National Reconnaissance Office under Eisenhower

In this chapter America's first two concrete human-spaceflight programs come to the forefront: NASA's Project Mercury and the Air Force's Dynasoar. Eisenhower's particular attitude toward human spaceflight, and its role (or lack thereof) in a competitive quest for prestige relative to the Soviets, largely determined the trajectory of both projects. As the DOD rendered invaluable assistance to NASA's Mercury, it also independently sought human access to space under its own devices. These programs developed simultaneously with the secretive construction of the third—and final—organizational leg of America's space program, the NRO, whose responsibility was to develop, launch, and operate America's reconnaissance satellites. While continuing classification challenges prevented a full investigation of the NRO during the Eisenhower and subsequent administrations, its creation is briefly touched upon in this chapter to roughly complete the picture of at least the organizational component of America's early space program.

Mercury's Antecedents

The story of human spaceflight in the Eisenhower administration—like many trends in his space policy—does not begin with Sputnik but rather has pre-Sputnik antecedents. In this case, one must look at the efforts of the Air Force to justify a human presence in space before the fall of 1957. In the months before NASA began operations in October 1958, the overriding question was whether Eisenhower would assign the human-spaceflight mission to the new NASA or to the Air Force. He gave it to NASA in August 1958, and Project Mercury officially came into existence that fall. Simultaneously, however, the Air Force continued to pursue its Project Dynasoar, also designed to place a human in orbit. Therefore, the Eisenhower administration laid the foundation for the complex NASA-DOD relationship concerning human-spaceflight projects that would fully emerge in the Kennedy administration.

Early Air Force Man-in-Space Activity

As early as November 1948, the Air Force's School for Aviation Medicine held a symposium on "The Medical Problems of Space Travel."[1] In 1949 the Air Force appointed Dr. Hubertus Strughold as the first professor of space medicine at its School for Aviation Medicine. One study of the evolution of space medicine concludes, "By the midfifties current thinking in the Air Force was increasingly oriented toward possible manned-space flight."[2] Indeed, numerous sources attest to the fact that by early 1956, the USAF was seriously studying the requirements of human spaceflight. Most of this work emanated from the Air Research and Development Command (ARDC), and more specifically its BMD under General Schriever. In March 1956, the BMD initiated a series of studies termed *Manned Ballistic Rocket Research System* to examine the technology of human spaceflight and create preliminary designs of spacecraft capable of being recovered from orbital conditions. The BMD secured the assistance of the NACA and industrial contractors in this effort.[3] These studies continued at a relatively low level throughout 1957 but began to pick up momentum in 1958—as the government started forging its response to Sputnik. For instance, on 31 January 1958, General Putt, Air Force deputy chief of staff for development, directed the ARDC to determine the quickest way to put a man into space and recover him. Putt also wrote Dryden, NACA director, and formally invited the NACA to participate in this effort, encompassing both a one-orbit human flight and a boost-glide research airplane (Dynasoar). In early February 1958, the NACA verbally informed the ARDC that NACA was preparing its own manned-capsule designs to be ready by late March; therefore, they could not cooperate with the USAF effort before that point.[4] One NACA insider stated that [Dryden] "had a very good ear to the ground. . . . he could see the handwriting on the wall, that probably the manned space program was going to be run by a civilian agency. . . . He didn't want to sign any papers with the Air Force at that time."[5] In contrast to the lack of cooperation in designing a ballistic capsule to orbit the earth, DOD-NACA cooperative work on Dynasoar was formalized and will be discussed below.

General LeMay, vice-chief of staff of the Air Force, did not want to delay initiating a capsule-type design and on 27 February 1958 ordered the ARDC to prepare and submit an official USAF man-in-space program as soon as possible. The next day Roy Johnson, director of the ARPA, recognized "The Air Force has a long term development responsibility for manned space flight capability with the primary objective of accomplishing satellite flight as soon as technology permits." Johnson authorized development of a test vehicle for experimental flights with laboratory animals with the goal of eventually orbiting a human.[6] The Air Force again turned to the NACA and indicated, "The Air Force would like NACA to participate in the examination of the man-in-space problem and to furnish guidance and experience in the logic of the program, and the feasibility from the technical point of view."[7] The NACA seemed more willing to cooperate in March and April than it had in January 1958. On 11 April Dryden signed an agreement to conduct a joint man-in-space program with the USAF.[8] This agreement was never implemented because the NACA quickly "tabled" it. It proposed "management of the design, construction, and operational phases of the project shall be performed by the Air Force."[9] During April, however, the NACA did at least supply inputs into the BMD report issued on 25 April in response to LeMay's February tasking, "Air Force Manned Military Space Systems Program." This report was the first official Air Force human-spaceflight plan and consisted of four phases:

1. Man-in-Space-Soonest would determine the functional capabilities and limitations of humans in space by means of earth-orbital flights.

2. Man-in-Space-Sophisticated would have a vehicle capable of 14-day orbital flights and conduct experiments essential for a lunar-exploration program.

3. Lunar Reconnaissance would explore the moon with a television camera and other instrumented packages.

4. "Manned Lunar-Landing-and-Return" would test the equipment for circumlunar flights before climaxing with a human lunar landing, brief surface exploration, and return.[10]

The estimated date of completion for the entire program was December 1965 with an estimated total cost of $1.5 billion.

The PSAC reported to Killian that "an NACA–Air Force cooperative effort on the manned-satellite program appears to be in high gear with every reason to believe a satisfactory working agreement in this field will continue."[11] This April 1958 plan was the first of seven such plans the Air Force would publish in 1958—most were scaled-back versions of this original plan designed to reduce expenses in the face of waning support for a military human-spaceflight program beyond Dynasoar.[12]

The May 1958 version of the Air Force's Man-in-Space Plan has been significantly declassified and reveals the importance the USAF attached to the program as well as the problems inherent in it: "The precise mission of the USAF in space with either uninhabited or manned vehicles cannot as yet be conclusively stated." The Air Force concurred with the PSAC that reconnaissance, communications, and early warning were the immediately available military uses of space but also stated, "These applications are merely the rudimentary ancestors of the sophisticated Air Force space weapons systems of the 1970–1980 era and beyond" because "man exceeds by many orders of magnitude the capabilities of present and prospective automata in perceptive acuity, level of judgment and decision making ability, and flexibility." The May 1958 plan stated that the USAF must gain approval for a human-spaceflight program in which "manned landing on the moon and return to earth has been chosen as the specific terminal mission." In the end, this would mean "the weapons systems designer of the future will have to him the bonus alternative of utilizing the moon as a base of Air Force operations."[13] As explained in the previous chapter, this type of rhetoric, forecasting the military control and use of space, was in no way attractive to the civilian space policy makers in the Eisenhower administration because it contravened the space-for-peace policy and could possibly endanger the free passage of reconnaissance satellites. Therefore, assigning the human-spaceflight mission to NASA became increasingly likely during the summer of 1958.[14] All indications from civilian Eisenhower administration officials "made clear that quick approval of a military man in [the] space program was not forthcoming. . . . [T]he military services and particularly the Air Force found their space prospects disheartening. Obviously the military services no longer controlled development of space vehicles and programs [and] the new fiscal year offered little hope for change."[15]

As its expansion into a larger and more powerful NASA became increasingly likely after Eisenhower submitted his Space Act in April 1958, the NACA had little reason to forge a cooperative human-spaceflight program with the military because it appeared there was a good chance the new NASA would be given the mission itself. This explains the quick turn of events in March, when Dryden and the NACA agreed to work with the Air Force in developing its man-in-space program as a "joint project for a recoverable manned satellite test vehicle." In April Eisenhower submitted his Space Act that "appeared likely to transform NACA into the focal point of the nation's efforts in space," and in May the NACA withdrew from the cooperative joint undertaking that Dryden had signed 11 April and tabled indefinitely its participation with the Air Force in human-spaceflight R&D outside of Dynasoar.[16] As a memo from Dryden's office files delicately stated, once Eisenhower submitted his Space Act in April, the NACA leaders discussed the Air Force's offer for a cooperative man-in-space project and agreed that "the prospective Agreement should be put aside for the time being. The matter may be taken up again when the responsibilities of ARPA and NASA have been clarified."[17]

The NACA had started independent investigations into human-spaceflight vehicle designs within its own laboratories, separate from the USAF. As Project Mercury's official history explains, "The research engineers at Langley and on Wallops Island were pushing their own studies. They could see the opportunity to carry out a manned satellite project coming their way." Throughout spring 1958, the NACA's labs "were urgently engaged in basic studies" of propulsion, spacecraft configuration, orbit and recovery techniques, guidance and control, and the myriad other details of a human-spaceflight program.[18]

The NACA's work progressed to the point that wind-tunnel experiments were conducted on various vehicle designs, and rockets were launched with models of assorted orbital vehicles. Overall, the NACA engineers "were steadily modifying the manned ballistic satellite design itself" and by late 1958 had settled on the design that became the basis of Project Mercury.[19] Therefore, while some in the Air Force would later lament that Project Mercury was simply a wholesale borrowing from the USAF Man-in-Space Plan, this appears not to be the case. The NACA was independently working on most aspects of spacecraft design, and when Eisenhower did award NASA the human-spaceflight mission in August 1958, the NACA's efforts could then incorporate Air Force designs and engineering resulting from man-in-space plans. Therefore, while the USAF could not claim Project Mercury was simply a redesign of their man-in-space design, neither were later NASA assessments entirely correct in stating, "Project Mercury had grown out of the pioneering work on manned space flight at Langley Research Center," thereby ignoring the Air Force's contributions to Mercury's designs.[20]

Even as the Eisenhower administration was poised in July and August to formally assign the human-spaceflight mission to the DOD or NASA, its opinion of the prestige-related importance of the program seemed low. A panel with representatives from all relevant agencies (ARPA, BOB, NACA, NSC, and PSAC) concluded on 2 July 1958 that the human-spaceflight program "in general was looked upon rather unfavorably. The amount of psychological effect was questioned and no scientific applications were advanced. It was generally agreed that man-in-space (orbit) should not be put on a crash basis. . . . The man-in-space program should be handled on a long-term basis."[21] This would remain the Eisenhower administration's position for the remainder of its duration.

Eisenhower Awarded NASA the Mission

In late July 1958, the ARPA/USAF and the PSAC/NACA drew up papers supporting their respective cases for being the organization given the human-spaceflight mission. The ARPA said the DOD had a pressing requirement to undertake an immediate R&D program in this field because "such a program will lead to a significantly improved capability to accomplish existing military missions, such as reconnaissance, navigation and communications. In addition, development of such a capability is inherently a component of necessary military programs of a future, entirely predictable character." A human being would be a "superior mechanism" in most space systems because unmanned satellites would have a limited life span and could not be repaired. Therefore, "It is quite likely that a single high sophistication and manned and recoverable vehicle system will be both more efficient and more economical."[22] The NACA/PSAC presentation emphasized that "NASA through the older NACA has the technical background, competence, and continuing within-government technical back-up to assume this responsibility with the cooperation and participation of the Department of Defense."[23] The PSAC summarized to its chairman Killian, "At the present time there is no seriously proposed weapon system, or military operation, which requires the development of a manned satellite. In addition, no reasons have been advanced which indicate that this research and development activity is 'necessary to make effective provision for the defense of the defense of the US' other than the 'feeling' that the military ultimately will require manned satellites or other space vehicles." The scientists concluded human spaceflight was essentially research—not operations—and so should be in NASA. In addition, if the mission were given to DOD, this "cannot help but set a precedent for future, more extensive manned projects."[24]

The ARPA seems to have sensed that the NACA and Killian's office had the stronger case for presentation to Eisenhower. The ARDC commander informed the CSAF and vice-CSAF that ARPA director Johnson explained to him that "the current prevalent view in the White House is that there is no requirement for 'Man in Space.'"[25] As the preeminent history of this period states, "But by August the Air Force's hopes for putting a man into orbit sooner than the Soviet Union, or than any other agency in this country, were fading rapidly before the growing consensus that manned space flight should be the province of the civilian administration."[26] Most scholars conclude it was on 20 August 1958 that Eisenhower decided to award the human-spaceflight mission to NASA and not the DOD. His decision was closely tied to his space-for-peace policy, plus the fact that there was no clear military justification to put humans into orbit. In addition, the human-spaceflight budget would be $40 million, to include $30 million from NASA and $10 million transferred from the ARPA.[27] As the NACA's official history explains, Eisenhower "did not want to hand over to any group in the Pentagon a large and potentially enormous new area of activity, especially when he seriously doubted the services' ability to handle their current missions."[28]

One of the first actions NASA took when it started operations on 1 October 1958 was to officially approve Project Mercury as its program for human spaceflight.[29] In March 1959, Glennan would testify to Congress, "Finally, despite reports to the contrary, there is only one U. S. manned-satellite program: NASA's Project Mercury. . . . And representatives of each of the services are regular working members of the Project Mercury team."[30] Glennan was making a fine distinction between Dynasoar and a manned-satellite program. Dynasoar was a sort of powered glider designed for orbital operations but presented by the Air Force as intended for suborbital R&D into hypersonic flight— equal to or exceeding five times the speed of sound. Through this semantic massaging, the Dynasoar was not technically a space vehicle because it would not complete an entire orbit. Realistically, however, the Air Force fully expected the Dynasoar to engage in orbital operations once it was perfected. In the Eisenhower administration, it was politically advisable to present it as a suborbital vehicle so it would not become mixed up in the complicated ARPA, NASA, OSD, PSAC, and USAF organizational give-and-take.

Eisenhower, Prestige, and Human Spaceflight

Before plunging into the programmatic details of Dynasoar and Mercury, however, it is necessary to examine the last relevant component of Eisenhower's space policy: What were his views on using human spaceflight as a competitive tool for winning prestige in the Cold War struggle with the USSR after NASA's establishment? Evidence presented thus far indicates he was not keen on the notion before October 1958. Nor would he endorse the idea of a human-spaceflight race after NASA began operations. This is not to say the Eisenhower administration was unaware of Soviet plans for human spaceflight. The CIA reported in August 1958, "We believe that the ultimate foreseeable objective of the Soviet space program is the attainment of manned space travel on an interplanetary scale. . . . While the Soviet space program was undoubtedly initiated to serve scientific purposes, an immediate aim was to achieve political and propaganda gain."[31] However, Eisenhower concluded he would not take the Soviet-competitive bait in the case of human spaceflight. While he did approve Project Mercury, he did not let its budget skyrocket, and he was ready to terminate NASA's human-spaceflight program at the conclusion of his administration because it represented everything he wanted to avoid in space policy: it was hugely expensive, was "driven almost entirely by the competition with Russia, and lacking in a compelling scientific rationale."[32]

In an NSC meeting in May 1959, Eisenhower reminded his staff, "If a program is being conducted for psychological reasons only, we must look at it with a jaundiced eye."[33] Eisenhower tried to maintain a delicate balance between a crash human-spaceflight program and completely ignoring such a program. He said later that year concerning human spaceflight, we need "some achievements that will encourage and hearten our people. At the same time we should seek to discover scientific principles that will be of use to our military forces."[34] Killian delivered a speech in September 1960, which, even though he had returned to MIT, still represented the advice Eisenhower received from his trusted scientists on the wisdom of a spectacular race for prestige using human spaceflight. Killian acknowledged that he believed that "our man-in-space program is on the way to becoming excessively extravagant and will be justified only as a competitor for world prestige with the Soviet man-in-space program. Many thoughtful citizens are convinced that the really exciting discoveries in space can be realized better by instruments than by man. . . . Unless decisions result in containing our development of man-in-space systems and big rocket boosters, we will soon have committed ourselves to a multibillion dollar space program."[35] Killian's successor, Kistiakowsky, stated with great displeasure that Project Mercury "would be only the most expensive funeral man has ever had"[36] and was a "scientific luxury that should not be allowed to divert national efforts from more urgent scientific challenges here on earth."[37]

It was in the final months of the Eisenhower administration that his conclusions, and those of his subordinates, concerning the inadvisability of pursuing human spaceflight for prestige-related reasons became most clear. At a conference with the president in October 1960, Glennan discussed the idea of sending humans to the moon. NASA had been exploring this possibility under its Project Apollo study program. Glennan said his conclusion was that this was a multibillion-dollar project "of no immediate value" and that he was "screwing up his courage to state publicly that this should not be done." Eisenhower then agreed that such a project was "useless at this moment and would not think it really worth the money." Eisenhower said, "He likes to see us go ahead on useful things but he is not much of a man on spectaculars." He added he had "little interest in the manned aspects of space research," having realized that "some stunts, such as the Lindbergh trip across the Atlantic, have some virtue," but Eisenhower "emphasized that he would not be willing to spend tax money to send a man around the moon." Eisenhower concluded by concurring with Glennan's suggestion that the whole issue should be left for the next president and "emphasized his desire to avoid crash programs. He said there is such a thing as common sense, even in research." Finally, Eisenhower tasked his science advisors, Kistiakowsky and Glennan, to form a panel to reach a position on the amount of effort that was appropriate for the human-spaceflight program after Mercury.[38]

One challenge in fully understanding Eisenhower's conception of not using human spaceflight as a competitive tool in the Cold War race for prestige is why did he then authorize the Saturn rocket as the next-generation space booster, with the ability to lift much higher payloads than the ICBM-based boosters? Glennan stated clearly in October 1960 that the Saturn program would cost at least $7 billion over 10 years and that "there is really not much use doing this unless we are aiming at placing a man on the moon," an effort which overall would cost $14 to $35 billion.[39] While Eisenhower did authorize the Saturn and increased its funding, there is no record of exactly which prestige-related payloads he believed it should launch. The record is, however, clear on the fact that extensive human-spaceflight missions were not the types of missions he envisioned for the superbooster. Eisenhower expressed this puzzling notion to National Security Advisor Gordon Gray in November 1960: "The President felt that the only place we ought to be even in a clandestine way contesting with the Soviet Union is the development of the big engine. He repeated his often expressed view that little would be accomplished by putting a man into space."[40]

The report Eisenhower tasked Glennan and Kistiakowsky with preparing, concerning the appropriate level for the human-spaceflight program, was prepared by a panel chaired by Donald Hornig, a PSAC member and chemistry professor at Princeton University.[41] The Report of the Ad Hoc Panel on Man-in-Space, delivered on 16 December 1960 declared, "We have been plunged into a race for the conquest of outer space. . . . The most impelling reason for our effort has been the international political situation which demands that we demonstrate our technological capabilities if we are to maintain our position of leadership." The report explained Mercury was by definition a "somewhat marginal effort, limited by the thrust of the Atlas booster," which, as a converted Air Force ICBM, had barely sufficient power for human-spaceflight payloads. Nevertheless, Hornig's panel concluded Project Mercury had to be pushed due to "political desire either to be the first nation to send a man into orbit, or at least to be a close second. This marginal capability cannot be changed substantially until the Saturn booster becomes available."[42] In fact, the report stated that even the Saturn rocket would probably not be enough to land humans on the moon and safely return them; a new Nova rocket, possibly featuring both chemical and nuclear propulsion, would be required.[43]

The ad hoc panel's conclusion was sure to shock the fiscally prudent president. It would cost a minimum of $350 million to orbit a human; $8 billion to circumnavigate the moon; and

$26–35 billion to land on the moon and return—perhaps around 1975. Saturn should be regarded as only an intermediate step because it "must be followed by a much bigger development before manned lunar landing is possible."[44] Glennan's initial reaction to such talk was that when the discussion turns to the prestige of the United States resting on the question of who lands a human on the moon first, then "It seems clear that all sense of perspective has gone out the window. Clearly, with the probability that at least ten years must elapse before we can accomplish the feat of putting a man on the moon, the leadership and stature of the United States will no longer be in question. Either we will be the leader or we will not."[45]

Eisenhower's reaction was even more caustic when the Hornig panel's report was briefed to him at an NSC meeting on space on 20 December 1960. Glennan introduced the presentation by explaining that somewhere between 1964 and 1966, "The United States would have to decide (1) whether to spend large sums of money to put a man on the moon, and (2) if a landing were to be attempted, what vehicle should be developed for this purpose?"[46] Glennan outlined NASA's long-term budget picture and even before the Hornig panel briefing, Eisenhower replied that "he had a thousand questions" because, "In the space field there appeared to be no practical test of the immediate usefulness of a program. . . . He was anxious to do whatever was necessary for security, but wished to avoid the development of a *SPUTNIK* complex. . . . He was not prepared to say that he would support a program of $2.4 billion for space activities in 1970."[47] Glennan pointed out, "He had already decided not to embark on a full-scale man-in-space program beyond MERCURY." Eisenhower seemed to agree because he said he had always thought $1 billion per year should be the ceiling on the space-program budget and yet that ceiling would apparently be breached in 1962: "The President said he was reluctant to spend sums of this magnitude on space activities. He had no hesitation in supporting vast programs designed to acquire specific scientific information, or programs which were necessary for psychological reasons," but he believed the budget amounts in Glennan's long-range projections for $1.9 billion were excessive.[48]

At this point Eisenhower received the actual ad hoc panel's briefing on the man-in-space program. To which, "The President said that, like Isabella, we were hocking our jewels for this purpose." The all-inclusive figure for a lunar effort was now presented as $33–46 billion. Kistiakowsky, who gave the briefing, recalls Eisenhower "just about blew a gasket. He was horrified."[49] The NSC minutes record his lamenting that "the SPUTNIK complex impelled us to do everything yesterday. . . . He had to think about the country as a whole, the economy, and the other demands on the budget. He believed it might be necessary to establish an annual budgetary ceiling for space activities." Kistiakowsky pointed out that "to a large extent the objectives of the space program must be charged to the cold war. The Soviets had succeeded by propaganda in instilling the idea that achievements in space were an accurate overall measure of a country's scientific and technological potential." To which Eisenhower replied, he could "use $1 billion to better advantage on some other aspect of the cold war."[50] Clearly space in a general sense was not Eisenhower's preferred tool for Cold War competition.

Much less did the idea of human spaceflight appeal to him as an appropriate instrument for prestige gathering. The 20 December 1960 meeting's minutes explain, "The President said he was ready to say that he saw no scientific or psychological reason for carrying the man-in-space program beyond the MERCURY program. He thought the idea of a man on the moon was sheer Buck Rogers fiction. . . . The President said we were facing a difficult fiscal problem because our rate of expenditure was increasing faster than our economic growth."[51] Glennan's diary entry concerning this December 1960 meeting records Eisenhower's response to such huge sums for a lunar landing: "He couldn't care less whether a man ever reached the moon."[52] The conclusion of the NSC as a whole was therefore that "further testing and experimentation will be necessary to establish whether there are any valid scientific reasons for extending manned space flight beyond the MERCURY program."[53]

Clearly, then, at the end of his tenure, Eisenhower was convinced that human spaceflight should, at best, continue after Mercury contingent upon obtaining further scientific justification, but not for prestige-related reasons. At worst, human spaceflight might very well end completely after Mercury's conclusion, if no persuasive scientific reason for its continuation could be found. Human spaceflight was not an arrow in Eisenhower's Cold War quiver. As John Logsdon has written, the situation in early 1961 for human spaceflight was "extremely gloomy."[54] One final piece of evidence supports this conclusion: Eisenhower's final budget message of 16 January 1961. In it he said that while Mercury components continued to be tested and hope existed for a human orbital flight in 1961, "Further testing and experimentation will be necessary to establish whether there are any valid scientific reasons for extending manned space flight beyond the Mercury program."[55] Again, Eisenhower felt human spaceflight needed additional scientific reasons for its continued existence; prestige was not a legitimate justification in his mind. In fact, even this budget message allowing for potential scientific justification was a tempered version of what Eisenhower had wanted to state, which was that human spaceflight would definitely end after Mercury's completion. NASA associate administrator Robert Seamans explained, "Eisenhower wanted to put in that there should be no commitment of any sort to any follow-on manned flight effort beyond Mercury. . . . It was a very, very negative statement."[56]

The historian must understand this general lack of enthusiasm pervading the Eisenhower administration concerning human spaceflight before delving into the specifics of the two relevant systems, Projects Mercury and Dynasoar. Without the context, it would be difficult to understand the deliberate and purposeful pace at which they both proceeded during the Eisenhower years. Understanding Eisenhower's beliefs concerning fiscal solvency and his antipathy toward competing for prestige via human spaceflight makes their relatively low level of effort, when compared to the Kennedy years, more comprehensible.

Dynamic Soaring

The best way to picture the Dynasoar is to imagine an isosceles triangle, the entire surface of which would act as a delta-shaped wing. On top and bisecting the triangle would be a cylindrical fuselage for the pilot in front and a payload bay in back. At the base of the two equal sides, to either side of the fuselage, would be smaller triangles attached perpendicularly to the main wing structure; these would provide additional aerodynamic surfaces for control and stability. The spacecraft would be launched vertically by means of a modified ICBM, or perhaps a first-generation Saturn, and separate from it upon reaching orbital velocity. After conducting its mission on an orbital glide path around the earth (reconnaissance was most often mentioned, although some sources speculated about delivering bombs from the Dynasoar) at perhaps 13,000 mph or more, the Dynasoar would reenter the earth's atmosphere by means of retro-rockets that would slow its velocity. Further, its assorted aerodynamic control surfaces permitted maneuverability upon atmospheric reentry and, thus, a selection of bases at which to land, within certain range limitations. Thus, the Dynasoar's concept of operations was often referred to as a boost-glide vehicle. The Dynasoar was radically different from NASA's Projects Mercury, Gemini, and Apollo, all of which shared the same and familiar basic design of a wide cone with a slightly rounded base with a cylinder (for the reentry parachutes) attached to the top. These ballistic capsules had only the most limited maneuverability when compared to that which Dynasoar's aerodynamic wings and control surfaces provided. While the Boeing Corporation did manufacture Dynasoar prototypes, the Dynasoar was never actually launched because President Kennedy's SECDEF McNamara canceled it in December 1963. Mercury, Gemini, and Apollo capsules, however, were launched with great fanfare and publicity throughout the 1960s and early 1970s.

A comprehensive early history of hypersonic flight in general and the Dynasoar specifically is not directly relevant to this study and is expertly covered elsewhere.[57] The most important early R&D, such as conceptual studies and preliminary testing of boost-glide vehicles, was done in Germany before and during World War II in a quest to develop a bomber capable of reaching the United States. After the war, key individuals such as Gen Walter Dornberger, formerly in charge of the Nazi's V-weapons program, immigrated to America. Dornberger worked for the Bell Aircraft Corporation and ceaselessly campaigned for support for a hypersonic aircraft such as that which would eventually become Dynasoar. One source calculates that he made 678 presentations to various groups before May 1958 to muster support for getting a Dynasoar-type aircraft off the drawing boards.[58] However, "Through lack of funds and high-level interest, this project was not even begun until 1956–57."[59] This is not to say activity was absent before Sputnik. The Air Force, its contractors, and the NACA conducted numerous feasibility studies on various hypersonic vehicles and their designs.

For instance, in February 1956 an ARDC document mentioned a Bomi (bomber-missile) "which has been extensively studied since 1951 by the Bell Aircraft Corporation [and] underwent formal evaluation last fall by the NACA. . . . It was concluded that this concept represents a potential major breakthrough."[60] An NACA meeting in February 1957 discussed what the next phase of the DOD-NACA flight research should be after the X-15 explored the upper reaches of the atmosphere. Dryden discussed a boost-glide vehicle similar to the Dynasoar, which would be boosted by a rocket and glide back down through the atmosphere. He said feasibility studies from the NACA, the USAF, and the Air Force's contractors "have indicated that with early, intensified research and study it would be possible to construct a manned airplane employing this principle . . . that would fly at tremendous speeds and have a range otherwise unobtainable in manned flight." He further reported, "Members of the NACA staff and of the staff of the Air Force have discussed this matter on several occasions and are of the opinion that it is timely from a technical point of view to start a project of this type now." At the end of this meeting, "It was agreed that the NACA staff should cooperate with the Air Force in connection with a new research airplane to follow the X-15."[61]

It is from this type of early, pre-Sputnik document that one discovers mention of several Dynasoar predecessors. Dornberger and Bell Aircraft presented Bomi to the Air Force as early as April 1952, but the period thereafter was filled with uncertainty, studies, reviews, and discussions with "little or no unanimity of opinion."[62] The USAF's Wright Air Development Center completed a contract with Bell Aircraft on 1 April 1954 calling for a study of an advanced bomber-reconnaissance system. By 12 May 1955, the Air Force issued a general-operational requirement for a hypersonic-bombardment system.[63] Other pre-Dynasoar boost-glide-vehicle concepts discussed in various settings were a Robo (rocket-bomber, really a redesignation of the Bomi idea), a reconnaissance vehicle called Brass Bell, and an NACA-USAF hypersonic R&D vehicle known as Hywards (Hypersonic Weapon and R&D System). By 30 April 1957, the USAF had consolidated the multitude of study efforts under the single name of Dynasoar.[64] The unified Dynasoar program immediately before Sputnik consisted of three stages: an experimental glider, a reconnaissance vehicle, and a bombardment vehicle.[65] Air Force expenditures on Dynasoar during the 1954 through mid-1957 study phase have been estimated at $3 million, with USAF contractors spending another $3.8 million.[66]

The most important point concerning a rocket-bomber or a bomber-missile during the formative period of the space-for-peace policy was that the civilian "DOD and [Eisenhower] administration officials did not believe a satellite should be employed as an offensive atomic weapon system or orbital bomb. Based on this policy, the closer BOMI's speed approached orbital velocity, ironically, the closer it would approach a mission the Eisenhower administration would be less likely to support."[67] This would cause long-term viability problems for Dynasoar if Dynasoar continued to be cast in a weapons-carrying role, thereby demonstrating the USAF's "proclivity for a manned strategic bomber to fulfill the fundamental mission inherent to achieving its independence from the Army in 1947—strategic nuclear bombardment."[68] If the Air Force insisted on assigning an offensive mission to Dynasoar as a space vehicle, "A day of atonement could be coming."[69] Nevertheless, in the swirl of post-Sputnik panic and response, the Air Force issued System Development Directive 464 in November 1957 that for many marked the official start of the Air Force's concrete Dynasoar program.[70] This meant the Air Force could start the preliminary process of requesting actual spacecraft designs from potential contractors and move in the direction of selecting a prime contractor to build the vehicle. The NACA's and later NASA's role throughout the Dynasoar program was largely limited to research advice and laboratory and wind tunnel assistance (see below and chap. 7 for more detail). In June 1958, the Air Force selected Boeing and a consortium headed by the Martin Corporation and Bell Aircraft for a design competition. In November 1959, the USAF selected Boeing as the primary system contractor while Martin would develop the rocket to launch Dynasoar.[71] Estimated Air Force spending on Dynasoar during the 1958–59 study and design competition phase was $18 million.[72]

It is important to understand that the Air Force after Sputnik saw the Dynasoar as a space vehicle and viewed it as a system to conduct both offensive missions such as bombardment and defensive missions such as reconnaissance. A briefing to the nation's highest-ranking generals by the Air Force in November 1957 explained Dynasoar "will represent a major technological breakthrough in performance and mission capability for manned bombardment and reconnaissance. As weapons systems, they will represent the first step in manned space flight." The USAF anticipated a conceptual test vehicle in Step I of the Dynasoar program by 1963; a reconnaissance system with a range of at least 5,500 miles in Step II by 1966; and a global-range bombardment system in Step III by 1971.[73] The Air Force justified Dynasoar to the PSAC as follows: "DYNASOAR will represent the first of a whole new generation of manned weapon systems that will succeed present day turbo-jet powered weapon systems and may eventually supplement unmanned ballistic weapon systems."[74] Others within the executive branch were aware of Dynasoar's proposed missions. PSAC staffer Robert Piland wrote Killian in February 1958 explaining the Dynasoar concept and stated, "The contemplated Air Force uses are reconnaissance and strategic bombardment." Piland also portended the difficulty Dynasoar would have reaching operational status when he speculated its costs could easily reach a billion dollars, that there would be "tremendous development problems," and that "it probably would have all the disadvantages of the present family of ballistic missiles such as vulnerability, long readiness time and generally complex operational procedures." Piland explained that manned or unmanned satellites could be designed to offer the Dynasoar's advantageous properties of recallability, maneuverability, and accuracy and would "give a much more desirable deterrent setup."[75] Therefore, the often-heard criticism that the Air Force did not have specific and well-defined missions in mind for Dynasoar is not correct. It quite clearly wanted to use the vehicle for space reconnaissance and space bombardment. The latter of these missions goes a long way toward explaining why it would encounter so much resistance during the Kennedy administration as it neared operational status.

The NACA/NASA Role in Dynasoar

The Air Force was quick to enlist the NACA's assistance in the Dynasoar program after its official go-ahead in November 1957. Its deputy chief of staff for development, General Putt, wrote NACA director Dryden in January 1958, "The Air Force is convinced that we must undertake at once a research vehicle program having as its objective the earliest possible manned orbital flight" and so has undertaken a design competition for a "hypersonic boost glide vehicle nicknamed Dyna Soar I." Putt told Dryden that the concept conformed closely to previous NACA recommendations and would be able to orbit as a satellite. Since both the Air Force and the NACA were "well along in investigations seeking the best approach to the design of a manned earth orbiting research vehicle," Putt invited the NACA "to collaborate with the Air Research and Development Command in this important task."[76] However, an internal NACA memorandum in early February revealed that the NACA had concluded, "ARDC did not consider us equal partners in the development of this vehicle in the sense that we are in the X-15 project. This was as suspected but had not known until recent meetings."[77] Therefore, the Air Force's basic proposal would be that "NACA enter into the Dynasoar I project in the role of consultant only. . . . In the role of a consultant only we would feel that our responsibilities would rest in mainly expressing opinions for which we had been asked." The author of the memo to Dryden said the NACA had to ascertain exactly what its responsibilities with Dynasoar were to be or else "we might find ourselves involved in something for which we had neither adequate finances nor manpower."[78]

This ambiguity meant that an official USAF-NACA memorandum of understanding (MOU) in response to Putt's January letter was not forged until May 1958. It stated, "Overall technical control of the project will rest with the Air Force, with the advice and assistance of the NACA. . . . Financing of the design, construction, and Air Force test operation of the vehicles will be borne by the Air Force. . . . Management of the project will be conducted by an Air Force project office within Headquarters ARDC." Therefore, the NACA's (and later NASA's) participation would be largely in the area of technical consultation "to maximize the vehicle's capabilities from both the military weapon system development and aeronautical-astronautical research viewpoints." All flight testing would be accomplished by an NACA-USAF-contractor committee "chaired by the Air Force."[79] The Air Force's leading role in all phases of Dynasoar's research and development was clearly predominant at the beginning of the project's NACA-USAF interface and would remain so as NASA was created and until the project's termination in December 1963.[80]

Tenuous Support for Dynasoar and the USAF Response

The USAF was politically savvy enough to realize that as a costly R&D project at least partially devoted to delivering weapons from space, the Dynasoar had a shaky political foundation at best within the Eisenhower administration. Only 28 days after NASA started operations, an internal BOB memo assessed Dynasoar and concluded, "The project as now conceived appears to be premature if not entirely impractical as a weapons project, and overly expensive as an experimental vehicle project. In the 1960 budget considerations, we are again recommending strongly that it be canceled."[81] The scientists so influential in the creation of Eisenhower's space policy continued to be skeptical. Piland told Killian after NASA's creation "The Air Force plans to use the glide missile [Dynasoar] for reconnaissance and bombardment. The coming of the reconnaissance satellite has brought the need of this vehicle for reconnaissance into question. . . . As a bombardment vehicle the glide missile must be compared with the ballistic missiles, including Minuteman."[82] Piland also relayed that NASA believed that the Dynasoar was "a reasonable extension of the research airplane concept" and as such would be valuable for studying and evaluating flight problems in the hypersonic regime, but nevertheless, "NASA maintains its usual position of not commenting on the military utility of the vehicles. NASA also has not commented on the relative priority of this project."[83] Nor would NASA do so over the course of Dynasoar's existence.

Piland reiterated, "It is hard to see how the system could be had for less than a billion dollars." He concluded PSAC's evaluation of Dynasoar by stating, "Its desirability as a weapon system has not been clearly established in comparison with reconnaissance satellites and ballistic missiles. The question of the need for a satellite vehicle capable of maneuvering and landing upon reentry appears to be confused with the need for a glide missile" (emphasis in original).[84] Later that month, BOB director Stans, in a meeting with Eisenhower, stated, "Since the program [Dynasoar] represents a space experiment, there is considerable question as to whether the program should be pursued with the Department of Defense or with NASA. The discussion of switching the program to NASA was not conclusive."[85] From the USAF perspective, Dynasoar seemed threatened from many sides. The powerful PSAC felt its missions were not justified when compared to unmanned satellites and missiles. The BOB felt it might more properly be under the organizational cognizance of NASA. Finally, the Dynasoar's budget was imperiled. The BOB had withheld $10 million of the approximately $18 million that Dynasoar was supposed to have received for FY 58. SAF Douglas had to ask SECDEF McElroy to intervene on 4 December 1958. After much discussion and intra-DOD wrangling, Deputy SECDEF Quarles issued a memo on 7 January 1959 releasing the $10 million but emphasized the funds were for R&D purposes only. They did not represent the DOD recognition of Dynasoar as a weapons system.[86]

The Air Force responded to these assorted threats to Project Dynasoar by subtly changing the way it presented the project. The USAF's director of Advanced Technology explained how the SAF disseminated guidance that "suborbital aspects of Dynasoar be emphasized. . . . It is recommended that the weapon system aspects and the capabilities of Dynasoar as a space vehicle be avoided." The Air Force should therefore present Dynasoar as a military test system, which "will explore and solve the problems of hypersonic flight including return from near orbital velocity."[87] The Air Force hoped that by emphasizing the testing or R&D functions the Dynasoar would do suborbitally and by downplaying its orbital military missions of reconnaissance and bombardment, it could attenuate some of the pressure coming from individuals questioning its purpose and from those desiring to reduce its budget. As one Air Force historian explained, the Air Force "had been successful in retaining control of Dyna-Soar by asserting that it has less than an orbital flight capability. . . . As a safeguard, the Air Force continued for some time to emphasize the suborbital rather than the orbital characteristics of Dyna-Soar while going forward with its development as rapidly as weak funding and strong opposition within OSD permitted."[88]

Internally, however, the Air Force continued to regard Dynasoar as a program leading to an eventual orbital operational weapon system. In a January 1959 document arranged in question-and-answer format and intended to serve as an internal institutional expression of the USAF's space policy, the Air Force asked itself, "Why shouldn't NASA be conducting development of Dyna Soar?" The answer was, "Because it is not a research vehicle, but an intermediate step to a weapon system." In trying to head off charges that Dynasoar duplicated Mercury, the Air Force stated, "Mercury is a soundly conceived project to meet its objectives which are to put a man in orbit as simply and quickly as possible. It will not give us the capability for controlled flight and precise landing after leaving an orbit as Dyna Soar will."[89]

For the most part, this Air Force tactic of reorienting Dynasoar's external focus as an R&D test vehicle while maintaining a continuing internal focus on preparing Dynasoar to serve as a weapons system was successful. Dynasoar survived the remainder of the Eisenhower administration, albeit at a low budgetary level when compared to other DOD projects. By April 1959, DDR&E York stated that the primary goal for Dynasoar would be suborbital exploration of hypersonic flight and that he considered the testing of military subsystems and the attaining of orbital velocities as secondary objectives. He therefore approved $14.5 million for FY 59 funding.[90] Near the end of 1959, there was another brief flurry of concern within the Air Force that Dynasoar might be transferred to NASA. The CSAF wrote to the undersecretary of the Air Force in late October, "The Air Force must not lose Dynasoar. Will you please put all of this in context for me?" (emphasis in original).[91] Three days later, Boushey, director of advanced technology fretted, "The loss of the DYNA SOAR project to NASA appears imminent." He reached this conclusion based upon a DOD budget-review session that tentatively removed all FY 61 funding for Dynasoar, "contemplating its elimination." Boushey also said he believed Kistiakowsky and York had discussed "and may have decided upon cancellation of DYNA SOAR, with NASA to pick up the pieces as experimental in-house work. NASA plans for MERCURY to include winged DYNA SOAR-like vehicles."[92]

But as NASA continued to focus on Mercury and made no overt attempts to capture managerial control of Dynasoar, this concern over a potential NASA takeover of Dynasoar receded and is not found in any 1960 primary sources. The overriding concern from the Air Force perspective seemed to be the continuing challenge of justifying the program to OSD and then to the executive branch as a whole.

The BOB's hostility to Dynasoar continued when BOB director Stans simply declared at a conference with the president in November 1959, "The Dyna-soar makes no sense at all." SECDEF McElroy retorted that he "had cut the Dyna-soar submission [for FY 61] from $150 million to $25 million." Eisenhower did not speak to Dynasoar directly, simply saying that "within five years we must be balancing our budgets, or we will be ruining our defense by swings of the pendulum. . . . He asked Mr. McElroy to go over the budget again minutely to make it a little leaner and tougher."[93] The BOB's bottom line was firm, "Since the nation is already committed to the Mercury project for scientific and prestige purposes, it would seem that the possible military benefits do not warrant the continuation of the Dynasoar project at this time."[94] Glennan certainly did not want NASA to become embroiled in a situation in which NASA's Mercury was used as a justification for eliminating Dynasoar. He wrote concisely, "There is no direct relationship between Mercury and DynaSoar."[95] The factor of perceived duplication between NASA's Mercury and Dynasoar was not significant during the Eisenhower administration. However, the perceived duplication between NASA's Gemini (the follow-on to Mercury) and Dynasoar would become a key factor in McNamara's cancellation of Dynasoar in 1963.

Top Air Force leaders continued to plan the long-term structure of the program, despite the lean budgets and uncertain high-level support. In November 1959, the Air Force estimated total program cost of $638 million by FY 66 and outlined a three-step program. Step I would feature a full-scale but unmanned Dynasoar vehicle for tests on the ground, from a B-52 bomber, and on a modified Titan ICBM. Step II tests would begin to incorporate the Dynasoar's internal equipment for global-range and orbital testing of military subsystems "and for initial operational test and use." During this step the Dynasoar would be launched by a larger booster and was expected "to achieve orbital velocity." Finally, Step III was designed to provide an operational military weapons system and use either the Saturn or another larger booster.[96] Despite these well-laid plans, Dynasoar's financial reality was a different matter. A status report of December 1959 complained the assistant SAF had refused to release more than $1 million of the $35 million programmed for Dynasoar in FY 60. The report also stated that the $58 million programmed for FY 61 was inadequate and would cause a delay of at least a year in the program.[97]

Another challenge by 1960 was the fact that the Air Force's plan to emphasize the suborbital R&D aspect of Dynasoar had succeeded so well that OSD began to consider that rationale the only reason for Dynasoar's existence, discounting what the Air Force felt was its real and ultimate, albeit downplayed, significance as a weapons system. The Air Force's Directorate for Advanced Technology reported on meetings with DDR&E York and Undersecretary of the Air Force Joseph Charyk (responsible within the Air Force context for its R&D activities). According to the directorate, York and Charyk now believed that "orbit is not an acceptable objective" for Dynasoar; that any vehicle designed to perform inspection of hostile satellites (another potential Dynasoar mission, along with bombardment and reconnaissance) should not have wings but should be like Mercury; that Dynasoar's only certain primary objective was "exploration of the hypersonic regime; that there is as yet no military requirement for winged reentry, however . . . it is acceptable to explore the hypersonic regime;" and that the Dynasoar should be like the X-15 in that it have no foreseeable military use or be made to lead to development of any later weapon system (emphasis in the original).[98] Since these two men were the ones most directly responsible for the USAF's R&D program, these conclusions did not bode well for converting, at some later time, the Dynasoar from a perceived R&D-only platform into an operational vehicle for reconnaissance, bombardment, or satellite inspection. Nevertheless, at a NASA-USAF conference in April 1960, the USAF representative stated, "The fundamental objective of the Dyna-Soar program is to establish a technological basis for the development of future military weapon systems. . . . Dyna-Soar must be able to test military equipment and the man-machine relationship. Dyna-Soar must achieve orbital capability" (emphasis in original).[99]

Therefore, at the end of the Eisenhower administration, Dynasoar's political status was stable in the sense that R&D was continuing but it was not hopeful from the Air Force's long-term perspective of fielding an operational weapon system. Support at the civilian OSD leadership level was at best lukewarm, and within the broader executive branch such as the PSAC and the BOB, Dynasoar could encounter outright opposition. For instance, the PSAC's Strategic Systems Panel monitored Dynasoar and in September 1960 concluded that while human spaceflight in the hypersonic realm was a legitimate research objective, "A program to develop these capabilities might more logically be a prime NASA responsibility rather than that of the Air Force, [but] at this late date it would probably be a mistake to shift responsibilities."[100] Nevertheless, Kistiakowsky was concerned Dynasoar "may develop into another gigantic program with emphasis on a poorly-defined or nonsensical strategic operational requirement. . . . He stressed that he thought the program must not be considered in isolation, but in the context of other programs such as those for strategic delivery systems, reconnaissance systems, and Mercury and Apollo."[101]

Whatever his subordinates' opinions may have been concerning Dynasoar, Eisenhower at the end of his presidency was not enthusiastic about the project. At an NSC meeting in December 1960, DDR&E York informed Eisenhower that the Dynasoar program would cost at least $700 million. Eisenhower replied that "Dynasoar would be a desirable project to play around with if unlimited funds were available. However, he was not in the least impressed by the usefulness of Dynasoar as a project which would compete with other defense programs for scarce funds." Eisenhower further explained that "his comments on Dynasoar had been based on his view of the national security race rather than the technological race. . . . The President believed that Dynasoar as well as a great many research and development projects were useful concepts but he was unable to understand what practical utility a great many of these concepts would have."[102]

Nevertheless, the Dynasoar's programmatic status at the end of the Eisenhower administration seemed relatively stable with a FY 61 budget of $87 million.[103] Contracts had been let for the glider, the launch vehicle, and the launch vehicle engines. Eisenhower's final Aeronautics and Space Report described NASA's continuing role in supporting Dynasoar: "NASA is carrying out a wide range of research activities in its laboratories and wind tunnels to determine configurations that can best stand the stresses of space flight."[104] NASA's Dynasoar involvement clearly continued to be supporting and consultative in nature and did not include policy decisions. However, the existence of other human-spaceflight programs, under NASA's active management, would soon become intertwined with Dynasoar's fortunes not at NASA's behest, but due to McNamara's drive for consolidation and efficiency.

Project Mercury and the DOD Role

NASA's human-spaceflight program, Project Mercury, requires less extensive coverage than Dynasoar in the context of this book for two reasons. First, an exhaustive history of the program already exists.[105] Second, after the decision was made to award the primary human-spaceflight mission to NASA in the summer of 1958, the NASA-DOD relationship in Project Mercury then became relatively straightforward and utilitarian. DOD, particularly the Air Force, provided absolutely critical support for the program and established a formal structure to manage such support. Policy-level concerns rarely intruded upon the process whereby NASA was essentially dependent on the DOD for much of the equipment, people, ranges, and tracking stations necessary to execute Project Mercury. The DOD and the Air Force provided the support that NASA required, knowing that by doing so, they would further America's experience in human spaceflight and assist the creation of an infrastructure that would, in an emergency, be available for national defense purposes.

Most Mercury operations and actual launches took place during the Kennedy administration; however, the Eisenhower administration laid the foundation. During the Eisenhower era, the only actual launches were unmanned and consisted of one Mercury-Atlas (MA) combination in which the capsule was mated with a modified Atlas ICBM and two Mercury-Redstone (MR) combinations in which the capsule was mated with the Army's Redstone ballistic missile. A bad omen was the failure of the first MA launch on 29 July 1960 when it exploded approximately 65 seconds after launch. This was especially distressing because the more powerful MA combination was programmed to launch all but the first two American astronauts into space. The first MR combination on 21 November 1960 "marked the absolute nadir of morale among all the men at work on Project Mercury" when it lifted off the launchpad almost four to five inches, settled back, jettisoned the escape tower, and deployed its parachutes. This MR failure "was the most distressing, not to say embarrassing, failure so far in Project Mercury. Critics waxed unrestrained."[106] Fortunately, the second MR had a successful suborbital flight, reaching an altitude of 135 miles, 4,200 mph, and impacting 235 miles down range on 19 December 1960. When Kennedy took office, the Mercury program had a 33 percent success rate in its first three test launches, far less than that required for a human-spaceflight system, and a 100 percent failure rate with the crucial MA combination.[107] Mercury's status was tenuous at best at the beginning of the Kennedy administration. One trio of analysts concluded after the initial Mercury test-launch failures, "Mercury looked horribly like another Vanguard."[108] The scholarly retort to this was, "If Project Mercury were on the verge of technological bankruptcy, as some critics claimed, the problem was that man was still landlocked by inadequate boosters."[109] Finally, Mercury was supposed to be a sort of "quick-and-dirty" way for America to put a human into orbit. The original cost estimates for the entire program were $200 million but expenditures by mid-1960 had already risen to $250 million.[110]

The NASA-DOD interface in Mercury began with the creation in mid-September 1958 of the Joint NASA-ARPA Manned Satellite Panel, with membership six to two in favor of NASA. Soon, NASA had ordered nine Atlas (the Air Force would eventually provide 14), eight Redstone, and two Jupiter missiles from the DOD to begin the Mercury program.[111] Glennan wrote Deputy SECDEF Gates in July 1959 to acknowledge, "We [in NASA] have recognized from the beginning the very considerable reliance that must be placed upon resources under the custody of the Department of Defense if the program is to be successful without an inordinate expenditure of time and money." He asked Gates to keep the organizational structure governing DOD's assistance to NASA in Mercury "as simple and straightforward as possible with delegation of authority to the field commander to the maximum extent possible."[112] Gates complied on 10 August 1959 by naming Maj Gen Donald Yates, USAF, commander of the Atlantic Missile Range (AMR) complex of facilities in Florida, as well as the DOD representative for Project Mercury, who would serve as the single point of contact for the NASA-DOD Mercury-operational interface.[113] Over the next few years this position was strengthened so that the DOD representative could exercise control not only over DOD tracking support, but also over the recovery, launch, booster, medical, and all other support activities.[114] Lt Gen Leighton Davis, USAF, replaced Yates as AMR commander and DOD Mercury representative in July 1960.

A sample of USAF accounting of its support for Mercury from November 1959 reveals four major categories:

1. Launch support, in which AMR personnel: prepared launch operations plans; provided the launch vehicle and the personnel required to launch it; provided the launch pads and hanger areas for the booster and the capsule; and provided standard launch services such as range safety and security.

2. Tracking support, in which the AMR: provided space for NASA data collection equipment in AMR facilities; collected data using AMR equipment in Florida; operated NASA equipment located at worldwide AMR tracking stations; provided logistical and maintenance support to worldwide NASA stations; and operated NASA stations located at worldwide AMR locations.

3. Recovery support, in which the Air Force provided assistance in planning recovery operations as well as provided search and rescue services. The Navy of course provided the surface vessel component of the recovery forces.

4. Transportation, in which AMR personnel and vehicles provided all short-notice, scheduled-passenger, and cargo-carrier services for NASA. FY 60 costs totaled $17.4 million at AMR, of which only $10.6 million was reimbursed.[115]

The general trend of assigning numerous military personnel to NASA was surveyed in chapter 3. One famous example of this was America's first seven astronauts, often collectively termed the *Mercury Seven*. They were all military officers. Glennan recalled his initial question to Eisenhower on this issue. Eisenhower simply responded, "Of course, you will use military men. They are in the service for matters of this kind."[116] Tom Wolfe's *The Right Stuff* tells the romanticized but nevertheless fascinating story of Scott Carpenter, L. Gordon Cooper, John Glenn, Virgil "Gus" Grissom, Walter Schirra, Alan Shepard, and Donald "Deke" Slayton.[117] A scholarly assessment states that while using military test pilots "greatly simplified the astronaut selection procedure" and thereby reduced required training time, "The fame of the astronauts quickly grew beyond all proportion to their current activities and their preflight mission assignments."[118] The military officers who served as astronauts should most properly be seen as perhaps the most visible component of DOD support to Project Mercury, but nevertheless only one small part of a much larger military effort to ensure NASA's human-spaceflight project proceeded as quickly as possible within the limits of available resources.

At the end of his tenure, Glennan was satisfied with the DOD's support of Mercury. He wrote the SECDEF to thank him for the "excellent cooperation and support you are giving us in this difficult research and development task. This well-integrated operation seems to me to speak for itself in elegant terms of the kind of cooperation that exists between the military and civil components of our space program."[119] Even when NASA undertook unpleasant tasks such as identifying why the 29 July 1960 MA launch exploded 65 seconds after launch, it made a point to include in the final report that "all Department of Defense support for the operation was very good."[120] Looking ahead to the end of the Mercury program in June 1963, one calculation showed total DOD support at $133 million ($67.6 million for launch vehicles), of which NASA reimbursed $99.8 million.[121]

Having discussed DOD's specific support of NASA in this chapter and generally in the preceding chapter, the question arises of whether there was any reciprocal support by NASA for the DOD. Concerning specific programs or services during the Eisenhower administration there was not. There existed only the general notion that by developing space technology, facilities, and experience, NASA was creating a national asset that could, in times of crisis, be made available for national security purposes. NASA deputy administrator Dryden explained, "NASA's role in the national security program is the development of space technology and the conduct of a program of scientific exploration in the atmosphere and in space."[122] Glennan told ARPA director Johnson, "My own belief [is] that all of the work of the National Aeronautics and Space Administration will eventually find military application and therefore has military implications."[123] During the Kennedy and Johnson administrations, NASA would render some very limited direct assistance to the DOD in assorted projects, but the standard presentation to congressional and executive branch leaders continued to be that NASA contributed to the nation's defense insofar as it was permitted to develop the technology, facilities, and operational experience of spaceflight.

Looking to the Future

Neither NASA nor the DOD was anchored solely in the contemporary realities of pursuing Mercury and Dynasoar and dealing with the numerous associated challenges. Both organizations looked to future and follow-on human-spaceflight projects such as space stations and lunar landings. As early as 10 July 1959, NASA held a conference "to study the various aspects of placing a manned space laboratory in operation. . . . This project is envisioned as one of the initial steps in the actual landing of a man on the moon in 10–15 years." The participants at this conference even reached preliminary design decisions such as that the station should have a one-year life, incorporate a two-person crew, have a near-equatorial 400-mile altitude orbit, be comprised of a rigid cylinder with a parabolic solar energy collector, and weigh 7,000 lbs.[124] NASA's Langley Laboratory continued to study the space station concept and produce preliminary designs throughout the Eisenhower and Kennedy administrations. One from October 1960 shows an inflatable space laboratory based on the Mercury spacecraft.[125]

The Air Force also had ongoing-study efforts of space stations/laboratories. Seven USAF contractors studied a Military Test Space Station (MTSS) from 1958 to 1961 and designed a small station to be available in the mid-1960s. These detailed MTSS studies provided the Air Force with the raw data needed to design the Military Orbital Development System (MODS) in 1962 that subsequently fed into the MOL design process in 1963.[126] It must be stated, however, that these studies were "at a relatively low level of effort" and that there was little progress toward any operational platform due to "the lack of a validated requirement for the presence of military man in space, particularly in view of the Nation's dedication to peaceful use of space."[127]

Concerning the drive to reach the moon, the Air Force's early study efforts in this area were discussed in chapter 3 in the context of interservice rivalry. These studies continued throughout NASA's early years, again at a low level and hampered by the perception of the lack of any firm requirement for

179

a military presence in the lunar environment created by the space-for-peace national policy. An Air Force colonel responsible for monitoring the Air Force's lunar study effort and coordinating it with NASA wrote in July 1960, "Although military requirements in the lunar area are not now fully defined, the moon clearly represents an area over which conflicts may arise." The officer then briefly described SR-183, the USAF's examination of a lunar observatory, and mentioned that a separate study of "the military aspects of the moon" was currently "undergoing evaluation."[128]

Much more important for the future of American space policy was NASA's institutional decision during the Eisenhower administration, despite Glennan's skepticism and long before Kennedy's lunar-landing decision, that a lunar landing would be NASA's primary long-range goal in space. The complicated process whereby NASA internally reached this decision is comprehensively and skillfully presented in Logsdon's seminal *The Decision to Go to the Moon.*[129] The details are not germane to this study's discussion but the bottom line is. "NASA planners, in mid-1959, chose a manned lunar landing as the appropriate goal of the second-generation NASA manned space flight program. That is, almost two years before the Kennedy political decision to attempt a manned lunar-landing program, NASA had chosen such a program on technological grounds as the logical successor to Project Mercury."[130] NASA could examine and incorporate not only the USAF's study effort, but also the Army's Project Horizon and the results of the many industrial contractors that had contributed to these studies. In addition, NASA's early decision to focus on the lunar landing meant it got a sort of "head start" on planning for the specifics of vehicle configuration, launch modes, propulsion requirements, and myriad other operational details associated with traveling to and returning from the moon without which, Logsdon says, "It is unlikely space experts would have told Kennedy in 1961 a lunar landing was possible by 1967."[131] This held true for other operational and technical questions, leading Logsdon to conclude that by the time of Kennedy's decision in 1961, "For some time and in some detail, Americans had been thinking about how to go to the moon."[132]

Even though Eisenhower did not support, endorse, or fund a lunar-landing effort, the R&D that NASA did carry out with internally available funds permitted it to address many technological and operational questions. NASA's final Eisenhower-era long-range plan dated 12 January 1961 simply stated, "Manned space flight is a key element in the overall NASA program. . . . The program for the next ten years is directed toward providing the means for manned flight to the moon. . . . When a national decision is made to proceed with a manned lunar exploration program, design and construction of a spacecraft for manned circumlunar flight will be undertaken. This plan assumes that a decision will be made to proceed."[133] While not officially sanctioned to begin Project Apollo under Eisenhower, career NASA employees planned for it as much as it could and were biding time until, it hoped, the next president gave it formal approval. Dryden stated quite honestly, "We were trying to get in a position to make proposals. . . . A new group was coming in and NASA needed a new sales pitch."[134] NASA's patrons in Congress urged even bolder action: "A high priority program should be undertaken to place a manned expedition on the moon in this decade. . . . NASA's 10-year program is a good program, as far as it goes, but it does not go far enough."[135]

Reconnaissance Satellites and the Creation of the National Reconnaissance Office

Not only NASA and the DOD conducted America's space program. A third organization emerged late in the Eisenhower administration to manage the reconnaissance satellite programs and eventually became the third major first-generation participant in the US space program. This body was originally called the Office of Missile and Satellite Systems in the fall of 1960, but approximately a year later during the Kennedy administration, it was renamed the National Reconnaissance Office. The NRO was then and still is under joint USAF-CIA management. Only in 1992 was even the very existence of the NRO officially declassified.

Before the NRO's creation, America's first reconnaissance satellite program was pulled out of the Air Force, reassigned to a joint USAF-CIA management team, and rechristened Corona. Subsequently, the Air Force continued in its attempt to design and construct its own independent reconnaissance satellite, called Samos. Thus, not only was there continuing tension between the DOD and NASA over assorted issues, there was also continuing strain and even resentment between the Air Force and the CIA over the direction and composition of the US reconnaissance satellite program. It was Corona, operating under its unclassified cover program called Discoverer, which conducted the first successful launch of an American reconnaissance satellite in August 1960. One analyst described this event as "perhaps the most important development in military technology since the atom bomb. The spy satellite revolutionized the intelligence business."[136]

Among the numerous questions Eisenhower's scientific advisors intensely studied after Sputnik were reconnaissance satellites.[137] After much discussion, early in February 1958 Eisenhower accepted a recommendation that a small part of the Air Force's WS-117L program featuring a satellite with a returnable film capsule would be taken from the USAF and placed under joint management of Brig Gen Osmond Ritland, USAF, and the CIA's Richard Bissell for accelerated development. Though designed as an interim program, Corona in fact "would become the backbone of our entire intelligence collection system for the next 12 years."[138] Despite the difficulties posed by numerous technical challenges and perfecting the procedure for aerial recovery of film capsules from space, the 13th Corona launch on 13 August 1960 involved the successful recovery of a capsule from space (without film). The 14th Corona launch on 18 August 1960 did carry a complete photographic system, took pictures, and had its film successfully recovered and processed.[139]

Before the NRO's establishment in late 1961, "The CORONA program operated under a loose, unstructured arrangement by which the CIA and the Air Force jointly ran the effort. . . . For a time the relationship worked well."[140] But concern grew within the Eisenhower administration that two separate reconnaissance satellite programs existed: the joint CIA-USAF Corona venture, and the independent Air Force effort called originally WS-117L (part of which had been extracted and rechristened Corona), then renamed Sentry, and finally Samos. Science advisor Kistiakowsky commented on the "unbelievable chaos among the highly classified projects—the piling up of one project on top of another without any effective mechanism for evaluating even the potential usefulness of each." In particular he said that the reconnaissance satellite area was "a very distressing situation," which by May 1960 involved "administrative chaos" and "technical troubles."[141] Eisenhower finally stepped in and ordered SECDEF Gates to recommend an overall management scheme for reconnaissance satellites. Gates in turn appointed a panel consisting of Kistiakowsky, Undersecretary of the Air Force Charyk (who would become the NRO's first director) and Deputy DDR&E John Rubel to conduct the actual investigation.[142]

Kistiakowsky wrote that his group recommended a direct line of command from the SAF to the officer in charge of the USAF's reconnaissance satellite program and that the joint CIA-USAF management of other reconnaissance satellite programs continue.[143] Eisenhower approved these recommendations on 25 August 1960 and the Air Force created an Office of Missile and Satellite Systems (OMSS) to manage Samos in September.[144] It was not until a year later that the OMSS was renamed the NRO with an organizational structure explicitly recognizing joint CIA-USAF management responsibility. The NRO's historian explained that

> on 6 September 1961, CIA and the Air Force officially signed a charter establishing a National Reconnaissance Program (NRP). Under that agreement, a covert NRO would finance and control all overhead reconnaissance projects. The NRO was to be managed by a joint directorship of the CIA and the Air Force reporting to the Secretary of Defense. . . . The Air Force provided the missiles, bases, and recovery capability for the reconnaissance systems. The CIA, in turn, conducted research and development, contracting, and security. The agreement also left the CIA in control of the collection program.[145]

Finally, one day after Eisenhower approved the Gates/ Kistiakowsky recommendations that led to the NRO's creation, he also issued a directive establishing a new and entirely separate security classification system for reconnaissance satellites: "I hereby direct that the products of satellite reconnaissance, and information of the fact of such reconnaissance . . . shall be given strict security handling under the provisions of a special security control system approved by me. I hereby approve the TALENT-KEYHOLE Security Control System for this purpose." Eisenhower emphasized that anyone with access to what became known as "TK" information was strictly prohibited from "imparting any information within this system to any person not specifically known to them to be on the list of those authorized to receive this material."[146] From this point forward, virtually no primary sources concerning the NRO as an organization or reconnaissance satellites themselves (except Corona) are available. Therefore, discussions of the NRO as the third organizational leg of the US space program (such as those speculating on continued Air Force–CIA managerial tension) or of reconnaissance satellites during the Kennedy and Johnson administrations rely almost wholly upon secondary, speculative, and largely conjectural sources.

This chapter has examined the final elements of Eisenhower's space policy: the actual human-spaceflight programs of NASA's Mercury and the USAF's Dynasoar; the relationship between these projects; and the creation of the NRO to supervise and direct the reconnaissance satellite program, which stands as the third institutional wing (after the DOD and NASA) of the initial American space edifice. Eisenhower clearly blazed the trail that his predecessors would follow in most aspects of the space program. The one glaring exception would of course be Kennedy's approval of Project Apollo, which would reverse Eisenhower's philosophy of not using human spaceflight as a competitive tool for international prestige. In fact, Kennedy's space policy would highlight beating the Soviets to the moon and back and move Project Apollo to the very center of American space policy. Nevertheless, in most other areas, Kennedy and Johnson continued in the same general direction that Eisenhower pointed them. Reconnaissance satellites remained paramount, and the overall tenor of NASA-DOD relations continued to be characterized by a complex mix of support, coordination, and rivalry. The next chapter will examine in detail the one major change Kennedy did make in Eisenhower's space program and philosophy—emphasizing human spaceflight for prestige purposes and thereby sending America on its way to the moon.

Notes

1. Futrell, *Ideas, Concepts, Doctrine*, vol. 1, 544.
2. Link, *Space Medicine in Project Mercury*, 12.
3. For more details on these early study efforts see, Air Force, "Outline of History of USAF Man-in-Space," August 1962.
4. Ibid., 1.
5. Gilruth, oral history interview, 27 February 1987, 242.
6. Air Force, *Chronology of Early Air Force Man-in-Space Activity*, 18.
7. Faget to NACA associate director, memorandum, 5 March 1958, 1.
8. Air Force, Outline of History of USAF Man-in-Space," August 1962, 2.
9. Wood, transmittal of copies of proposed MOU between Air Force and NACA for joint NACA-Air Force, 1.
10. Air Force, *Chronology of Early Air Force Man-in-Space Activity*, 23.
11. Piland to Killian, memorandum, 14 April 1958, 1.
12. Piper, *The Space Systems Division*, 2.
13. Remaining ambiguities could be resolved by expedited R&D and experimentation, which "will lead directly to maximized military mission applications and to space superiority. The USAF further believes that a national capability to control space is a fundamental requirement since on it will depend the future position, prestige, and welfare of the U.S.A." USAF/ARDC/BMD, *USAF Manned Military Space Development Plan*, vol. 2, II-1–II-3.
14. Air Force, *Chronology of Early Air Force Man-in-Space Activity*, 32–33.
15. Ibid.
16. Swenson et al., *This New Ocean*, 81–82.
17. Wood, memorandum for files, 11 April 1958, 1.
18. Ibid., 93–94.
19. Ibid., 94.
20. Emme, "Historical Origins of NASA," 20.
21. Piland to Killian, memorandum, 8 July 1958, 5–6.
22. Clark to Killian, memorandum, 23 July 1958, 1–3.
23. The NACA/PSAC also believed the exception clause of the Space Act supported NASA's claim to the human-spaceflight mission: "The assignment of the direction of the manned satellite program would be consistent with the President's message to Congress and with the pertinent extracts from the National Aeronautics and Space Act of 1958." Dryden to Killian, Manned Satellite Program, letter, 18 July 1958, 1–2.
24. Piland to Killian, memorandum, 8 July 1958, 1–2
25. Brown, memorandum for record, 25 July 1958, 1.
26. Swenson et al., *This New Ocean*, 101.

27. Ibid., 101–2. Other sources concurring in the 20 August 1958 time frame and discussing budgetary figures include Air Force, "Outline of History of USAF Man-in-Space," 3; and Emme, *Chronology of Man-in-Space*, 6.

28. Roland, *Model Research*, 298–99.

29. Hall, "Instrumented Exploration and Utilization of Space," in Emme, *Two Hundred Years of Flight in America*, 189.

30. Glennan, Statement before the Subcommittee on Governmental Organization for Space Activities, 24 March 1959, 14.

31. CIA, National Intelligence Estimate, 20 December 1955, 26.

32. Greenstein and Callahan, *The Myth of Presidential Leadership*, 40.

33. NSC, 406th Meeting of the NSC, memorandum of discussion, 13 May 1959, 8.

34. Goodpaster, memcon, 21 September 1959, 4.

35. Killian (speech to the Dallas Council on World Affairs, 23 September 1960, 4).

36. Hansen, *Spaceflight Revolution*, 51.

37. Holmes, *America on the Moon*, 193.

38. John Eisenhower, memcon, 13 October 1960, 1–3. Glennan interjected that while Mercury was "moving ahead under a full head of steam . . . if we fail to place a man on the moon before 20 years from now, there is nothing lost." Glennan, *The Birth of NASA*, 245.

39. Glennan, *Birth of NASA*, 255.

40. Gray, memorandum for record, 8 November 1960, 3.

41. Hornig would later become Lyndon Johnson's science advisor and then president of Brown University.

42. PSAC, Ad Hoc Panel on Man-in-Space, report, 16 December 1960, in Logsdon et al., vol. 1, 408–9.

43. Ibid., 3. The intricate relationships between various types of vehicles within the Saturn family are best presented in Bilstein, *Stages to Saturn*.

44. PSAC, Ad Hoc Panel on Man-in-Space, report, 16 December 1960, in Logsdon et al., vol. 1, 7.

45. Glennan, *Birth of NASA*, 269.

46. NSC, 470th Meeting of the NSC, memorandum of discussion, 20 December 1960, 2–3.

47. Ibid., 3.

48. Ibid., 4.

49. Kistiakowsky, oral history interview, 22 May 1974, 37.

50. NSC, 470th Meeting of the NSC, memorandum of discussion, 20 December 1960, 4–5.

51. Ibid., 5–6.

52. Glennan, *Birth of NASA*, 292.

53. NSC, 470th Meeting of the NSC, memorandum of discussion, 20 December 1960, 6.

54. Logsdon, *The Decision to Go to the Moon*, 37.

55. Eisenhower, *Public Papers of the Presidents, 1960–61*, no. 414, 16 January 1961, 972.

56. Seamans, oral history interview, 8 May 1968, 14.

57. See Houchin, *The Rise and Fall of Dyna-Soar*; and Hallion, *The Hypersonic Revolution.*

58. Gibney, "The Missile Mess," 39. See also House, *The National Space Program*, H. R. 1758, 7.

59. House, *The National Space Program*, H. R. 1758, 7.

60. Colchagoff to Anderson, memorandum, 16 February 1956, 2.

61. NACA, Minutes of Meeting, 21 February 1957, 8.

62. Air Force, *Commander's Congressional Policy Book*, vol. 2, 3.

63. Futrell, *Ideas, Concepts, Doctrine*, vol. 1, 543–44.

64. Bowen, *Threshold of Space*, 34.

65. Futrell, *Ideas, Concepts, Doctrine*, vol. 1, 544.

66. DOD, DDR&E, Report to Define a DOD Position on DYNA SOAR, 20 February 1962, 1.

67. Houchin, *Rise and Fall of Dyna-Soar*, 67.

68. Houchin, "Why the X-20 Program was Proposed," 4.

69. Houchin, *Rise and Fall of Dyna-Soar*, 100.

70. Stares, *The Militarization of Space: U.S. Policy*, 129.

71. Air Force, *Air Force Information Fact Sheet, X-20 Dyna-Soar*, 6.

72. DOD, DDR&E, Report to Define a DOD Position on DYNA SOAR, 1.

73. Air Force, briefing, Armed Forces Policy Council, 5 November 1957, 6.

74. Piland to Killian, memorandum, attach. 1, 3 November 1958, 1.

75. Piland to Killian, memorandum, 28 February 1958, 1–2.

76. Putt to ARDC commander, memorandum, 31 January 1958.

77. Soule to NACA Headquarters, memorandum, 10 February 1958, 1–2.

78. Ibid.

79. White and Dryden, memorandum of understanding, 20 May 1958.

80. For instance, on 14 November 1958, Glennan and White signed a new MOU, "Principles for Participation of NASA in Development and Testing of the Air Force System 464L Hypersonic Boost Glide Vehicle (Dyna Soar I)." However, this document was simply a reiteration of the May 1958 MOU's language but substituted NASA for the NACA.

81. Shapley to BOB director, memorandum, 28 October 1958, 1.

82. Piland to Killian, memorandum, 3 November 1958, 2.

83. Ibid., 2.

84. Ibid., 2–3.

85. Goodpaster, memcon, 28 November 1958, 8.

86. Douglas to SECDEF, memorandum, 4 December 1958, 1. See also Houchin, *The Rise and Fall of Dyna-Soar*, 124.

87. Boushey to director of requirements, memorandum, 8 January 1959, 3.

88. Bowen, *An Air Force History of Space Activities*, 168.

89. Air Force, "Air Force Policy on Space," 28–29 January 1959, 12, 14.

90. Houchin, *The Rise and Fall of Dyna-Soar*, 131.

91. White to Charyk, memorandum, 27 October 1959, 1.

92. Boushey to the CSAF, memorandum, 1. Boushey elaborated, "In a 29 October 1959 high level OSD-NASA presentation on the SATURN (now scheduled for transfer to NASA) Dr. Von Braun justified the choice of 220 inches as the diameter of the second stage SATURN booster (rather than 160 inches) entirely on the basis of assumed DYNA SOAR requirements" (ibid.).

93. Goodpaster, memcon, 16 November 1959, 8–9.

94. BOB to Dr. Reid, memorandum, 23 November 1959, 1.

95. Glennan to NASA leaders, memorandum, 2 January 1960, 5.

96. Air Force, office of the CSAF, record of decision, 17 November 1959, 1.

97. LeMay to CSAF White, status report, 1.

98. Ferer, memorandum for record, 14 March 1960, 1.

99. Moore, Dyna-Soar Program Status, remarks, 11–14 April 1960, 3–4.

100. PSAC to Kistiakowsky, report on Dynasoar, 17 September 1960, 1. However, panel member George Rathjens demurred from the panel's endorsement of human spaceflight in the hypersonic field and stated unmanned vehicles could gather sufficient information: "I have not attempted to make the arguments that the man is needed, because I do not understand them. In fact, I am inclined to believe he is not needed and that the technology can and should be developed largely with wind tunnel and other work on the ground, and with instrumented drones." Rathjens proposed a radical reorientation of the Dynasoar program that eliminated the human presence and emphasized technological development; such a change would permit quicker technological development and cost only $100 million, compared to a billion dollars for the manned version. Rathjens, PSAC Strategic Systems Panel, memorandum for other members, 23 September 1960, 1–2.

101. Rathjens, PSAC Strategic Systems Panel, memorandum for other members, 23 September 1960, 1–2.

102. NSC, 469th Meeting of the NSC, memorandum of discussion, 8 December 1960, 1–24.

103. Air Force, Development Directive for System 620A—DYNA SOAR, 12 October 1960, 1.

104. Eisenhower, "Year One of the Space Age," 11, 17.

105. Swenson et al., *This New Ocean*, n.p.

106. Ibid., 297.

107. Concise launch data can be found in NASA, *NASA Pocket Statistics*, while a full accounting is in Swenson et al., *This New Ocean*, 133; and appendix D, Flight Data Summary, 638–39.

108. Young et al., "Why We Went to the Moon," 31.

109. Swenson et al., *This New Ocean*, 272.

110. Elliott, *Finding an Appropriate Commitment*, 114.

111. Swenson et al., *This New Ocean*, 110, 123.

112. Glennan to Thomas Gates, letter, 22 July 1959, 1, 3.

113. Gates to the service secretaries, memorandum, 10 August 1959, 1.

114. For a detailed look at this process, see Clements, *The Coordination of Manned Spaceflight Operations between DOD and NASA*.

115. Martin to secretary of the Air Force, report, 10 November 1959, 1–4.

116. Glennan, oral history interview, 5 April 1974, 1.

117. Wolfe, *The Right Stuff*, 1979.

118. Swenson et al., *This New Ocean*, 160.

119. Glennan to Thomas Gates, letter, 19 December 1960, 1.

120. Cited in Swenson et al., *This New Ocean*, 278.

121. Cantwell, *The Air Force–NASA Relationship in Space*, 23.

122. Dryden to Lay, memorandum, 16 September 1959, 1.

123. Glennan to Roy Johnson, letter, 17 November 1958, 6.

124. Henry to NASA associate director, memorandum, 5 October 1959, 1.

125. Newkirk et al., *Skylab: A Chronology*, 15.

126. Air Force to the House DOD Subcommittee on Appropriations, insert for the record, 1 April 1965, 2.

127. Coulter and Loret, "Manned Orbiting Space Stations," 37.

128. O'Neill to Heaton, memorandum, 30 July 1960, 1.

129. Logsdon, *The Decision to Go to the Moon*, 40–62.

130. Ibid., 40.

131. Ibid., 58.

132. Ibid., 62.

133. NASA, Office of Program Planning and Evaluation, 12 January 1961, 8.

134. Dryden, oral interview, 26 March 1964, 10.

135. House, Committee on Science and Astronautics, *Space, Missiles, and the Nation*, 5 July 1960, 55–56.

136. Richelson, "From CORONA to LACROSSE: A Short History of Satellites," B1.

137. Haines, "The National Reconnaissance Office," 143–56.

138. Wheelon, "Lifting the Veil on CORONA," 251.

139. Schriever, oral history interview, 2 July 1996, 2.

140. Haines, "The National Reconnaissance Office," 147.

141. Kistiakowsky, *A Scientist at the White House*, 45, 196, 245.

142. Eisenhower to Gates, letter, 10 June 1960, 1. See also Richelson, *America's Secret Eyes in Space*, 45.

143. Kistiakowsky, *A Scientist at the White House*, 387.

144. Sharp to the CSAF, memorandum, 1–2.

145. Haines, "The National Reconnaissance Office," 149–50. Numerous secondary sources, however, have discussed the basic facts surrounding the NRO's creation and have pointed out the fact that the NRO sprang from the Office of Missile and Satellite Systems, see Richelson, *Secret Eyes*, 47; David Spires, chap. 2, "From Eisenhower to Kennedy: The National Space Program and the Air Force's Quest for a Space Mission, 1958–1961," in *Beyond Horizons: A Half Century of Air Force Space Leadership*; R. Hall, "The Eisenhower Administration and the Cold War: Framing American Astronautics to Serve National Security," 27, 68; and Stares, *The Militarization of Space*, 46.

146. Eisenhower to the secretaries of State and Defense, memorandum, 26 August 1960, 75.

Chapter 5

Kennedy, Prestige, and the Manned Lunar Landing Program

This chapter will examine three primary points. It will start with a brief look at Kennedy's general approach to the Cold War in an attempt to lay the background for how his space policy fit into his larger philosophy. The bulk of the chapter will detail his space policy, how it differed from Eisenhower's, and how Kennedy brought the notion of using human space-flight as a competitive tool for prestige in the Cold War to the forefront. Finally, Kennedy's proposals for cooperating with the Soviets in space projects will be analyzed in an attempt to determine if, near the end of his term, he began to turn away from the competitive framework in which he viewed human space-flight and towards a more détente-oriented, internationalist philosophy.

A noteworthy historiographical point governs the analysis of the Kennedy and Johnson administrations' space policies. The plethora of primary source documentation available from the Eisenhower administration becomes a relative dearth from the Kennedy and Johnson era. Two reasons appear to explain this difference. First, Kennedy's decision-making process did not feature an extensive and rigidly structured staff system similar to Eisenhower's. The copious documentation created by the NSC and its subsidiary groups, the PSAC panels, and numerous other bodies from the Eisenhower administration declined dramatically during the Kennedy administration. Instead of Eisenhower's military-derived hierarchical staff system, Kennedy appears to have relied more on ad hoc groups and informal consultations to gather the information he needed to reach a conclusion.[1] One analyst explains, "Kennedy eschewed broad policy declarations as futile. Instead he approached each issue from an action perspective and organized special interagency task forces to deal with them."[2] This method of collecting and using information leaves behind a much less distinct paper trail. One of the consequences of Kennedy's aversion (and later, Johnson's) to numerous, long official policy documents was "there was no comprehensive, presidentially approved statement of national space policy while John Kennedy or Lyndon Johnson were [sic] president, as there had been under Eisenhower."[3] In fact, Eisenhower delivered at least five.[4] In summary, "The ad hoc, collegial style preferred by Kennedy generally produced far fewer written descriptions of policy-making deliberations from the NSC and elsewhere than did Eisenhower's more rigid and formalized structures for the NSC and other bodies."[5] Kennedy biographer Richard Reeves explains that Kennedy was determined not to be trapped by procedures: "He liked a certain disorder around him, it kept his people off balance, made them try a little harder. He dismantled Eisenhower's military-style national security bureaucracy, beginning with the Operations Coordinating Board [the NSC's OCB]. . . . His use of the National Security Council itself was casual enough that when Gen Earle Wheeler, the chief action officer of the Joint Chiefs of Staff, was handed National Security Action Memorandum 22 . . . he realized he had never seen numbers 5 to 21." Wheeler commented to his staff, "The lines of control have been cut. But no other lines have been established." Reeves explains Kennedy believed the lines of power should be like spokes of a wheel, all coming and going from him: "He preferred hallway meetings and telephone calls to desk officers." Kennedy was asked early in his administration why he had not convened the NSC. He replied, "These general meetings are a waste of time. Formal meetings of the NSC are not as effective, and it is much more difficult to decide matters involving high national security if there is a wider group present." Kennedy explained he preferred one-on-one meetings or small-group gatherings. Reeves concludes, "Short conversations and long hours substituted for organization." Indeed, by April 1961, Kennedy had called only two Cabinet meetings, then stopped them altogether, declaring, "They're a waste of time."[6]

The second point relevant to the relative lack of primary source documents from the Kennedy and Johnson administrations is tied closely to the first. An informal style of policy making often involved fewer official memoranda, letters, and official policy statements available for later analysis. In the case of Kennedy's lunar-landing decision, good documentation does survive concerning the process whereby the decision was made to go to the moon, but the process whereby this decision was implemented over the next several years is more thinly documented. Not only did presidential-management style not lend itself to the production of such documents, no single figure or body devoted itself during the Kennedy and Johnson administrations to taking virtually verbatim notes from every meeting the president attended and later translating those notes into an official memorandum of conference which was then placed in the historical record. During the Eisenhower administration, the Office of the Staff Secretary produced hundreds, if not thousands, of such memoranda of conference for virtually every meeting in which the president participated. A brief look at this book's bibliography will reveal the importance of individuals from that office such as Andrew Goodpaster and L. A. Minnich. Bodies such as the NSC and the Cabinet also had individuals that produced detailed records of each meeting. The NSC series at the Eisenhower Library contains almost 500 memoranda of separate NSC meetings.[7] In the end, historical analysis of questions concerning general space policy and the human-spaceflight story in particular is not impossible for the Kennedy and Johnson administrations but is, quite frankly, currently based on much less primary-source documentation than is a similar analysis of the Eisenhower administration.[8]

Kennedy and the Cold War

Kennedy's Cold War philosophy shares many characteristics with Eisenhower's. Both men believed containing the Soviet Union was necessary. Both men believed the USSR posed a genuine threat to America. Nonetheless, both men also believed pursuing an active containment strategy did not preclude searching for means to reduce tensions, slow down the arms race, and reach some kind of détente. Elements of the sword and olive branch were not mutually exclusive in the way each man structured his Cold War policies. One noted Cold

War historian explains that Kennedy's worldview and the policies flowing from them "differed in no important essential from the Eisenhower policies after 1954. The new Administration was only more efficient and determined in carrying them out."[9]

Kennedy's inaugural address, in a not-so-subtle reference to Eisenhower, stated, "The torch has been passed to a new generation of Americans—born in this century." He further declared, "Let every nation know, whether it wishes us well or ill, that we shall pay any price, bear any burden, meet any hardship, support any friend, oppose any foe to assure the survival and success of liberty." He touched upon the dichotomy of his Cold War aims: "Let us never negotiate out of fear. But let us never fear to negotiate." Nevertheless, "In the long history of the world, only a few generations have been granted the role of defending freedom in its hour of maximum danger. I do not shrink from this responsibility—I welcome it."[10] Early in his administration, Kennedy rarely shrank from the following type of rhetoric: "We are opposed around the world by a monolithic and ruthless conspiracy that relies primarily on covert means for expanding its sphere of influence—on infiltration instead of invasion, on subversion instead of elections, on intimidation instead of free choice, on guerrillas by night instead of armies by day. Its preparations are concealed, not published."[11]

In more private settings, Kennedy was not quite as much of an alarmist, but still firm. For instance, he wrote Soviet leader Nikita Khrushchev on 21 February 1961 to propose a summit, saying, "You may be sure, Mr. Chairman, that I intend to do everything I can toward developing a more harmonious relationship between our two countries."[12] When he met with Khrushchev in Vienna during the first week of June 1961 for their initial summit, Kennedy said that since the two countries were "competing with each other in different parts of the world," the two men had to "find during his Presidency ways and means of not permitting situations where the two countries would be[come] committed to actions involving their security and endangering peace, to secure which is our basic objective." When Khrushchev stated, "He did not want to conceal that the USSR was challenging the United States; it wants to become richer than the United States," Kennedy disagreed with this economic motive and said his own interpretation of the situation was that "the Soviet Union was seeking to eliminate free systems in areas that are associated with us. . . . This is a matter of very serious concern to us." Khrushchev of course denied this, and afterwards Kennedy said people and governments must have free choice and that the real problem was "how to conduct this disagreement in areas where we have interests without direct confrontation of the two countries and thus to serve the interests of our people." As Khrushchev continued to deny any culpability, Kennedy started to become flustered and interjected that "Mao Tse Tung had said that power was at the end of the rifle." Khrushchev said he did not believe this.[13] The DOS record of the summit ends with, "The President concluded the conversation by observing that it would be a cold winter."[14] Entire books can be, and have been, devoted to Kennedy's overall Cold War policy and how it was or was not instantiated in particular crises. The salient points for the space policy discussion are simply that: one, Kennedy was willing to be firm with the Soviet Union and dramatically increase defense spending; two, he did see the United States as engaged in a competitive struggle with the USSR; but, three, he was also willing to negotiate measures to reduce tensions and move toward a détente, though one must guard against overemphasizing this final trend.

Kennedy, the Cold War, and Defense Spending

Throughout 1961 and 1962 a succession of Cold War crises plagued the Kennedy administration: the Bay of Pigs; Laos; the Congo; Berlin; and, most serious of all, the Cuban missile crisis. With only slight exaggeration, one historian states, "The thousand days of the Kennedy administration resonated with the constant sound of alarm bells."[15] The details are not germane to this chapter's focus, but several overarching points are. First, Kennedy did not hesitate to significantly increase defense spending as part of the Cold War competitive environment. Before the Vienna summit, he had already recommended increasing US defense spending by $650 million.[16] After Vienna he requested (and Congress appropriated) an ad-

195

ditional $3.24 billion for defense.[17] Further increases meant that in January 1962 Kennedy requested a $51.6-billion defense budget for FY 63 (the total federal budget that year was $92.5 billion).[18]

Theodore Sorensen was one of Kennedy's closest personal advisors, and he recalled, "Kennedy believed in arming the United States to provide bargaining power and backing for disarmament talks and diplomacy." Kennedy's basic instruction on defense spending was "Under no circumstances should we allow a predetermined arbitrary financial limit to establish either strategy or force levels." Sorensen says in his three years Kennedy conducted "the largest and swiftest [defense] buildup in this country's peacetime history, at a cost of some $17 billion in additional appropriations" that provided the United States with a versatile arsenal "ranging from the most massive deterrents to the most subtle influences."[19] McNamara concurred: "I would say that a major instruction which I received from President Kennedy was to develop a defense program that would assure the security of our Nation without regard to arbitrary budget ceilings."[20] Therefore, it seems unlikely that Kennedy would balk at significantly boosting space spending due to financial concerns if he believed an accelerated space program would somehow contribute to America's overall well-being and to the United States' Cold War struggle with the Soviet Union. Such was the case with Project Apollo.

Kennedy and Competing with the Soviets

The second point from Kennedy's Cold War approach relevant to space policy is that in general he seems to have had no aversion to competing with the Soviets. After the Soviets broke the voluntary US-USSR-Great Britain moratorium observed since November 1958 and resumed testing nuclear weapons in the atmosphere on 31 August 1961, McGeorge Bundy recorded, "The President's patience is at an end." Bundy added that Kennedy said, "The world is being subjected to threats and terror. We have to show both our friends and our own people that we are ready to meet our own needs in the face of these new Soviet acts."[21] Part of this competitive dynamic involved Kennedy making it absolutely clear to Khrushchev that

Kennedy knew the American nuclear arsenal was superior to the Soviet Union's, that Khrushchev should not press his demands on issues like Berlin too far, and that America would prefer peaceful competition in areas like space instead of an escalating arms race.[22]

The chosen vehicle for communicating this competitive resolve to Khrushchev was a speech on 21 October 1961 by Deputy Secretary of Defense Gilpatric to the National Business Council. Reeves states, "Kennedy appointed himself Gilpatric's editor, going through the text line by line and number by number."[23] Gilpatric later concurred that the speech was coordinated "all the way up to and including the president."[24] The tone of the speech clearly seems to have been intended to impress upon the Soviets that the United States was ready, willing, and able to compete:

> The total number of our nuclear delivery vehicles . . . is in the tens of thousands, and, of course, we have more than one warhead for each vehicle. . . . Our forces are so deployed and protected that a sneak attack could not effectively disarm us. The destructive power which the United States could bring to bear even after a Soviet surprise attack upon our forces would be as great as, perhaps greater than, the total undamaged force which the enemy can threaten to launch against the United States in a first strike. In short, we have a second-strike capability which is at least as extensive as what the Soviets can deliver by striking first.[25]

This speech, as one Cold War historian summarized, marked the "final expression of Kennedy's determination to overturn his predecessor's method of dealing with the Soviet Union."[26] Kennedy would not refrain from competing against the Soviets in their chosen field, be it nuclear arms or space. Kennedy would not shrink from pointing out America's areas of superiority and those areas in which America needed to catch up. A race to the moon would be one competitive mode that Kennedy embraced and the one directly relevant to this study. As Kennedy commented in his first State of the Union message, America did not want to compete militarily with the USSR if it had a choice. However, "Open and peaceful competition—for prestige, for markets, for scientific achievement, even for men's minds—is something else again. For if Freedom and Communism were to compete for man's allegiance in a world at peace, I would look to the future with ever increasing confidence."[27]

Thawing

The third Kennedy Cold War principle relevant to space policy is—after the brinkmanship of the Cuban missile crisis in October 1962 forced Kennedy and Khrushchev to directly face the possibility of nuclear war—some movement toward détente. One scholar says that while the missile crisis did not mark the end of the Cold War, it "signified the end of that acute phase of Soviet pressure and attempted blackmail" that so distressed Kennedy.[28] At a minimum, Kennedy incorporated conciliatory language into his speeches. John Lewis Gaddis explains that while Kennedy made no significant alteration of his earlier policy of seeking agreement on negotiable issues while taking care not to convey any sense of weakness to the Soviets, it was Khrushchev who made most of the obvious movement toward détente because he "now abandoned his obviously counterproductive strategy of seeking to bully the West into an easing of antagonisms."[29] Whoever moved and how much is not the issue. There appeared to be a greater willingness to tone down the rhetoric and take concrete actions to lessen tensions. As Kennedy said after the missile crisis, "The achievement of a peaceful solution to the Cuban crisis might well open the door to the solution of other outstanding problems."[30] One must not stretch the reconciliation point too far, however. As Kennedy said in January 1963, "Here hope must be tempered with caution . . . . I foresee no spectacular reversal in Communist methods or goals." Kennedy foresaw a continuously rising defense budget because "there is no substitute for an adequate defense, and no 'bargain basement' way of achieving it." Nevertheless, "We do not dismiss disarmament as merely an idle dream. For we believe that, in the end, it is the only way to assure the security of all without impairing the interests of any. . . . In short, let our adversaries choose. If they choose peaceful competition, they shall have it."[31]

The most famous example cited for a sense of budding rapprochement was Kennedy's American University speech of 10 June 1963. One biographer reports Kennedy ordered the speech's drafts kept away from the DOS and DOD officials who normally coordinated on presidential foreign policy and national security addresses. It so impressed the Soviets that *Izvestia* reprinted it in full, and the Soviets turned off the thousands of transmitters normally jamming signals from the Voice of America so it could be heard in Eastern Europe and the USSR.[32] In it Kennedy said, "Total war makes no sense in an age when great powers can maintain large and relatively invulnerable nuclear forces and refuse to surrender without resort to those forces." Therefore, peace was "the necessary rational end of rational men. . . . We have no more urgent task." While Kennedy said he hoped Soviet leaders would "adopt a more enlightened attitude" toward the pursuit of peace, he added, "I believe we can help them do it." He warned Americans against falling into the same trap of the Soviet leaders in which they actually start to believe the propaganda they write about Americans. Kennedy emphasized, "No government or social system is so evil that its people must be considered as lacking in virtue. As Americans, we find communism profoundly repugnant as a negation of personal freedom and dignity. But we can still hail the Russian people for their many achievements." Kennedy pointed out that the hard reality was that both sides "have a mutually deep interest in a just and genuine peace and in halting the arms race."[33]

But even among these words of conciliation, the competitive dynamic was not far from Kennedy's mind. He also stated in the American University speech, "We are unwilling to impose our system on any unwilling people—but we are willing and able to engage in a peaceful competition with any people on earth."[34] Too much can also be made of the spirit of détente in this speech; one Kennedy insider says its effect "was to redefine the whole national attitude toward the cold war."[35] Historians often ignore the closing section of his address in which he emphasized, "The Communist drive to impose their political and economic system on others is the primary cause of world tension today. For there can be no doubt that, if all nations could refrain from interfering in the self-determination of others, the peace would be much more assured."[36] Nevertheless, it is difficult to find such words of conciliation in Kennedy's rhetoric before the Cuban missile crisis. One perceptive commentator explains that after the Cuban missile crisis, "The change was not in Kennedy but in what he perceived to be his political environment."[37] Whatever its origins, this very nascent détente is relevant to space history because it is in this context at the end of his term that Kennedy suggested the lunar-landing program could be made a joint US-Soviet effort. This in turn undermined the competitive, prestige-oriented dynamic in the minds of many, including some in Congress.

Kennedy, Space Policy, and Prestige

Having sketched the aspects of Kennedy's Cold War orientation that were applicable to his space policy, the next logical question is, exactly what was Kennedy's space policy? The answer forms the heart of this chapter and is an important determinant of the NASA-DOD relationship in human spaceflight detailed in the next two chapters. Before he became president, and perhaps even during the first few weeks of his administration, Kennedy appears not to have devoted any great effort to contemplating space policy. During the 1960 presidential campaign, it was an issue that helped him support his general theme that America was somehow trailing the USSR and required a new leader who would get the country back on its feet and moving again. Before the campaign Kennedy viewed space as an issue only inasmuch as it supported his assertions that Eisenhower and Nixon had permitted the United States to fall behind the USSR's military and that a dangerous missile gap was opening.

Prepresidential Attitudes and Statements

In an address to the District Democratic Meeting in Topeka, Kansas, 7 November 1957, a month after Sputnik, Kennedy said the United States was losing the satellite-missile race with the USSR because of "complacent miscalculations, penny-pinching, budget cutbacks, incredibly confused mismanagements and wasteful rivalries and jealousies." Kennedy called for Eisenhower to "tell us exactly where we stand today and where we go from here. The people of America are no longer willing to be lulled by paternalistic reassurances, spoon-fed science-fiction predictions or by pious platitudes of faith and hope."[38] Kennedy linked space concerns, the missile gap, and national security throughout the post-Sputnik period and during his presidential campaign. Perhaps most well known was his missile-gap speech delivered on the Senate floor on 14 August 1958. His campaign later reprinted this speech in booklet form for widespread distribution. In it he claimed the United States was "about to lose the power foundation that has long stood behind our basic military and diplomatic strategy" because in the past, "We have possessed a capacity for retaliation so great as to deter any potential aggressor from launching a direct attack upon us. . . . The hard facts of the matter are that this premise will soon no longer be correct." He explained the United States was "rapidly approaching that dangerous period called the 'gap' or the 'missile lag' period, which is . . . a period in which our own offensive and defensive missile capabilities will lag so far behind those of the Soviets as to place us in a position of great peril. . . . the deterrent ratio might well shift to the Soviets so heavily, during the years of the gap, as to open to them a new shortcut to world domination." Kennedy claimed their "sputnik diplomacy" was an example of this process through which "the periphery of the free world will slowly be nibbled away."[39]

Kennedy's usual suggestions for remedies included vastly increased spending on missiles and nuclear aircraft. When space was mentioned, Kennedy placed it in the national security context. In a February 1960 speech, he said the Soviet satellites meant that "for the first time since the War of 1812, foreign enemy forces potentially had become a direct and unmistakable threat to the continental United States, to our homes and to our people. . . . But only belatedly were sufficient time and attention given to our missile program. And even then sufficient funds were not forthcoming." Kennedy concluded it was easier to gamble with survival, "But I would prefer that we gamble with our money—that we increase our defense budget this year—even though we have no absolute knowledge that we shall ever need it. . . . That is the harder alternative."[40]

201

On those occasions in which Kennedy, Johnson, or the Democratic Party did specifically address space issues, their concerns were linked with either the US-USSR competitive dynamic or with the missile gap. On 18 December 1959, Johnson declared, "We cannot concede outer space to communism and hold leadership on earth."[41] The Democratic platform for 1960 used an extract from R. E. Lapp, Frank McClure, and Trevor Gardner's, *Position Paper on Space Research* of 31 August 1960.

> The Republican Administration has remained incredibly blind to the prospects of space exploration. It has failed to pursue space programs with a sense of urgency at all close to their importance to the future of the world. It has allowed the Communists to hit the moon first, and to launch substantially greater payloads. . . . The new Democratic Administration will press forward with our national space program in full realization of the importance of space accomplishments to our national security and *our international prestige*. We shall reorganize the program to achieve both efficiency and speedy execution.[42] (emphasis added)

Kennedy told a Veterans of Foreign Wars convention in August 1960, "The world's first satellite was called a *Sputnik*, not *Vanguard* or *Explorer*. The first living creatures to orbit the earth were Strelka and Belka, not Rover and Fido. Now let me make it clear that I believe there can be only one defense policy for the United States, and that this is summed up in the word 'first.' I do not mean 'first, but.' I do not mean 'first, when.' I do not mean 'first, if.' I mean first period."[43]

For Kennedy being first, in space or elsewhere, was part of what he perceived as a contest for the "hearts and minds" of people worldwide, particularly the developing nations. As he stated in a September 1960 campaign speech,

> The hard, tough question for the next decade is whether we or the Communist world can best demonstrate the vitality of our system. Which system, the Communist system or the system of freedom is going to be able to convince the watching millions in Latin America and Africa and Asia, who stand today on the razor edge of decision and try to make a determination as to which direction the world is moving? I think it should move to us. I think ours is the best system. I do not agree with Mr. Khrushchev when he says he is going to bury us. I think

we can demonstrate in the next ten years, in the next 40 years, that our high noon is in the future, that our best days are ahead.[44]

Kennedy's most pointed attack on Eisenhower's space policy during the 1960 campaign came in an article published under his name for the aerospace trade magazine *Missiles and Rockets*.[45] In it he declared, "We are in a strategic space race with the Russians, and we have been losing. . . . Control of space will be decided in the next decade. If the Soviets control space they can control earth." Therefore, the United States "cannot run second in this vital race. To ensure peace and freedom, we must be first." Kennedy nodded towards civilian space pursuits by saying that goals like space laboratories and Americans on the moon were possible, though their target dates "should be elastic. All these things and more we should accomplish as swiftly as possible. This is the new age of exploration; space is our great New Frontier." He may also have given encouragement to those within the Air Force who concluded the USAF's space role would increase under Kennedy: "The United States must have preeminence in security as an umbrella under which we can explore and develop space for the benefit of all mankind. Reorganization of the cumbersome, antique and creaking machinery of the Department of Defense is high on the agenda of the new Democratic administration." Even in this article designed to specifically address space issues, Kennedy presented in detail his plan for augmenting defense spending, increasing the number of strategic missiles, and expanding and modernizing conventional forces.[46]

While this article was indeed "full of the clash and clamor of the space race," Logsdon points out it is uncertain if it actually represented Kennedy's thinking, given the fact that it stands "in rather direct contrast to some more cautious statements on the space program made soon after his inauguration."[47] Nevertheless, if nothing else, it demonstrates Kennedy's willingness to use space and missile concerns as a political issue in the 1960 campaign. Summarized one scholar, "Kennedy was successful in magnifying the salience of the space issue and in linking the issue to his overall 'New Frontier' theme."[48]

Two problems arise with Kennedy's use of the missile-gap issue (with space matters linked to it) during the period before the 1960 election. First, "The problem was that there was no missile gap." Eisenhower knew from U-2 and particularly from early reconnaissance satellite information that the United States was firmly in the lead in ICBM production. In fact, Eisenhower administration officials briefed Kennedy and Johnson, but they "persisted in exploiting the issue . . . with cartoonish simplicity."[49] CIA Director Dulles reported to Eisenhower on 3 August 1960 that in accordance with Eisenhower's instructions he had briefed both Kennedy and Johnson for over two hours on, among other issues, "an analysis of Soviet strategic attack capabilities in missiles and long-range bombers and of Soviet nuclear testing prior to the moratorium."[50] Yet, the accusations of a missile gap continued to fly.

Second, there are questions concerning whether Kennedy actually believed America's supposed lagging in space exploration was an important issue (although he may have been genuinely concerned with the missile questions, at least until briefed by the Eisenhower administration). An interviewer asked Kennedy in the spring of 1960 if he favored combining the civil and military space-development programs under an overall commissioner, similar to the AEC arrangement. He replied, "Both civilian and military agencies can make a contribution to the development of space technology. We must not be bemused by neatly drawn organizational charts. . . . Nor do I believe that a Manhattan-type project is necessary." He added that combining the military and civilian programs "will inevitably dilute the fundamental responsibility of the Department of Defense for this country's military security. I do not believe that any such dilution is either wise or necessary."[51] In addition, before his election, Kennedy never defined exactly what he had in mind for the American space program. He was silent on the specific changes he would make, never elaborating beyond charging the Eisenhower administration with fiscal neglect of the program and linking the program to a missile gap.

Finally, others recall Kennedy displaying a distinct lack of interest in the space program when not campaigning or making speeches in the Senate. Charles Stark Draper was director of MIT's Instrumentation Lab and often briefed, both formally and informally, Kennedy on science, technology, and R&D issues. Draper recalled meeting Kennedy and his brother Robert at a restaurant-bar in Boston after Sputnik but before the 1960 election and watching the maitre d' hang soda straws by a cross pin, light one end and insert it into a bottle, and watch them pop to the ceiling. A "rather heated argument" ensued between the Kennedys and Draper over the usefulness of rockets. Draper said John Kennedy "could not be convinced that all rockets were not a waste of money, and space navigation even worse."[52] Kennedy's science advisor Wiesner said concerning the space program before Kennedy came to office, "He hadn't thought much about it." As one *Apollo* history concludes, "Certainly Jack Kennedy the senator hadn't been interested in space. . . . He really wasn't convinced that manned space flight had a place in his vision of the *New Frontier*." As he took office, human spaceflight was not "on the agenda at all."[53]

After the Election

Ambiguity, conflict, and uncertainty concerning American space policy and the role of human spaceflight within it characterized the Eisenhower-Kennedy interregnum and early 1961 because, "For the first few months of his administration, Kennedy did not actively involve himself in space policy."[54] However, by 25 May 1961, and probably several weeks earlier, Kennedy had decided to send America to the moon and back in quest of Cold War prestige. What conditions changed, and why did Kennedy make this decision? The historian need go no further in answering these questions than Logsdon's *Decision to Go to the Moon*. This deservedly classic treatment details every facet of Kennedy's decision and its ultimate impact. The present author does not pretend to offer new insights beyond Logsdon's theses but will attempt to summarize the important developments and to highlight the role of the DOD in the process.

A few days after the election, presidential transition team member Walt Rostow wrote Kennedy to raise the kind of space-related questions he believed Kennedy's administration would need to address. The fact that these questions still required resolution illustrates the indeterminate nature of Kennedy's space thinking at that time. Rostow said the key issues requiring resolution included:

> Should we stick with NASA and a continued split between scientific and military space programs; or should we go for a space AEC? . . . What should be the objective of the scientific space program? In light of these objectives, is [the] present program big enough? Too big? Shall we proceed with Project Mercury? If so, at what pace and with what objectives? . . . How and when should we internationalize the scientific space efforts with other nations of the Free World? With the Russians?[55]

Kistiakowsky recalls that after Eisenhower was briefed on and rejected Project Apollo and the PSAC's Hornig's Ad Hoc Panel on Man-in-Space report late in December, Kennedy was also given the panel's report "and had then a negative reaction to the moon-landing proposition"[56] and even that Kennedy said, "Project Apollo was for the birds."[57] One prescient presidential advisor early on stated that the heart of the space problem facing Kennedy was the question of "pressing achievement for the sake of psychological effect, regardless of concrete scientific or military utility." This official even foresaw Lyndon Johnson's eventual role when he wrote Kennedy, "You wanted something you could give him to work on and worry about. I hope this meets the purpose." *This* being the interrelated complex of questions concerning, Should the United States get out of the space-for-prestige race and focus on space applications which have tangible value? Or should the United States press the space-for-prestige angle? If so, what particular "firsts" were most appealing and dramatic?[58]

The most-often-discussed early Kennedy administration space document is the report from a group headed by another MIT professor and Kennedy confidant, Jerome Wiesner. Wiesner would soon be named and serve as Kennedy's science advisor. One of his responsibilities during the transition period was to examine America's space program and make recommendations to the president-elect. The Wiesner committee's "Report to the President-Elect of the Ad Hoc Committee on Space" of 10 January 1961 has been characterized as "hastily prepared" and offering Kennedy "no new options," but if nothing else it "did make explicit the beliefs of many influential scientists."[59]

Wiesner began by emphasizing that ICBMs were "the most important of all space programs" and that "for the near future the achievement of an adequate deterrent force is much more important for the nation's security than are most of the space objectives," but that there were five other motivations for a vital, effective space program. First was prestige because, "during the next few years the prestige of the United States will in part be determined by the leadership we demonstrate in space activities." The report also cited national security, scientific observation and experimentation, practical nonmilitary applications, and possibilities for international cooperation. Most of the rest of the report was devoted to explaining what it felt were the "serious problems within NASA, within the military establishment and at the executive and other policy-making levels of government."[60] The Wiesner report charged that in addition to the lack of large-capacity space launch vehicles, one of the major handicaps for the American space program

> has been the lack of a strong scientific personality in the top echelons of its organization. . . . There is an urgent need to establish more effective management and coordination of the United States space effort. . . . Neither NASA as presently operated nor the fractionated military space program nor the long dormant space council have been adequate to meet the challenge that the Soviet thrust into space has posed to our military security and to our position of leadership in the world. . . . Many inexperienced people have been placed in positions of major responsibility.[61]

This was a stinging criticism of some levels of NASA's leadership as well as the overall structure of the space program.

While the report left open the possibility that human spaceflight could be justified by the prestige motive, it concluded, "A crash program aimed at placing a man into an orbit at the earliest possible time cannot be justified solely on scientific or technical grounds." Further, Mercury had to be carefully evaluated because of the problems in its test-launch program and whatever was decided about human spaceflight,

> We should stop advertising MERCURY as our major objective in space activities. . . . It exaggerates the value of that aspect of space activity where we are less likely to achieve success, and discounts those as-

pects in which we have already achieved great success. . . . Indeed we should make an effort to diminish the significance of this program to its proper proportion before the public, both at home and abroad.[62] (emphasis in original)

The Wiesner report was most certainly *not* a ringing endorsement for either the current American human-spaceflight effort or the idea of competing with the USSR for prestige in space.

Throughout January 1961, the mood within NASA continued to be uncertain due to the critical nature of the Wiesner report, which Dryden claimed "was the only knowledge which President Kennedy on coming into office had about the NASA space program."[63] Due to the fact that Glennan had resigned and left town, there was no contact with the Kennedy administration until it finally nominated a new NASA administrator, James E. Webb, on 31 January.[64] In a diary entry for 3 January 1961, Glennan commented, "To my surprise, not one single word or hint of action has been forthcoming from the Kennedy administration."[65] Associate Administrator Seamans said the feeling in NASA was "Why would anybody turn it down? It must mean that the plans for NASA are being pulled in."[66] The fact that Kennedy did not endorse the Wiesner report in toto at a press conference when he said, "I don't think anyone is suggesting their views are necessarily in every case the right views" was perhaps some small solace.[67]

Things began to look up for NASA after Webb assumed the reins. With perhaps only slight exaggeration, one source states, "From that moment on, NASA seems to have been watched over by a solicitous Providence."[68] For instance, on 21 February 1961, a week after Webb was sworn in, there was the first completely successful Mercury-Atlas test launch. Webb has been described as the prototypical politician-manager who knew where all the bodies were buried, could play congressional appropriations committees with finesse, and was willing to employ hard-eyed calculation and deviousness when required. Whatever one's opinion of his methods, there is little doubt that from the moment of his appointment, "The role he played from then until his resignation in the fall of 1968 was indispensable."[69]

Kennedy's conversion to an ardent space racer and competitor was not immediate however. In early February he said, "We are very concerned that we do not put a man in space in order to gain some additional prestige and have a man take disproportionate risk. . . . Even if we should come in second in putting a man in space, I will be satisfied if when we finally put a man in space his chances of survival are as high as I think they must be."[70] Logsdon explains Kennedy's hesitancy to make any basic changes to Eisenhower's space framework continued until Kennedy "became convinced that space achievement was linked closely to the power relationships between East and West, and was a symbolic manifestation of national determination and vitality." When he finally did make that connection, then there was a dramatic reversal of Eisenhower's policy and a decision to go to the moon.[71] Perhaps Kennedy's first intense exposure to the space program was a meeting on 22 March 1961 with Johnson and officials from NASA who were requesting a supplemental appropriation.

Before asking for an acceleration of the space program, Webb sought and obtained new SECDEF McNamara's opinion. Webb recorded, "With respect to the question of accelerating our present program, Secretary McNamara feels that a most careful review should be made, that this should be done about four weeks from now if we can wait that long, and has a general feeling that we should accelerate the booster program." Webb stated their meeting's flavor "was clearly one in which he [McNamara] at this time would generally support the kind of items" Webb was considering submitting to the BOB.[72] Accordingly, Webb formally requested BOB director David Bell consider a NASA request for a supplemental appropriation of $308 million, increasing NASA's FY 61 budget to $1.42 billion; the two main items were $173 million for the Saturn superbooster project and $42 million to officially begin Project Apollo, a step Eisenhower had specifically prohibited in December 1960.[73]

When Bell was initially hesitant to forward NASA's request to Kennedy, Dryden perceptively replied, "Well, he may not feel he has the time, or you may not feel he has the time, but whether he likes it or not, he's going to have to consider it.

Events will force this."[74] Kennedy, Johnson, NASA officials, and others did finally gather for the 22 March meeting, which Logsdon says began Kennedy's close involvement in space policy that was to culminate two months later in his lunar-landing speech.[75] Webb's main point was that "we cannot regain the prestige we have lost without improving our present inferior booster capability, and doing it before the Russians make a major breakthrough in the multimillion pound thrust range. . . . The extent to which we are leaders in space science and technology will in some large measure determine the extent to which we, as a nation, pioneering on a new frontier, will be in a position to develop this emerging world force." The next day, after a supplemental meeting with Johnson, new NASC executive secretary Edward C. Welsh, Wiesner, and Bell, Kennedy decided he would grant most of the funds required to accelerate the Saturn booster and other launch vehicles but would *not* authorize the millions requested for the official commencement of Project Apollo. Clearly, at the end of March 1961, Kennedy "had not made up his mind at this time what his general attitude toward manned flight would be."[76] Kennedy approved $125.7 million of NASA's $308-million request.[77]

To the Moon

The major event that seems to have forced Kennedy's hand was another spectacular Soviet first in space. On 12 April 1961, the Soviets launched the first human into space, Yuri Gagarin, who flew for 108 minutes in his *Vostok* spacecraft. Any number of historians cites "the enormous reaction of the public and the press to the Soviet man-in-space achievement," with striking parallels to the furor that erupted after Sputnik three and one-half years earlier.[78] Khrushchev reportedly exclaimed, "Let the capitalist countries catch up with our country!" while the Central Committee of the Communist party claimed the Gagarin flight "embodied the genius of the Soviet people and the powerful force of socialism."[79]

On the day of the Gagarin flight, Kennedy held a press conference during which he stated, concerning the string of Soviet space firsts since Sputnik, "However tired anybody may be, and no one is more tired than I am, it is a fact that it is going to take some time" to catch up with the USSR. The United States was behind and "the news will be worse before it is better, and it will be some time before we catch up."[80] Privately, Kennedy reportedly remarked, "Russian housing is lousy, their food and agricultural system is a disaster, but those facts aren't publicized. Suddenly we're competing in a race for space we didn't even realize we were in."[81] Congressmen demanded a response. Rep. James Fulton declared, "I believe we are in a race, and I have said many times, Mr. Webb, 'Tell me how much money you need and this committee will authorize all you need.'"[82]

Within two days, by 14 April, it appears Kennedy "reluctantly came to the conclusion that, if he wanted to enter the duel for prestige with the Soviets, he would have to do so with the Russians' own weapon, space achievement."[83] A key meeting took place on that date with Sorensen, Bell, Wiesner, Webb, Dryden, and Kennedy. Also in attendance was journalist Hugh Sidey, who later recorded Kennedy's main problem with catching up with the Soviets in space: "The cost. That's what gets me. . . . When we know more, I can decide if it's worth it or not. If someone can just tell me how to catch up. Let's find somebody—anybody. I don't care if it's the janitor over there, if he knows how. There's nothing more important. . . . I'm determined to get an answer."[84] It appears then that while Kennedy had not made his final decision, the stage was set for a full-scale inquiry that would supply Kennedy with specific available options from which he could select his precise plan.[85]

Kennedy initiated the information-gathering process by tasking his vice president. He charged Johnson as chairman of the NASC "to be in charge of making an overall survey of where we stand in space" and to answer numerous questions "at the earliest possible moment,"

1. Do we have a chance of beating the Soviets by putting a laboratory in space, or by a trip around the moon, or by a rocket to land on the moon, or by a rocket to go to the moon and back with a man? Is there any other program which promises dramatic results in which we could win?

2. How much additional would it cost?

211

    3. Are we working 24 hours a day on existing programs? If not, why
       not? . . . Are we making maximum effort?[86]

Johnson surveyed numerous individuals in the scientific, business, and military communities for their inputs on Kennedy's questions; however, the task of actually writing the response to the president fell to Seamans, Rubel, and Shapley. Webb and McNamara would sign the document. Within a day, McNamara gave a partial response: "Dramatic achievements in space, therefore, symbolize the technological power and organizing capacity of a nation. It is for reasons such as these that major achievements in space contribute to national prestige. . . . Our attainments constitute a major element in the international competition between the Soviet system and our own."[87]

As Johnson was gathering information and opinions, Kennedy tipped his hand at a press conference on 21 April 1961 when he said, "We have to make a determination whether there is any effort we could make in time or money which could put us first in any new area. . . . If we can get to the moon before the Russians, we should. . . . I think we face an extremely serious and intensified struggle with the Communists."[88]

Eight days after Kennedy's 20 April memo, Johnson gave Kennedy a preliminary response. He explained he had consulted with such luminaries as NASA's Wernher von Braun, Gen Bernard Schriever, deputy chief of Naval Operations Vice Adm John Hayward, NASA leaders, Wiesner, and BOB senior officials, along with members of the business community. Johnson said the emerging consensus was that the "Soviets are ahead of the United States in world prestige attained through impressive technological accomplishments in space." While the United States has greater resources than the USSR to devote to attaining space leadership, it has so far "failed to make the necessary hard decisions to marshal those resources to achieve such leadership." In addition, "Dramatic accomplishments in space are being increasingly identified as a major indicator of world leadership," and if the United States does not act soon, "the margin of control over space and over men's minds through space accomplishments will have swung so far on the Russian side that we will not be able to catch up, let alone assume leadership."[89]

Johnson telegraphed Kennedy's conclusions and reinforced his conclusions at a meeting on 3 May 1961 when he said, "Free men are losing real estate to the Communists, and we are behind the Communists in the race for space. I believe it is the position of every patriotic and knowledgeable American that past policies and performances in space have not been enough to give this country leadership. That is the conclusion of the President. Moreover, that is, and has long been my conclusion." Johnson added that Kennedy was determined to move the United States into its proper position in space, one of leadership: "There is no other place for our country." Johnson closed by remarking Kennedy appeared ready to expand the total program from $22 billion over 10 years to $33 billion.[90] The remaining task was simply for McNamara and Webb to submit a detailed plan.

Two days later, on 5 May 1961, the first American finally went into space. Alan B. Shepard had a 15-minute, 116-mile spaceflight from Cape Canaveral, Florida. Kennedy apparently had considered the space-for-prestige question in some detail both before and after Shepard's flight. Attempting to get a sense of the Third World's perspective, Kennedy asked Tunisian president Habib Bourguiba after Shepard's flight if he would rather have an extra billion dollars a year in American foreign aid or have the United States mount a lunar-landing effort. "Bourguiba stood silent for several moments. Finally Bourguiba said, 'I wish I could tell you to put it in foreign aid, but I cannot.' "[91] The question of America's prestige in the international community clearly weighed heavily on Kennedy after the Gagarin flight. America's first human in space just three weeks later only reinforced the idea that space was indeed the Cold War competitive arena of the future. Logsdon explains that Shepard's flight was "one final event [which] helped ensure that an accelerated space program would be accepted by the president and the country. . . . The unqualified success of the flight swept away any of Kennedy's lingering doubts with regard to the role of the man in space flight."[92]

Important DOD Input into the Decision

Before the final Webb/McNamara position paper of 8 May 1961 was prepared, Johnson received final written replies from individuals with whom he had spoken earlier.[93] Schriever's reply is particularly important because it highlights why the Air Force's space-oriented officers supported the lunar-landing effort both before and after Kennedy's impending decision. Schriever said it was his "strong conviction that achievements in space in the critical decade ahead will become a principal measure of this nation's position in world leadership—a world in which it is becoming increasingly obvious that there will be no second." Schriever felt the main obstacle in America's space program was "the artificial and dangerous constriction of 'space for peaceful purposes' and 'space for military uses.'" When coupled with an "attitude of defeatism and a seeming resignation to second place in the space competition with the Soviets," a dangerous condition results, which "places at serious and unacceptable risk both our national prestige and our military security." Schriever said America's past space policy had failed to recognize "the military potential of space and the fact that achievements in space have been the single most important influence in the world prestige equation." Schriever concluded that a manned lunar landing and return would be the appropriate centerpiece of "a greatly expanded and accelerated space program [which] must reflect a singleness of purpose, a sense of urgency, a full acceptance of the Soviet challenge, and a refusal to admit there is any place for the United States but first."[94]

Schriever also explained to Logsdon that the USAF's space community supported the lunar-landing program because "it would put a focus on our space program. . . . I felt that we needed a major national space program for prestige purposes, for those things we could see as having national security implications and because of the need for advancing technology." Logsdon also notes this was the same basic idea the Air Force had supported since 1958: using a lunar landing as a central feature to give focus and lend global impact to the American space program. As to whether, by early 1961, anyone thought the Air Force should manage the lunar-landing program,

Schriever told Logsdon, "That never came up. At that point, there was no argument who was going to run the program."[95] In a perfect world, the Air Force certainly would have preferred to direct the lunar-landing program. But a NASA-directed program was infinitely preferable to the Air Force than no program at all because of the facilities, technology, and experience it would create for America and make available for potential defense applications.

Though Schriever and his corps of space-oriented officers provided important input to Johnson, the most important figure in the DOD input to the lunar-landing decision was SECDEF McNamara. Over the weekend of 6–7 May 1961, a group (consisting of Webb, McNamara, and various subordinates such as Dryden and Seamans for Webb, Gilpatric and DDR&E Harold Brown and his deputy Rubel for McNamara, along with BOB representative Willis Shapley), hammered out the final decisions. McNamara clearly had no problem with NASA pursuing an extensive human-spaceflight program for prestige purposes. In fact, at one point in the lunar-landing discussion that weekend, NASA Associate Administrator Seamans recalled McNamara remarked, "Well, are you sure that is a bold enough step?" He wondered, "Now are you sure we shouldn't take an even bigger bite and consider manned planetary [travel]?" Seamans said the NASA personnel were "very strong in the view that this was too big a step to commit the country to."[96] Seamans averred that the only thing DOD brought up was the question of large, solid rocket motors, believing the Air Force should be granted additional funding to pursue this project. Other than that, McNamara was receptive to the NASA staff's ideas for establishing a lunar landing and return as America's primary space goal in the 1960s.[97] The 8 May Webb/McNamara final memo appears to have been drafted primarily by Seamans and Rubel, based on a report Rubel had previously drafted, with last-minute editorial input from Webb.[98] Again, in all the final discussions and drafts, "It was absolutely accepted that this was NASA's responsibility, to take this on, and there was no question of, say, the DOD wondering if we should do it or in any way doing anything but saying, 'This is your responsibility. Jim Webb, you and NASA have got to do this.'"[99]

Kennedy Committed

When all was said and done, the 8 May 1961 Webb/McNamara recommendations, over 25 pages long, are the most important space policy document of the 1960s. Webb and McNamara recommended a $626-million add-on to the 1962 space budget, all of which would go to NASA except for $77 million, which was directed to the DOD for the solid-rocket engine R&D. The objective was "manned lunar exploration in the latter part of this decade." The men explained that space projects could be undertaken for four reasons: scientific knowledge, commercial-civilian value, military value, or national prestige. The United States was ahead in the scientific and military categories and had greater potential in the commercial arena but trailed in the space-for-prestige field. Therefore,

> This nation needs to make a positive decision to pursue space projects aimed at enhancing national prestige. Our attainments are a major element in the international competition between the Soviet system and our own. . . . The non-military, non-commercial, non-scientific but 'civilian' projects such as lunar and planetary exploration are, in this sense, part of the battle along the fluid front of the cold war. Such undertakings may affect our military strength only indirectly if at all, but they have an increasing effect upon our national posture. . . . It is vital to establish specific missions aimed mainly at national prestige.[100] (emphasis in original)

The Webb/McNamara package endorsed a lunar landing before the end of the decade because it "represents a major area in which international competition for achievement in space will be conducted. . . . It is man, not merely machines, in space that captures the imagination of the world." They acknowledge a lunar landing "will cost a great deal of money" and require "large efforts for a long time." Nevertheless, given "The Soviets have announced lunar landing as a major objective of their program," the United States has little choice if it wants to compete: "If we fail to accept this challenge it may be interpreted as a lack of national vigor and capacity to respond. . . . Perhaps the greatest unsurpassed prestige will accrue to the nation which first sends a man to the moon and returns him to earth. . . . The exploration of space will not be complete until man directly participates as an explorer."[101] Johnson quickly endorsed the Webb/McNamara conclusions and forwarded them to Kennedy because Kennedy had dispatched Johnson on a fact-finding tour of Southeast Asia. Logsdon records that on 10 May 1961 Kennedy met with his close advisors to ratify the Webb/McNamara package forwarded by Johnson. Bundy recalled, "The President had pretty much made up his mind to go" and was not particularly interested in hearing arguments to the contrary. Kennedy approved the package exactly as McNamara and Webb had laid it out.[102] On 25 May 1961, Kennedy announced his decision to the nation in a special message to Congress.

Kennedy said all the actions he proposed related to the responsibility of America to be "the leader in freedom's cause" because "the adversaries of freedom plan to consolidate their territory—to exploit, to control, and finally to destroy the hopes of the world's newest nations. . . . It is a contest of wills and purposes as well as force and violence—a battle for the minds and souls as well as lives and territory. And in that contest, we cannot stand aside." Accordingly, Kennedy actually proposed many initiatives before detailing his lunar-landing plan. He discussed measures "to turn recession into recovery," to aide the economic and social progress of the developing nations, to increase NATO's strength, to increase the American strategic deterrent, to triple US civil defense expenditures, and to strengthen the Arms Control and Disarmament Agency.[103] In fact, the lunar-landing decision was the final major point in his speech. Kennedy explained:

> Finally, if we are going to win the battle that is now going on around the world between freedom and tyranny, the dramatic achievements in space which occurred in recent weeks should have made clear to us all, as did the sputnik in 1957, the impact of this adventure on the minds of men everywhere who are attempting to make a determination of which road they should take. . . . It is time to take longer strides—time for a great new American enterprise—time for this Nation to take a clearly leading role in space achievement, which in many ways may hold the key to the future on earth. . . . For while we cannot guarantee that we shall one day be first, we can guarantee that any failure to make this effort will make us last. . . . We go into space because what-

ever mankind must undertake, free men must fully share. . . . I believe that this nation should commit itself to achieving the goal, before this decade is out, of landing a man on the moon and returning him safely to the earth.[104]

Later that day at a NASA press conference featuring Webb, Dryden, and Seamans, a reporter asked, "Is this an accelerated effort predicated on the assumption that we want to beat Russia to the moon?" NASA leaders replied simply, "Yes."[105] Kennedy had clearly concluded that national prestige was an important element in national power because what other nations and people thought about American power "was as important, if not more important, than the reality of that power. . . . A basic reason for the lunar landing decision was Cold War politics, phrased in terms of containing Soviet political gains from their space successes."[106] Human spaceflight became, under Kennedy, one expression of that power. Johnson's earlier conclusion: "Failure to master space means being second best in every aspect. . . . In the eyes of the world first in space means first, period; second in space is second in everything" became the Kennedy administration's guiding space policy.[107] Logsdon summarizes that the lunar-landing decision "is perhaps the ultimate expression of 'technological anticommunism' in terms of which way of life can best master nature, not control men."[108]

Webb had earlier emphasized to Johnson the central importance of the financial question in not only the lunar-landing decision, but also the long-term execution of the program. Webb wrote the vice president,

> I feel it imperative that you and the President understand we [Webb and McNamara] will need the assurance that the Nation is committed to this and that every effort will be made to put something between us and a situation in which we might be running like two foxes before two packs of hounds (Congress and the press), dependent only on our own skill and cunning to evade the pursuers and still carry on the work. . . . I want to make clear that we can only succeed if you are strongly with McNamara and me over the months and years ahead to do the really tough things we are going to have to do.[109]

In this passage Webb identified the foremost space-policy question of Kennedy's remaining term and Johnson's entire presidency: what level of financial support was appropriate for the overall NASA program and the lunar-landing program within it?

Some Consequences of the Decision

In the short term, the budgetary impact of Kennedy's decision was tremendous. NASA's FY 62 budget was increased $549 million; when coupled with the already-approved March supplemental, Kennedy had increased Eisenhower's final NASA budget of $1.1 billion by 61 percent in six months. In this process, "Congress approved his requests, almost without a murmur."[110] Kennedy had a radically different economic philosophy than the fiscally cautious Eisenhower. Logsdon explains Kennedy preferred "to use fiscal and monetary policy as tools for managing the national economy according to the tenets of the new [Keynesian] economics. Kennedy preferred government expenditures for needed programs instead of tax cuts as a means of injecting spending power into the economy."[111] Therefore, Kennedy did not have a visceral disdain for large new spending proposals such as the lunar-landing program, especially when this particular new venture meshed so nicely with his competitive Cold War philosophy.

It is fortunate Kennedy was amenable to new spending because "Project Apollo grew like a baby Paul Bunyan, and within two years consumed more than 50 percent of the entire NASA research and development budget."[112] Webb reorganized NASA by abolishing Glennan's all-inclusive Office of Space Flight Programs and creating two subdivisions, the Office of Manned Space Flight (OMSF) and the Office of Space Science and Applications. The OMSF disproportionately benefited from the subsequent Apollo-induced massive NASA-budget increases and was soon the dominant force within the NASA hierarchy, as the OMSF's directors determined "NASA's choice of future goals, controlled completely most of its budget, and preserved assiduously the separation of the space agency between manned and unmanned space flight constituencies."[113]

This concentration on human spaceflight was crushingly expensive. From Eisenhower's recommended level of $1.1 billion for NASA in FY 62, NASA's actual budget skyrocketed for the three years over which Kennedy had direct control: FY 62,

$1.8 billion; FY 63, $3.7 billion; and FY 64, $5.1 billion.[114] Of the FY 62 figure, 50.7 percent was for human spaceflight, which increased to 65.8 percent of the dramatically increased FY 64 total.[115] NASA employees went from 10,000 in 1960 to 34,000 in 1966, and NASA contractor employment grew tenfold from 37,000 to 377,000 in the same period.[116] Former NASA historian Launius estimated at the peak of its employment, one in 50 Americans worked on some aspect of Project Apollo.[117] Most estimates of the overall cost of the lunar-landing program are between $20–25 billion, a figure that translates to $91–114 billion in 1989 dollars.[118] The lunar-excursion-module portion of the Apollo spacecraft cost literally 15 times its weight in gold.[119] One assessment is that NASA's mobilization for Project Apollo was "comparable, in relative scale, to that undertaken by the U.S. to fight World War II."[120] NASA, by size of budget, was the fifth largest federal organization, after Defense; Treasury; Agriculture; and Health, Education, and Welfare.[121]

Truly impressive technology resulted from these outlays, however. The Saturn V rocket that would take the Apollo spacecraft to the moon had greater than eight million parts,[122] and the explosive potential of a million pounds of TNT,[123] (a megaton, which is more than most nuclear warheads) and 7.5 million pounds of thrust,[124] far in excess of any ICBM. The Saturn V at 363-feet tall was six stories higher than the Statue of Liberty, weighed six million pounds, and was the approximate size and weight of a Navy destroyer.[125] The United States spent an estimated $2.2 billion just constructing the infrastructure at Cape Canaveral/Kennedy, Houston, Texas, and Huntsville, Alabama, to support Apollo; tracking and communications facilities alone cost another $300 million.[126]

By the end of the Kennedy administration, however, the era of blank checks for NASA budgets appeared to be over. For FY 64 Kennedy actually asked Congress for a $5.7-billion NASA budget, but Congress approved only $5.1 billion. The $600-million difference "was the largest, both in absolute and relative terms, ever made on a NASA budget request."[127] NASA had problems with cost estimates, as Mercury was originally budgeted for approximately $200 million but cost almost $400 million.[128] The question that arises is, Did Kennedy's commitment to competing for prestige via human spaceflight and a lunar landing falter before his assassination in November 1963? Or, did he remain firmly committed to a space race with the Soviets?

Did Kennedy's Commitment Hold Firm?

Kennedy continued throughout 1961 to support his earlier decision. In November he stated, "I say this with complete conviction, there is no area where the United States received a greater setback to its prestige as the number one industrial country in the world than in being second in the field of space in the fifties. . . . And while many may think that it is foolish to go to the moon, I do not believe that a powerful country like the United States, which wishes to demonstrate to a watching world that it is first in the field of technology and science . . . want[s] to permit the Soviet Union to dominate space."[129] In National Security Action Memorandum (NSAM) 144 of 11 April 1962, Kennedy awarded Apollo the "DX" rating, signifying it was among those projects "being in the highest national priority category for research and development and for achieving operational capability" and thus, had first call in case of shortages of material or labor.[130]

Nevertheless, Webb reported a discussion he had with Kennedy in mid-1962 in which Kennedy "wanted to talk a little about the relation of this plan [Apollo] to that of the Russians. . . . He said he still thought the Russians were ahead in terms of world opinion." But, "He was quite concerned about the high level of expenditures involved in our program, plus the military program, and urged that everything be done that could possibly be done to see that we accomplish the results that would justify these expenditures and that we not expend funds beyond those that could be thoroughly justified."[131] While in no way implying Kennedy was questioning his original commitment to a lunar landing, his statements to Webb do at least indicate a level of concern with the high level of expenditures required for Project Apollo a year after his decision. Kennedy's friend and science advisor Wiesner recalled that by August 1962 Kennedy expressed "great irritation" with the ever-increasing cost trend in the space program.[132]

221

Concern over financial pressures did not modify Kennedy's public rhetoric on the space issue. At Rice University in September 1962 he delivered his second famous space-related address and wholeheartedly endorsed the lunar goal.

> The exploration of space will go ahead, whether we join in it or not . . . and no nation which expects to be the leader of other nations can expect to stay behind in this race for space. . . . We mean to be part of it—we mean to lead it. . . . We shall not see it [space] be governed by a hostile flag of conquest, but by a banner of freedom and peace. . . . The vows of this nation can only be fulfilled if we in this nation are first, and, therefore, we intend to be first. . . . Our leadership in science and in industry, our hopes for peace and security, our obligations to ourselves as well as to others, all require us to become . . . the world's leading space-faring nation. . . . Only if the United States occupies a position of pre-eminence can we help decide whether this new ocean will be a sea of peace or a new terrifying theater of war. . . . We choose to go to the moon in this decade and do the other things, not because they are easy, but because they are hard, because . . . that challenge is one that we are willing to accept, one we are unwilling to postpone, and one which we intend to win. . . . We do not intend to stay behind, and in this decade we shall make up and move ahead.[133]

1962 Review

Within the White House, however, there were limits to Kennedy's acceptance of Apollo's budget increases. A dispute arose within NASA between Webb and the person he chose to head the OMSF, D. Brainerd Holmes. Holmes was directly responsible for day-to-day management of the Apollo program. Holmes believed Apollo was of such critical importance that it should proceed on an all-out-crash basis, with access to virtually unlimited funds. He wanted a $400-million supplemental appropriation for Apollo so he could actually accelerate the schedule to permit a lunar landing in 1967. Webb's position was that Apollo should be in some kind of relative balance with NASA's other responsibilities such as space science.[134] By November (just after the Cuban missile crisis), the dispute reached Kennedy, who asked for Webb's opinion. Webb argued, "The objective of our national space program is to become preeminent in all important aspects of this endeavor." In Webb's mind, this meant that "the manned lunar landing program, although of highest national priority, will not by itself create the preeminent position we seek." Webb believed the broader US interests in science "demand we pursue an adequate, well-balanced space program in all areas, including those not directly related to the manned lunar landing."[135] Kennedy sided with Webb: there was no $400-million supplemental; Apollo did not proceed on a "blank-check" basis; and Holmes soon departed NASA. Logsdon stated that "the president's acceptance seemed to indicate that across-the-board preeminence was indeed his guiding policy objective for the United States in space," although the pursuit of this objective would proceed within a reasonable financial framework.[136]

1963 Review

By 1963 Kennedy felt the need for a second review of the space program. In contrast to the 1962 review, which was generated primarily by forces within NASA wanting an even higher priority for Apollo, the 1963 review "appears to have been stimulated by increasing external criticism of the priority being given to the space program rather than other areas of science and technology, and was focused on those aspects of the program not linked to Apollo."[137] By 1963 many within the scientific community felt the human-spaceflight program was too expensive and siphoned off resources that could be usefully employed by other scientific disciplines. Eisenhower continued to believe Apollo was a waste of resources. Finally, congressional Republicans, among others, criticized Kennedy for ignoring military space requirements.

A few examples must suffice to represent the rising chorus of criticism by 1963. Eisenhower wrote, "By all means, we must carry on our explorations in space, but I frankly do not see the need for continuing this effort as such a fantastically expensive crash program. . . . Why the great hurry to get to the moon and the planets?. . . I think we should proceed in an orderly, scientific way, building one accomplishment on another, rather than engaging in a mad effort to win a stunt race."[138] The criticism that Kennedy neglected military space projects was seen in a January 1963 Republican Congressional Committee report: "The Kennedy administration's failure to build up a strong military space capability is perhaps the most disastrous blunder by any government since the last World War."[139] Vannevar Bush, who is given credit for harnessing the scientific R&D community in service of the government in World War II, represented the opinion of many within the scientific community when he wrote Webb in April concerning the lunar-landing program: "The program, as it has been built up, is not sound. The sad fact is that the program is more expensive than the country can now afford; its results, while interesting, are secondary to our national welfare. . . . This is no time at which to make enormous—and unnecessary—expenditures. . . . This program has never been evaluated objectively by an adequately informed and disinterested group, and I fear it never will be."[140]

Accordingly, Kennedy asked Johnson to conduct another careful review on 9 April 1963 because he felt "the need to obtain a clearer understanding of a number of factual and policy issues relating to the National Space Program which seem to arise repeatedly in public and other contexts." Kennedy's five specific questions included inquiries concerning the differences between his program and Eisenhower's; principal benefits flowing from the program; major problems resulting from the space program; what reductions in the program could take place without compromising the lunar-landing timetable; and was there adequate NASA-DOD coordination.[141] In his capacity as NASC chairman, Johnson gathered inputs much as he had in the spring of 1961.

The DOS was critical of the continuing race posture. Its response said, "Continuing emphasis on a crash program for a manned lunar landing, particularly in the Cold War context of a race with the Soviets, will strengthen the impression abroad that our program is motivated by political and security considerations. It will tend to reduce the credibility of our program as a balanced, rationally-paced undertaking for essentially scientific and beneficial purposes." The author concluded, "By the time a manned lunar landing has been accomplished our success may well have a less advantageous impact abroad than we expect."[142] Webb, not surprisingly, disagreed and supported the current effort in space. He said the criticisms "arise from a narrow view of the progress required to achieve the lunar goal, and a tendency to evaluate the program only in terms of immediate objectives. This attitude fails to recognize that the Apollo program is not an end in itself, but rather an initial major objective on which to focus our efforts. . . . The skill and knowledge gained and the resources developed in the Apollo program will provide the basis for space power required to carry out necessary tasks in space for many years to come."[143] Johnson was clearly in Webb's camp on this issue. He declared shortly before giving his official report to Kennedy, "I do not believe that this generation of Americans is willing to resign itself to going to bed each night by the light of a Communist moon."[144]

Johnson's official response to Kennedy's tasking was on 13 May 1963. He explained that Eisenhower's space program through 1970 would have cost $17.9 billion while Kennedy's featuring Apollo had a price of $48.1 billion. Eisenhower's plan was that of a "second-place runner" while the Kennedy plan was designed "to make this country the assured leader before the end of the decade." The benefits included not just prestige but also economic and national security returns. Johnson concluded no major problems would result from the space program since it employed only 3 percent of the nation's engineers. Johnson did not cite any portions of the NASA program amenable to reduction, and he offered up no major NASA-DOD problems. His fundamental conclusion was to stay the course.

> The space program is not solely a question of prestige, of advancing scientific knowledge, of economic benefit or of military development. . . . Basically, a much more fundamental issue is at stake—whether a dimension that can well dominate history for the next few centuries will be devoted to the social system of freedom or controlled by the social system of communism.
>
> The United States has made it clear that it does not seek to "dominate" space. . . . But we cannot close our eyes as to what would happen if we permitted totalitarian systems to dominate the environment of earth itself. For this reason our space program has an over riding urgency that cannot be calculated solely in terms of industrial, scientific, or military development. The future of society is at stake.[145]

This report is important because not only does it appear to have been Kennedy's fundamental position for the remainder of his term but also because it represents Johnson's thinking only six months before he would become president.

Only five days later, Kennedy declared, "I believe the United States of America is committed in this decade to be first in space."[146] On 16 November 1963, less than a week before his assassination, Kennedy toured Cape Canaveral and at one point insisted on standing directly beneath a giant Saturn rocket and asked, "Now, this will be the largest payload that man has ever put in orbit?" When told it was, he replied, "That is very, very significant." In the helicopter ride back from watching a Polaris submarine missile launch, Kennedy made NASA associate administrator Seamans repeat the entire briefing on the Saturn and asked Seamans if the Saturn's capabilities were greater than those of the Soviets' largest rocket. When assured they were, Kennedy said, "That's very important. Now, be sure that the Press really understands this." Before exiting the helicopter, Kennedy reminded Seamans, "Now, you won't forget, will you, to do this?"[147] In his final days Kennedy continued to regard the space program as a competitive race with the Soviets for worldwide prestige. As Sorensen testified, Kennedy was not "deterred by a swelling chorus of dissenters at home."[148] Robert Rosholt explained that by 1963 "NASA and the space program had already gained a momentum that was not easily deflected."[149]

In the speech Kennedy would have delivered in Dallas on the afternoon of 22 November, Kennedy was prepared to explain, "The [space] effort is expensive—but it pays its own way, for freedom and for America. . . . There is no longer any doubt about the strength and skill of American science, American industry, American education and the American free enterprise system. In short, our national space effort represents a great gain in, and a great resource of, our national strength."[150] Finally, his speech for that evening contained this assessment of the American space effort: "We are not yet first in every field of space endeavor, but we have regained worldwide respect. . . . And we have made it clear to all that the United States of America has no intention of finishing second in outer space. . . . This is still a daring and dangerous frontier; and there are those who would prefer to turn back or to take a more timid stance. But Texans have stood their ground on embattled frontiers before, and I know you will help us see this battle through."[151] Even if one allows for rhetorical flourish, these hardly seem the words of a man about to pull back from a drive for preeminence or one preparing to abandon a competitive effort.

Kennedy, the Soviet Space Program, and a Joint Lunar Landing

One potentially puzzling sequence of events remains, however. If one holds that Kennedy's commitment to the human-spaceflight-for-prestige equation remained firm until his final days, how does one account for his offer in September 1963 to transform the lunar-landing program into a joint US-Soviet effort? Would this not indicate a significant withdrawal from the competitive ethos? A necessary precursor to exploring these questions is to survey the sequence of events during the Kennedy administration concerning cooperating in space with the Soviets as well as how offers of cooperation related to the overall American estimates of the Soviet space program and whether the Soviets were even in a race with the United States to the moon. It will be recalled from previous chapters that during the Eisenhower administration there were initiatives in this field but little progress—a factor attributed by most participants and scholars to Soviet intransigence. Much the same pattern persisted during the Kennedy administration. It is possible to point to more concrete initiatives and results from Kennedy's term, however, to include a preliminary UN agreement to ban the stationing of weapons of mass destruction in space—a precursor to the 1967 Outer Space Treaty. In addition, while there were some assertions that the Soviets had dropped out of the lunar race, Kennedy appeared to either discount them completely or at least not view them as credible enough to undermine his commitment to Apollo.[152]

Assessing the Soviet Space Program

The CIA's input to Kennedy on the Soviet space program reinforced the competitive dynamic. Representative was the CIA's National Intelligence Estimate 11–1–62 from December

227

1962 that concluded the Soviets were likely to conduct "a space program of much broader scope than in the past, but attempts to accomplish spectacular 'firsts' will continue. . . . Dramatic manned space flights are likely in the course of the next few years. . . . Some Soviet statements indicate that a program for a manned lunar landing is under way in the USSR. . . . We estimate that with a strong national effort the Soviets could accomplish a manned lunar landing in the period 1967–1969." Of vital importance to understanding Kennedy's later offer to make the lunar-landing program a joint one with the Soviets is the CIA's conclusion that, from the Soviets' perspective, "the political prestige at stake in a lunar race is likely to preclude cooperation in this area, even though it is by far the most costly of the possible new programs. The Soviets would seek a significant degree of international cooperation only if the economic burden of their space program becomes so heavy that this program or key economic and military programs were jeopardized. Under such conditions the Soviets would prefer cooperation to competing unsuccessfully or at too high a price." However, for the foreseeable future, the CIA stated, "We believe that the Soviet leaders are committed to a continuing space program of sizable proportions as an element of national power and prestige."[153] It seems likely that even if Kennedy genuinely desired space cooperation with the Soviets, up to and including a lunar landing, and made legitimate proposals for such joint endeavors, there had to be at least an element in his calculations cognizant of the fact that the Soviets would be extremely unlikely to accept these offers.[154]

Therefore, when a flurry of speculation arose in 1963 that the Soviets had withdrawn from the moon race, Kennedy was skeptical. Renowned British astronomer Sir Bernard Lovell, director of Britain's Jodrell Bank Experimental Station, returned from a trip to the USSR in July 1963 and reported that the president of the Soviet Academy of Sciences, M. V. Keldysh, had told him that the USSR had rejected, for the time being, any plans for manned lunar landings due to insurmountable problems of radiation in space.[155] Lovell explained he had visited "all the major Soviet optical and radio observatories" and had concluded, "I don't think that there is any priority at the moment for the manned moon program—definitely not in their budget anyhow. . . . I got an astonishing impression during my visit there that the ice was rapidly cracking, and that there was a really genuine desire for cooperation."[156] This caused immediate and intense excitement in the press, with rampant speculation that the United States could now slow down the pace of the Apollo program and save money. However, Kennedy seemed not to take Lovell's charges seriously, stating there was still "every evidence that they are carrying on a major campaign and diverting greatly needed resources to their space effort. With that in mind, I think that we should continue . . . with our own program and go on to the moon before the end of this decade."[157]

Shortly thereafter, Khrushchev reopened the issue by declaring at a Third World meeting of journalists, "We are not at present planning flights by cosmonauts to the moon. Soviet scientists are working on this problem. It is being studied as a scientific problem, and the necessary research is being done. . . . We do not want to compete with the sending of the people to the moon without careful preparation. . . . Much work will have to be done and good preparations made for a successful flight to the moon by man."[158] These statements sent American space officials into damage control mode. NASC executive secretary Welsh explained,

> There is nothing in Mr. Khrushchev's statement which warrants concluding: (1) They are abandoning a lunar project, (2) They are lessening or slowing down their space program, or (3) They won't in the near future try a manned flight around the moon and back. It appears that Mr. Khrushchev has taken this means of encouraging a space slowdown in the United States and thereby trying to maintain a competitive advantage from our slower pace rather than from his speeding up.[159]

The DOS corroborated Welsh's interpretation by stating what Khrushchev meant was that while the Soviets were not working on short-range, operational plans for a lunar landing, they were working on the problem in general. Therefore, Khrushchev did not announce the end of the Soviet lunar program: "All told Khrushchev has committed himself to nothing."[160] This was apparently true because Khrushchev quickly reversed himself. He told a group of visiting American businessmen, "We in the Soviet Union have never given up our goal of placing a man upon the surface of the moon at the proper time. We have never said we had given it up. This is an interpretation which the Americans have given to my statement. Again, I will say that the Soviet Union has an active program in space research with specific orientation to landing a Soviet man on the surface of the moon when the time is proper and our capabilities have been developed."[161] Once again, this is not a sequence of events likely to have created within the Kennedy administration the impression that the Soviets were going to accept United States offers of joint lunar landings or other space cooperation projects.[162] Kennedy dismissed the whole brouhaha when he said, "The fact of the matter is that the Soviets have made an intensive effort in space, and there is every indication that they are continuing and that they have the potential to continue. I would read Mr. Khrushchev's remarks very carefully. . . . I think we ought to stay with our program. I think that is the best answer to Mr. Khrushchev."[163]

Cooperating with the Soviets in Space

These twin discussions of intelligence information available to Kennedy on the Soviet space program and of Soviet attempts to persuade the United States that there was no moon race help set the stage for this chapter's final topic: US-Soviet space cooperation and Kennedy's September 1963 offer of a joint lunar program. Hopeful rhetoric concerning cooperation was present in Kennedy's speeches from his first day in office. In his inaugural address he said, "Let both sides seek to invoke the wonders of science instead of its terrors. Together, let us explore the stars."[164] One perceptive scholar explains that Kennedy made these, and subsequent, offers of US-USSR space cooperation "knowing full well that there was little likelihood that Khrushchev would accept his offer" because if Khrushchev did, "It would tacitly be recognizing the equality of the United States in space activities."[165]

A footnote to the Vienna summit of June 1961 was an informal Kennedy-Khrushchev exchange on a joint lunar-landing program. Apparently during lunch on the first day, Kennedy suggested combining the lunar-landing efforts (less than two weeks after his famous 25 May speech announcing his decision). The DOS memo recorded, "With regard to the possibility of launching a man to the moon, Mr. Khrushchev said that he was cautious because of the military aspect of such flights. In response to the President's inquiry whether the United States or the USSR should go to the moon together, Mr. Khrushchev first said no, then said 'all right, why not?' "[166] Khrushchev's final remark was probably in jest because the next day he reversed himself.

> Mr. Khrushchev said he was placing certain restraints on projects for a flight to the moon. Such an operation is very expensive and this may weaken Soviet defenses. Of course, Soviet scientists want to go to the moon, but the U.S. should go first because it is rich and then the Soviet Union will follow. In response to the President's inquiry whether perhaps a cooperative effort could be made in that direction, Mr. Khrushchev said that cooperation in outer space would be impossible as long as there was no disarmament. The reason for this is that rockets are used for both military and scientific purposes. The President said that perhaps coordination in timing of such efforts could be achieved in order to save money. . . . Mr. Khrushchev replied that might be possible but noted that so far there had been few practical uses of outer space launchings. The race was costly and was primarily for prestige purposes.[167]

Once again, the historian of these events is hard pressed to avoid the conclusion that as much as Kennedy may have hoped differently, he had to be aware of the fact that Khrushchev was not going to be receptive to American offers of large-scale cooperation throughout Kennedy's administration. One can argue that Khrushchev's reluctance was due to financial reasons, disarmament concerns, and worries about military-technology transfer. Also he felt that by competing with the United States he would grant legitimacy to the American program. Whatever the case, the fundamental point remains: Kennedy almost certainly knew there was little chance Khrushchev could or would seriously respond to American offers of cooperative or joint space projects.

There was no reason, then, why Kennedy could not deliver pleas, such as at the UN in September 1961 that "the new horizons of outer space must not be riven by the old bitter concepts of imperialism and sovereign claims. The cold reaches of the universe must not become the new arena of an even colder war." Kennedy also declared the United States would support any UN effort toward "reserving outer space for peaceful use, [and] prohibiting weapons of mass destruction in space or on celestial bodies, and opening the mysteries and benefits of space to every nation."[168] One concrete result of Kennedy's speech was that the USSR did agree to expand the UN Committee on Peaceful Uses of Outer Space to 23 members, and so the COPUOS had its first official meeting with a full contingent of countries in March 1962; it began work on a resolution that would ban the deployment of weapons in space.[169] This effort would culminate in one of the two concrete results of the international space cooperation efforts during Kennedy's term. On 17 October 1963, the UN General Assembly adopted Resolution 1884, "Stationing Weapons of Mass Destruction in Outer Space." This resolution did exactly what its title implied; it prohibited the orbiting of weapons of mass destruction (nuclear, biological, chemical weapons) around the earth or on other celestial bodies such as the moon.[170]

The second identifiable product from the international cooperation initiatives of the Kennedy era began with America's first orbital flight of a human. On 20 February 1962, John Glenn in his Mercury capsule *Friendship 7* made three orbits of the earth and flew in space for four hours and 55 minutes. Besides making him an instant hero, it generated a congratulatory message from Khrushchev that read, "If our countries pooled their efforts . . . to master the universe, this would be very beneficial for the advance of science and would be joyfully acclaimed by all peoples who would like to see scientific achievements benefit man and not be used for 'cold war' purposes and the arms race."[171] Kennedy immediately responded, "I welcome your statement that our countries should cooperate in the exploration of space. . . . I am instructing the appropriate officers of this Government to prepare new and concrete proposals for immediate projects of common action."[172] Kennedy issued NSAM 129 instructing that NASA, the NASC, and Wiesner cooperate with the DOS in developing these proposals because "the President does require that there be a prompt and energetic follow-up of his message to Chairman Khrushchev."[173] More important were Kennedy's private instructions to Webb, delivered through National Security Advisor Bundy. Bundy wrote Webb that Kennedy "knows that there are lots of problems in this kind of cooperation, and he knows also that you have a great head of steam in projects which we do not want to see interrupted or slowed down. At the same time, there is real political advantage for us if we can make it clear that we are forthcoming and energetic in plans for peaceful cooperation with the Soviets in this sphere." Therefore, Kennedy hoped NASA's staff could "go a little out of their way to find good projects."[174] The overall tone of Kennedy's instructions gives the distinct impression that he was not overly concerned with any possible cooperative projects in and of themselves (he didn't mention any specific initiatives) but rather the "real political advantage" that could be extracted from the image of a peaceful, cooperative America.

What followed was another exchange of Kennedy-Khrushchev letters and then further talks by their designated representatives, NASA deputy administrator Dryden and Soviet academician and scientist Anatoly Blagonravov. Dryden and Blagonravov met nine times between March 1962 and May 1965.[175] The concrete cooperative actions resulting from these negotiations have best been collectively referred to as "only token results."[176] Various levels of cooperation eventually took place in four areas: meteorological satellite systems and the exchange of their data, using the passive American communications satellite *Echo II* for cooperative experiments, satellites for studying and mapping the earth's magnetic field, and a joint review of information gathered in the areas of space biology and medicine. As Khrushchev freely admitted in his memoirs, the USSR simply was not interested in genuinely extensive space cooperation because this would have given America access to Soviet space and missile technology and by doing so "we would have been both giving away our strength [space technology] and revealing our weakness [lagging ICBM development]."[177] Congress correctly concluded, "Khrushchev seemed to be concerned less with cooperating in space than with making a concrete political reality of the abstract Soviet claim that a shift in the balance of world power against the

West had occurred, and that this was attributed, among other factors, to Communist superiority."[178]

NASA's director of International Programs emphasized that the assorted projects and data exchanges resulting from the Dryden-Blagonravov talks in the early and mid-1960s provided for coordination and not integration, "a kind of arm's length cooperation in which each side carries out independently its portion of an arrangement without entering into the other's planning, design, production, operations, or analysis. No classified or sensitive data is exchanged. No equipment is to be provided by either side to the other. No funds are to be provided by either side to the other."[179] Kennedy himself wrote Rep. Albert Thomas in September 1963 and explained, "Our repeated offers of cooperation with the Soviet Union have so far produced only limited responses and results."[180] Given this limited progress by 1963 in developing concrete US-Soviet space cooperation, it seems unlikely Kennedy concluded that he had much to lose by rhetorically offering Khrushchev a joint lunar-landing effort because Khrushchev would almost certainly either reject or ignore the proposal.

In the summer of 1963, simply making such a grand proposal—during an address to the UN General Assembly—had distinct appeal to Kennedy. After the Cuban missile crisis in October 1962, there had been at least some thawing in US-Soviet relations. Some even spoke of a nascent détente. The clearest piece of evidence was that after US-USSR talks for a complete banning of nuclear tests had failed, the countries did work out a Limited Test Ban Treaty (LTBT) in July 1963 that banned the testing of nuclear weapons in space, the atmosphere, and under water.[181] As movement within the UN framework toward a resolution banning the stationing of weapons of mass destruction in outer space gained momentum, Kennedy may very well have seen the offer of a joint lunar landing as one which would provide America with an even brighter image as a peaceful nation enthusiastically embracing all types of disarmament and weapons control. As Sorensen recalled, Kennedy "did not think it possible to achieve in his administration a sweeping settlement of East-West divisions. But he did hope that small breakthroughs could lead to larger ones, and that brick by brick a détente could be built, a breathing space, a 'truce to terror' in which both sides could recognize that mutual accommodation was preferable to mutual annihilation."[182]

Accordingly, when Kennedy spoke to the UN on 20 September 1963, he indirectly referred to the Cuban missile crisis when he said, "The clouds have lifted a little so that new rays of hope can break through." Kennedy pointed to the LTBT, the easing of tensions over Berlin, and resolution of the Congo and Laos crises as evidence of the fact that "we meet today in an atmosphere of rising hope." Kennedy offered several proposals for maintaining and augmenting the momentum towards peace and said,

> I include among these possibilities a joint expedition to the moon. Space offers no problems of sovereignty. . . . Why, therefore, should man's first flight to the moon be a matter of national competition? Why should the United States and the Soviet Union, in preparing for such expeditions, become involved in immense duplications of research, construction, and expenditure? Surely we should explore whether the scientists and astronauts of our two countries . . . cannot work together in the conquest of space, sending some day in this decade to the moon not the representatives of a single nation, but the representatives of all of our countries.[183]

Taken at face value, Kennedy's speech would certainly appear to have been a legitimate, good-faith offer for a joint lunar-landing program. But making a legitimate, good-faith offer is not mutually exclusive with holding out little realistic hope that a positive response to that offer will be forthcoming. The evidence in the case of Kennedy's joint lunar-landing offer appears to support the interpretation that while Kennedy may very well have not been acting or speaking disingenuously, he also may not have been at all optimistic, based upon past Soviet/Khrushchev behavior, that his offer would be taken seriously, much less elicit a favorable response. Analysts should remember Kennedy's statement earlier that summer in the midst of the Lovell episode: "The kind of cooperative effort which would be required for the Soviet Union and the United States together to go to the moon would require a breaking down of many barriers of suspicion and distrust and hostility which exist between the Communist world and ourselves. There is no evidence as yet that those barriers will come down. . . . I would welcome it, but I don't see it as yet, unfortunately."[184]

Nevertheless, the historian must also avoid dismissing entirely Kennedy's sincerity in making his September 1963 offer. Only 10 days before his assassination, he signed NSAM 271, *Cooperation with the USSR on Outer Space Matters*. In it Kennedy addressed Webb,

> I would like you to assume personally the initiative and central responsibility within the government for the development of a program of substantive cooperation with the Soviet Union in the field of outer space, including the development of specific technical proposals. . . . These proposals should be developed with a view to their possible discussion with the Soviet Union as a direct outcome of my September 20 proposal for broader cooperation between the United States and the USSR in outer space.[185]

A formal presidential NSAM is more than a continuing wish. Kennedy clearly wanted his administration to press forward with the exploration of potential US-Soviet Union cooperative space projects. Quite possibly the only sure statement the analyst can make is that the tensions that had been present within Kennedy's space policy from the beginning of his presidency between racing competitively for prestige in space and cooperating internationally in space continued until his death.[186]

It is possible that Kennedy found himself almost whipsawed between conflicting advisors within his administration. On the one hand, Johnson and Webb seemed inclined to support as low a level of cooperation with the USSR as possible. On the other hand, elements within the DOS and Arms Control and Disarmament Agency, represented by individuals such as national security advisor Bundy, wanted to achieve as much space cooperation with the Soviets as quickly as possible. NSAM 271 may represent the continuing ambivalence within Kennedy's mind as to which pursuit was paramount: competition or cooperation. Perhaps near the end of his presidency, the proponents of cooperation had the upper hand, given the tenor of NSAM 271. Whatever the case, and absent additional evidence, one can safely state that no firm resolution or conclusion is possible; the ambivalence in Kennedy's space policy continued throughout his tenure. Janus continued to gaze in both directions.

This bidirectional space-policy orientation in one sense reflected the continued ambivalence one finds in Kennedy's overall Cold War policy. For instance, one must balance the indications of détente and Kennedy's inspiring American University speech of June 1963 with other Cold War statements he made *after* that address. In Berlin, Kennedy declared,

> Ich bin ein Berliner . . . There are many people in the world who really don't understand, or say they don't. What is the great issue between the free world and the Communist world. Let them come to Berlin. . . . Freedom has many difficulties and democracy is not perfect, but we have never had to put up a wall to keep our people in. . . . The wall is the most obvious and vivid demonstration of the failures of the Communist system [and] an offense not against history but an offense against humanity.[187]

One returns again to the image of Janus looking in both directions. Kennedy's Cold War policy and his space policy considered as a subset of it were clearly an amalgam of "accommodative and confrontational policies" because "Kennedy was, above all, a pragmatist who viewed the Cold War . . . as a conflict of interests rather than of ideologies."[188] For him there was not necessarily any conflict in signing an atmospheric and space nuclear-test ban and continuing to test underground, or in being willing to sell the Soviets surplus wheat while refusing to sell them strategic, defense-oriented items, or even in exploring the possibilities of disarmament while maintaining a stockpile of arms. Kennedy's Cold War policy, with the space program clearly a part of it, "was marked by heterogeneous features: on the one hand, an obsession not to appear soft on the Soviets and a distinct preoccupation with conveying a tough and virile image; and, on the other hand, a penchant for stressing the common interests brought about by the 'dark forces of destruction' unleashed by science."[189] Kennedy himself said, "Let us always make clear our willingness to talk, if talk will help, and our readiness to fight, if fight we must. . . . When we think of peace in this country, let us think of both our capacity to deter aggression and our goal of true disarmament."[190]

A final point concerning Kennedy's joint lunar-landing proposal of September 1963 bears mentioning. Whether or not Kennedy believed the suggestion was likely to elicit an affirmative response from the USSR, the very fact that he made the offer seems to have cost Apollo a measure of congressional support. At the same time Kennedy was making the offer he was asking that NASA's FY 64 budget be approved at the level of $5.7 billion. However, on 10 October 1963, the House voted 125:110 to forbid spending any federal funds for "participating in a manned lunar landing to be carried out jointly by the United States and any Communist-controlled, or Communist-dominated country." The House language would force the president to seek special approval for any part of the space program used in a joint-lunar-exploration program. In addition, Congress was beginning the appropriations process that would result, as described earlier in this chapter, in the reduction of Kennedy's NASA budget request by $600 million to $5.1 billion.[191] A Republican congressman explained the cut as resulting from the fact that the Russians were focusing on earth orbital space in their space program, not the lunar environment and because of "the President's suggestion made recently before the world that lunar programs in technology, operation and objective be shared with the Soviet Union. . . . The mere fact that the President has suggested such a possibility infects the entire Apollo program with fiscal uncertainty."[192] At a minimum from this point forward in the realm of forging space policy, "Congress could no longer be taken for granted."[193] Given this adverse congressional reaction, it was unlikely Lyndon Johnson would, during his presidency, risk any of his political capital (which he wanted to use to jump-start his Great Society initiatives but which ended up being rapidly depleted by the Vietnam war) on bold propositions for US-Soviet space cooperation. In fact, he did not. US-Soviet space cooperation during the Johnson administration was simply the continuation of the Kennedy-era initiatives, specifically the decreasingly fruitful Dryden-Blagonravov talks and transforming the UN Resolution banning weapons in space into the Outer Space Treaty in 1967.

Logsdon provides the most important conclusion for this chapter. He summarizes, "In terms of its political underpinnings, it is more appropriate to place the Apollo decision in the 1950's than in the 1960's. Apollo was one of the last major political acts of the Cold War; the moon project was chosen as a symbol of the head-to-head global competition with the Soviet Union." As a symbolic undertaking Apollo was "intended to demonstrate to the world that the United States remained the leading nation in technical and social vitality. Almost equally important, though not as clearly articulated, Kennedy saw Apollo as a means of restoring American pride and self-confidence, which appeared to have been badly damaged by the Soviet Union's surprising demonstration of technological and strategic strength through its series of space firsts." The foundation for Kennedy's space policy was the simple fact that as a political leader Kennedy "found unacceptable the notion of the United States taking second place to the Soviet Union in a critical area of human activity." The contrast with the Eisenhower administration could not be starker. Overall, "Kennedy himself was much more interested in the political payoff of Apollo than he was in the across-the-board acceleration of the space program, but he had little choice but to approve the whole package."[194] Harvey Brooks points out another aspect of the Apollo decision that Kennedy found appealing: Apollo provided a highly visible and easily understandable demonstration of American technological prowess "without directly threatening the USSR or raising public fears of a military confrontation. It was like a challenge between the champions of two medieval armies, the race for the moon serving as a partial surrogate for more threatening forms of competition."[195]

Another analyst makes the telling point that, "In a very real sense, the final U.S. response to the Sputnik challenge was not complete until Neil Armstrong and Buzz Aldrin walked upon the Sea of Tranquillity on 20 July 1969. . . . The moon race completely overshadowed all other U.S. space activities such as the continuing attempts of the Air Force to build a manned military space mission."[196]

The next chapter will detail the institutional climate that developed between the DOD and NASA during the Kennedy administration. It will include the crucial factor of tension within the DOD between the OSD and the corps of Air Force space enthusiasts that hamstrung the latter's aspirations for military human spaceflight.

239

Notes

1. For a fuller explanation, see Gaddis, *Strategies of Containment*, 198.

2. Firestone, *The Quest for Nuclear Stability*, 81.

3. Logsdon, "The Evolution of U.S. Space Policy and Plans," 382. Logsdon also explains that the NASC did draft such a space policy document, but "it never received presidential sanction" (ibid.).

4. They were NSC 5520, *Draft Statement of Policy on U.S. Scientific Satellite Program*, 20 May 1955; PSAC, *Introduction to Outer Space*, 26 March 1958; *National Aeronautics and Space Act of 1958*; NSC 5814/1, *Preliminary U.S. Policy on Outer Space*, 8 August 1958; NSC 5918, *U.S. Policy on Outer Space*, 26 January 1960.

5. Hayes, *Struggling Towards Space Doctrine*, 161.

6. Reeves, *President Kennedy*, 88.

7. The staffs of the Kennedy and Johnson Libraries have informed this author there are very few such equivalent extensive records corresponding to those presidents. A related point concerning primary sources relates simply to the passage of time. More time has elapsed since the end of the Eisenhower administration, and therefore many more documents have been declassified. Declassification is indispensable to the space historian because the space arena, particularly the military space field, tends to be one of the most heavily classified research topics. More raw data is available from the Eisenhower administration simply because the staffs of various archives have had a few more years to sift through, consider, and declassify Eisenhower-era documents when compared to the Kennedy and Johnson material from the 1960s (which, as mentioned above, is much less in quantity to begin with). The author has reached this conclusion after discussions with the declassification officials at not only the three presidential libraries in question but also other facilities such as the AFHRA (Air Force Historical Research Agency), AFHSO (Air Force History Support Office), LOC (Library of Congress), and NARA (National Archives and Records Administration).

8. It appears that only the passage of time and additional declassification authority, such as researchers' extensive use of the 17 April 1995 presidential EO 12958, as amended by EO 13292, 25 March 2003, will help rectify this situation. EO 13292 states that by 31 December 2006 all classified records more than 25 years old and with "permanent historical value" shall be automatically declassified whether or not the records have been reviewed. However, the EO also lists nine reasons why agency heads may exempt their records from automatic declassification, and, as with any governmental decree, agencies can apply for special waivers from the EO's requirements. EO 12958, *Classified National Security Information*, 17 April 1995. As amended by EO 13292, *Further Amendment to Executive Order 12958, Classified National Security Information*, 25 March 2003, Washington, D.C.: GPO.

9. LaFeber, *America, Russia, and the Cold War, 1945–1966*, 229.

10. Kennedy, *Public Papers of the Presidents*, no. 1, 20 January 1961, 1–3.

11. Kennedy, *Public Papers of the Presidents*, no. 153, 27 April 1961, 334–38.

12. Beschloss, *The Crisis Years*, 70.

13. DOS, memcon, 3 June 1961, 2–5. The two men continued at loggerheads over the Berlin situation, the nuclear-test ban question, the crises in Congo and Laos, nuclear disarmament, and the two countries' general relationship during the remainder of the summit, which was their only face-to-face meeting. Their exchanges concluded with Khrushchev exclaiming, "The U.S. wants to humiliate the USSR and this cannot be accepted. He said that he would not shirk his responsibility and would take any action that he is duty bound to take." Khrushchev continued by stating that if the United States did not sign a peace treaty with East Germany ceding control of West Berlin to East Germany, then "the USSR will have no choice other than to accept the challenge; it must respond and it will respond. The calamities of a war will be shared equally. War will take place only if the U.S. imposes it on the USSR." DOS, memcon, 4 June 1961, 1–3.

14. DOS, memcon, 4 June 1961, 1–3.

15. Kunz, "Introduction: The Crucial Decade," 3.

16. This would augment US counterguerrilla warfare special forces such as the Green Berets; *increase* Polaris ballistic-missile submarines from 19 to 29; double the production of Minuteman ICBMs; and increase air and ground alert of bombers. Kennedy, *Public Papers of the Presidents*, no. 99, 28 March 1961, 230–35.

17. This increment increased the Army from 875,000 to a million, increased the Navy by 29,000, and the Air Force by 63,000, and doubled draft calls and call-ups of reservists. During his first six months in office, Kennedy increased Eisenhower's defense budget by $6 billion total, to $47.5 billion. Reeves, *President Kennedy*, 201. Kennedy boasted at an 11 October 1961 news conference that this $6-billion (14%) growth over Eisenhower's defense budget had increased: the number of Polaris submarines by 50 percent; the number of bombers on 15-minute strategic alert by 50 percent; the production capacity for Minuteman missiles by 100 percent; airlift capacity by 75 percent; antiguerrilla forces by 150 percent; and production of M-14 infantry rifles from 9,000 to 14,000 per month. Kennedy, *Public Papers of the Presidents*, no. 415, 11 October 1961, 658. Kennedy, Johnson, and McNamara, in various settings and throughout the course of Kennedy's administration, would frequently use these figures and others for similar increases in tactical aircraft procurement, active duty Army divisions, aircraft carriers, civil defense, and many other measurements of the vast increases in spending for nuclear and conventional forces. An interesting footnote, however, is that due to the rapid economic growth during Kennedy's administration, defense spending as a percentage of GNP actually declined from 9.1 percent to 8.5 percent. Gaddis, *Strategies of Containment*, 226. As of 2005, America's defense expenditures stood at less than 3 percent of GDP.

18. Booda, "Kennedy Asks $51.6 Billion for Defense," 26.

19. Sorensen, *Kennedy*, 608–9.

20. Futrell, *Ideas, Concepts, Doctrine*, vol. 2, 23.

21. DOS, memcon, 5 September 1961, 163.

22. By late 1960, American reconnaissance satellites were regularly returning imagery from the Soviet Union. Early in Kennedy's administration, high officials from the president down were convinced by this imagery that the so-called missile gap, an important issue in the just-completed election campaign, did not, in fact, imperil America. The only missile gap that did exist was actually in reverse: America's strategic superiority was so vast that the USSR was actually the victim of a missile gap when comparing its strategic capabilities to America's. The best one-volume treatment of the complex history of the missile gap is Bottome, *The Missile Gap*, 1971. At the time of the Cuban missile crisis, two years after the presidential campaign, the United States had over 5,000 deliverable nuclear weapons while the Soviets had approximately 300. Reeves, *President Kennedy*, 375.

23. Reeves, *President Kennedy*, 246.

24. Gilpatric, oral history interview, 30 June 1970, reel 5, 71.

25. Beschloss, *The Crisis Years*, 330.

26. Ibid., 350.

27. Kennedy, *Public Papers of the Presidents*, no.11, 30 January 1961, 23.

28. Ulam, *The Rivals*, 337.

29. Gaddis, *Russia, the Soviet Union, and the United States*, 242.

30. Kennedy, *Public Papers of the Presidents*, no. 515, 20 November 1962, 831.

31. Kennedy, *Public Papers of the Presidents*, no. 12, 14 January 1963, 17–18.

32. Reeves, *President Kennedy*, 507, 514.

33. Kennedy emphasized: "Let us not be blind to our differences, but let us also direct attention to our common interests and to the means by which those differences can be resolved" (ibid.).

34. Ibid.

35. Schlesinger, *A Thousand Days*, 900.

36. Kennedy, *Public Papers of the Presidents*, no. 232, 10 June 1963, 460–63.

37. Beschloss, *The Crisis Years*, 600.

38. Miller, *Statements of John F. Kennedy on Space Exploration*, 1957 section.

39. Kesaris, *Presidential Campaigns*, pt. 2, 1986, reel 11.

40. In a Senate speech on American defense policy, 29 February 1960, Kennedy blamed the Eisenhower administration for this turn of events because the missile and space gap was "but another symptom of our national complacency, our willingness to confuse the facts as they were with what we hoped they would be, . . . our willingness to place fiscal security ahead of national security." Branyan and Larsen, *The Eisenhower Administration*, vol. 2, 1228, 1231. The editors include representative samples of Kennedy's nu-

merous missile-gap speeches in their Eisenhower volumes as examples of Eisenhower's opponents' use of the missile gap as a political issue.

41. Kesaris, *Presidential Campaigns*, pt. 1, 1986, reel 4, 7–8.

42. Ibid.

43. Miller, *Statements of John F. Kennedy on Space Exploration*, 1960 section. In other iterations of this speech Kennedy added that the first country to place its national emblem on the moon was Russia, not America. Sorensen recalls that while Kennedy's opponent, Vice Pres. Richard Nixon, would often highlight how he shook his finger in Khrushchev's face during their "kitchen debate" and proclaimed, "You may be ahead of us in rocket thrust but we are ahead of you in color television," Kennedy responded, "I will take my television in black and white. I want to be ahead in rocket thrust." Sorensen, *Kennedy*, 182–83.

44. Kesaris, *Presidential Campaigns*, pt. 2, 1986, reel 8, 2. In a speech in Elmhurst, IL, 25 October 1960, he espoused a standard theme of most of his campaign speeches, "American prestige, essential to our influence and security, has declined these last eight years even more sharply than we realized. . . . I do not say that the balance of power is determined by a popularity contest. But I do say that our prestige affects our ability to influence these nations, to strengthen the forces of freedom within them, to convince them of which way lies peace and security. . . . If we are to save the peace and rebuild our security, we must remold the symbol of Uncle Sam as the forceful spokesman of a great and generous nation." Kesaris, *Presidential Campaigns*, pt. 2, reel 10, 1, 3. This particular speech lamenting America's loss of prestige was released by the Democratic Party as News Release B-2783. A report from Johnson's staff (in late October 1960, after Kennedy and Johnson were on the same ticket and were therefore no longer overt rivals) concluded, "It is hardly an overestimate to say that space has become for many people the primary symbol of world leadership in all areas of science and technology. . . . Our space program may be considered as a measure of our vitality and ability to compete with a formidable rival, and as a criterion of our ability to maintain technological eminence worthy of emulation by other peoples." Lehrer to Johnson, memorandum, 31 October 1960, 6.

45. Actually, Edward C. Welsh, who would soon be named Kennedy's executive secretary for the National Aeronautics and Space Council, explained, "I was asked and did prepare some materials for speeches and articles on both defense and space for nominee Kennedy." He said he wrote the 10 October 1960 *Missiles and Rockets* piece. Welsh, oral history interview, 20 February 1969, 2, 25. This practice is, of course, not unusual for politicians in general.

46. Kennedy, "If the Soviets Control Space, They Can Control Earth," 12–13.

47. Logsdon, *Decision to Go to the Moon*, 65. "Clash and clamor" is Logsdon citing Diamond, *The Rise and Fall of the Space Age*, 31.

48. Depoe, "Space and the 1960 Presidential Campaign," 227.

49. Beschloss, *The Crisis Years*, 25–27.

50. Dulles to the President, memorandum, 3 August 1960, 1.

51. NASA, *Selected Statements of President Kennedy on Defense Topics*, 201. The latter portion is another example of the kind of Kennedy statements that may have given the Air Force the idea he was amenable to a larger military role in space.

52. Draper, oral history interview, 2 June 1974, 1.

53. Murray and Cox, *Apollo*, 60–61.

54. Logsdon, *Decision to Go to the Moon*, 64.

55. Rostow to Kennedy, memorandum, 7 November 1960.

56. Kistiakowsky, *A Scientist at the White House*, 409.

57. Kistiakowsky, oral history interview, 22 May 1974, 38.

58. Neustadt to Senator Kennedy, memorandum, 20 December 1960, 1. It should be noted that one of the few space-related actions Kennedy did take early on was to have Welsh draft and sign an amendment to the Space Act on 25 April 1961 that made the vice president, instead of the president, chairman of the NASC. Johnson then assumed an important role in the long and difficult task of finding someone willing to serve as NASA administrator in an environment of uncertainty and ambiguity. In April and May, Johnson would spearhead the effort that recommended a lunar landing and return as the best way to beat the Soviets in space.

59. Levine, *Managing NASA in the Apollo Era*, 17.

60. Wiesner et al., "Report to the President-Elect of the Ad Hoc Committee on Space," 10 January 1961, 1–4.

61. Ibid., 4, 6, 14.

62. Ibid., 14–15, 17.

63. Dryden, oral history interview, 26 March 1964, 1.

64. Johnson claimed he had to interview "about 20" candidates before he found one. Webb was willing to take the job in the face of uncertainty over NASA's future and the perceived threat of a possible DOD takeover of the space program. Johnson, *The Vantage Point*, 278. Another source maintains Johnson stated he interviewed 28 individuals. Emme, "Presidents and Space," 39. Whatever the specific number, after perfunctory Senate confirmation hearings, Webb was sworn in on 14 February 1961 and is the key figure in NASA throughout the rest of the period this book covers until his resignation became effective in October 1968.

65. Glennan, *The Birth of NASA*, 93. In further diary entries until he departed Washington on 19 January, Glennan makes clear NASA was still completely in the dark as to Kennedy's plans for NASA specifically or the space program in general.

66. Murray and Cox, *Apollo*, 69.

67. Kennedy, *Public Papers of the Presidents*, no. 8, 25 January 1961, 15.

68. Murray and Cox, *Apollo*, 70.

69. Ibid., 71. For a full biography of Webb, see Lambright, *Powering Apollo*, 1995.

70. Kennedy, *Public Papers of the President, 1961*, no. 25, 8 February 1961, 70.

71. Logsdon, *Decision to Go to the Moon*, 93.

72. Webb, memorandum for record, 24 February 1961, 1.

73. Webb to David Bell, letter, 17 March 1961, 1.

74. Logsdon, oral history interview of Seamans, 5 December 1967, 4.

75. Logsdon, *Decision to Go to the Moon*, 91.

76. Ibid., 97–99.

77. Lambright, *Powering Apollo*, 91.

78. Emme, "Historical Perspectives on Apollo," 378. One team of scholars says Gagarin's flight was a "crushing disappointment to many Americans," that Congress was "stampeded" by the flight, and that the flight "provided a tremendous impetus to the desires of Americans . . . to become first once again." Swenson et al., *This New Ocean*, 334–35.

79. Lambright, *Powering Apollo*, 93. Khrushchev further gloated about Gagarin, "This victory is another triumph of Lenin's idea, confirmation of the correctness of the Marxist-Leninist teaching. . . . This exploit marks a new upsurge of our nation in its onward movement towards communism." Holmes, *America on the Moon*, 84.

80. Kennedy, *Public Papers of the Presidents*, no. 119, 12 April 1961, 262–63.

81. Beschloss, *The Crisis Years*, 114.

82. Representative Anfuso remarked, "I want to see our country mobilized to a wartime basis. . . . I want to see what NASA says it is going to do in 10 years done in 5. I want to see some first coming out of NASA, such as the landing on the moon." Logsdon, *Decision to Go to the Moon*, 103, 105.

83. Ibid., 105.

84. Sidey, *John F. Kennedy, President*, 122–23. See also Logsdon, *Decision to Go to the Moon*, 105.

85. Logsdon, *Decision to Go to the Moon*, 107. The Bay of Pigs fiasco began the next day, 15 April. It remains undetermined exactly what influence this event may or may not have had on Kennedy's lunar-landing decision. Scholars differ in their assessments. While no explicit evidence exists linking it directly to Kennedy's thinking on his response to Gagarin, Lambright's conclusion seems reasonable in that the Bay of Pigs "created an atmosphere at the White House in which the president felt he had to assert leadership right away." Lambright, *Powering Apollo*, 94–95. Logsdon concurs, stating, "The fiasco of the Bay of Pigs reinforced Kennedy's determination, already strong, to approve a program aimed at placing the United States ahead of the Soviet Union in the competition for firsts in space. It was one of the many pressures that converged on the president at that time, and thus its exact influence cannot be isolated." Logsdon, *Decision to Go to the Moon*, 112.

86. Kennedy to Vice President Johnson, memorandum, 20 April 1961, 424.

87. McNamara to Johnson, memorandum, 21 April 1961, 424–25.

88. Kennedy, *Public Papers of the Presidents*, no. 139, 21 April 1961, 310–11. The second portion of the citation is the first and only time the author has been able to discover in which Kennedy stated very explicitly the

concept of beating the Russians to the moon. When Kennedy signed the amendment to the Space Act on 25 April making the vice president the head of the NASC, Kennedy said it was a "key step toward moving the United States into its proper place in the space race. . . . I intend that America's space effort shall provide the leadership, resources, and determination necessary to step up our efforts and prevail on the newest of man's physical frontiers." Kennedy, Statement upon signing HR 6169, 25 April 1961 (ibid., 321–22.).

89. Johnson to Kennedy, memorandum, 28 April 1961, 427–29. Johnson said manned exploration of the moon would be an achievement of not only great propaganda value but may be the one space spectacular that America could accomplish before the USSR. He recommended that if more resources and efforts were quickly put into the American space program, America could conceivably be first in 1966 or 1967 to circumnavigate the moon and perhaps even accomplish a lunar landing. However, at the present time, "We are neither making maximum effort nor achieving results necessary if this country is to reach a position of leadership" (ibid.).

90. Johnson, Opening Statement for the Vice President's Ad Hoc Meeting on Space, 3 May 1961, SPI document 1121, 1; and transcript of the meeting itself, 12. *Exploring the Unknown*, vol. I, reprints the transcript of the meeting, 433–39, but not Johnson's opening statement.

91. Murray and Cox, *Apollo*, 83.

92. Logsdon, *Decision to Go to the Moon*, 121, 123.

93. Logsdon makes an important point concerning the overall process leading to the lunar-landing decision and how the decision was being justified in nonscientific terms: "At no time during the consultations was PSAC as a body asked for its opinion on the choice of a lunar landing as a central feature of an accelerated space program." Logsdon, *Decision to Go to the Moon*, 118. The PSAC's influence, and that of Wiesner as head of the Office of Science and Technology within the White House, did not disappear during the Kennedy administration, but the scientists' input into the space program's direction and overall space policy definitely waned when compared to Killian, Kistiakowsky, and the PSAC under Eisenhower.

94. General Schriever to Johnson, memorandum, 30 April 1961, 1–4.

95. Logsdon, *Decision to Go to the Moon*, 114–115, based on Logsdon's oral history interview of Schriever, 3 November 1967.

96. Oral history interview of NASA associate administrator Robert C. Seamans Jr., by the author, 5 July 1996. See also Logsdon, oral history interview of Seamans, 5 December 1967, 11. See numerous other sources verifying the McNamara general disposition and interplanetary suggestion, most of which are based on participants' interviews. For example, see Mandelbau, "Apollo: How the United States Decided to Go to the Moon," 651; Seamans summarizes his involvement in this decision and other key NASA events in his biography, Seamans, *Aiming at Targets*, chap. 2.

97. Oral history interview of NASA associate administrator Robert C. Seamans Jr., by the author, 5 July 1996. It should be noted that the author repeatedly contacted Mr. McNamara with requests for an oral history inter-

view to explore not only his role in the lunar-landing decision but in all the major issues of this study's remaining chapters. McNamara finally responded by saying, "I would like to help but I do not wish to rely on my memory to discuss events of 30 plus years ago and I do not have time to do the necessary research work." McNamara, note to the author, 15 October 1996.

98. Seamans, *Aiming at Targets*, 89. See also Logsdon, *Decision to Go to the Moon*, 125.

99. Logsdon, oral history interview of Seamans, 5 December 1967, 12.

100. Webb and McNamara to Vice President Johnson, memorandum, 8 May 1961, 16.

101. Ibid. In addition to the lunar-landing proposal, the package also recommended that the United States develop a worldwide operational satellite communications capability; a worldwide satellite weather prediction system; and the large-scale boosters, both solid- (by the DOD) and liquid-fueled (by NASA) because of their potential military use and their obvious necessity in the lunar-landing effort. These large rockets were the DOD's only real nonprestige-related interest in the accelerated program: "It is certain . . . that without the capacity to place large payloads reliably into orbit, our nation will not be able to exploit whatever military potential unfolds in space" (ibid., 16). However, even in the context of this document devoted to laying out a plan for increasing America's prestige via space projects, the authors felt necessary to highlight the crucial role of reconnaissance. They further stated, "The existence of the Iron Curtain creates an asymmetry in military needs between the U.S. and the Soviet Union which compels us to undertake a number of military missions utilizing space technology that would appear to be unneeded by the USSR. We have in the past and are likely in the future to continue to feel the need for reconnaissance. The SAMOS project is intended to fill this need." McNamara and Webb stated that Samos, the Midas program for the "earliest possible warning of ballistic missile attack," and the Discoverer program made for a three-way American investment in reconnaissance satellites exceeding a billion dollars (ibid., 24).

102. Logsdon, *Decision to Go to the Moon*, 126.

103. Kennedy, *Public Papers of the Presidents*, no. 205, 25 May 1961, 396–403.

104. Ibid. Kennedy made it perfectly clear that this would be "a course which will last for many years and carry heavy costs. . . . If we are to go only half way, or reduce our sights in the face of difficulty, in my judgment it would be better not to go at all. . . . I believe we should go to the moon. But I think every citizen of this country as well as the members of Congress should consider the matter carefully in making their judgment . . . because it is a heavy burden" (ibid.).

105. NASA, News Release no. 61-115, 25 May 1961, 5–6.

106. Logsdon, *Decision to Go to the Moon*, 134, 162.

107. McDougall, "Technocracy and Statecraft in the Space Age," 1025.

108. Logsdon, *Decision to Go to the Moon*, 164.

109. Webb to Johnson, letter, 4 May 1961, 1.

110. Logsdon, *Decision to Go to the Moon*, 126, 129.

111. Ibid., 155.

112. Hall, "Instrumented Exploration and Utilization of Space," 190.

113. Hall, "Thirty Years into the Mission," 135.

114. NASA, *Aeronautics and Space Report of the President*, A-30. NASA's budget would peak at $5.25 billion in FY 65 and decline steadily thereafter. The NASA historian explains that this FY 65 figure was 5.3 percent of the federal budget, which would have equaled $65 billion in FY 92's budget, a year in which NASA's actual budget stood at less than $15 billion. Launius, *NASA*, 68.

115. Rosholt, *An Administrative History of NASA*, 245.

116. Hirsch and Trento, *The National Aeronautics and Space Administration*, table, p. 57.

117. Launius, *NASA*, 70.

118. Alex Roland, "The Lonely Race to Mars," 37. Others citing similar figures are Lambright, *Powering Apollo*, 2; and Hall, "Project Apollo in Retrospect," 155.

119. Hirsch and Trento, *The National Aeronautics and Space Administration*, 115.

120. Kraemer, "NASA and the Challenge of Organizing for Exploration," 91.

121. Van Dyke, *Pride and Power*, 27.

122. Ibid.

123. Murray and Cox, *Apollo*, 88.

124. Lambright, *Powering Apollo*, 187.

125. Ibid.

126. Hirsch and Trento, *The National Aeronautics and Space Administration*, 115.

127. House, Committee on Government Operations, H. Rep. 445, 4 June 1965, 74.

128. Ibid. A NASA meteorological satellite, *Nimbus*, had cost overruns of $9 million by mid-1963 in a total contract of $22 million. An orbiting astronomical observatory had cost overruns of $34 million in a $92-million contract.

129. Kennedy, *Public Papers of the Presidents*, no. 477, 18 November 1961, 18.

130. NSC, NSAM 144, 11 April 1962, 1.

131. Webb to Dryden and Seamans, letter, 4 May 1962, 1.

132. Wiesner, oral history interview of Jerome, 24 July 1974, 4. A BOB document from that same month attests to the fact that "the President's desire [is] that the space programs be given an especially critical review in view of the prospective large increases in expenditures." BOB, *Draft Staff Report*, I-1.

133. Kennedy, *Public Papers of the Presidents*, no. 373, 12 September 1962, 669.

134. For background on the Holmes-Webb controversy, see Seamans, *Aiming at Targets*, 103; and Logsdon, "The Evolution of U.S. Space Policy and Plans," 381.

135. Webb, Report for Kennedy, 30 November 1962, *Exploring the Unknown*, vol. 1, 461, 465–66.

136. Logsdon, "Evolution of U.S. Space Policy and Plans," 381.

137. Ibid., 381–82.

138. Eisenhower, "Are We Headed in the Wrong Direction?" 24. Eisenhower reiterated the same points a year later, asking, "But can we best maintain our over-all leadership by launching wildly into crash programs on many fronts?. . . This racing to the moon, unavoidably wasting vast sums and deepening our debt, is the wrong way to go." Later in 1963 Eisenhower declared, "Anybody who would spend $40 billion in a race to the moon for national prestige is nuts." Eisenhower, "Spending into Trouble," 19. See also Loory, "Project Mercury Comes to End," 1. Into the Johnson administration, Eisenhower continued his criticism of Apollo: "This program has been blown up all out of proportion. With hysterical fanfare our space research has been presented as a crash effort, as a 'race to the moon' between the United States and Russia which we must win at all costs. . . . We are breezily assured that the cost and dislocation brought about by this moon race are worthwhile for the new 'prestige' they will bring us" but the only sure return from a lunar voyage is that it "will set a new record for a trip taken on borrowed money." Eisenhower, "Why I Am A Republican," 19. Similarly, Republican congressman H.R. Grosse from Iowa quipped, "It would be my hope that if and when we get to the Moon, we will find a gold mine up there, because we will certainly need it." Hechler, *The Endless Space Frontier*, 124.

139. McDougall, *The Heavens and the Earth*, 391. Rep. Louis C. Wyman elaborated on Republican reasons for supporting reductions in Apollo spending and increasing military space expenditures, "A manned trip to the moon, far from being a crash program, should have a lower priority than assurance of continuing American military control of inner space. . . . If the world is to stay at peace, what it needs and what this country must have, is an American policeman in space. Not a civilian climbing a moon crater with a handful of moon dust. This can come later when we can afford it. Right now we need a manned, armed space vehicle with a hunter-killer capacity." House, Committee on Appropriations, Report no. 824, 88th Cong., 1st sess., 7 October 1963, 20, 22.

140. Bush to Webb, letter, 11 April 1963, 2–3. Other scientists criticized NASA's emphasis on human spaceflight. Dr. Philip Abelson, editor of *Science* magazine said this overemphasis "is having and will have direct and indirect damaging effects on almost every area of science and technology and . . . may delay the conquest of cancer and mental illness." Levy, "Conflict in the Race for Space," 205.

141. Kennedy to Johnson, memorandum, 9 April 1963, 467–68.

142. Packard to the executive secretary, NASC, memorandum, 9 April 1963, 2.

143. Webb to Johnson, letter, 3 May 1963, 1, 3.

144. "Johnson Doesn't Want Red First on Moon," *Washington Sunday Star,* 12 May 1963, 1.

145. Johnson to Kennedy, memorandum, 13 May 1963, 468–73.

146. Kennedy, *Public Papers of the Presidents,* no. 194, 18 May 1963, 412.

147. Seamans, oral history interview, 27 March 1964, 43–45.

148. Sorensen, *Kennedy,* 527. Another scholar concurs, "In the end, the debate of 1963 was clearly won by the advocates of the manned lunar landing." Levine, *The Future of the U.S. Space Program,* 89.

149. Rosholt, *An Administrative History of NASA,* 282.

150. Coughlin, "The Wall of Space," 48.

151. Kennedy, *Public Papers of the Presidents,* no. 478, 22 November 1963, 897. The day before he was murdered, in one of his final speeches, Kennedy declared that when the Saturn was launched the next year, it would be "for the first time, the largest booster in the world, carrying into space the largest payload that any country in the world has ever sent into space. I think the United States should be a leader. A country as rich and powerful as this which bears so many burdens and responsibilities, which has so many opportunities, should be second to none. . . . This nation has tossed its cap over the wall of space, and we have no choice but to follow it. Whatever the difficulties, they will be overcome. Whatever the hazards, they must be guarded against." Kennedy, *Public Papers of the President,* no. 472, 21 November 1963, 883.

152. John Logsdon and Alain Dupas offer a cogent and succinct examination of the Soviet lunar-landing program. In it they explain that in fact the Soviet government did not give preliminary approval to a Soviet lunar-landing plan until December 1964 and a final go-ahead did not come until November 1966. Logsdon and Dupas, "Was the Race to the Moon Real?" 20. For a comprehensive treatment of the Soviet space program and their perspective on a space and lunar-landing race during this period, see Siddiqi, *Challenge to Apollo,* 2000.

153. CIA, NIE 11-1-62, 1–3, 23–24.

154. DOS intelligence information (it is unknown if it reached the presidential level or not) would have reinforced the idea that the Soviets were engaged in an active program to reach the moon. One of State's intelligence reports stated there was an American student in the USSR who was friendly with a Soviet citizen and physicist who was working on the Soviet lunar project. The physicist had told the student that "plans for the Soviet project are well advanced and that a launching should take place 'soon.'" The DOS report stated this information could not be disseminated outside the US government due the risk of identifying its sources. See DOS, *Airgram,* 22 March 1963, 1.

155. Harvey, "Preeminence in Space," 977.

156. "Is U.S. Running Alone in the Race to the Moon? Interview with Sir Bernard Lovell," *US News and World Report,* 12 August 1963, 70–71.

157. Kennedy, *Public Papers of the Presidents*, no. 305, 17 July 1963, 568. Given the fact that, according to Logsdon and Dupas, and Siddiqi, the Soviet government did not give preliminary approval to its lunar-landing program until December 1964, two months after Khrushchev was deposed, (see above); Khrushchev's statements in October 1963 were technically correct.

158. Welsh, "Khrushchev, Address to Third World Meeting of Journalists," report attachment, 25 October 1963, 3.

159. Welsh, "Premier Khrushchev's Statement re Moon Project," report, 29 October 1963, 1.

160. DOS, Khrushchev's Obscure and Noncommittal Statements About Moon Shots, note, 5 November 1963, 1.

161. Senate, *Khrushchev's Statement to American Businessmen*, staff report, 7 November 1963, 1.

162. Recent scholarly evaluations of the Soviet space program tend to agree with the conclusion that the Soviets did, in fact, pursue a serious lunar-landing program at least throughout the 1960s. McDougall states, "But there is enough technical evidence . . . to suggest that the Soviets were in the race for the moon despite their disclaimers after the fact." McDougall, *Heavens and the Earth*, 289. In 1989 three MIT faculty members visiting the USSR stumbled upon an actual Soviet lunar lander at a technical institute in Moscow. In addition, Soviet engineers told them the Soviet spacecraft was ready to go to the moon in 1968. The *New York Times* concluded, "After years of denial by silence and misinformation, the Soviet Union has now disclosed that in the 1960s it was indeed racing the United States to be first to send men to the moon. . . . The Soviets disclosed that repeated failures of a booster rocket delayed the program and eventually caused its cancellation in the early 1970s. . . . American authorities on Soviet space activities said the disclosures were the most definitive evidence yet that there had been a 'Moon race.'" Seamans, *Aiming at Targets*, 323–24. Logsdon describes recently declassified information as well as testimony from 1960s-era Soviet space officials that proves "the moon race was indeed real." He states that photographs and engineering descriptions of Soviet lunar hardware mean scholars have "a much clearer picture of just how extensive the Soviet lunar program was." Logsdon and Dupas, "Was the Race to the Moon Real?" 16, 18. Again, the most comprehensive treatment of these Soviet space program questions is Siddiqi, *Challenge to Apollo*.

163. Kennedy, *Public Papers of the Presidents*, no. 448, 31 October 1963, 832.

164. Kennedy, *Public Papers of the Presidents*, no. 1, 20 January 1961, 2. Several days later in his State of the Union address he said he intended "to explore promptly all possible areas of cooperation with the Soviet Union" to include weather satellites, communication satellites, and probes to Mars and Venus because "both nations would help themselves as well as other nations by removing these endeavors from the bitter and wasteful competition of the Cold War." Kennedy, State of the Union Address, 30 January 1961; and ibid., 26–27.

251

165. Launius, *History of the U.S. Civil Space Program*, 58.

166. DOS, memcon, 3 June 1961, Vienna, reel 24, 1.

167. DOS, memcon, 4 June 1961, Vienna, 1–2.

168. Kennedy, *Public Papers of the Presidents*, no. 387, 25 September 1961, 622.

169. Frutkin, *International Cooperation in Space*, 144.

170. For the full text, see DOS, *Documents on Disarmament, 1963*, 538.

171. Portree, *Thirty Years Together*, 1.

172. Kennedy to Khrushchev, letter, 22 February 1962, 1.

173. NSC, NSAM 129, 1.

174. Bundy to Webb, memorandum, 23 February 1962, 1.

175. For a good succinct account of the content of each meeting and the resulting agreements/memoranda of understanding, see Malloy, "The Dryden-Blagonravov Era of Space Cooperation," 40–45.

176. Kohler, "An Overview of US-Soviet Space Relations," xxiv.

177. Khrushchev, *Khrushchev Remembers*, 54.

178. Senate, *Soviet Space Programs: 1962–1965*, report, 89th Cong., 2nd sess., December 1966, 62–63.

179. Frutkin, *International Cooperation in Space*, 100–101.

180. Kennedy to Rep. Albert Thomas, letter, 23 September 1963, 950. McDougall's quip is appropriate: "The Dryden-Blagonravov negotiations have been described more often than their results warrant." McDougall, *Heavens and the Earth*, 516, n. 28.

181. For text and details, see DOS, *Documents on Disarmament, 1963*, 291–93. The treaty was initialed by US, Soviet, and British representatives on 5 August 1962. The US Senate ratified the treaty by a vote of 80:19 on 24 September 1963. Smaller testimonials to a growing Soviet-American thawing in 1963 were the "Hot Line" agreement of 20 June 1963 establishing a direct communications link between Moscow and Washington, the commencement of the sale of $250 million of surplus American wheat to the USSR, the initiation of negotiations to begin direct air service between New York and Moscow, and the opening of new consulates in both countries. See any one of a number of sources for these developments such as Firestone, *The Quest for Nuclear Stability*, 38; or Gaddis, *Russia, the Soviet Union, and the United States*, 244.

182. Sorensen, *Kennedy*, 517.

183. Kennedy, *Public Papers of the Presidents*, no. 366, 20 September 1963, 693, 695.

184. Kennedy, *Public Papers of the Presidents*, no. 305, 17 July 1963, 567–68. At any rate, the United States never received any kind of Soviet reply to Kennedy's September 1963 joint lunar-landing offer.

185. NSC, NSAM 271, 12 November 1963, 1. Webb's final response to NSAM 271 of course had to be delivered to Johnson. On 28 January 1964 he wrote Johnson to suggest four potential areas of American-Soviet cooperation. These were projects for the determination of: micrometeoroid density in space between the earth and moon; radiation and energetic particle environment be-

tween the earth and moon; character of the lunar surface; and selection of lunar-landing sites. Webb to Johnson, letter, 28 January 1964, 1.

186. I am indebted to Prof. John Logsdon of the Space Policy Institute of George Washington University for his thoughts concerning Kennedy's posture toward space cooperation with the USSR contained in this paragraph and the next.

187. Kennedy, *Public Papers of the President,* no. 269, 26 June 1963, 524–25. Addressing NATO headquarters a week later he stated, "Communism has sometimes succeeded as a scavenger but never as a leader. It has never come to power in any country that was not disrupted by war, internal repression or both. . . . They [Communists] cannot look with confidence on a world of diversity and free choice, where order replaces chaos and progress drives out poverty." Kennedy said the growing strains within the Communist bloc "make it increasingly clear that this system, with all its repression of men and nations, is outmoded and doomed to failure." Kennedy, *Public Papers of the Presidents,* no. 291, 2 July 1963, 551.

188. Firestone, *The Quest for Nuclear Stability,* 60–61.

189. Beukel, *American Perceptions of the Soviet Union as a Nuclear Adversary,* 37.

190. Kennedy, *Public Papers of the Presidents,* no. 426, 19 October 1963, 996–97.

191. Loory, "House Rebuffs Kennedy's U.S.-Red Moon Trip in Limiting Space Funds," 1.

192. House, Committee on Appropriations, Report no. 824, 88th Cong., 1st sess., 7 October 1963, 22.

193. Lambright, *Powering Apollo,* 121.

194. Logsdon, "The Apollo Decision in Historical Perspective," 4–5.

195. Brooks, "Motivations for the Space Program," 10.

196. Hays, *Struggling Towards Space Doctrine,* 173.

NASA Photo

PHOTO SECTION

Apollo 9 Command/Service Modules photographed from Lunar Module

Conceptual renderings of Dynasoar (Reproduced by permission from danroam@ deepcold.com)

Gemini 6 views Gemini 7

Mercury 8 in Hangar S, Cape Canaveral, Florida

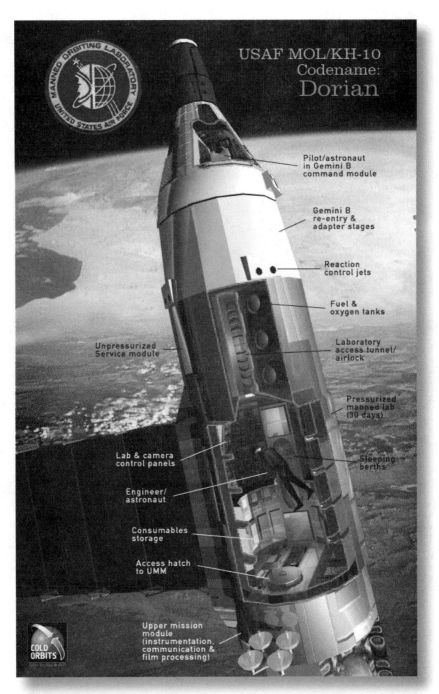

USAF MOL/KH-10
Codename:
Dorian

Pilot/astronaut
in Gemini B
command module

Gemini B
re-entry &
adapter stages

Reaction
control jets

Fuel &
oxygen tanks

Unpressurized
Service module

Laboratory
access tunnel/
airlock

Pressurized
manned lab
(30 days)

Lab & camera
control panels

Sleeping
berths

Engineer/
astronaut

Consumables
storage

Access hatch
to UMM

Upper mission
module
(instrumentation,
communication &
film processing)

Conceptual renderings of Dynasoar (Reproduced by permission from danroam@
deepcold.com)

NASA Photo

Skylab and Earth Limb

Chapter 6

NASA, DOD, McNamara, and the Air Force under Kennedy

An account of the NASA-DOD relationship during the Kennedy administration must begin by examining the election campaign and transitional period in which many believe the Air Force was waging a "campaign" of its own to secure more responsibility in the space program, almost certainly at the expense of NASA.

This chapter examines that hypothesis and discusses the climate within the DOD itself that was so important to how it related to NASA. Finally, the chapter closes with a look at the specifics of support, coordination, and rivalry that continued to characterize the NASA-DOD relationship from 1961–63.

Air Force "Campaign" to Usurp NASA Responsibility

The idea that the Air Force was waging a concerted effort to somehow usurp NASA came from three sources. First, an Air Force–formed committee under Trevor Gardener issued a report containing language that fueled such speculation. Second, speeches and briefings delivered by certain Air Force officers added to media conjecture. Third, internal Air Force documents were open to the interpretation that something was amiss with the NASA-DOD relationship. Together these supposedly comprised an intra-USAF and public relations offensive designed to convince the incoming Kennedy administration and the Congress that the Air Force had the institutional capabilities required for, and therefore should be awarded, a greater role in space.

The Gardner Committee

General Schriever established the Gardner study committee on 11 October 1960. Gardner had been an energetic and dynamic special assistant for R&D to the Air Force from February 1953 to February 1955, during which time he was the driving force behind the Air Force's successful push to accelerate the ICBM effort. He then served as the USAF's first assistant secretary for R&D but resigned in February 1956 because he felt the Eisenhower administration was not devoting adequate resources to the ICBM crash effort.[1] He maintained close contacts with the service's space and missile community, and Schriever asked him in October 1960 "to review current ARDC space development objectives and resources and to recommend a program which would enable the Air Force to effectively meet its development responsibilities in space in the 1960–1970 time period."[2] Schriever wrote outgoing NASA administrator Glennan to reassure him that the Air Force had no designs on taking over NASA, that the report was designed for internal Air Force planning uses (he provided Glennan with a copy of his tasking letter to Gardner, cited above), and emphasized that if the committee made any recommendations in areas "clearly scientific or commercial in nature" such recommendations would be highlighted in the report so they could be dealt with by the appropriate agency.[3] While the committee did not release its report until late March 1961, it was generally known that the Air Force was making some kind of internal assessment as to its future space plans, and many assumed NASA was thereby somehow threatened.

As it turned out, the Gardner committee's report—when released in March—had no great impact because DDR&E Brown "just gave it short shrift. Nothing ever came of it."[4] The 64-page report did not directly challenge either NASA's institutional existence or its specific missions in space. In fact, NASA was scarcely mentioned. The report did lament, "Our insistence on classifying space activities as either 'military' or 'peaceful' has exposed us to unnecessary international political problems. . . . The Air Force must improve its organization and procedures so that its actions in this new field of endeavor will reflect a full understanding of these complex facts, particularly as they relate to other agencies and governments. . . . National security considerations alone justify a major increase in the DOD space effort. . . . Unless we meet the Soviet challenge with a dramatically invigorated space program, our international prestige will be further damaged."[5] Such nebulous statements—neither indicting NASA directly nor stating the Air Force should take it over, yet ignoring NASA while criticizing the current state of affairs—continued throughout the report: "While the role that the Air Force is to play in the U.S. exploration of space is not yet determined, both past experience and existing resources indicate that this role should be a major one, and should be established in the near future. . . . The Air Force should take the lead in improving our international position resulting from space actions, plans and events."[6] The Air Force should have foreseen the public relations danger inherent in statements such as this one:

> The challenge of the unknown and of the unoccupied will make manned space exploration inevitable—first in orbit, then of the moon and afterwards of the planets. The Department of Defense, through the Air Force, should prepare to play a major role in this difficult exploration. The Air Force should urgently develop the fundamental capability to place and sustain man in orbit. . . . It is essential that the Air Force play a major support role in manned exploration of the moon and planets.[7]

Speeches and Briefings

A Schriever speech of 21 November 1960 represents the oratorical component of the perceived campaign. Schriever began with the standard Air Force line, "For the first time in the history of our Nation, we are open to a destructive nuclear surprise attack." As part of deterring such an attack, "The importance of satellites and other space systems as essential elements of our military strength is not fully appreciated." However, Schriever then specifically *denied* that the United States should have a single, unified space program and pointedly called for close cooperation between the civilian and military space programs so that facilities could be used to their fullest. He elaborated that there was a clear divergence between the DOD's and NASA's space roles: the DOD's was to exploit space for the security and survival of the United States, while NASA's was to investigate space for scientific and other peaceful purposes. Therefore, the two organizations would require

265

different types and numbers of space systems and vehicles. In the next decade, DOD would require larger numbers of vehicles for its defensive missions than NASA would for its scientific exploration. In addition the military vehicles would require longer life and higher reliability, simple operational and maintenance procedures, and have the ability to be quickly launched.[8]

Schriever set off alarm bells by emphasizing the Air Force's current capabilities in space. He explained, "Within the USAF there exists a great array of facilities capable of projecting the Air Force into the aerospace age." The Air Force's BMD "constitutes the greatest single collection of space age managers in the free world." Together with the USAF's rocket-testing laboratories and launch facilities, its tracking stations and satellite-test centers, its scientific laboratories and its bioastronautics laboratories, the Air Force's facilities were a valuable national asset: "I haven't mentioned all of the Air Force facilities for space nor even all of those which we have in ARDC oriented toward that vast arena. . . . The Air Force has the resources for the space age."[9] In no way did Schriever directly compare the USAF's capabilities to NASA's. Conversely, he did not mention NASA's contributions to America's space infrastructure.

Internal Documents

The internal document universally pointed to as evidence of an Air Force campaign was the *Air Force Information Policy Letter for Commanders* for December 1960. This publication was regularly used by the Air Force to explain current policy on particular issues and give commanders guidance in establishing local policies, composing speeches, and so forth. The four-page, December 1960, edition was subtitled "Air Force Competency in Space Operations" and concluded both Nixon and Kennedy had displayed "a realization at the highest levels of our Government that military supremacy in space is as essential to our security as military supremacy at altitudes near Earth." The pamphlet continued,

> During the past 20 years the broadest base for current US programs in aerospace has been largely developed by the Air Force—plus the aerospace industry, research institutions, and government agencies, such as NACA and NASA, which have helped make the Air Force the

world's leading aerospace arm. The know-how and facilities that have sprung from this military effort are a national resource of immeasurable value not only to Free World survival, but to scientific and technological advances for the welfare of mankind. . . . From its start, NASA's Project Mercury has been nourished by Air Force aerospace medical skills and people.[10]

The newsletter went on to describe in detail the assorted Air Force space vehicles, launchers, facilities, and installations giving the Air Force the "unparalleled competency to assume an even more stronger [sic] supporting role in gaining and maintaining general aerospace supremacy for our nation."[11] Again, the letter did not have an overtly imperialistic or hegemonic quality to it. On the other hand, like Schriever's speech, it did not specifically discuss NASA's facilities and capabilities, nor did it give a perspective of what the Air Force was doing in space and what NASA did in space; NASA was essentially ignored. This *Policy Letter* reflected, in a general sense, the briefings (such as Schriever's discussed above) that the Air Force was giving to congressmen and representatives of the aerospace industry.

The Press Weighed In

The Schriever-type speeches and the policy-letter-type documents were enough to send the press into a frenzy of speculation long before the March release of the Gardner report. *Aviation Week* confidently declared four days after the *Policy Letter*'s release:

> The Air Force is preparing a major political offensive to bring about changes in national space policy and law that would let it proceed with detailed, specific plans for space weapons involving 'tens of thousands, perhaps hundreds of thousands' of satellites in a continuing series of technical briefings to industry leaders and groups within the service. . . . Not only does [the] Air Force expect to invade a province of NASA by proposing manned space vehicles and large booster development, but it intends to enter the communications satellite area, now monitored by the Army.[12]

The venerable *New York Times* picked up on this and added:

> The Air Force has drafted a publicity offensive to stake out a major role for itself in the nation's space program. The offensive is clearly keyed to the change in administrations. It is the openly expressed belief of the Air Force that the Kennedy administration will look more favorably

upon military operations in space than does the Eisenhower administration. . . . The [*Air Force Information Policy*] letter serves to point up the probability of a major battle between the military services and between the Defense Department and the National Aeronautics and Space Administration over which agency should play the major role in the space program.[13]

The *Policy Letter*, in fact, said absolutely nothing about the "probability" of any kind of institutional battle. John Finney wrote this *Times* article and covered aerospace affairs for that newspaper. He later told NASA associate administrator Seamans why he rarely wrote anything noncontroversial or even positive about either NASA or the Air Force space programs: "OK, I write a good article and if I'm lucky it will be on page 33. If I write something controversial, I have a chance of getting it on page 1. It's as simple as that. I'm paid by what page I get my articles on."[14]

One cannot completely dismiss assertions of an Air Force campaign as media fabrications. There was, at a minimum, concern within NASA's congressional patrons—the House and Senate space committees—that something was afoot. Kenneth BeLieu was staff director for the Senate Committee on Aeronautical and Space Sciences, chaired by Vice President-elect Johnson, and wrote Johnson in December 1960, "The Air Force can be expected—and apparently already has started—to make a basic power play to grab the entire Space program. This would involve eliminating NASA. . . . The Air Force would have the entire aerospace industry behind it."[15] Outgoing DDR&E York was evidently concerned enough to inquire from the SAF what was going on with these speeches, briefings to industry, and so forth. The acting SAF replied, "The philosophy underlying the briefings was to present systems, concepts, and studies to industry. Many of the topics were not presented as approved programs or as authorized." The purpose of the Air Force presentations to the aerospace industry was "to give industry the benefit of AFBMD [Air Force Ballistic Missile Division] thinking about possible courses of missile and space research and development in the future. . . . It is Air Force policy to give this type of briefing to industry rather than have industry attempt to predict future Air Force research and development efforts on incomplete and fragmentary information." Furthermore, "It has become evident that the Air Force program of study requirements is especially susceptible to misunderstanding [because] enthusiasm on the part of industry and the press frequently describes these requirements out of context. The Air Force is determined to minimize such occurrences."[16]

Resolution

By February 1961, the Air Force had suffered enough adverse publicity from these accusations of waging a campaign and was attracting enough unwanted attention from high-level civilian leaders that it stated in another *Policy Letter for Commanders*:

> From NASA's beginning, in 1958, Air Force–NASA cooperation has been close and mutually beneficial. . . . The Air Force agrees with NASA that there should be a clear realization, both in this country and throughout the world, that the United States has a single space exploration program administered by NASA; and that activities in the space environment related to national defense devolve from the responsibilities of the Defense Department for the defense of the Nation, and clearly must be managed by the Department of Defense. Air Force activity in space projects is devoted solely to the latter. . . . Although each agency has a different sphere of responsibility, both NASA and the Air Force contribute to each other's program.[17]

The flap over an Air Force campaign steadily died down from this point forward, not only due to Air Force pledges of cooperation with NASA, but also because it became absolutely clear that Congress would not tolerate any significant alteration in the division of space responsibilities. Seamans recalls voicing his concerns over NASA's general situation, and especially the possibility that the Air Force might take over part of NASA's programs, to NASA general counsel Johnny Johnson early in 1961. Seamans asked, "Do you suppose they're even thinking of absorbing NASA back into the Department of Defense?" Johnson's perceptive reply was, "There is no chance. The political situation would never permit. . . . They may be thinking about it, but if they should try, they won't get away with it."[18]

By mid-February Webb was also onboard as NASA administrator and could not only boost NASA morale but could also exercise his political skills on NASA's behalf against any Air

Force initiatives, perceived or otherwise. Webb later explained that the Air Force never had any chance of expanding its space responsibilities into, for instance, the lunar-exploration area because of powerful members of the space committees such as Brooks in the House and Robert Kerr in the Senate. Webb said the Air Force "could fuss for it [more space responsibility]. They could get the newspapers saying they ought to have it. But the power structure was not oriented so that they could prevail. I was perfectly happy for them to float around, and make the noise, and make the bids. I knew where the power was, and where the votes lay, so I wasn't bothered by that. . . . And, you notice I never bothered to answer. . . . But I was still in very close touch with the people who held the balance of power" in Congress.[19] On the very day he was nominated, CSAF White wrote Webb, "Contrary to some published reports, the Air Force and NASA have enjoyed a very close and cooperative relationship. We in the Air Force will do our utmost to maintain this cooperative spirit."[20] Additional documentary evidence bears out Webb's assessment of the fundamental political situation.

The key figure in the rest of the correspondence—related to assessing the Air Force "campaign"—was Representative Brooks, chairman of the House Committee on Science and Astronautics (hereafter referred to as the House space committee). He zealously guarded NASA's responsibilities against any encroachment. When CSAF White, heard Brooks was concerned about supposed Air Force moves to gain increased responsibility at NASA's expense, he wrote Brooks, "I can assure you that any action or statements by any Air Force individual or groups which tend to create such impressions are in direct contradiction to the established beliefs and policies of the Air Force." General White stressed that "the excellent spirit of teamwork that characterizes the cooperation between that agency and the Air Force . . . has not changed and, in fact, our close cooperation with NASA at both the policy and working levels has never been stronger than it is today." White closed by asking Brooks for any help Brooks could offer in "specifically identifying the 'pressure groups within the USAF' to which you refer and the specific actions taken by these groups toward 'degrading the position of NASA.'"[21] Outgoing NASA administrator Glennan also attempted to persuade Brooks there was no grand conspiracy taking place to undermine NASA. He wrote Brooks shortly after White did in January 1961 to emphasize, "Surely, in the early days of NASA, much strain and competition was in evidence. However, with the great assistance of Secretary Gates [Defense], Secretary Douglas [Air Force], and General White, I believe there has been eliminated from the scene the sort of competition which is destructive in nature. Arguments there will be, but these are now conducted with good will on both sides."[22]

Brooks was not mollified by either Glennan's or White's reassurances. He wrote Glennan in mid-February that he was happy with the private assurance of Glennan and White that all was well but, "I have been waiting with great interest for a public renunciation of these charges. Although both you and General White have given me private assurances in this matter, neither the U.S. Air Force nor NASA has specifically attempted to set the public record straight. . . . I am concerned about the 'end runs' which tend to circumvent the spirit of the agreements which constitute the foundation for the operation of the AACB."[23] Top Air Force leadership appeared to be at a loss at this point as to how to placate Brooks. One Air Force internal memo could only conclude, "There is no 'power struggle' afoot! . . . It is possible that someone is giving Mr. Brooks some faulty advice which has the net effect of keeping alive an erroneous public impression of NASA–Air Force waste, duplication, and unhealthy competition. It is interesting to note that both NASA and Air Force informally have agreed that not only is there no waste or duplication, but that the national interest demands the application of further resources to the U.S. national space effort."[24] Chief of Staff White could only resolve to meet with new Administrator Webb, along with NASA deputy administrator Dryden and Seamans, to try to "determine how we may, with finality, lay the ghost of this alleged NASA–Air Force dissension and duplication to rest. . . . The first order of business is to get Congressman Brooks on our side."[25]

271

Apparently Brooks was in such a state that only direct word from the president would reassure him that NASA was not imperiled. Therefore, he wrote Kennedy on 9 March 1961:

> I am seriously disturbed by the persistence and strength of implications reaching me to the effect that a radical change in our national space policy is contemplated within some areas of the executive branch. In essence, it is implied that United States policy should be revised to accentuate the military uses of space at the expense of the civilian and peaceful uses. . . . The voluminous rash of such reports appearing in the press, and particularly in the military and trade journals, is, it seems to me, indicative that more than mere rumor is involved.[26]

Kennedy's reply made his position crystal clear:

> It is not now, nor has it ever been, my intention to subordinate the activities in space of the National Aeronautics and Space Administration to those of the Department of Defense. I believe, as you do, that there are legitimate missions in space for which the military services should assume responsibility, but that there are major missions, such as the . . . application of space technology to the conduct of peaceful activities, which should be carried forward by our civilian agency.[27]

If the speculation of an Air Force takeover had begun to dwindle during February with the Air Force's official denial, then in March and April it quickly faded after Kennedy's letter. After his lunar-landing speech on 25 May 1961 announcing that NASA would be the agency primarily responsible for this ambitious goal, the whole question virtually disappeared. As one participant in the lunar-landing decision deliberations recalled when asked if there were ever any discussion of *not* having NASA manage the effort, "There was never the slightest."[28]

It appears Kennedy only addressed the NASA-DOD balance-of-power issue one other time during the rest of his presidency. In a 14 June 1962 press conference, he was asked if there were any plans for a major realignment of the American space program to give the military a bigger role (Finney in the *New York Times* had just published another series of articles speculating this was the case). Kennedy responded, "The military have an important and significant role, though the primary responsibility is held by NASA and is primarily peace, and I think that the proportion of that mix should continue."[29] Kennedy seemed quite convinced, even before his lunar-landing decision, that the fundamental NASA-DOD division of effort was appropriate. After giving NASA responsibility for Project Apollo, the central element of Kennedy's expanded space effort, there was even less of a chance he or the congressional space committees would permit any fundamental alteration in the managerial responsibilities of the American space program.

Several conclusions emerge from this sequence of events associated with charges of an Air Force campaign for a heightened space role. First, if one accepts the evidence offered by proponents of such an offensive, it must naturally follow that the campaign was waged *not* by the highest civilian or uniformed levels of the Air Force but by the cadre of space enthusiasts headed by Schriever. The correspondence between White and Glennan clearly indicates they had an amiable personal relationship and were puzzled as to why there was such concern over the NASA-DOD relationship.

Second, the evidence offered as supporting the campaign notion is itself open to divergent interpretations; it contains no overt references to taking over NASA as an institution or any of its programs. To extract subtleties of bureaucratic hegemony from the speeches and *Air Force Information Policy Letters* requires reading between the lines and imputing what the Air Force meant, if anything, by *not* discussing NASA as part of America's space program.

Third, at most, any such campaign did not go beyond speeches to civic groups and briefings to the aerospace industry. The worst reasonable interpretation of Air Force actions is that they comprised a clumsy, ill-timed, and poorly executed public relations effort. Any Air Force attempt to take substantial action encroaching on NASA's territory would have been firmly resisted by a coalition of NASA's dynamic new leader Webb, the congressional space committees, Kennedy, and possibly new SECDEF McNamara, who may have viewed a healthy and powerful NASA as a way to check the power of an overly ambitious Air Force.[30] As will be seen below, Webb quickly formed agreements with McNamara on assorted issues within the NASA-DOD relationship, effectively countering any Air Force moves.

Fourth, even if one assumes a powerful Air Force move to take over NASA, once Kennedy awarded NASA the lunar-landing mission, any such campaign had no chance of success and quickly would have died. As Logsdon concludes, "It is unlikely that the Kennedy administration could have, or would have, agreed to the Air Force demands for a larger space role at the expense of NASA."[31]

The discussion of the Kennedy-era NASA-DOD relationship may now move from the realm of speculation and conjecture into areas in which more concrete historical evidence is available. The central preliminary question is in fact quite complex. By 1961 what did the Air Force as an institution believe about its role in space? Did the civilian leadership such as new SAF Eugene Zuckert concur? What kind of reception did the Air Force position receive in the OSD now headed by McNamara? Finally, how do the answers to these questions play into the NASA-DOD relationship? Such is the task for the remainder of this chapter.

DOD Directive 5160.32

The Air Force believed it detected a hopeful sign concerning McNamara's stance on military space when he issued DOD Directive 5160.32, *Development of Space Systems*, on 6 March 1961. With this decree, McNamara consolidated the USAF's role in the military space realm. He declared that while each military service could conduct undefined "preliminary research to develop new ways of using space technology," all space technology proposals beyond "preliminary research" had to be submitted to the DDR&E for consideration and eventual SECDEF approval. Then, "Research, development, test, and engineering of Department of Defense space development programs or projects, which are approved hereafter, will be the responsibility of the Department of the Air Force." Only the SECDEF could make exceptions to the assigning of space developmental responsibilities to the Air Force and only then "in unusual circumstances."[32] In effect, this directive "made the Air Force the DOD executive agent for all space development programs, regardless of service of ultimate use. It enabled the Air Force to determine the shape of space developments to best suit its own requirements."[33] In addition, it effectively ended the interservice competition for space once and for all; the only programs remaining outside the Air Force were the Navy's *Transit* navigation satellite and the Army's *Advent* communication satellite.[34]

The origins of this directive consolidating Air Force control of DOD space projects apparently were found in a review McNamara ordered of the military space program after the Wiesner Report called it "fractionated." He assigned the review to his new Office of Organization and Management Planning Studies, which quickly discovered the Air Force was already responsible for more than 90 percent of the DOD's space R&D and, in the remaining 10 percent, it still provided the boosters and launch facilities.[35] Schriever recalled that at that point Deputy SECDEF Gilpatric approached him and said, "Look, General, you straighten out the situation between ARDC and AMC [Air Materiel Command] and we'll direct that the space research and development and space activity within the Department of Defense be assigned to the Air Force."[36] Schriever assigned the task of creating a new intra–Air Force organizational structure for space activities to a small working group that prepared, coordinated, and obtained the necessary approval to create a new Air Force Systems Command (ultimately commanded by Schriever) that would combine the functions and organizations of the old ARDC and AMC. The new systems command had responsibility for the R&D, design, testing, procurement, and delivery to the operational commands of every new weapons system in the Air Force to include space systems.[37] This new organization and Schriever as its commander were the key players in the Air Force and military space program for the remainder of the Kennedy and Johnson administrations, although Schriever retired in 1966.

While the Air Force chose to emphasize the aspects of the directive that centralized its control of the military space R&D realm, the more important clauses were the ones granting the DDR&E and the SECDEF final-approving authority for all space projects. McNamara's key watchdog within the DOD for waste and duplication emphasized that the real reason for the directive was to prevent interservice conflicts and to centralize

OSD control over Air Force space proposals "by further restructuring the independent freedom of action of the three military services . . . [and] by limiting the latitude of the military departments to increase emphasis and funding for various projects."[38] McNamara would use the DDR&E as a strong staff arm to exercise firm control over all Air Force space proposals during his tenure. The Air Force could do very little developmental work on any space system without explicit DDR&E approval. The best assessment of DOD Directive 5160.32 was provided by Zuckert, who said that in fact it "was solely jurisdictional. It just gave us jurisdiction in the space field. There was the question of how much support we would get. . . . It was like getting a franchise to run a bus line in the Sahara desert."[39] Zuckert's point, borne out by future events, was that just as a franchise to traverse the Sahara is basically meaningless due to the lack of traffic demand, so would be the responsibility for space R&D if McNamara and his DDR&E, Harold Brown, refused to sanction such work. The Air Force actually had very little independence as a result of the directive; the OSD through the DDR&E would exercise tight control over USAF activities for the rest of the decade and beyond.

The fundamental clash that developed turned out not to be between NASA and the Air Force or between NASA and the DOD but in fact between the Air Force and the OSD. McNamara, Brown, and Brown's deputy Rubel repeatedly quashed the Air Force's grand plans for putting humans in space and extending the American deterrent shield into orbit. The key link to the NASA-DOD relationship is that not only did the OSD insist that Air Force space proposals offer a definite and identifiable increase in American security, but they also must not duplicate in any way NASA's work in space R&D. If NASA were working on a project that could possibly fulfill the requirements of a system the Air Force were proposing, then the OSD would almost certainly reject the Air Force proposal and order a cooperative venture with NASA. Therefore, an accurate portrayal of the NASA-DOD institutional relationship should be characterized more precisely as the NASA-OSD-USAF relationship. It is necessary first to examine what the space community within the Air Force felt was the proper role for the military in space. Next, what did McNamara and the OSD conclude concerning the military in space? Finally, what sorts of tensions arose due to clashing interpretations and how did this begin to feed into the NASA-OSD-USAF relationship?

The Air Force's Space
Philosophy and Space Plan

High-ranking Air Force officers often emphasized the important role they perceived the new domain of space playing in national security. When he retired on 30 June 1961, CSAF White remarked, "I make this prediction, in the future the people who control space will control the world."[40] The most enthusiastic space officers, such as Schriever, continued to believe that an artificial distinction between "peaceful" and "military" activities in space inhibited the Air Force's ability to operate in the space medium. Schriever's complaint was not with NASA as an institution but rather the space-for-peace policy and philosophy behind the creation of NASA. He told a Senate committee in July 1961, when asked if the military space programs were being adequately and properly supported, "No sir, I think we have been inhibited in the space business through the 'space for peace' slogan. I think that there has been too arbitrary a division made between the Department of Defense and NASA in this area . . . when in fact no technical and little other distinction between the two exists." Schriever recommended that "the sense of urgency that exists across the whole front of space projects should be injected into the manned military space program. . . . If the artificial division between peaceful and military space programs is removed" then the United States could surpass the Soviet Union."[41]

The Air Force Space Plan of 1961

The clearest expression of Air Force sentiments in the Kennedy administration was the first, and only, full-blown Air Force Space Plan, released in September 1961. In this 88-page document, the USAF detailed exactly what it believed it should do in space, what programs were required, and what these would cost.

The Space Plan explicitly stated, "The prestige value of spectacular achievements, such as has been enjoyed by the Soviets, is recognized as having significant importance in the cold war struggle between two opposing ideologies." In any American effort to respond, "It is appropriate that the Air Force become completely involved in, and carry a major share of, this effort. . . . Whereas the Air Force strongly advocates an aggressive military space program, it recognizes that to arbitrarily separate military from nonmilitary space-development responsibilities is fundamentally unsound. The capabilities and facilities of the Air Force will be used to support the entire National Space Program, not just the distinctively military portions."[42]

The bulk of the Space Plan went on to detail the specific missions and systems the Air Force believed America required to ensure the Soviets could not pose a threat from space. Throughout the document, the Air Force emphasized the role humans must play in space systems. For instance, "Man is unique in his ability to make on-the-spot judgments. He can discriminate and select from among alternatives which have not been anticipated. He is unusually adaptable to rapidly changing situations. Thus, his inclusion in military space systems, if feasible, can be expected to increase significantly their flexibility as well as the probability of mission success." The Air Force believed satellite inspection and neutralization would be performed by a manned system, Dynasoar.[43] The anticipated Air Force space program included everything from the development of rendezvous, docking, and reentry techniques to "a permanent, manned, military test space station [for] evaluating operational concepts and hardware possibilities for: space command posts; permanent space surveillance stations; space resupply bases; permanent orbital weapon-delivery platforms; subsystems and components."[44] The Air Force even called for a space shuttle to be the next vehicle constructed beyond Dynasoar in 1965 and used as a space station resupply vehicle.[45] Deputy CSAF for R&D Lt Gen James Ferguson told Congress that implementation of the Air Force Space Plan would require increasing FY 63 funding from OSD's programmed $826 million to $1.3 billion and FY 64 from $1.3 billion allocated by the OSD to $1.86 billion.[46]

In his public advocacy for the Air Force Space Plan and the accompanying increased funding, General Schriever apparently went too far. In an 11 October 1961 speech to the American Rocket Society, Schriever remarked, "I have been, am being, and, if the situation is not changed, will continue to be inhibited if our space efforts continue to be carried out under an unnecessary, self-imposed national restriction; namely, the artificial division between space for peaceful purposes and space for military purposes."[47] This generated a pointed memo from McNamara to SAF Zuckert: "Gene, if such statements were made, they seem inappropriate. What do you plan to do?"[48] Zuckert reminded Schriever, "It must be clear that Air Force people do not publicly complain about Government decisions, in this case the assignment of space development responsibilities to a special agency created for that purpose." And Air Force spokesmen must "avoid giving the impression in the public press that the Air Force is 'shackled,' or 'inhibited,' or not getting adequate support from NASA."[49] Zuckert later commented that Schriever's speech "didn't make the President very happy either"[50] and that "I chewed him [Schriever] out about that speech because Jim Webb complained to me and we gradually worked things out."[51]

Two factors seem to have deflated any Air Force effort to increase its space budget through its Space Plan. First, on 20 February 1962, John Glenn became the first American to orbit the earth. One analyst said, "A great feeling of relief and euphoria swept the nation as the feat brought an outpouring of international acclaim and good will to the United States, not only for the achievement itself, but for the public manner in which it had been conducted." Any interest that may have existed within Congress for expanding the Air Force space program in accordance with the USAF's new Space Plan quickly dissipated.[52] Second, the OSD was not in late 1961 or early 1962 amenable to Air Force requests for greater space funding. When the Air Force did request an immediate $400-million supplement for the FY 63 budget to begin implementing its Space Plan, DDR&E Brown informed White's successor as CSAF, General LeMay, on 20 August 1962 that it would be difficult "to justify any blanket increase in funding for space programs at this time."[53] This became a standard response from the OSD throughout the Kennedy administration to Air Force requests for increased space funding.[54]

1962 and Beyond

The space enthusiasts within the Air Force continued to desire a larger Air Force role in space, but as the Kennedy administration continued, it became increasingly clear this was not likely, given the close supervision OSD exercised over Air Force space initiatives. After the situation described above with the 1961 Space Plan, future Air Force space planning documents were not publicized nor was there any effort to garner congressional support for them. The OSD's firm managerial control ensured there would be no replays of the events surrounding the release of the 1961 plan. By October 1962, the Air Force had completed a revision of its 1961 Space Plan. It reiterated the two purposes of the Air Force in space: "To enhance the general military posture of the United States through military use of space" and "To provide a military patrol capability within the space region." Together the systems that would provide these capabilities would deny to any hostile power "the uninhibited military exploitation of space, and to provide a system of protection for U.S. scientific activities in space." The revised plan continued to call for better space boosters; space weaponry; development of reliable rendezvous, docking, and transfer procedures; and maneuverable reentry and precision recovery.[55]

LeMay's case to Zuckert to support the revised Space Plan of 1962 included figures to demonstrate the increasing budgetary discrepancy between what Air Force officers felt was required for a proper military role in space and what the OSD was permitting. Table 4 shows the plan's proposed military space expenditures compared to DOD's budgeting baseline for space, in millions of dollars. These figures illustrate the gulf in thinking between the Air Force and the OSD; the Air Force was planning for a space program more than three times larger than the OSD seemed willing to authorize. As an official Air Force history explains, "Unfortunately, the five-year program—which served the useful function of crystallizing Air Force thinking on its space goals—made no great impact upon the

Table 4. Military space expenditures proposed Space Plan of 1962 compared to DOD's budgeting baseline for space (in millions of real-year dollars)

Year	1963	1964	1965	1966	1967
USAF	1,210	1,995	2,410	2,398	2,302
DOD	919	992	857	776	688

Adapted from Vice-CSAF Curtis LeMay to Eugene Zuckert, 19 October 1962, folder: 6-1962, Box B128, Curtis LeMay Papers, Library of Congress, 4.

OSD. McNamara for all practical purposes ignored the document." Deputy DDR&E Rubel's response "was very discouraging" as he indicated in a late October 1962 meeting that the plan would receive little support in the OSD because OSD had concluded the plan failed to justify the requirement for proposed programs.[56] In a speech that month, Rubel declared, "proposals which served abstract doctrines about the military role in space would not be entertained."[57]

The OSD's Perspective

McNamara stated clearly, concisely, and often the criteria an Air Force space proposal would have to meet before he would even consider approving it. "First it must mesh with the efforts of the National Aeronautics and Space Administration in all vital areas. We must ensure that the Defense and NASA programs, taken together, constitute an integrated national program, and that knowledge and information flow freely between the two. Second, projects supported by the Defense Department must promise, insofar as possible, to enhance our military power and effectiveness."[58] If there were even a hint of possible duplication with any ongoing NASA program, the Air Force would have an extremely difficult time justifying a project to the OSD. If the Air Force could not show quantifiably and specifically exactly how the proposed space project was able to enhance military power and effectiveness, then once again, OSD approval was extremely unlikely. Behind these criteria there was an oft-expressed OSD

281

skepticism concerning the necessity for military personnel in space and for any increased military space budget. DDR&E Brown's deputy Rubel had day-to-day responsibility for monitoring the military space program. Rubel summarized in October 1962, "Our expenditures on space developments have been remarkably high in relation to viable concepts for military applications in space. In fact, despite extraordinary efforts we have not evolved any very new ideas for military applications in space during the past several years. This is especially true of manned military applications."[59]

The OSD termed its overall orientation to the military space question the "building block" approach. The DOD divided military space into two broad areas. First were those missions currently deemed viable and able to support existing and concrete DOD requirements such as robotic satellites for meteorology, navigation, geodesy, communications, and early warning against ballistic missile attack. These military space programs would be "integrated with the overall military program, supplementing or complementing other military activities."[60] The second category was certain developments in basic technology, "the building blocks necessary for a flexible capability to move rapidly into systems needed in the future as specific defense requirements and missions are defined. These building blocks include structures, guidance and control systems, maneuverable reentry vehicles, propulsion, and man himself."[61] The OSD tended to regard these potential future building blocks as an insurance policy against a Soviet surprise in the use of space for military purposes. DDR&E Brown explained, "At this point in time it is difficult to define accurately the specific characteristics that future military operational systems of many kinds ought to have. We must, therefore, engage in a broad program covering basic building blocks which will develop technological capabilities to meet many possible contingencies. In this way we will provide necessary insurance against military surprise in space by advancing our knowledge on a systematic basis so as to permit the shortest possible time lag in undertaking full scale development programs as specific needs are identified."[62]

Brown obliquely referred to Dynasoar when he gave the example of rendezvousing with a satellite and returning to earth as a building-block capability being worked on: "Again, while a firm military requirement for all such systems does not now exist, we are following the 'building block' approach."[63]

McNamara's Management Philosophy and Systems Analysis

A short discussion of McNamara's underlying management philosophy, which compelled him to enforce efficiency, consolidation, and the elimination of duplication from all DOD programs, the military space program included, is necessary to fully understand the NASA-DOD relationship. McNamara's quest for cost reduction and single, efficient programs guaranteed that if NASA had some kind of a program (such as Gemini) exploring a particular capability (such as rendezvous and docking in space), it was extremely unlikely that he would approve a DOD program exploring those same capabilities (Dynasoar), even if Dynasoar were also going to explore other capabilities.

McNamara later wrote that one of his core conclusions about leading the DOD was that "the dynamics of efficient management in so complex an institution as the Defense Department necessarily require the use of modern managerial tools and increasing efforts to determine whether the 'cost' of each program and each new project is justified by the 'benefit' or strength it adds to our security." He described himself as the type of SECDEF who was a real leader who "immerses himself in his operation, leads and stimulates an examination of the objectives, the problems and the alternatives" and not just a judge who "waits until subordinates bring him problems for solution, or alternatives for choice." His diagnosis of the Pentagon's main problem was not that the SECDEF lacked authority but rather "the absence of the essential management tools needed to make sound decisions." McNamara then began "applying strict standards of effectiveness and efficiency to the way we spend our Defense dollars. . . . These reforms would necessarily change traditional ways of doing things, and limit the customary ways of spending defense money."[64]

McNamara believed the primary problem was that, in the past, "the three military departments had been establishing their requirements independently of each other. The results could be described fairly as chaotic." For instance, the Army planned for a long war of attrition and therefore stockpiled months and sometimes years of supplies while the Air Force assumed future conflicts would be short nuclear exchanges and so maintained only a few days of supplies. McNamara therefore insisted the DOD budget "for the first time grouped together for planning purposes units which must fight together in the event of war." So the Navy's Polaris submarine and the USAF's bombers and ICBMs would be compared with and evaluated in terms of each other, not in terms of other intraservice priorities. Which would be most cost-efficient in destroying Soviet targets? What was true within the DOD would, in the case of military space, also be true for the Air Force in relation to NASA. Air Force space proposals would be judged not only in terms of what they could add to America's deterrent, but also in terms of whether they duplicated NASA capabilities. McNamara emphasized, "Adding a weapon to our inventory is not necessarily synonymous with adding to our national security," and so the process of approving a new system "must begin with solid indications that a proposed system would really add something to our national security. The United States cannot even seriously consider going ahead with a full-scale weapons-system development until that basic requirement has been met. . . . We need to keep the number of new systems as low as possible consistent with security"[65] The Air Force suffered particularly hard in this evaluating process, losing not only its sole human-spaceflight project, Dynasoar, in December 1963, but also its nuclear-powered airplane, the B-70 bomber, Saint (an unmanned satellite interceptor), and the Skybolt missile to McNamara's drive for efficiency and centralization between 1961 and 1963. The Air Force said the minimum acceptable number of Minuteman missiles would be 3,000, but only 1,000 were eventually approved.[66] One analyst summarized, "During the first couple of years of the Kennedy Administration, the Air Force could not win a single battle with McNamara."[67]

The specific mechanism whereby McNamara evaluated one system in the context of other systems designed to provide similar capabilities was called Planning, Programming, and Budgeting System (PPBS), often referred to simply as systems analysis. This book cannot hope to provide complete details of the labyrinthine operational principles of this process.[68] Fortunately, McNamara did summarize, "Major program priorities can be meaningfully determined only in terms of the total program, and a proper balancing of all the elements of the defense effort can only be achieved at the Department of Defense level."[69] He added, "It provides the mechanism through which financial budgets, weapons programs, force requirements, military strategy and foreign policy objectives are all brought into balance with one another." The result was an annual Five-Year Defense Program that was backed by the "full range of analytic support with operations research and other modern management techniques" which, in turn, "allowed us to achieve a true unification of effort within the Department without having to undergo a drastic upheaval of the entire organizational structure."[70] Systems analysis "offered McNamara the natural quantifier with which he could gain control of this sprawling empire."[71]

Quantitative analysis was the key. A proposed new project had to include detailed mathematical justification on: (1) exactly how it would add to America's national security; and (2) why it was more cost-effective to provide that particular capability with this new system rather than with older existing systems or with other competing proposals for new systems to provide the same capability. Deputy Assistant SECDEF for Systems Analysis Alain Enthoven explained that PPBS compared programs to determine "the degree of military effectiveness that can be achieved with a particular capability for a given expenditure." Pentagon comptroller Charles Hitch continued, "In each case we are interested not only in the military worth of the proposed requirement but also in its cost. In our view, military effectiveness and cost are simply two sides of the same coin. . . . Properly applied analytical techniques help to minimize the areas in which unsupported judgment must govern in the decision-making process."[72] The new budgeting techniques resulted in a DOD budget with 620 subcategories. When the military services appealed his decisions in the fall of 1961 in every single subcategory, McNamara made a point of confirming all 621 of his decisions in a single day.[73] For the Air Force and its space proposals, NASA's R&D was part of the PPBS equation in the sense that if the OSD concluded an Air Force space proposal duplicated or had the potential to duplicate a NASA project, it was highly unlikely the USAF system would win OSD approval.

The Air Force's conundrum, of course, was that since so little was known about the space environment in the early 1960s, it had very few hard facts and almost no concrete numbers to incorporate into the systems-analysis computers. The Air Force was asked to prove its requirements in a realm, space, for which little information existed, but was not permitted to build the systems required to operate in that realm and thereby gather the requisite information.[74] This dilemma has been termed the *requirements merry-go-round*.[75] One government report said the USAF's dilemma under McNamara was that "space experimentation was restricted unless it could prove beforehand that hard requirements existed," which, of course, could not be done without the information the experiments were designed to gather.[76]

McNamara's biographer amplified that, at the core of PPBS, was McNamara's "unshakable faith in the importance of financial controls, in the 'truth' as discoverable through statistics, and in the importance of using this kind of information as the basis for organizational planning and control."[77] Of special concern for McNamara was the rapidly growing R&D field, which included most Air Force space expenditures. He told Congress in 1963, "Research and development expenditures, whether measured in budget terms or in program terms, have been mounting steadily over the years, but too much of this effort is not producing useful results. What we want are weapons and equipment that the fighting man can use. We are not interested in supporting the intellectually challenging but militarily useless, engineering 'tour de force.'"[78] This was not a hopeful sign for a robust military human-spaceflight program in general or Dynasoar in particular. One observer said that during the

Kennedy years, "The Air Force was caught in a bind which threatened to grow tighter than anything it had known and mastered during the Eisenhower period."[79]

McNamara's drive for efficiency, to eliminate duplication, and to enforce commonality in systems as much as possible certainly had the laudable goal of providing America with the most capable defense at the lowest possible level of expenditure.[80] However, even scholars sympathetic to McNamara and his "whiz kids" agree that, "A principal result of McNamara's administrative reform was to install a decision-making system that had the effect of increasing the centralization of authority in and around the Secretary of Defense. . . . McNamara's administrative innovations substantially increased the influence of civilian advisors on questions relating to matters of military strategy.[81] Walter McDougall stated less delicately, "In every functional pyramid, new layers of centralized, civilian bureaucracy splayed out from the organizational box of OSD in 1961." McDougall also discussed "the managerial shift from the uniformed services to the civilian bureaucracy fanning out from the Office of the Secretary of Defense," which "pulled all strings into OSD." He concluded, "McNamara's whiz kids were everywhere, removing every vestige of independent authority and, with it, much of the pride of career officers."[82]

In addition to the obvious loss of institutional power the military services suffered as a result of McNamara's managerial reforms, the way in which these reforms were implemented also contributed to a sense of dismay among many in the Air Force. McNamara's biographer Deborah Shapley explains that McNamara's "reign was colored by moral righteousness and arrogance" because he and the PPBS cadre he emplaced within the OSD "were young, fresh, and convinced that history was on their side. Their mistake was to appear contemptuous of the military institutions whose follies they sought to reform." Shapley concluded that McNamara's treatment of the military services "reveals a basic flaw in his revolution—his disdain for the military institutions and culture he was presuming to change."[83] She added, "McNamara's analytic strengths were coupled with a limited personal capacity to understand

and empathize with the culture and traditions of the organizations he commanded."[84]

The Air Force Reacted

The tension between top-level Air Force leadership and McNamara's OSD grew quickly to a high level. After White retired as CSAF in June 1961, he could honestly express himself: "I am profoundly apprehensive of the pipe-smoking, tree-full-of-owls type of so-called professional 'defense intellectuals' who have been brought into this nation's capital. I don't believe a lot of these overconfident, sometimes arrogant professors, mathematicians, and other theorists have sufficient worldliness or motivation to stand up to the kind of enemy we face." LeMay succeeded White and commented on the 34-year-old Brown, DDR&E, exercising control over Air Force R&D efforts, "Why, that [SOB] was in junior high school while I was out bombing Japan!" LeMay reportedly asked, "Would things be much worse if Khrushchev were Secretary of Defense?"[85]

Schriever's Systems Command was probably most affected by the OSD's new procedures and philosophy because it was responsible for the R&D process leading up to the USAF's fielding of new weapons. He noted, "I never once had a session with McNamara relating to a single major program decision, not one time in all the five-and-a-half or six years that we overlapped." Yet Schriever reported McNamara's "completely undisciplined staff . . . would go charging around all over the country. . . . Most of the time we didn't even know that they were wandering about. In no circumstances were we ever provided with copies of their reports when they came in, so we didn't even know what the hell was going on." Schriever reported that essentially the OSD "usurped all this authority, but they had no responsibility" so that Air Force officers were being whipsawed by the ever-changing requirements for thousands and thousands of pages of documentation the OSD demanded. Schriever concluded, "Mr. McNamara had no concept of management. . . . He demanded all kinds of loyalty, but he dispensed no loyalty down. . . . So if I seem to have little respect for Mr. McNamara, that's precisely correct. I didn't have while I was on active duty, and I don't have today. I think that he did many things that we're still suffering from and will suffer from for many, many years to come."[86] Schriever described his long-term efforts to convince McNamara that sometimes the Air Force had to undertake cutting-edge R&D to generate the technology necessary to maintain American military superiority: "I have tried and tried but he won't listen to me."[87]

One should not view such thoughts as simply the bile resulting from military officers having lost autonomy and influence. The USAF's top civilian, SAF Zuckert, stated, "I would have to say that my batting average for getting my views adopted by McNamara was very low. . . . I think McNamara saw the Air Force as a very powerful force with the Congress and with the people, by reason of its size and its missions. I think he felt that one way he could control the Air Force was to keep it off balance. He wanted the Air Force to know at all times who was the boss. . . . Even on little things he would get involved" such as when McNamara ordered that no more Naval Academy graduates could transfer to the Air Force. Zuckert concluded, "When McNamara dealt with me, nearly ninety percent of the time he was completely arbitrary. 'The Air Force does not know what they are doing. No, that is not the way it is going to be.' He was very rough."[88]

The official Air Force history of this era confirms, "The Secretary of Defense continued to rigidly control funding and insisted on absolute program definition. . . . Frequently, USAF projects were submerged in cooperative ventures with other national agencies. This situation resulted in part from the efforts of the Secretary of Defense to assure that the most efficient and economical use was made of the nation's space resources" and from the national space-for-peace policy which "placed the greater emphasis on devoting space to peaceful and scientific purposes, with responsibility vested in a civilian space agency." In fact, recounted the Air Force history, "It was becoming clear by 1963 that there really was no such thing as an 'Air Force Space Program'—that Air Force space activities would be conducted within the context of an overall 'DOD Space Program.'"[89] This meant that during the Kennedy administration there was a "fundamental schism" between the USAF and the OSD on how to get a space project started or

Table 5. The 1961 situation from the Air Force perspective

Air Force Spacemen	OSD Spacemen
Enthusiastic, zealous	Sober, cautious, conservative
Long experience in military space work	New to military space
Eager to sponsor multiple solutions to a single space problem	Determined to select the single best solution in advance
Advocates of a total space systems concept	Believers in an R&D demonstration concept

Adapted from Col Paul E. Worthman, USAF, "The Promise of Space," *Air University Review* 20 (January–February 1969): 120–27.

how to continue to manage one already underway. Table 5 above shows the contrast as viewed by the Air Force. One Air Force colonel said, "Communication between the two agencies was frequently strained, and relations were complex. Following its own convictions rigorously, the OSD began to cancel or slow down a number of Air Force 'pre-Kennedy' programs. The cases took on a dreary similarity, with a regular pattern of review, revision, de-emphasis, or elimination."[90]

The OSD Held Firm

OSD skepticism toward increased military space spending in general and toward the Air Force's perceived requirement for military officers in space continued. DDR&E Brown told Congress in June 1962, concerning manned military space systems, "I cannot define a military requirement for them. I think there may, in the end, turn out not to be any."[91] Brown added the DOD was relying heavily on NASA to develop the technology of human spaceflight, stating, "We have no intention to preempt those areas which are the proper pursuit of NASA, and, as a sign of this, their planned effort for next year in space is very much larger than those within the Department of Defense. . . . We are not attempting nor do we have any intention or any reason to compete or duplicate the large variety of orbital missions which are planned as part of the national space program by NASA."[92] June 1962 was the same month

Kennedy declared, "The military have an important and significant role, though the primary responsibility is held by NASA
and is primarily peace, and I think that the proportion of that
mix should continue." Given presidential satisfaction with the
civil-military mixture in the US space program, there was little
reason for the OSD to augment the USAF's space budget or
approve proposals for new space projects.

Perhaps the clearest expression of DOD's orientation by late
1962 was Deputy DDR&E Rubel's speech on 9 October 1962
on military space, which the DOD disseminated as an official
press release to emphasize its nature as official DOD policy.
The highlights of Rubel's address were:

> Extensive programs and projects under NASA supervision will prove
> equally applicable to systems and devices in space whether these are
> used for military or non-military purposes. . . . Despite extraordinary
> efforts we have not evolved any very new ideas for military applications
> in space during the past several years. This is especially true of
> manned military applications. Most manned military missions in space
> still, after years of study, seem little or no more viable than they ever
> did. . . . Nevertheless, we are anxious to build a base on which future
> systems could, if needed, be constructed. We are not yet ready to de-
> sign the building, but we want the building blocks at hand. . . . Doc-
> trinal abstractions such as 'sea power' or 'air power' or 'aerospace
> power' are often useful for analysis. . . . But these doctrinal abstrac-
> tions do not translate well into new programs and projects. Here, tech-
> nology takes over. . . . If you are going around with your head in the
> clouds, you'd better keep you [sic] feet on the ground.[93]

The *New York Times* commented, "Pentagon authorities made
clear that Mr. Rubel's speech was intended as a rebuttal to
members of Congress and some Air Force leaders who have
been campaigning for increased military space expenditures.[94]
Concerning military men in space, McNamara explained less
than two weeks later,

> I am not prepared to say that we will or will not need to have manned
> spacecraft. I cannot read the future. . . . At this time I see no clear re-
> quirement for manned satellites for military purposes. Trying to put a
> man into the space vehicle leads to complications and delay. . . . At
> present, we can do almost everything we need to do without a man in
> the satellite. Much of what we need to do now we can do better with-
> out a man—and sooner. But we must be prepared to put man in space
> in the future should new requirements develop."[95]

The OSD fully supported Kennedy's emphasis on continuing Eisenhower's space-for-peace policy, albeit with a new emphasis of human spaceflight for prestige via Apollo. OSD officials regularly quashed any speculation about offensive uses of space such as orbital bombardment. Deputy SECDEF Gilpatric on 5 September 1962 emphasized in a speech, "We have no program to place any weapons of mass destruction into orbit. An arms race in space will not contribute to our security. I can think of no greater stimulus for a Soviet thermonuclear arms effort in space than a U. S. commitment to such a program. This we will not do. We will of course take such steps as are necessary to defend ourselves and our allies, if the Soviet Union forces us to do so."[96] The report resulting from the Kennedy-mandated major review of the space program conducted in the fall of 1962 (described last chapter) stated, "The Secretary of Defense and his assistants have taken a restrictive approach in their reviews, based on the conclusion that there are no valid new military requirements which justify at this time a major expansion in the military space programs."[97]

Accordingly, in January 1963 the OSD disapproved Air Force space budget requests to start a space station program called the Military Orbital Development System and to purchase NASA Gemini capsules and use them for military experiments (Blue Gemini). Those two systems were among the 13 new programs in space that the Air Force asked the OSD for permission to start, of which the OSD allowed none. The memo summarizing the military space situation to the CSAF after OSD disallowed almost all of the Air Force's proposals said, "In terms of the *Five-Year Military Space Program*, DOD action is short of Air Force proposals by 1.3 billion dollars. For FY 64, DOD is providing 55 percent of the level recommended by the Air Force." The memo explained that for spacecraft projects the numbers for Air Force proposals and OSD approval were, in millions of dollars: FY 63—587 vs. 537; FY 64—1,032 vs. 367; Total—1,619 vs. 804.[98] McNamara's rejoinder was that the FY 64 military space budget overall was $1.65 billion, which was $50 million greater than FY 63 and almost $400 million greater than FY 62. McNamara added that military space represented 20 percent of the entire DOD R&D budget, an amount greater than that devoted to developing strategic weapons.[99]

Congress and the President on the OSD-USAF Relationship in Space

One reporter summarized the OSD-USAF situation: "While the issue is occasionally constructed as competition between NASA and the Air Force for authority and funds, the argument is basically between the Air Force and the upper echelons in the Department of Defense."[100]

Rep. George Miller, who became chairman of the House space committee after Overton Brooks died in September 1961, agreed, "The problem is that the military space enthusiasts have not been able to obtain all the green lights they want from their bosses."[101] Miller's speech also indicates one of several reasons why the OSD's space policy was likely to prevail over any Air Force attempts to increase military space spending or programs: by the time of the Kennedy administration, NASA had developed powerful congressional patrons. Miller supported McNamara's idea that new military space projects must be justified before they were approved,

> But the space critics are vague about what they want. Something really good, they say, is bound to turn up. That's fine. I agree. And as it does, I say, 'let's go.' I cannot understand, however, initiating a program when the requirement it must meet is unknown or can be better met by another system. The balanced program we are following is the one devised by the President after meticulous study of the Nation's needs, resources, and aspirations. . . . Our defense officials are not dolts.[102]

Beyond the OSD and certain important NASA patrons in Congress, the most important determinant of whether or not the OSD's decisions concerning military space and military officers in space would prevail was, of course, Kennedy. As alluded to previously, he had no great inclination to disturb McNamara's policies or the general division of responsibilities in the NASA-DOD equation he inherited. Numerous sources have observed that "McNamara's actions had the full support of the president,"[103] that Kennedy "was enamored of McNamara's brilliance, almost always backed him up,"[104] and that "because of his standing with Presidents Kennedy and Johnson, McNamara had more influence than any of his predecessors"[105] as SECDEF. Clearly, the Air Force had no choice but to accept McNamara's formulation of the proper scope, scale, content, and pace of the military space program, including his reluctance to authorize military personnel in space.[106]

Herbert York was DDR&E during the Eisenhower administration and for the first few months of Kennedy's. His impression was that "[Kennedy] was already of the view that the Air Force was much too 'gung ho.' I think or I have the impression that he already felt that the Air Force was what some people might call a little bloodthirsty."[107] And yet Kennedy did reappoint LeMay as CSAF. Kennedy commented, "LeMay's like Babe Ruth. Personally he's a bum, but he's got talent and the people love him."[108] He also stated, "It's good to have men like Curt LeMay . . . commanding troops once you decide to go in. . . . I like having LeMay head the Air Force. Everyone knows how he feels. That's a good thing."[109] Therefore, one should not overdo Kennedy's animus toward the military or the Air Force. The fundamental point is simply that he was unlikely to overrule McNamara by embracing any Air Force proposals for altering the military space program that McNamara concluded were inappropriate, nor was he likely to endorse any significant shifts in the NASA-DOD relationship in favor of the Air Force.

One Space Program, Not Two

The final outcome of the entire complicated issue of NASA-OSD-USAF interaction was a reversal of the Eisenhower proposition that a single, unified space program was an impossible goal and should not be pursued. Kennedy's administration in fact concluded a single, national space program did exist and that what NASA and the DOD did in space must be carefully coordinated so as to avoid waste, duplication, and fruitless effort. This reinforces the conclusion that the Air Force was unlikely to receive approval for any program which, when evaluated in light of any NASA project, could be accused of overlap or redundancy. Any number of senior administration officials made the same point Deputy SECDEF Gilpatric made when he observed in June

1962, "Some people have the erroneous impression that there are two space programs—a NASA program and a DOD program. What this nation has in fact is a National Space Program, part of which is funded and directed by NASA and part of which is funded and directed by DOD."[110]

An ancillary point that complemented the idea that America had a single, coordinated space program was the idea that American military activities in space were, in fact, peaceful activities, just like NASA's. Therefore, one should not speak of the "military" use of space and the "peaceful" use of space but rather of the aggressive and nonaggressive use of space. Behind this conclusion was of course the commitment to ensuring that reconnaissance satellites would continue to enjoy unmolested transit through space. A State Department official explained, "The test of the legitimacy of a particular use of outer space is not whether it is military or nonmilitary, but whether it is peaceful or aggressive. . . . The United States has military space programs, but all of our space activities will continue to be for peaceful, i.e., defensive and beneficial purposes."[111] One of the strongest administration exponents of the "military space is space for peace" proposition was NASC executive secretary Welsh. He delivered many speeches in which he explained, "We do not have a division between peaceful and nonpeaceful objectives in our space program. All the objectives are peaceful. It should be clear, however, that projects to help keep the peace are just as peaceful as any other space projects."[112] One of the instructions Kennedy issued to American representatives to the UN Outer Space Committee and General Assembly in August 1962 was to forcefully explain and defend the notion that "the distinction between peaceful and aggressive uses of outer space is not the same as the distinction between military and civilian uses, and that the U. S. aims to keep space free from aggressive use and offers cooperation in its peaceful exploitation for scientific and technological purposes."[113] Again, the point to carry forward in discussing the intricacies of the NASA-DOD relationship and into the next chapter's discussion of Dynasoar-Gemini-MOL is that at the highest administration levels, up to and including the president, there was a strong desire to avoid any suggestion whatsoever that the United States had any aggressive intent in space. While it actively encouraged the notion that the defensive military uses of space were in fact peaceful uses, the administration simultaneously insisted that the OSD ensured that no offensive uses of space taint the American space program.

The NASA-DOD Relationship: An Overview of Support, Coordination, and Rivalry

While the historian can safely state that during the Kennedy administration there were no *major* problems in the NASA-DOD institutional relationship, there were nevertheless some notable difficulties. In particular, the top officials of each organization, Webb and McNamara, did clash over several top-level policy issues. Shortly after he was sworn in as NASA administrator, Webb and his deputy Dryden met with senior DOD officials (McNamara, Gilpatric, and outgoing DDR&E York) to discuss how the two organizations would keep in touch. Webb reported, "It was agreed that Mr. Gilpatric and I would meet from time to time for lunch and would bring others as needed."[114] With that began extensive interaction between Webb and assorted senior DOD officials that continued throughout the Kennedy administration. Important points relating to the lunar-landing decision's NASA-DOD component were surveyed in the previous chapter. Other important groundwork can be found in this chapter's discussion of the supposed Air Force campaign to gain a larger role in the space arena. The remainder of this chapter examines the specifics of the support, coordination, and rivalry that comprised the NASA-DOD institutional relationship from 1961 to 1963.

Generally, No Major Problems

By the summer of 1961, shortly after Kennedy's lunar-landing decision, the NASA-DOD situation seemed to be well under control. The minutes from a July AACB meeting note that "the Vice President was astonished and delighted at the unanimity of the NASA-DOD recommended program objectives and approach."[115] Schriever emphasized to his boss that he desired to assist NASA in the lunar-landing program in every way possible, regardless of what his personal opinion might be on the underlying space-for-peace policy. "Our relationship has been very good," he noted. "We have worked out at the working level a very good relationship. I think that . . . there are many things in the Lunar Program that will have military applications."[116] An NASC meeting that included Johnson, Secretary of State Dean Rusk, McNamara, and Webb recorded in August, "Coordination between NASA and DOD is excellent, with every intention to keep it that way."[117] As NASA's lunar-landing program gained momentum, the Air Force began to realize that while its lunar-landing role was primarily supportive in nature, it was nonetheless important because Apollo would also create the "building blocks" of spaceflight experience and infrastructure. Zuckert said, "The NASA has a massive program to acquire a capability to operate in space. The Air Force is supporting it to the limit of our abilities. We need what NASA will learn. . . . We have an excellent working relationship with the NASA and feel that we, NASA, and the Nation benefit from this relationship."[118] While the OSD and the USAF had their differences over military space policy, OSD officials such as Rubel agreed that the basic NASA-DOD situation and level of cooperation "has been one of continuing improvement since the creation of NASA, as operating procedures evolved, as policies were established, as relative responsibilities were defined, as personnel became better acquainted and familiar with each other's problems, as internal organizations were improved and as the Aeronautics and Astronautics Coordinating Board mechanism evolved."[119]

In fact, *Missiles and Rockets*, one of the trade magazines most critical of the perceived slighting of military space, commented in April 1962 on a "gleeful conspiracy" between NASA and the Air Force, which consisted of "the growing cooperation—both in spirit and deed—between the National Aeronautics and Space Administration and the U. S. Air Force." The magazine said there were "some top NASA officials, possessed of a vision broader than the confines of their own agency, who are acutely aware of the need for major manned and unmanned military space programs. . . . To bring these military capabilities to fruition at the earliest possible moment, they are aiding and abetting the movement to bring the Air Force, to a certain degree, under NASA's strong financial shelter in the Apollo program."[120] Shortly thereafter, a NASA official wrote to the magazine to say, "I thoroughly enjoyed reading your perceptive and well-written editorial."[121] McNamara added, "Increasingly, the space efforts of Defense and NASA have become interwoven and more effective. . . . I am determined . . . to ensure the continuation of this excellent relationship."[122] These institutional encomiums could be cited ad infinitum, but the point remains, at least for public consumption and often in private meetings, high administration officials displayed no sense of alarm or even concern over potentially serious NASA-DOD conflict.

NASA-DOD Difficulties

This is not to say, however, that tension, rivalry, and conflict were absent. Most of the clashes are directly associated with the management of the Gemini program, its relationship to Dynasoar, and the repercussions of both the Gemini and Dynasoar programs on the space station/MOL issue. These are detailed in the next chapter, which examines specific programmatic issues. However, some hints of a more general, institutional strain are also detectable. NASA's deputy associate administrator for Defense Affairs W. Fred Boone correctly pointed out, "When there are two government agencies that have responsibilities and areas of activity which to some extent overlap, that in a sense compete for budget dollars, and that are headed by two such dynamic, strong-willed, articulate men as Mr. Webb and Mr. McNamara, one should not be surprised to find conflicting policies and opinions between them."[123] For instance, Boone said that in the opinion of NASA leaders, McNamara unreasonably and "consistently avoided any acknowledgment that the NASA R&D program was making a contribution to national security." This attitude of the Pentagon top management toward NASA "filtered down through all echelons of the Defense establishment," and as a result, "Some key officials in OSD and the Services . . . appeared to be inhibited from laying before us their needs for new technology and from exploring opportunities for cross-support for fear of bringing down on their heads the ire of the Secretary of Defense."[124]

Webb wrote Johnson in May 1963 with three suggestions for improving coordination between NASA and the DOD, which he felt was good but could be better. First, he called for "Earlier coordination in the study phase of advanced projects to eliminate unwarranted duplication." Webb said that "cross-fertilization" of research and technology should be strengthened so as to "reveal additional applications of NASA discoveries and advancements to some of the most critical military problems." In addition, he desired "greater participation by the DOD in NASA projects to enhance the knowledge and capability of the services in space and space-oriented applications."[125] During the summer of that year, Boone wrote an extensive report surveying the "divergent philosophies, attitudes, and interpretations of the Department of Defense and the National Aeronautics and Space Administration" and concluded that specific problems appeared to be centered in the areas of national policy, planning, ground-support operations (ranges, tracking stations, and data-collection centers), and aeronautical research. The first two are immediately relevant to this study. Concerning national policy, Boone stated, "DOD sees the civilian and military space programs as one program which should be jointly conducted to attain both civilian and military objectives." Therefore, the military should have a stronger voice in shaping the direction of the total space program. In turn this had made for DOD attempts to achieve greater roles in some NASA programs such as Gemini. Boone added,

> The desire to control is especially strong within the Air Force because many within it consider space operations "simply an extension of flight operations in the atmosphere," and therefore should be under Air Force control. Lacking greater support for this position at the DOD level, the Air Force has made 'end runs' to members of Congress and the White House staff, and has launched an intensive and well organized public relations campaign to convert the public to the Air Force point of view. The Air Force is inclined to look upon NASA as a competitor rather than a partner in the field of space.[126]

Boone recommended that McNamara and Webb conduct a vigorous effort to indoctrinate their subordinate staffs and agencies to the facts. First, it was and is the intent of Congress for the United States to maintain in the eyes of the world a peaceful image for the US space program, and so NASA will remain an independent, civilian agency. Second, certain advantages accrue to the DOD from civilian management such as international cooperation and the R&D issuing forth from civilian scientific organizations and universities.[127]

In terms of planning, there appeared to Boone to be a difference of opinion concerning the desirability of joint programs versus coordinated programs. DOD seemed to desire the former where both participating agencies receive equal management and decision-making responsibilities; no major decisions are made without the concurrence of both agencies. NASA preferred the latter because in the coordinating process it maintained managerial and decision-making control while fully recognizing the DOD's interests in major NASA programs, such as Gemini, and keeping it informed concerning their progress. For instance, NASA did not want to be limited by having to specifically tie various concepts to military operational requirements concerning its long-range studies for space stations. At the same time, NASA desired to be "ever alert to discern those areas of research which appear to offer the most promising potential for the solution of military problems." If NASA had to obtain DOD concurrence to conduct studies of future concepts, this would "seriously obstruct NASA's ability to discharge its statutorily assigned functions." Nevertheless, the DOD strongly believes that all planning related to NASA programs that were of interest to DOD should be jointly conducted from its inception, Boone explained. "This view has led DOD to seek inflexible agreements concerning the manner in which NASA's advance exploratory studies may be initiated, including signoff authority for DOD."[128] The ramifications of these policy and planning differences will become evident in the next chapter's Gemini-Dynasoar-MOL discussion.[129]

Webb-McNamara Difficulties

Despite Boone's dispassionate discussion of the general disagreements between NASA and the DOD, most of the nonprogrammatic, leadership/headquarters-level tension appeared to have resulted from direct clashes, related to personality conflicts and otherwise, between Webb and McNamara. For instance, records from a meeting McNamara attended in March of 1963 to discuss Dynasoar contain the following puzzling observation: "Mr. McNamara raised the question of what would be an optimum test bed [for hypersonic R&D] during the NASA briefing. Someone at the NASA briefing raised the point that the Space Act provided that space be used for 'peaceful purposes.' Mr. McNamara was very scornful, saying that he was prepared to get the law changed."[130] This supposed-McNamara remark must remain a mystery because no further evidence exists of McNamara attempting to have the Space Act amended. At a minimum, however, it does indicate that McNamara had some type of negative feelings ("scornful") toward the general concept of space for peace, and possibly even NASA per se, although this certainly did not translate into any amenability toward Air Force space proposals.

More concrete evidence exists documenting the McNamara-Webb personal difficulties. McNamara's deputy, Gilpatric, recounted that his boss "took a dislike to Webb because Webb took so long in getting to the point. And so I think he mishandled Webb. He sort of goaded him into taking extreme positions. The result would be that Webb would go up to the Hill and see his good friends like Bob Kerr [chairman, Senate space committee] and Clint Anderson [chairman, House space committee] and didn't do McNamara any good. . . . It was just an unnecessary bit of exacerbation to take him on in such a militant fashion."[131]

Webb's biographer, W. Henry Lambright, reporting on the overall situation states:

> In the early period after the Apollo decision, Webb and McNamara met regularly for lunches, accompanied by aides, to facilitate coordination. At one of these lunches, McNamara lectured Webb, so offending the NASA administrator that he and Seamans walked out, and the regular lunches were discontinued. Although the two senior officials dealt with one another as little as possible thereafter, they had to cooperate to some extent for common interests. Webb used Seamans as a surrogate, and McNamara used similarly appropriate substitutes.[132]

An interview with Seamans confirmed this account of what Seamans called "The Black Luncheon." At this particular luncheon, McNamara told Webb there was no point in their having meetings "just for pleasantries." Webb agreed and McNamara stated, "I just happen to have a piece of paper here" and proceeded to read from it. Seamans recalled, "Well, boy, you never heard such a scathing denunciation of NASA. It was about a page-and-a-half or two pages on how we'd agreed to things and hadn't carried through on them. Jim's face was getting red and he was getting madder and madder and madder. He practically exploded. And that was the last meeting we ever had." Seamans dated this incident to between the spring and fall of 1962. Seamans said that he and Rubel had to handle most direct NASA-DOD communication from that point forward because Webb and McNamara would not speak to each other.[133] Webb testified that he had to remind McNamara when they were speaking that he should not and could not treat Webb like he treated his service secretaries and other subordinates. "I did tell him that, on an occasion when I felt that improper pressures were being applied. I said, 'You are not going to get NASA under your thumb, as you have the Air Force. . . . There was always this feeling, if NASA joins with the Air Force, then it makes a lot of problems for the Secretary of Defense. And I always made clear to him, we wouldn't do that. But they still never were quite sure."[134]

In setting the stage for the discussion of the specific support, coordination, and rivalry that is to follow, it is necessary to note that not all was sweetness and light between the two organizations, or at least between the organizations' leaders. There was an undercurrent of tension between McNamara and Webb that could and did erupt, most particularly with Project Gemini (explored in the next chapter).

The NASA-DOD Relationship: Tension and Rivalry Specifically

Tensions over exactly how to rectify the problems that surfaced early during the Mercury program demonstrate that there were specific points of contention between NASA and the DOD. Other working-level difficulties were present, and the two organizations had to finally and completely put to rest rumors and misinterpretations associated with supposed sentiment toward combining NASA and the DOD.

Shortly after being sworn in as NASA administrator, Webb had to choose between alienating the Air Force and alienating his own staff. The Mercury-Atlas test flight had failed during the Eisenhower administration due to a catastrophic explosion. Investigations revealed the most likely cause to have been weakness of the metal where the Mercury capsule was mated with the Atlas ICBM. NASA proposed that for the next test launch—scheduled for 18 February 1961, four days after Webb's confirmation—this section of metal be strengthened with the addition of an eight-inch-wide steel corset or "belly band" until the thicker-skinned Atlases that NASA had on order could be delivered. Schriever and the Air Force protested vehemently because another Atlas failure would reflect very badly on the US ICBM deterrent force, then based on the Atlas. Schriever wrote, "It is my recommendation that no more thin-skin[ned] Atlas boosters should be flown in the Mercury program because of the high risk of failure. . . . The only sensible approach is to delay the next Mercury/Atlas flight until approximately 1 April 1961 when a thick-skin[ned] Atlas will be available." Schriever further explained, "Since failure of the Atlas booster during launch would reflect unfavorably on the prestige of the United States and would be incorrectly interpreted by many agencies as a weakness in the Atlas weapon system, I do not concur with the proposed launch of the field modified (restraining band) booster."[135]

Webb supported NASA's decision to launch and refused to budge even when the Air Force took its protest to the White House by appealing to Kennedy's science advisor Wiesner. Webb said of his new organization and staff, "I knew that if I turned their advice down and took advice from outside of NASA, I would have a very hard time building the confidence of the staff."[136] The launch went ahead on 21 February and was successful. Webb called his choice to back NASA a "critical decision" because it set the tone of his supporting NASA over the Air Force, even when pressure was applied at the presidential level.[137] As his biographer states, Webb's fortitude "proved an auspicious beginning for the new administrator," earned him the gratitude of NASA's professional cadre, and "won the grudging respect of the air force, which knew Webb could not be intimidated." And, since NASA's technical judgment had proven correct, "The air force would not be so quick next time to challenge Webb and those advising him."[138] Webb recalled his first weeks at NASA as a time when he and Air Force leaders were "like two strange animals . . . sparring around, smelling each other, seeing what could be done, testing each other out."[139] The director of the Mercury program said Webb's decision saved four–five months on Mercury's schedule, compared to waiting for the thicker-skinned missiles before restarting testing.[140]

Combine NASA and the Air Force?

Chapter 3 noted that CSAF White had written several of his subordinate commanders on 14 April 1960:

> I am convinced that one of the major long-range elements of the Air Force future lies in space. It is also obvious that NASA will play a large part in the national effort in this direction and, moreover, inevitably will be closely associated, *if not eventually combined with the military.* It is perfectly clear to me that particularly in these formative years the Air Force must, for its own good as well as for the national interest, cooperate to the maximum extent with NASA, to include the furnishing of key personnel even at the expense of some Air Force dilution of technical talent. . . . I want to make it crystal clear that the policy has not changed and that to the very limit of our ability, and even beyond it to the extent of some risk to our own programs, the Air Force will cooperate and will supply all reasonable key personnel requests made on it by NASA.[141] (emphasis added)

Individuals who want to prove the Air Force was campaigning to take over NASA almost always cite this highlighted passage, out of context. This Eisenhower-era letter is relevant to the Kennedy-era NASA-DOD rivalry/tension discussion because it was not until then that it was extensively discussed. Congress held hearings to discuss McNamara's DOD Directive 5160.32, and the general question of the DOD's intentions toward NASA surfaced, as did White's letter in particular.

When asked if he believed the DOD should take over NASA, Gilpatric emphatically replied, "I certainly do not. We have plenty of problems today. We don't need any more." When asked, "And you say now you have no intention of infringing upon any of the rights of NASA?" Gilpatric replied, "That is correct." White explained his sole purpose was to "make it crystal clear that the policy is we will cooperate with NASA," even at some risk to Air Force programs. When asked if there were any planning at any level within the Air Force to take over NASA, White replied, "Absolutely not. None then, none now, and I know of no one else who has contrary views in the Air Force. I would like to point out that this is not a statement of advocacy, but a statement of possible fact— . . . No planning whatsoever."[142] White closed by assessing NASA-DOD relations as "optimum, both in the past, present, and I am certain for the future. . . . The job is plenty big for all of us. . . . The idea of a combination is so remote to my own thinking that I haven't seen that particular specter."[143]

Next to testify concerning White's letter was Schriever, who allowed that he was probably largely responsible for White feeling compelled to pen it. Schriever explained how he had expressed reluctance at giving up officers working on Air Force space systems and transferring them to NASA, "I knew it would hurt ARDC considerably to turn these people over to NASA, so I resisted their assignment, not because I didn't want NASA to have them, but because of the effect it would have on ARDC." Therefore, White issued his letter making it clear the Air Force would support NASA personnel requests."[144] Chairman Brooks asked Schriever, "There is no effort on the part of the Air Force to encroach on the normal fields of NASA activity, is there?" Schriever replied, "No sir. . . . I see no reason why we cannot work shoulder to shoulder in the most cooperative manner and there is plenty to do for both, I can assure you."[145]

When the House space committee issued its report, it summarized:

> Witnesses from the Department of Defense have disavowed any designs on NASA, and have renewed promises to work in full cooperation with NASA. The committee is happy to have these assurances from the proper officials in DOD. However, the committee has a large bulk of printed material which derogates NASA in relation to the DOD. This would seem to throw the responsibility for slurring remarks about the importance or the efficacy of NASA on nongovernmental sources; but whatever the source, the committee regrets such attacks as unwise.[146]

Apparently the committee was hopeful that the USAF-NASA situation was under control but left the impression that congressional vigilance would continue. Therefore, any Air Force attempts at making inroads into NASA's responsibilities, however unlikely they might be, would be met with firm congressional resistance at the hands of NASA's congressional patrons.

A Sample of Working-Level Difficulties

Beyond the headquarters McNamara-Webb level of tensions in the policy-making realm, there were also problems at the working level where policies were supposed to be executed. Perhaps the most persistent problem area was the national launch ranges specifically which organization should control what portions and functions of the ranges. A full examination would require a separate chapter at a minimum, but a brief survey adds some working-level detail to the story of high-level policy making.

The main US launch facility was at Cape Canaveral on Florida's east coast; the many and diverse Air Force facilities, including tracking stations, associated with the Florida range were collectively termed the Atlantic Missile Range and later the Eastern Test Range (ETR). The range and its support components had been developed after World War II primarily by the Air Force, which operated them for all agencies who used them. When Kennedy tasked NASA with Project Apollo, NASA knew it would have to assume a much greater role at the Cape because of the huge size of the Saturn family of boosters required to take three humans to the moon and back. One historian explained that if all stages of the *Saturn V* were to explode simultaneously, "The force of the detonation would approach that of a small atomic bomb."[147] This being the case, NASA would require a large amount of undeveloped land near Cape Canaveral to construct its own separate launch facilities for Apollo. In the meantime, it would call even more heavily upon the Air Force's range infrastructure for the interim launches. By August 1961, NASA had announced its plans to purchase 324 square kilometers (111,000 acres) north of Cape Canaveral, centered on Merritt Island. From this point forward, there were at least two years of constant bickering between the Air Force and NASA over myriad questions associated with the new Merritt Island Launch Area (MILA). Who would buy which portions of land? Where would the Saturn launch sites and their required buffer zones be located on the new land? Could the Air Force place any launch sites for its new, large booster, the *Titan III*, on NASA's parcel? Could the rockets launched by one agency overfly the other agency's facilities? What role would each agency play in the administration and management of the new MILA and its facilities, and how would this impact upon current practices at the AMR? Which agency would fund which range activities and based upon what formula?[148]

On the one hand, "The Air Force quite simply viewed the new area as an extension of Cape Canaveral Missile Test Annex" and thus largely under its control.[149] On the other hand, NASA wanted to have a much higher degree of autonomy at the MILA facility than it had at AMR, where NASA was essentially a client of the Air Force required to formally request the use of launch stands, tracking stations, and so forth through the Air Force hierarchy. By mid-1962, "The bureaucratic infighting reached a draw." The Air Force was allowed to construct its *Titan III* launch sites on the south end of MILA. In return NASA retained jurisdiction over the entire complex and received permission to acquire 60 more square kilometers at the north end of the MILA because of the Air Force facilities on the south end. The first NASA-DOD MILA agreement, in January 1963, agreed MILA would be considered a NASA installation, separate and distinct from the AMR. There would, in essence, be two launch ranges in Florida, to cover the Atlantic Ocean, Africa, and part of the Indian Ocean. In sum, "NASA had established its status as more than a tenant of the Air Force. . . . The decision finally came down, NASA alone—not NASA and the Air Force—would put a man on the moon."[150]

As one reporter observed, "Crossing from Air Force installations into NASA's . . . is like going from one country into another."[151] The story of USAF-NASA tension over the MILA specifically and the national range question in general was far from over. Over the course of the Kennedy and Johnson administrations, innumerable subissues were constantly being discussed at one level or another. Who would reimburse whom and at what level for which services rendered? Who was responsible for and would pay for the aircraft and ships that helped track spacecraft after launching? Who would be in charge of which of the many overseas tracking stations? Should these worldwide facilities be combined and operated in a colocated manner for both NASA and DOD? Under whose control? The list goes on and on and on. Boone's memoirs are probably the easiest access to this complex web of issues.[152] There was no shortage of working-level tension as these myriad questions were negotiated and settled, sometimes over several years.

When all was said and done, Zuckert expressed what was important from the headquarters, policy-making perspective. "At the top level, we know it's absolutely necessary for progress in both the military program and in the NASA program that we get along. We can't afford to be played off one against the other. . . . There has been a maturing of the relationship. Sure, there'll be difficulties and the difficulties will generally be exaggerated."[153] Perhaps some of these rumblings of NASA-DOD tension/rivalry even reached Kennedy. One of his questions to Lyndon Johnson in the April 1963 space program review was, "Are we taking sufficient measure to insure the maximum degree of coordination and cooperation between NASA and the Defense Department in the areas of space vehicles development and facility utilization?"[154] Johnson replied that the NASC, AACB, numerous coordinating arrangements within the agencies, and more than 50 joint-written agreements were all operating or in effect to ensure the maximum degree of coordination in the National Space Program. "However," he added, "it is inevitable that controversies will continue to arise in any field as new, as wide-ranging, and as technically complicated as space. . . . It must be kept in mind that no mechanical application of a formula will insure maximum cooperation and coordination and a minimum of duplication and waste. Continuous monitoring at a high level is essential at every stage of the development of the space program."[155] Therefore, while Johnson did not ignore the unavoidable tension and rivalry that existed between two large bureaucracies, he was confident that it was under control and that it could be kept under control if policy makers maintained proper vigilance.

The NASA-DOD Relationship: Coordination Specifically

There were few concrete results from the perceived rivalry and tensions existing between NASA and the DOD during Kennedy's term, but Webb did create a special office called the Office of Defense Affairs (ODA) within NASA in November 1962. Retired Navy Admiral Boone was in charge until January 1968. Officially, his duties, and those of his staff, were "to strengthen the flow of technical and management information between NASA and the Department of Defense"[156] and "to improve working relationships between NASA and the DOD; to expedite the flow of information; and to promote coordination on matters of mutual interest."[157] Unofficially, he was supposed to "take the heat off Seamans on the military interface."[158] The importance of Boone's office in policy making was relatively limited; one source posited, "As with the AACB, its establishment was more notable as an expression of policy than for any immediate accomplishment."[159] Nevertheless, the fact that Webb felt the need for such an organization illustrates both the perception of tension that existed as well as the constant efforts to alleviate nascent rivalry through coordination at multiple levels and by numerous bodies (ODA, NASC, AACB, and working-level committees).

Another overarching point about the NASA-DOD coordination efforts is that McNamara probably used the extensive body of agreements between the OSD and NASA "as a check on the air force."[160] McNamara and Gilpatric both "wished to bring the services under tighter control." Such agreements were just as valuable for Webb because they "undercut the Air Force's attempt to take over the space program."[161] Chapter 3 described briefly the government report from 1965 that listed 88 separate "major" NASA-DOD agreements[162] and the comprehensive NASA accounting from 1967 that described 176 NASA-DOD accords.[163] A government accounting in 1965 determined that NASA, at the headquarters level alone, was involved in 203 interagency coordination and advisory bodies.[164] What is important is the degree to which almost every possible facet of the NASA-DOD relationship was legalistically and contractually delineated.[165] Zuckert referred to "a numerous series of peace treaties between NASA and ourselves [USAF],"[166] while another source said, "Much of the cooperation between NASA and DOD occurs on the basis of formal written agreements, somewhat suggestive of treaties between sovereign powers."[167]

While some may dismiss this proliferation of bodies, committees, boards, panels, and groups as inevitable bureaucratic accretion, it did ensure that despite the delicacy and potentially explosive nature of NASA-DOD relations, "There has never been a disagreement that could not be resolved by the Administrator and the Secretary of Defense."[168] McNamara concurred and added, "Because we have two agencies, and because it is difficult to categorize in advance the project as either civilian or military, and yet because we have the two agencies, we have to assign management responsibility to one or the other, it means there must be a rather formal and really quite an intricate relationship between these agencies, and that is what we are building up."[169] Therefore, Dryden explained that the emergence of the AACB in the spring of 1960 was not the be-all and end-all of the NASA-DOD coordination process. He said it was "only one of the channels for coordination. . . . Not all questions and problems relating to the activities of DOD and NASA, of mutual interest to both, will be resolved as a result of consideration of the matter by the Aeronautics and Astronautics Coordinating Board. Some matters are handled directly by the Administrator and the Secretary of Defense; others are settled at the level of the managers of specific programs and projects."[170] The AACB remained, however, the most visible symbol of NASA-DOD coordination. McNamara concluded, "The functions and work of this Board provide one of the best examples of continuing and effective cooperation between Government agencies engaged in parallel and interacting fields of activity."[171]

A Case Study: Launch Vehicles and the Large Launch Vehicle Planning Group

The functions of and subjects addressed by the myriad coordinating bodies, groups, boards, panels, and committees were as varied as the organizations themselves. One area that was particularly important, because of its direct applicability to the human-spaceflight projects, was the coordinating effort concerning launch vehicles. In fact, coordinating the NASA and DOD launch vehicle families was one of the first matters to which Webb and McNamara turned their attention. On 14 February 1961, Webb and Gilpatric signed an agreement stating, "It is hereby agreed that neither the DOD nor the NASA will initiate the development of a launch vehicle or booster for space without the written acknowledgment of the other agency that such a development would be deemed consistent with the proper objectives of the National Launch Vehicle Program."[172] They hoped this would prevent a proliferation of launch vehicles, with the attendant cost escalation.

Carefully coordinating the national fleet of launch vehicles developed by NASA and DOD became even more important just three months later with Kennedy's lunar-landing decision. An entirely new and larger class of vehicle would be required to launch humans and their associated equipment to the moon and ensure their safe return. While the vehicle that would eventually be built to do this was called the Saturn V, it was not in fact a direct descendant of the vehicle that had been accelerated during the Eisenhower administration. Certainly there were technological elements in the Saturn V that descended from the initial work performed by von Braun's team at the ABMA and then the Marshall Space Flight Center. But the Saturn V was actually closer in configuration and characteristics to the Nova vehicle that NASA and the PSAC had speculated about in the latter stages of the Eisenhower administration.

The Large Launch Vehicle Planning Group's (LLVPG) antecedents are found in a Webb letter to McNamara in July 1961. Webb said that, given NASA's new responsibilities in Apollo, "Formulation of detailed planning for the specification and development of large launch vehicles consistent with both NASA and DOD objectives" was imperative. He proposed the agencies establish a joint LLVPG to accomplish this task. Nicholas E. Golovin, technical assistant to the associate administrator of NASA, would direct it; its deputy director would be Lawrence Kavanau, special assistant to the DDR&E for space. The LLVPG would report to the NASA associate administrator Seamans and

311

Deputy DDR&E Rubel.[173] The foundation for Webb's letter was a proposal from Seamans (dated the same day as Webb's letter to McNamara) to establish the LLVPG. Seamans more clearly described exactly what the LLVPG was to accomplish: "To determine the large launch vehicle configurations and operational procedures which will best meet the needs of the DOD and NASA." The LLVPG was to not only specify the particular configuration of the vehicles required to travel to the moon, it was also to determine the "operational procedures" necessary to do so. In spelling out the guidelines the LLVPG should consider in designing the launch vehicles, Seamans stated, "Both direct ascent and rendezvous options should be considered."[174] The LLVPG was to operate under the AACB's Launch Vehicle Panel. The tasking to determine operational procedures is important because the assessment that the LLVPG essentially failed rests on the fact that not only did it not recommend a specific vehicle configuration, it also did not outline a particular operational mode for reaching the moon.

A memo from Golovin acknowledged Seaman's tasking memo of 7 July described the LLVPG's 15 members, said the LLVPG had held its first meeting on 24 July 1961, and said that it expected to complete its work by November 1, barring any "substantial changes."[175] Little beyond the summary volume of the LLVPG's final report has been declassified.[176] However, speculation in the trade press by September 1961 said, "Bitter controversy is understood to be raking the top policy group charged with working out a national space vehicle program." *Missiles and Rockets* speculated that the LLVPG was divided over the relative merits of solid versus liquid-fueled big boosters and that DOD representatives were complaining that the deliberations were wasting time that should be spent getting the lunar program initiated because "the problems involved in building lunar rockets already have been studied to death."[177] Golovin's personal diary indicates significant dissension between NASA and DOD representatives concerning the DOD's proposal of the Titan III as the DOD's next-generation heavy-lift booster.[178] NASA representatives apparently believed such a vehicle would be redundant to the launcher that would take Apollo to the moon, soon to be known as Saturn.[179] In another entry Golovin records a lunch meeting with his deputy Kavanau, who in turn reported, "McNamara had told him [Kavanau] that the Air Force had rail-roaded through the Titan III recommendation by the LLVPG."[180] Seamans recalled that the DOD introduced the Titan III question into the LLVPG only late in the summer, in part "related to Rubel's very great concern that the Saturn would never work. . . . You get a tremendous 'flexing of interests,' in effect the DOD wanted us to endorse the Titan III. . . . And we weren't just about to endorse it."[181]

By November NASA was already proposing an internal group that would make "a finer cut of the Golovin recommendations" that would be "more specific with regard to the content and emphasis of a program." Apparently NASA felt the LLVPG would not soon recommend a concrete large-launch vehicle program that would "1. Meet the requirements of manned space flight, and 2. Have broad and continuing national utility (for other NASA and DOD missions)," and that NASA would have to consider unilaterally making such a determination for the specific vehicle for Project Apollo.[182] The tentative nature of the LLVPG's conclusions was evident in an AACB Launch Vehicle Panel meeting of 5 January 1962. Golovin briefed the LLVPG's preliminary conclusions, "The Group was of the opinion that earth orbit is probably the best approach from the point of view of reliability and human safety but that the lunar orbit might be attained earlier. The Group concluded that no specific approach should or could be selected at this time and established three classes of boosters according to required payload placement capabilities."[183]

LLVPG's Final Report

The LLVPG's final report was not published until September 1962, over a year after the first meeting for what was thought would be a 90-day project. Its principal recommendations did little to clarify the large launch vehicle situation and seemed to provide little concrete basis from which to plan America's future family of large launch vehicles. Golovin's recommendations basically said to develop everything that was currently being considered, and more—the Saturn C-1, Titan III, Saturn IVB, and a new vehicle called the class B vehicle. Concerning the specific mission mode, the same pattern prevailed; the

LLVPG recommended making a major engineering effort to develop both the earth-orbit and the lunar-orbit techniques as approaches for the lunar-landing mission but also to concurrently develop the direct-ascent capability.[184] In the end, as Seamans stated, the LLVPG involved "a lot of churning around, a lot of effort expended,"[185] but with few final or definite recommendations from which to proceed. One NASA history concluded, "Golovin's group did get mired in the mode issue, leaving the choice of an Apollo launch vehicle still unsettled. . . . Once again nothing was settled. . . . The committee's conclusions—or lack of them—reflected compromises and conflicting opinions."[186]

As in so many questions of space policy, Logsdon ably summarized the LLVPG bottom line. "Despite these analyses and the extensive efforts of the LLVPG, the group reached the end of the study with a relatively large number of critical questions unresolved," Logsdon observed. "As a result, the LLVPG recommendations were somewhat of a compromise and did not provide the basis for the development of an integrated national launch vehicle program, based on a 'building block' program, as had been hoped."[187] The LLVPG case study serves to illustrate that despite the extensive network of NASA-DOD coordination efforts, they did not automatically result in a smoothly efficient and intricately meshed national space program. Institutional interests and personality conflicts still played a part in a coordinating process involving two extremely large bureaucracies that were at times successful and at times a failure.

The NASA-DOD Relationship: Support Specifically

The type and nature of support that the DOD, particularly the Air Force, provided NASA during the Eisenhower administration (described in chap. 3) continued under Kennedy. Nevertheless, there was in some areas a greater movement toward independence. While the Air Force continued to supply launch vehicles in the sense of converted Atlas and Titan ICBMs for the Mercury and Gemini programs respectively, NASA constructed its own Saturn family of launch vehicles for Apollo. The Air Force continued to provide hundreds of officers for transfer to NASA but began to bristle at some NASA personnel requests. Finally, the area of exactly how much NASA would reimburse the Air Force for the multitude of services it provided emerged during the early 1960s as a contentious issue.

The Air Force supervised and administered many of NASA's contracts for hardware procurement. That meant NASA did not have to station contract administrators across the country; the contracts were handled by the preexisting nationwide network of the Air Force Systems Command (AFSC) procurement officers. However, as NASA's budget mushroomed after Kennedy's lunar-landing decision, the demand on these officers grew correspondingly. The AFSC reported by August 1961 that Air Force manpower used to administer NASA contracts was "taken out of the hide" of its officer corps and that "support of regular Air Force programs plus a vital role in the site activation of Atlas, Titan and Minuteman missiles have strained our manpower resources to the breaking point. Additional requirements without increased manpower authorizations can only result in a diluted contract management effort."[188] Though NASA continued to use the DOD regulations and procedures for procurement and contract administration, it began to assume more and more of the burden of administering its own contracts. This is an example of how, over the course of time, NASA moved away from an overt dependence on the military and toward greater institutional autonomy and bureaucratic competence. The same trend held true in many other areas.

The Air Force provided such a preponderance of the DOD support to NASA that McNamara in February 1962 issued DOD Directive 5030.18, *Department of Defense Support of the National Aeronautics and Space Administration*, that officially declared, "It is in the national interest for the Department of Defense, to the extent compatible with its primary mission, to make its resources available to NASA, in the form of facilities and organizations, in order to employ effectively the nation's total resources for the achievement of common civil and military space objectives." The directive also made clear, "Except as the Secretary of Defense may otherwise direct, the Secretary of the Air Force is assigned responsibility for the research, development, test and engineering of satellites, boosters, space probes, and associated systems necessary to support specific NASA projects and programs arising under basic agreements between NASA and DOD."[189] What had been de facto true was now de jure established—the Air Force was the primary provider of DOD support to NASA, though of course still subject to OSD supervision and control. It illustrated the trend described earlier in this chapter whereby McNamara encouraged centralization of military space responsibilities under the Air Force, probably so that OSD could tightly manage military space affairs. AFSC responded to this directive by establishing and filling a new position within NASA headquarters, AFSC deputy commander for Manned Space Flight, Maj Gen Osmond J. Ritland. Ritland was responsible for the direct USAF/AFSC-NASA interface, most of which dealt with human spaceflight, and for coordinating the Air Force's support to NASA.[190]

A Key Issue: Personnel

Perhaps the most valuable type of support the Air Force provided NASA was assigning talented managers from its pool of officers. Table 6 shows the annual number of military officers assigned to NASA. From 1966 on the numbers leveled off and gradually declined because NASA had existed long enough to begin developing its own pool of experienced and capable managers. In addition, "from 1966 on positions were not filled with [military] detailees until a reasonable effort had been made to obtain a civilian." But for 10 years after NASA's inception, Air Force personnel filled a managerial void in NASA

Table 6. Annual number of military officers assigned to NASA

1958	1959	1960	1961	1962	1963	1964	1965	1966	1967	1968	1969
66	67	77	117	161	239	249	280	323	318	317	268

Adapted from Jane Van Nimmen, Leonard C. Bruno, and Robert Rosholt, *NASA Historical Data Book*, vol. 1, *NASA Resources 1958–1968*, NASA SP-4012 (Washington, DC: USGPO, 1988), 80. Ihor Gawdiak and Helen Foder, 1969 figure, *NASA Historical Data Book, Volume IV: NASA Resources, 1969–1978*, NASA SP-4012 (Washington, DC: GPO, 1994), 68.

that could not have been obtained from any other source; Air Force officers were the only class of individuals experienced in conceiving, initiating, developing, and managing large air and space projects.[191] Seamans wrote CSAF White early in the Kennedy era, "We [NASA] are benefiting tremendously from the generous exchange of Air Force personnel now engaged in our projects."[192] In 1963 a NASA official wrote Webb that the Air Force personnel working for NASA had made it "possible for NASA to obtain the services of many fine officers with skills and experience not obtainable from other sources. The cooperation on the part of the Department of Defense has contributed materially to the success of NASA's efforts." In fact, this official urged Webb to try to modify the agreements with the DOD so that these officers could serve significantly longer than the normal three-year tour of duty with NASA.[193] Webb confirmed to the NASC executive secretary Welsh that Air Force personnel "possess certain skills and experience which are not available to NASA from any other source," and if they were ever withdrawn, this "would create a situation in the NASA manning structure which would seriously disrupt the momentum of the national space program."[194]

Perhaps the single most important officer the Air Force loaned to NASA was Brig Gen Samuel Phillips. The person responsible for this transfer was the new OMSF director George Mueller. Mueller had worked for the Space Technology Laboratory of the TRW Corporation in the 1950s when it was heavily involved with providing systems integration for the USAF ballistic missile effort. During the late 1950s, Phillips was the program director for the Minuteman ICBM and impressed Mueller with his performance. One of Mueller's first acts after arriving at NASA headquarters in September 1963 and surveying the situation was to write Webb and urge even greater integration of skilled Air Force personnel at even higher levels within NASA. Mueller explained, "The management of the very large contracts which are characteristic of the lunar program requires a set of skills and background experience which are not now a part of the present and past NASA structure." The solution was that "the national interest would be best served if we could bring to bear upon the management of the lunar program some of the specific program management experience and skills which were developed in the Department of Defense during the conduct of the Polaris, Atlas, Titan, and Minuteman development programs." Mueller closed by mentioning General Phillips as a perfect candidate to direct the Apollo program under his supervision as OMSF director.[195]

In December Webb officially requested General Phillips's transfer to NASA, stating, "We do not have NASA people with the requisite background in program management, nor have we been able to find in industry available people, qualified to carry out these responsibilities." Webb said Phillips was "uniquely qualified to carry out the responsibilities" of Apollo deputy program director and that "his talent is not available either within NASA or in industry."[196] The Air Force immediately complied with Webb's request, and Phillips reported to NASA on 31 December 1963. After a brief stint as deputy director, Phillips served as Apollo program director from October 1964 through the first lunar landing in July 1969, and until September 1969, exercising direct and day-to-day management and control over America's drive to the moon. He later became a four-star Air Force general. Referring to General Phillips and the other Air Force officers, Seamans said, "I don't know if we could have done the project without them."[197]

One joint Mueller/Phillips contribution from their military experience stands out as key to Apollo's success within the decade of the 1960s. The prevailing theory at NASA on how to test space launchers, with their numerous subsystems and assemblies, derived from the methodical work of von Braun and his German rocket scientists at NASA's Marshall Space Flight Center. They tested virtually every item connected with the rocket and its spacecraft separately "and with painstaking detail." This German model meant "long sequences of launches testing various parts of the Apollo configuration in space."[198] The alternative that the Air Force had developed in its ballistic missile program, under Schriever and others, was called "all up" testing. A number of components were tested together and launched together as complete systems, eliminating many tests, albeit with a higher risk of failure at each level of testing. As Phillips explained, "In the simplest terms, the all up concept means build it all and fly it in its final configuration the first time you fly it. . . . It very clearly is the concept that had been established and used in the Minuteman program."[199] Mueller and Phillips discovered that the Apollo program was structured in accordance with the laborious and time-consuming stage-by-stage testing method. Further, if this tactic were followed, America would not reach the moon by the end of the decade. They boldly ordered that the *Saturn V* be tested all up with all its stages and the spacecraft in working order on the first test flight.[200] Without this time compression generated by the all up decision, it seems unlikely that NASA could have reached the moon on schedule, especially considering the yearlong delay caused by the tragic fire that killed three Apollo astronauts on 27 January 1967.[201]

DOD's level of assistance, especially its personnel support, was so extensive and so key to NASA's success that some individuals were convinced there had to be a conspiracy whereby the Air Force was quietly infiltrating NASA in an attempt to take it over from the inside. Hall explained, "Liberal canting underscored the improved relations. So many Air Force line officers held management positions in NASA, those on the left declared, that the nation's space program was now being militarized from the inside out."[202] Whatever the case, Air Force personnel indisputably made a vital contribution to NASA's success in the 1950s and 1960s.

Money Trouble

Within the general topic of DOD support to NASA, one problem area started to emerge during the Kennedy administration but did not blossom into a seriously contentious issue until the Johnson era—exactly how much would NASA reimburse the DOD for services rendered? Chapter 3 noted that the November 1959 agreement on this subject basically stated that if the DOD received an order from NASA that had to be subcontracted out, NASA would only have to reimburse the direct cost of the subcontract; there would be no overhead or administrative charges. If the DOD had the capability to fulfill the contract at one of its facilities, NASA's costs would be limited to the costs directly attributable to performance of the contract;

319

there would be no charges for depreciation, rent, overhead, and so forth.[203]

Several thorny questions arose during the Kennedy presidency. Perhaps the stickiest was how was DOD to separate the costs peculiar to NASA programs, particularly at Cape Canaveral AMR, from the total cost of running the range? In addition, McNamara began to insist on cost sharing of common expenses, contrary to the November 1959 agreement. NASA replied that if it had to pay on a cost-sharing basis, it wanted a management voice commensurate with its share of the funding of common overhead expenses. Since the AMR (renamed Eastern Test Range during the Kennedy administration) was a national range used by several agencies, it was not practical to charge each agency on a cost-sharing basis. This question was negotiated, discussed, renegotiated, and rediscussed without resolution, until NASA and the DOD referred it to the BOB director in 1967 for arbitration.[204] The reimbursement question will be discussed, primarily in the context of the Johnson administration, in chapter 8. Boone expressed the central difficulty: "There was no sound, simple method by which a reasonably accurate estimate of a NASA share of range costs could be made, primarily because the accounting procedures in effect were inadequate to permit making a breakdown of costs associated with the individual segments of workload. Those areas in which direct NASA and DOD costs could be identified constituted only a very small percentage of the total workload and costs."[205] Seamans said the complicated reimbursement issue boiled down to "a mare's nest of accounting."[206]

Mercury and DOD Support of Mercury

NASA's official history lists the total cost of Project Mercury as $384 million.[207] Its two ballistic parabolic flights and four orbital flights ended with Cooper's journey aboard MA-9, *Faith 7*, on 16 May 1963. The DOD's integral role in Mercury involved providing everything from the astronauts to the launch vehicles, from the launch facilities to the recovery forces. "Providing support to Mercury flights has contributed greatly to the Department of Defense's knowledge and experience in areas of launch, network, recovery, communications, and medical space operations. Future spaceflight operations can be effectively supported by applying the experience and procedures derived during Project MERCURY," according to General Davis, the DOD representative for Project Mercury Support Operations.[208] A congressional report shows that for Cooper's flight the DOD provided "28 recovery ships, 171 aircraft, and 18,000 people serving in various capacities."[209] The DOD had to support 32 planned landing areas and 51 contingency landing areas for this final Mercury mission.[210]

The USAF reaffirmed its commitment to continued post-Mercury support after Kennedy greatly expanded NASA's responsibilities with Apollo. Zuckert wrote Webb, "I would like to again reaffirm the Air Force intention to provide the maximum possible assistance to NASA in the discharge of its important responsibilities for this program [Apollo]."[211] In October 1961, NASA and the DOD worked out a detailed, 40-page document specifying exactly how the DOD would support the lunar-landing program. It had separate sections on what the DOD would contribute in management; budgeting and funding; procurement and contracting; bioastronautics; technical support; global communications and instrumentation; technical facilities; range operations; civil engineering; logistical support; personnel; public information; technical information; and foreign technical data. This agreement concluded, "Integration of effort, rather than competition is mandatory." One of DOD's goals was, "Shaping the MLLP [Manned Lunar Landing Program] as feasible to expedite the attainment of basic military capabilities to operate in space."[212]

Of the almost $400 million total cost of Mercury, the DOD provided $133 million in support, or almost one-third of the project's budget, of which NASA reimbursed $100 million.[213] A breakdown of this $133 million is explained in table 7.[214] Not only did DOD personnel render valuable assistance to the NASA program, DOD physical resources such as ships, aircraft, and ICBMs converted to space-launch vehicles played a key role in the success of NASA's first human-spaceflight project. Generally, the DOD absorbed the cost of approximately 25 percent of this physical assistance. These unreimbursed expenses help explain McNamara's drive to establish the new cost-sharing precedent for reimbursement described above.

Table 7. A breakdown of DOD support for Project Mercury (in millions of dollars)

Service	Reimbursed	Absorbed	Total
USAF	83.8	10.4	94.2
Navy	12.2	19.8	32.1
Army	2.3	.7	3.0
Bioastronautics	1.5	2.4	3.9
Total	99.87	33.37	133.24

Adapted from Loyd S. Swenson Jr., James M. Grimwood, and Charles C. Alexander, *This New Ocean: A History of Project Mercury*, NASA SP-4201 (Washington, DC: GPO, 1966), 644–46.

The Reciprocal: NASA's Contributions to National Security

In the Eisenhower administration, NASA developed the idea that it was making a contribution to US national security and the mission of the DOD because it was developing the infrastructure, vehicles, and experience required to operate in space. These capabilities and facilities could be used in times of national emergency. This proposition continued to be NASA's position under Webb, during both the Kennedy and Johnson administrations. Webb said in July 1961, "I think it would be a very brave man who would say that the capacity to operate with large, manned vehicles in space would have no military value."[215] McNamara and the OSD did not seem to be in any rush to endorse this notion, but they did not make any effort to publicly dispute it either.

Seamans explained, "There is an important interchange of components and vehicles between the NASA and DOD programs. United States mastery of space is essential insurance against finding ourselves with a technology inferior to that the Russians will develop as they press forward on the space frontier. If we allow them to surpass us, their space technology in its military aspects will be used to jeopardize our security."[216] Finally, Webb often reiterated, "Our national security demands that we act to insure that no hostile power will use space as an unchallenged avenue of aggression against us. The scientific knowledge and technological skill developed in our program of lunar exploration will give us that assurance, and will form the basis for any military applications which the national interest may require."[217]

As part of his Kennedy-mandated review of the space program in 1963, Vice President Johnson asked both Webb and McNamara to estimate how much of NASA's program was militarily useful. Webb replied, "All of it can be directly or indirectly militarily useful" because everything from launch vehicles to tracking stations "can, in time of need be converted to, or can be utilized to handle military requirements. . . . All those [components] in the program could become indispensable elements of military power. . . . The capability to operate safely and reliably in space is necessary for military control. This capability is being developed both in space and on the ground through NASA programs." Webb concluded, "Therefore, as insurance against surprise and as the building of the necessary underlying capability, I believe this program is completely justified."[218] In his reply to Kennedy, Johnson basically endorsed and forwarded Webb's view on this particular question.[219] Shortly thereafter, and only a few days before he was assassinated, Kennedy explained at a press conference that the United States was spending $5 billion for the space program "of which at least a good percentage has a military implication in the sense of national security."[220]

McNamara's response to Kennedy, however, was not nearly as generous concerning the applicability of NASA's contribution to national security. He wrote that of NASA's budget, expected to be $5.7 billion for FY 64, only the following amounts in the listed categories "would be undertaken by DOD in the absence of a NASA program:" space research—$20 million; exploratory and advanced development—$100 million; Gemini-type program—$150–200 million; mission applications such as meteorological and communications satellites—$25–50 million. McNamara specifically pointed out, "Most of the increase in the augmented NASA effort . . . reflects the lunar program directly and has no demonstrable military value. . . . Based upon what we presently foresee, the Defense Department would not pay for the large augmented management and support effort, or any appreciable fraction of it, if NASA did not." McNamara's bottom line was that of NASA's requested FY 64 budget of $5.7 billion, "I have identified approximately $600–675 million of NASA effort which appears to have direct or indirect value for military technology."[221]

Privately, McNamara was reportedly even more insistent that national security not be used as a major justification for NASA's space program. Seamans recalled that when Webb asked the OSD if the DOD had a preference between the earth orbit rendezvous (EOR) and lunar orbit rendezvous (LOR) lunar mission modes, "The answer came back, 'Look, we're responsible for national security. Sure, you've got your program, we've agreed to your program, but don't try to build it under the umbrella of national security.' Because if it had been otherwise, then McNamara would not have wanted it to be run by anybody other than the Department of Defense. McNamara was very clear on that."[222] Webb was apparently cognizant of the fact that he could not push the national security justification of NASA's program too far or he would risk a more intrusive McNamara presence. Beyond the general statements cited above, Webb never clarified exactly how NASA's R&D was relevant to the DOD; he never progressed beyond saying NASA's abilities and facilities simply would be available for purposes of national defense. Webb later stated, "I never did want to particularly clarify that. . . . McNamara wanted to take the view that only the money that fed the projects under his control contributed to defense."[223]

While containing an element of exaggeration, there is also some truth to the statement that "by 1963, however, the Air Force needed NASA almost as much as NASA needed the Air Force."[224] At a minimum, NASA began to achieve a degree of emancipation from the high levels of dependence it had on the DOD during the Eisenhower administration. During its first few years, NASA had no choice—DOD was the only organization that had the facilities, the experience, the managerial expertise, and the rockets NASA required to do its job. Over time, however, NASA would develop its own resources in each of these categories and began to move away from its close reliance on the DOD; this process started during the Eisenhower administration and gained momentum during Kennedy's. As one scholar explained, "While the Air Force's participation in NASA activities was consolidated during the Kennedy administration, its influence actually declined" because of the rapid increase in NASA appropriations following Kennedy's lunar-landing decision. This decision not only increased NASA's political constituency but also "sealed the primacy of NASA's manned space flight programme over the Air Force's."[225]

One must not take this too far, as did one scholar who declared, "The important point is that *the military and the civilian space programs are gradually being integrated into one plan, and NASA is becoming part of the evolving United States 'Space Force.'.* . . . A combination of interagency politics and accounting maneuvers allows the Air Force increasing penetration into the space program without the nation's giving it a clear go-ahead. . . . [NASA is] an embryonic fourth military space service, sometimes rival, sometimes partner of the Air Force, in astronautical maneuvers in the capital." (emphasis in original)[226]

As is often the case, Arnold S. Levine represents a calmer and more rational perspective on NASA-DOD relations in general and for the Kennedy administration specifically:

> The essence of the NASA-DOD relationship had far more to do with mutual need than with philosophical arguments concerning the existence or the desirability of one space program or two. . . . The principles underlying the U.S. space program resulted less from anything enunciated in the Space Act than from President Kennedy's May 1961 decision to assign the lunar-landing program to NASA. But this decision was preceded by earlier moves by NASA and DOD officials and by Congress to prevent an Air Force takeover. . . . With the backing of the President and much of Congress and the acquiescence of McNamara, NASA, on the one hand, staked out its position as an independent agency while, on the other, waging a quiet behind-the-scenes battle with DOD to maintain that independence. . . . NASA would cooperate with the DOD, but never to the point of giving away its authority to meet its needs. The history of NASA from its establishment to the mid-1960s can be charted in terms of NASA's ability to design its own programs, procure its hardware, and support its spacecraft without overt interference from the military.[227]

The next chapter explains how the general principles of the NASA-DOD relationship set forth in this chapter came into play with the human-spaceflight projects of Dynasoar, Gemini, and the MOL.

Notes

1. Neufeld, *Ballistic Missiles in the United States Air Force*, 103.

2. Gardner, *Report of the Air Force Space Study Committee*, i.

3. Finally, Schriever pledged, "You may be assured that a copy of this document [the committee's final report, when it was available] will be made available to NASA." Schriever to Glennan, letter, 11 January 1961, 1.

4. Oral history interview of Schriever by the author, 2 July 1996. In another context, Schriever said the report was simply "put on the shelf." Schriever, "Comments," in Needell, *The First 25 Years in Space*, 28. Secretary of the Air Force Zuckert confirmed, stating, "I thought the program was much too ambitious" as presented in the committee's final report. Zuckert, oral history interview, 25 July 1964, reel 12, 127.

5. Gardner, *Report of the Air Force Space Study Committee*, 4, 6.

6. Ibid., 7, 14.

7. Ibid., 27, 32.

8. Schriever (speech, Allegheny Conference on Community Development, 21 November 1960, A-93–A-94).

9. Ibid., A-94.

10. Air Force, "Air Force Competency in Space Operations," 1 December 1960, 1–3.

11. Ibid.

12. Booda, "Air Force Outlines Broad Space Plans," 5 December 1960, 26.

13. Finney, "Air Force Seeks Top Role in Space," 11 December 1960, 68.

14. Seamans, *Aiming at Targets*, 97.

15. BeLieu to Johnson, memorandum, 17 December 1960, 2–3.

16. Garlock to DDR&E Herbert York, letter, 12 January 1961, 1.

17. Air Force, "NASA-USAF Cooperation," 1 February 1961, 1.

18. Robert Seamans, interview by the author, 5 July 1996. See also Seamans, series of oral history interviews, 2 November 1987.

19. Webb, oral history interview, 11 April 1974, 42–43.

20. White to Webb, letter, 31 January 1961, 1.

21. White to Brooks, letter, 19 January 1961, 1. White was citing from a letter Brooks had written him expressing Brooks' concerns. White was apparently quite forthcoming in his desire to help.

22. Glennan to Brooks, letter, 27 January 1961.

23. Brooks to Glennan, letter, 14 February 1961, 1.

24. Wilson's response to Brooks, internal memorandum, 1–2.

25. White to Glennan, letter, 21 March 1961.

26. Logsdon et al., *Exploring the Unknown*, vol. 2, 315–17.

27. As for the DOD role in space, Kennedy cleared the air: "I have been assured by Dr. Wiesner that it was not the intention of his space task force

to recommend the restriction of the NASA to the area of scientific research in space." Kennedy to Brooks, letter, 23 March 1961, 317.

28. Shapley, oral history interview, 14 December 1967, 26.

29. Kennedy, *Public Papers of the Presidents*, no. 245, 14 June 1962, 495.

30. Lambright, *Powering Apollo*, 90.

31. Logsdon, *The Decision to Go to the Moon*, 79. Another team of scholars concurred, "The so-called 'military-industrial complex' had failed, if indeed it had ever tried, to reduce NASA." Swenson et al., *This New* Ocean, 338.

32. DOD, Directive 5160.32, 6 March 1961.

33. Wolf, *The United States Air Force Basic Documents on Roles and Missions* , 361.

34. However, McNamara's cover letter to the Directive 5160.32 made clear it did not automatically predetermine the assignment of operational responsibility for each and every space system to the Air Force. He said operational responsibility of a particular space system would be done project by project and "will take into account the competence and experience of each of the services and the unified and specified commands." McNamara, cover letter to DOD Directive 5160.32, 6 March 1961, 1.

35. Gilpatric, testimony before House Committee on Science and Astronautics, March 1961, 11–12.

36. Schriever, oral history interview, 20 June 1973, 27–28.

37. Ibid. See Air Force, *The Genesis of the Air Force Systems Command*, for the complete history. The AFSC had four divisions: Space Systems, Aeronautical Systems, Ballistic Systems, and Electronic Systems.

38. Hitch, testimony before House Committee on Science and Astronautics, March 1961, 82.

39. Zuckert, oral history interview, 25 July 1964, 125.

40. Futrell, *Ideas, Concepts, Doctrine*, vol. 2, 215.

41. Berger, *The Air Force in Space, Fiscal Year 1962*, 3–6.

42. The Air Force pledged full support of NASA's lunar-landing effort: "The lunar program will provide valuable data for military activities in space. It is expected that the civil and military efforts in space programs during the next decade will continue to complement each other. . . . The Air Force will provide the fullest possible support to the lunar program." But the USAF also emphasized the threat the Soviet space program posed: "It is clear that the Soviets have the technical capabilities to develop a serious military space threat to the nation. The Air Force believes that these growing technical capabilities will be developed into a threat." Therefore, there existed "the definite possibility of a surprise action which could result in Soviet military dominance of space." Air Force, *Air Force Space Plan*, September 1961, 1, 3, 7, 11, 13.

43. Ibid., 21–22, 27.

44. Ibid., 37.

45. Ibid., 40, 44.

46. Testimony of Ferguson, *DOD Appropriations for 1963*, hearings pt. 2, 476–77.

47. NASA, *Aeronautical and Astronautical Events of 1961*, 54.

48. McNamara to Zuckert, memorandum, n.d., 1. Document is now declassified.

49. Zuckert to Schriever, memorandum, n.d. but sometime after 19 October 1961, 1. Document is now declassified.

50. Zuckert, oral history interview, 25 July 1964, reel 12, 128.

51. Zuckert, series of oral history interviews, December 1986, 54.

52. Cantwell, *The Air Force–NASA Relationship in Space, 1958–1968*, 32.

53. Berger, *Air Force in Space, Fiscal Year 1962*, 24.

54. Zuckert made a point to emphasize, "The National Aeronautics and Space Administration has a massive program to acquire a capability to operate in space. The Air Force is supporting it to the limit of our abilities. We need what NASA will learn. Necessarily, the Air Force also has space programs of its own." Zuckert, "The Secretary of the Air Force Speaks on Space Programs," 4.

55. LeMay to Zuckert, letter, 19 October 1962, 1–3.

56. Cantwell, *The Air Force in Space, Fiscal Year 1963*, 7.

57. Rubel clarified the level of DOD space spending "as close to the optimum size as we can make it in the light of all the uncertainties that must accompany such a program. In fact, we probably err on the side of allowing too generous a margin of safety for the effects of these uncertainties. Henceforth the DOD would emphasize hard military requirements." House, *Government Operations in Space*, 77–78.

58. McNamara, *Military Posture*, 483.

59. Rubel, "Military Space," 9 October 1962, 1.

60. Kennedy, *U.S. Aeronautics and Space Activities, 1961*, 33.

61. Ibid.

62. Brown, "Statement to the Senate Committee on Aeronautical and Space Sciences," 14 June 1962, 1.

63. Ibid.

64. McNamara, *The Essence of Security*, x, 87–88.

65. Ibid., 90–91, 93.

66. Shapley, *Promise and Power*, 107.

67. Kaplan, *Wizards of Armageddon*, 255.

68. The implementers of systems analysis in the Pentagon under McNamara have written entire books explaining the extraordinary intricacies of PPBS. See Enthoven and Smith, *How Much is Enough*, 1971; and Tucker, *A Modern Design for Defense Decision*, 1966. Charles Hitch was intimately involved in the systems-analysis process as McNamara's assistant SECDEF, comptroller. He provides a full-length treatment of PPBS in *Decision Making for Defense*, 1965 and a concise treatment in "Plans, Programs, and Budgets in the DoD," *Operations Research*, 1–17.

69. Kaufmann, *The McNamara Strategy*, 172–73.

70. McNamara, *Essence of Security*, 95.

71. Shapley, *Promise and Power*, 101.

72. Both Enthoven and Hitch cited in Kaufmann, *The McNamara Strategy*, 179–80.

73. Shapley, *Promise and Power*, 103.

74. Witze, "How Our Space Policy Evolved," 83–92.

75. Levine, *Managing NASA in the Apollo Era*, 219.

76. House, Committee on Government Operations, *Government Operations in Space*, 84.

77. Shapley, "Robert McNamara: Success and Failure," 418.

78. McNamara, *Military Posture*, 462.

79. Nieburg, *In the Name of Science*, 49.

80. Entire monographs have been written not only to evaluate the success of McNamara's overall PPBS approach, but also its application to specific weapons systems. The most famous example of enforced commonality was the Tactical Fighter Experiment, or TFX, which became the Air Force's FB-111 fighter-bomber. For TFX case studies, as well as overall systems-analysis evaluations, see Art, *The TFX Decision*, 1968; and Coulman, *Illusions of Choice*, 1977.

81. Donovan, *The Cold Warriors*, 155.

82. McDougall, *The Heavens and the Earth*, 332.

83. Shapley, *Promise and Power*, 240, 246.

84. Shapley, "Robert McNamara: Success and Failure," 420.

85. White from the *Saturday Evening Post* and LeMay quotations cited by Kaplan, *Wizards*, 255–56. LeMay likened McNamara to a hospital administrator who dabbled in brain surgery. Watson, *The Office of the Secretary of the Air Force*, 215.

86. Schriever, oral history interview, 20 June 1973, 35–37.

87. "A Quiet Retirement," *Time*, 9 September 1966, 25 (referring to Schriever's retirement from the USAF).

88. Watson, series of oral history interviews of Zuckert, 9 December 1986, 3–5. One should remember that Zuckert was relatively close to McNamara, having been on the faculty with him at the Harvard Business School shortly before World War II and that McNamara personally asked Zuckert to serve as secretary of the Air Force.

89. Cantwell, *Air Force in Space, Fiscal Year 1963*, 1, 8.

90. Worthman, "The Promise of Space," 124.

91. Senate, *NASA Authorization for Fiscal Year 1963*, 343.

92. Ibid., 342–43. That same month Brown told a trade magazine, "We cannot visualize or define now a military mission for man-in-space." Coughlin, "Speak Up, Mr. Secretary," 46.

93. Rubel, "Military Space," 9 October 1962, 1, 4–7.

94. "Pentagon Shows Caution on Space," *New York Times*, 10 October 1962, 25.

95. McNamara, interview in *Missiles and Rockets*, 22 October 1962.

96. Gilpatric, "In Gilpatric Speech Military Space Move Left to Russians," 16.

97. Gordon to the president, memorandum, 13 November 1962, 456.

98. Whisenand to LeMay, memorandum, 30 January 1963, 1.

99. "DOD Space Position Defended," *Missiles and Rockets*, 4 February 1963, 12.

100. Sickman, "The Fantastic Weaponry," 224.

101. Miller, *Congressional Record* (speech, 6 September 1962, 18674).

102. Ibid., 18671–72. Miller pointed out that of all the money America had so far spent in space, 43 percent was for military space: "I find it difficult to view this record as flagrant disregard for the military's interests." (ibid.).

103. Meier, *Politics and the Bureaucracy*, 147.

104. Kaplan, *Wizards*, 256.

105. Watson, *Office of the Secretary of the Air Force*, 242.

106. Bradlee, *Conversations with Kennedy*, 112. One Kennedy biographer posits that Kennedy "despised" Air Force chief of staff LeMay: "In the White House, Kennedy had walked out on LeMay, more than once. Walking out on generals was a Kennedy specialty. . . . 'I don't want that man near me again,' Kennedy said after one of his walk-outs on LeMay. McNamara and his men learned not to bring the general's name up. 'He had a kind of fit if you mention LeMay,' Roswell Gilpatric warned one of his assistants." Richard Reeves, *President Kennedy*, 182. During the Cuban missile crisis, LeMay reportedly pounded a table and exclaimed, "It's the greatest defeat in our history, Mr. President. . . . We should invade today!" Kennedy later mused, "It's lucky for us that we have McNamara over there." Beschloss, *The Crisis Years*, 544.

107. York, oral history interview, 16 June 1964, 9.

108. Reeves, *President Kennedy*, 183.

109. Schlesinger Jr., *A Thousand Days*, 912.

110. Gilpatric, statement before the Senate Committee on Aeronautical and Space Sciences, 13 June 1962, Congressional Appearances, 191. "Defense Department and those of the NASA shall be conceived, planned, and executed to insure that the totality of our space efforts adds up to a single program in the national interest." Gilpatric to the Senate Space Committee, November 1963, 21350. Kennedy's 1962 report to Congress stated, "It is national policy to maintain a viable national space program, not a separate program for NASA and another for Defense." Cited in House, Committee on Government Operations, *Government Operations in Space*, 63.

111. Gardner, "Cooperation in Outer Space," 359.

112. Welsh to the American Legion, message 6 October 1962, 1. Vice President Johnson declared, "The United States does not have a division between peaceful and nonpeaceful objectives for space, but rather has space missions to help keep the peace and space missions to improve our ability to live well in space." Kennedy, *U.S. Aeronautics and Space Activities, 1961*, introduction, 6. Johnson emphasized later in 1962 that "all of the U.S. space projects are peaceful, including those which help us maintain the peace. . . . So far as the U.S. is concerned, there is not a distinction between peaceful and

nonpeaceful purposes. They are all peaceful purposes. I wish I could say the same with confidence about the plans and objectives of the USSR." Johnson, "The Vision of a Greater America," 8.

113. NSC, NSAM 183,1.

114. Webb, memorandum for record, 24 February 1961, 1.

115. AACB, Minutes of the 9th Meeting, 2.

116. Schriever to CSAF LeMay, letter, n.d., 1.

117. NASC, Summary Minutes, 18 August 1961, 2.

118. Zuckert, Statement in the *GE Forum*, 10 January 1962, 2. In March 1962, Zuckert emphasized that the peacetime role of NASA and the defense role of the Air Force in the national space program "must advance in harness, and they do. They are interdependent. One cannot move without the other." Zuckert (speech, 6 March 1962, 1).

119. Rubel to Welsh, letter, 10 April 1962, 14. For instance, Seamans recalled that Webb had a self-imposed rule that appointments to sensitive NASA positions like Associate Administrator for Manned Space Flight were to be cleared with the DOD. Levine, *Managing NASA in the Apollo Era*, 138.

120. Coughlin, "The Gleeful Conspiracy," 46.

121. Lloyd, "Letter to the Editor," 14 May 1962, 7.

122. McNamara, statement before the Senate Armed Services Committee on the FY 1964–68 Defense Program, 21 January 1963, 136.

123. Boone, *NASA Office of Defense Affairs*, 8.

124. Ibid.

125. Webb to the vice president, memorandum 10 May 1963, 17.

126. Boone to Webb, report, 12 July 1963, 348–56.

127. Ibid.

128. Ibid.

129. Boone's discussion of the two remaining points, ground support operations and aeronautics, while interesting, is not directly relevant to this discussion and in fact any useful treatment of the NASA-DOD relationship in each of these areas merits at least a chapter-length treatment, possibly even an entire monograph.

130. McMillan to Zuckert, memorandum, 15 March 1963, 2.

131. Gilpatric, oral history interview, 30 June 1970, reel 5. W. Henry Lambright, Webb's biographer, interviewed McNamara and reported, "Webb talked too much for him and was too 'political.'" Seamans believed, "McNamara was more powerful than Webb. But Webb had more guile." Lambright, *Powering Apollo*, 120.

132. Lambright, *Powering Apollo*, 240, n. 56.

133. Oral history interview, 5 July 1996, by the author. It must be reiterated that the author made repeated attempts to secure an interview with McNamara but they were all rebuffed. McDougall cites an oral history interview with Willis Shapley, who was responsible for both the NASA and DOD budgets within the BOB before he became a NASA deputy associate administrator in September 1965, in which Shapley confirmed that by late 1962

Webb and McNamara were "not speaking to each other." McDougall, *The Heavens and the Earth*, 513, n. 55.

134. Webb, oral history interview, 11 April 1974, 33–34. In another interview Webb recalled meeting with Kennedy three weeks before Kennedy's death to relate to him that space might become an issue in the election campaign because "McNamara will not say that this program has military advantage. I will say that every bit of the things we're doing contributes to the military." Kennedy replied, "Well, you're not going to let this get personal, are you?" Webb said, "No. Just the fact that that's the way it is." Kennedy concluded by telling Webb, "Go ahead and do what you think is right." Webb, oral history interview, 15 October 1985, 226.

135. Schriever to General Curtin, memorandum, 13 February 1961, 5–6.

136. Webb, oral history interview, 15 March 1985; and NASM, 88.

137. Webb, oral history interview, 11 April 1974, 38.

138. Lambright, *Powering Apollo*, 90.

139. Ibid., 91.

140. Gilruth, oral history interview, 27 February 1987, 247.

141. White to Generals Landon and Wilson, letter, 14 April 1960, 1.

142. House, Committee on Science and Astronautics, hearing, March 1961, 35–36, 92–93.

143. Ibid., 97, 101.

144. Ibid., 101.

145. Ibid., 105–06.

146. House, Committee on Science and Astronautics, H. Rep. 360, 5 May 1961, 36.

147. Bilstein, *Orders of Magnitude*, 69.

148. The best treatment of this complex bureaucratic wrangling is Benson and Faherty, *Moonport*, 80–105.

149. Ibid., 95.

150. Ibid., 104.

151. Smith, "Canaveral, Industry's Trial by Fire," 204.

152. Boone, *NASA Office of Defense Affairs*.

153. Zuckert, oral history interview, September 1965, 40.

154. Kennedy to Johnson, memorandum, 9 April 1963, 468.

155. Johnson to the president, report, 13 May 1963, 472.

156. NASA, News Release no. 62-249, 21 November 1962, 1.

157. Webb to the vice president, memorandum, 10 May 1963, 16.

158. Hale to Welsh, memorandum, 19 November 1962, 1. Hale was quoting an unnamed source he had spoken with inside NASA.

159. Levine, *Managing NASA*, 219.

160. Lambright, *Powering Apollo*, 91.

161. Levine, *Managing NASA*, 18.

162. House, *Government Operations in Space*, 123–132.

163. NASA, "Inventory of NASA Interagency Relationships," 13 October 1967.

164. House, *Government Operations in Space*, 101.

165. Van Dyke, *Pride and Power*, 205. DOD coordinating process would be the Tri-Service Working Group on Solid Propellants, the NASA-DOD Space Science Committee, the NASA-DOD Working Group on Planetary Observatories, the DOD-NASA Wind Tunnel Study Group, and the Large Solid Motor Technical Assessment Committee. These particular examples, of which scores more could be listed, are from an internal Air Force document, *USAF/NASA Coordination in Space Problems*, 16 March 1961.

166. Zuckert, oral history interview, 25 July 1964, 125.

167. Van Dyke, *Pride and Power*, 204.

168. Stephen I. Grossbard, *The Civilian Space Program: A Case Study in Civil-Military Relations*, 167.

169. McNamara, testimony before the Senate Armed Services Committee, 22 January 1962, excerpted in *Air Force Information Policy Letter, Supplement for Commanders, Special Issue: Military Mission in Space, 1957–1962*, 18.

170. Dryden, statement to the House Committee on Science and Astronautics, 17 May 1962, 4–5.

171. McNamara to the Senate Armed Services Committee, 1963, cited by Van Dyke, *Pride and Power*, 202.

172. Webb and Gilpatric joint memorandum, 14 February 1961, cover letter, 1.

173. Webb to McNamara, letter, 7 July 1961, 1.

174. Seamans to Webb, memorandum, 7 July 1961, 2. The most likely scenario to explain the simultaneous dates is that Seamans prepared a package for Webb that included not only Seamans' memo but also a letter to McNamara already drafted and ready for Webb's signature and transmission. "Direct ascent" is a reference to one theory on how best to reach the moon: a gigantic multistage rocket would be launched from the earth's surface and after jettisoning its spent stages, it would proceed on to the moon. Retrorockets would slow its descent to the lunar surface. After completion of lunar exploration, the remaining stages would be reignited, and the astronauts would proceed back to earth. A truly mammoth vehicle would be required for this mode. "Rendezvous options" referred to another theory, earth orbit rendezvous (EOR) whereby relatively smaller, multiple, and separate rockets would be launched into earth orbit, rendezvous for assembly, and then proceed on to the moon, whereupon much the same procedure outlined above would take place. The actual mode selected for and used in the Apollo program was a third, hybrid option: lunar orbit rendezvous (LOR). In it a large rocket, but not as huge as envisioned for direct ascent, would blast off, leave the earth's atmosphere, jettisoning spent stages as necessary. However, only a small lunar excursion module (LEM) would descend to the lunar surface. After exploration, an even smaller subsection of the LEM would lift off from the lunar surface, rendezvous and dock with the command module in orbit above the moon, and proceed back to earth. For a full explanation see James R. Hansen, *Enchanted Rendezvous: John C. Houbolt and the Genesis of the Lunar-Orbit Rendezvous Concept*.

175. Golovin and Kavanau to the Launch Vehicle Panel of the AACB, memorandum, 31 August 1961, 1, 5.

176. Golovin et al., "Summary Report: NASA-DOD Large Launch Vehicle Planning Group," 24 September 1962. Document is declassified.

177. "Policy Split over Boosters Reported," 94.

178. The basic configuration of the Titan III was that a standard liquid-fueled Titan II USAF ICBM would have attached to it two large solid-fueled rocket engines, one on each side of the liquid-fueled core. This meant, of course, that the standard Titan II ICBM serving as the vehicle's core would have to be substantially modified to be able to withstand the added weight and thrust of the solid-fueled additional engines.

179. Golovin, Chronological File Entry, 30 October 1961, 1.

180. Golovin, Chronological File Entry, 22 November 1961, 1.

181. Seamans, oral history interview, 26 May 1966, 3–4.

182. Rosen to Holmes, memorandum, 6 November 1961, 1–2.

183. AACB, minutes of the 9th meeting, 5 January 1962, 2.

184. Golovin et al., "Summary Report: NASA-DOD Large Launch Vehicle Planning Group," 24 September 1962, 318–37.

185. Seamans, oral history interview, 15 December 1988, 389.

186. Brooks et al., *Chariots for Apollo*, 48–49. Another NASA source concurred, stating that when the LLVPG finished its work, "Too many questions remained open, too many answers equivocal, pleasing neither NASA nor Defense, and the committee had failed to produce the integrated national launch vehicle program it had been created for." Hacker and Grimwood, *On the Shoulders of Titans*, 68.

187. Logsdon, *NASA's Implementation of the Lunar Landing Decision*, 33. Another space scholar concurred and added that in the LLVPG process "the different requirements and institutional interests of NASA and the DOD became clear. Both agencies distanced themselves from the contents of the report." By the time of the report's release in September 1962, "It had been obvious for some time that there would be little cooperation between NASA and the DOD on large launch vehicles. The result was a further solidification of entirely separate and redundant rocket development programs in the civil and military spheres." Day, "Invitation to Struggle," 258–59.

188. Sidders and Bickett, *Air Force Support of Army, Navy and NASA Space Programs*, 9.

189. DOD, Directive 5030.18, 24 February 1962, 1–2.

190. Montgomery to multiple USAF recipients, memorandum, 23 April 1962, 1–2.

191. Levine, *Managing NASA in the Apollo Era*, 121–22.

192. Seamans to White, letter, 28 February 1961, 1.

193. Siepert to Webb, memorandum, 8 February 1963, 673–74.

194. Webb to Welsh, letter, 7 May 1963, 1. In fact, NASA internally expressed concern that elements within NASA were trying too hard to "recruit" military personnel to apply for transfers to NASA. NASA's Executive Officer wrote to the NASA personnel director that NASA's facility in Cleveland, the

Lewis Research Center, had been urging "interested military personnel, such as graduating seniors in ROTC programs, to write to Lewis Research Center if they want to be assigned for work there. . . . It seems to me that LRC is misinterpreting the spirit and intent of the NASA-DOD agreement. . . . I do not think it should be interpreted as a license for NASA to proselyte service personnel on a wholesale basis. If the Lewis announcement comes to the attention of the military departments, it can prove embarrassing to NASA." Young to Lacklen, letter, 10 September 1963, 1.

195. Mueller elaborated, "I have thought that the actual Air Force ballistic missile program management experience would be most appropriate. . . . I believe the Air Force experience would be most valuable to us, and it would fill what I believe to be our greatest void of capability. It is particularly worth noting that the Air Force, over a period of years, has developed the capability of managing and controlling the very contractors upon whom we have placed our primary dependence for the lunar program." Mueller to Webb, letter, 26 September 1963, 1–3.

196. Webb to Zuckert, letter, 11 December 1963, 1–2.

197. Seamans, oral history interview, 15 December 1988, 398. Secondary sources agree, stating that in Project Apollo, "the Air Force influence was pervasive, from the Headquarters level on down." Bilstein, *Stages to Saturn*, 289.

198. Lambright, *Powering Apollo*, 116–17.

199. Phillips, oral history interview, 22 July 1970, 25–26.

200. Ibid., 27–28; See also R. Hall, "Project Apollo in Retrospect," 158; and Lambright, *Powering Apollo*, 117–18.

201. Launius, *NASA*, 87–88.

202. R. Hall, "Civil-Military Relations in America's Early Space Program," 31.

203. DOD, "Agreement between the DOD and NASA: Concerning the Reimbursement of Costs, 12 November 1959, 293–96.

204. Levine, *Managing NASA*, 222–23.

205. Boone, *NASA Office of Defense Affairs*, 126.

206. Seamans, oral history interview, 15 December 1988, 384.

207. Swenson et al., *This New Ocean*, 508.

208. Davis to the SECDEF, report, 11 September 1963, 78.

209. House, Subcommittee on NASA Oversight, *The NASA-DOD Relationship*, 26 March 1964, 8.

210. Davis, *DOD Support of Project Mercury*, 26.

211. Zuckert to Webb, letter, 25 August 1961, 1. The program intervening between Mercury and Apollo was Project Gemini. It was not officially approved until December 1961, and so Zuckert could not have included it in this particular pledge for continued Air Force post-Mercury support.

212. Gilpatric, "Agreements for Support of Manned Lunar Landing Program," 2 October 1961, 4–5. One point occasionally mentioned concerning DOD and the Apollo program is that the DOD would have preferred NASA to select the EOR method over the LOR method because EOR would develop the near-earth rendezvous and docking capabilities the DOD estimated

would be crucial to missions such as satellite interception. When asked if DOD was upset when NASA chose LOR, Seamans replied that Webb specifically asked McNamara to review the EOR versus LOR question and make clear the DOD's preference: "They didn't even come up with anything that indicated that it made much difference to them." Seamans, oral history interview, 15 December 1988, 390. Seamans told this author that when the OSD replied to NASA's question on the OSD's preference for LOR versus EOR, "In effect, they came back and said it's immaterial to us." Seamans, oral history interview, 5 July 1996.

213. Davis, *DOD Support of Project Mercury*, 2.

214. Swenson et al., *This New Ocean*, 644–46.

215. Webb, "Why Spend $20 Billion to Go to the Moon?" interview, 3 July 1961, 60.

216. Seamans (speech to the Aerospace Corporation, 29 August 1961, 5).

217. Webb, "NASA-Industry Program Plans Conference," 11–12 February 1963, 86. Dryden went so far as to declare that the lunar-landing effort had two fundamental purposes: "(1) Insurance of the Nation against scientific and technological obsolescence in a time of explosive advances in science and technology; and (2) Insurance against the hazard of military surprise in space. The manned lunar exploration program constitutes essential insurance against finding ourselves with a position in the new technology inferior to that of a possible enemy." Dryden (speech to the American Aeronautics and Astronautics Society, 30 December 1961, 7–8).

218. Webb to the vice president, memorandum, 30 July 1963, 1–2, 4–5.

219. See Johnson's summary of his response to Kennedy in his memoirs, Johnson, *The Vantage Point*, 282.

220. Kennedy, *Public Papers of the President, 1963*, no. 459, 14 November 1963, 847.

221. McNamara to Lyndon Johnson, report, 3 May 1963, 342–47.

222. Seamans, oral history interview, 19 January 1988, NASM, 260.

223. Webb, oral history interview, 15 October 1985, 226.

224. Levine, *Managing NASA in the Apollo Era*, 236. Levine was referring in particular to the fact that the Air Force frequently used such NASA facilities as its 16 different wind tunnels, its ground-based flight-motion simulator, and its 18.3-meter vacuum-environmental sphere.

225. Stares, *The Militarization of Space*, 61–62. See also House, *Government Operations in Space*, 135, for a similar conclusion.

226. Etzioni, *The Moon-Doggle*, 136–37, 142.

227. Levine, "Management of Large-Scale Technology," 47, 50.

Chapter 7

Gemini, Dynasoar, and the Manned Orbiting Laboratory

This chapter attempts to delineate the complex relationship between the three primary human-spaceflight projects of the Kennedy administration that were relevant to the NASA-DOD relationship—Gemini, Dynasoar, and the MOL. This chapter first briefly describes the genesis of NASA's Project Gemini, then moves on to McNamara's attempt to wrest management control of Gemini from NASA for the DOD, and then discusses the actual role of DOD in the project that emerged in early 1963.

After Webb and McNamara defined the DOD's role in Gemini in January 1963, Gemini began to influence McNamara's thinking about the requirement for Project Dynasoar. By the end of the year he canceled Dynasoar, believing that a combination of the Gemini capsule and a module attached to it called the MOL could best fulfill DOD's human-spaceflight requirements. The exact specifications of MOL became clear, however, only after another significant period of NASA-DOD give-and-take to ensure that the MOL was not considered a space station, thereby infringing on a mission area in which NASA felt it should play the primary role.

Project Gemini and the DOD

Project Gemini is often lost in the shuffle between America's first human steps into space with Mercury and its successful drive to the moon with Apollo. Besides serving as a vital developmental bridge between Mercury and Apollo, Gemini is also of crucial importance within the NASA-DOD human-spaceflight framework. The capabilities it offered eventually convinced McNamara to cancel Dynasoar and initiate a new DOD human-spaceflight project based on the Gemini capsule, with a cylindrical laboratory attached to it—the MOL.

Overview of Gemini and the DOD's Role

On 14 April 1961, NASA offered a study contract to the McDonnell Corporation for an improved version of the Mercury spacecraft. This Mercury Mark II would increase the size of the original Mercury capsule by approximately 50 percent, so it could carry two astronauts instead of one. In addition, significant hardware modifications to the capsule would enable it to conduct advanced missions such as rendezvous, docking, and transfer of humans and material, as well as extravehicular activity (EVA) or "space walking." In a related development in May 1961, the Martin Company, the manufacturer of the Air Force's Titan missile, briefed NASA on ICBM's possible applications to the next level of NASA's human-spaceflight program. Therefore, on 7 December 1961, NASA officially approved a development plan for the Mercury Mark II program involving the larger and more capable capsule and the Titan rocket. On 2 January 1962, the program was given its official name—Project Gemini.[1]

From Gemini's earliest moments, there was disagreement over the exact role DOD should play. NASA's Fred Boone, deputy associate administrator for Defense Affairs and a retired admiral, said that from its inception Gemini was "visualized as a program in which the Air Force would be deeply involved."[2] A 7 December 1961 memo that explained Gemini was written by Seamans, NASA associate administrator, and Rubel, deputy DDR&E, and addressed to both Webb and McNamara stated that as a result of "extensive studies, it is believed that the development of an earth orbital rendezvous capability is most important for the timely accomplishment of the manned space flight and manned lunar missions." Therefore, Mercury Mark II (soon to be renamed Gemini) had been formulated "with the objective of achieving manned rendezvous and relatively long duration earth orbital flight on a schedule considerably earlier than possible for the Apollo spacecraft." Seamans and Rubel continued, "The overall management and direction for the Mercury Mark II/Agena rendezvous development and experiments is the responsibility of the NASA as part of the manned space flight program. However, it is recognized that it is highly desirable that the resources of the DOD, especially the Air Force, be utilized in a contractor relationship by the NASA to the maximum degree practicable, both in order to facilitate the attainment of project objectives and to permit

DOD organizations to acquire useful design, development and operational experience."[3]

They concluded by outlining the initial Air Force role in Gemini, which "should include that of being the NASA contractor for the Titan II launch vehicle of the Mercury Mark II spacecraft and for the Atlas-Agena vehicle used in rendezvous experiments. DOD responsibilities should also include assistance in the provision and selection of astronauts and the provision of launch, range and recovery support, as required by NASA."[4] The government's official description of Gemini said its goals were to "develop and fly at an early date, a two-man spacecraft capable of rendezvous and being brought together (docking) with another vehicle in orbit around the earth, and carry out orbital flights lasting from a few days to a week to study how man functions under prolonged conditions of weightlessness to carry out a variety of scientific investigations of space."[5] Internally, in the context of McNamara's attempt in late 1962/early 1963 to have DOD take over Gemini, NASA emphasized that Gemini was a critical link and essential step between Mercury and Apollo: "The experience to be gained in Gemini, both in hardware and in operations, is needed in order to proceed with the current Apollo program." If Apollo had to proceed without the benefit of Gemini, "This alone would cause a substantial delay in the achievement of a manned lunar landing, and would increase the Apollo program costs."[6] Nevertheless, the DOD's interest in Gemini continued because it offered two potentially valuable defense-related capabilities. First, its enlarged capsule offered a possible platform from which to gather reconnaissance information in which humans could screen and exercise some kind of discrimination over incoming data. Second, if the rendezvous and docking of spacecraft could be mastered successfully, Gemini could serve as a system with which to conduct manned inspection of possibly hostile satellites and potentially even the neutralization or destruction of such satellites.

In addition to their memo of December 1961, Seamans and Rubel signed an agreement the next month attempting to delineate exactly what NASA and the DOD would do in Gemini. NASA would be responsible for overall program management, planning, direction, system engineering and operation; development of the Gemini spacecraft and development of the interface, rendezvous, and docking equipment for the Gemini-Agena combination; Titan II-Gemini systems integration; overall mission responsibility for launch, flight, and recovery operation; overall command, tracking, and telemetry during orbital operations; and providing reciprocal support for any DOD space projects and programs within the scope of the Gemini project. The DOD would be charged with: developing and procuring the modified Titan II required to launch the Gemini capsule; procuring the Agena-target vehicles, as well as the Atlas boosters required to launch them; performing Atlas-Agena system integration; launching the Titan II and Atlas-Agena vehicles; and range support and recovery.[7] Over the remainder of 1962, however, McNamara concluded the DOD's role should be greater. By the end of the year, he took action.

The Air Force and Space Stations, 1962

The background for McNamara's assertion in November 1962 that the DOD should take over Gemini program management was Air Force efforts throughout 1962 to obtain the OSD's permission to begin a space-station project. While McNamara rebuffed these efforts, it seems likely that the Air Force made enough of a case concerning the requirement for military earth-orbital operations to convince McNamara that the DOD should at least have greater control of Gemini to assure Gemini met DOD requirements for the building blocks of developing earth-orbital techniques and equipment.

The Air Force had conducted low-level studies of space-station feasibility throughout the Eisenhower and into the early Kennedy administrations under the rubric Military Test Space Station. These efforts intensified in 1962 and moved toward specific designs of a program called Military Orbital Development System. Behind all these efforts (and behind the future MOL) lurked the reconnaissance requirement. As a DDR&E report to McNamara explained in February 1962, "In the near future it may become necessary to conduct optical surveillance from high altitude orbits. Very large optics will be required if good resolution is desired. Use of such optics may be quite feasible. . . . However, the practicability of such a system would almost certainly depend on the use of man for system adjustment and continued operation of equipment."[8] The second possible use OSD seemed to allow was the use of an orbital platform for the inspection and possible neutralization of hostile satellites. Lt Gen James Ferguson, deputy CSAF for Development, represented the Air Force space community's viewpoint when he declared on 12 February 1962, "We are convinced that a manned, military test space station should be undertaken as early as possible."[9]

Therefore, in the midst of the general OSD skepticism concerning the requirements for military men in space described in chapter 6, a small, experimental, DOD-manned orbital platform seemed to be the one tiny ray of hope the Air Force sensed in McNamara's otherwise negative attitude toward military human spaceflight. McNamara's explanation to Zuckert in February 1962 of his position on the DOD's human-spaceflight program opened with the standard caveat, "In the absence of a clearly defined military manned space mission, present military efforts should be directed to the establishment of the necessary technological base and experience upon which to expand, with the shortest possible time lag, in the event firm military manned space missions and requirements are established in the future." McNamara also added the standard stipulation that Air Force space efforts must be meshed with NASA's: "Space technologies primarily related to military applications must be advanced concurrently with those being exploited primarily for scientific applications," of which one example was the "establishment of comprehensive plans for cooperative DOD-NASA programs covering manned rendezvous." But then McNamara also allowed, "It may be necessary that the Air Force conduct a complementary experimental program of manned rendezvous directed at Defense requirements for docking and transfer involving uncooperative targets." Mostly, however, McNamara emphasized working with NASA and its Gemini program, suggesting that the Air Force study the feasibility of combining and adopting Gemini hardware with any emerging Air Force space platform. In addition, he closed his instructions to Zuckert by emphasizing, "It is recognized that

341

a space laboratory to conduct sustained tests of military man and equipment under actual environmental conditions impossible to duplicate fully on earth would be most useful. . . . The possible adaptation of GEMINI and DYNA SOAR technology and hardware to meet initial military experimental requirements for preliminary experimentation with a manned orbital test station is also worthy of study."[10]

Although this was by no means a ringing endorsement of a large, independent, highly capable Air Force space station, the Air Force saw it as at least a display by McNamara that he was willing to consider some type of presence for military officers in space. The Air Staff took this as official guidance and undertook an intensive planning effort.[11] AFSC's Space Systems Division drew up new plans and perfected old ones that described basically two different programs. A "Blue Gemini" would allow Air Force pilots to fly on six Gemini missions so that the Air Force could gain experience, train astronauts, and generally become oriented for the later MODS missions.[12] Sometime later Alexander Flax, assistant SAF for R&D, described Blue Gemini as "simply the idea that the Air Force would take over or follow on some of the NASA flights with Gemini with purely Air Force flights and Air Force experiments on these flights." Seamans added Blue Gemini "was really just a continuation of the present NASA Gemini, but under Air Force auspices. It did not as originally reviewed and studied include a laboratory module."[13] The best description of the still-classified Blue Gemini was "Blue Gemini was neither clearly defined nor officially sanctioned."[14] Blue Gemini appeared not to enter Air Force planning until August 1962. MODS would be an actual military space station using Gemini as a ferry vehicle.[15]

The USAF completed its development plan for MODS on 4 June 1962. This large and detailed package included separate chapters describing the particulars of operations, intelligence estimates, program management, scheduling, acquisition, civil engineering, logistics, manpower and organization, personnel training, financial, requirements, authorizations, security, and a program summary. It described a four-person space station with "an optimum design which takes maximum advantage of

GEMINI." According to the Air Force: "MODS will provide a manned long-duration orbital base which will enable the conduct of military tests and experiments under laboratory conditions in the space environment. . . . It is a significant step toward a long-duration manned space capability. Once developed, this technology will provide an extremely flexible capability to meet future military requirements. In this sense, MODS is not an end itself, but a means to an end." The system itself consisted of a permanently orbiting station module, an earth-based spacecraft comprised of a modified-Gemini capsule for ferry purposes, and a new launch vehicle, probably the Titan III. The crew of four could remain in the 1,700-cubic-foot-station module for 30 days without resupply, while the station itself would remain in orbit for at least a year. The USAF fully expected MODS to grow: "Ultimately, as MODS is expanded through modular extension, it will serve as a base from which experimental military space vehicles can be developed, tested and employed."[16]

Secondary sources have determined that the Air Force believed MODS could begin operations by March 1967 and cost $733 million.[17] In addition, MODS's primary missions have been listed as: general reconnaissance; request reconnaissance of given areas or targets; poststrike reconnaissance; continuous surveillance of an area; and ocean surveillance.[18] One problem with the USAF's MODS plan was that NASA desired a very similar station. One source said MODS was "in well-known competition with the NASA Manned Orbiting Space Station (MOSS) which the agency has tentatively scheduled for about 1966."[19] NASA's Langley Research Center had drawn up detailed plans for a Manned Orbital Research Laboratory (MORL) very similar to MODS in that it was also a medium-sized, zero-gravity station using much the same hardware and many of the same contractors as the Air Force proposed for MODS.[20] The MORL was significant because for the first time NASA was permitted to let contracts for study and design of a space station, whereas before such work had been done by NASA in-house.[21] NASA's Long Range Plan of January 1962 said NASA could launch a manned earth-orbiting laboratory as early as 1964, and by 1966 NASA could launch a much larger and more capable station based on Apollo spacecraft, hardware, and Saturn launch vehicles.[22] The possibility for future duplicative NASA-DOD programs was clearly present.

BOB, ever watchful for wasteful spending, pointed out the crux of the problem for NASA and the Air Force when it described the national space-station effort: "The presence of this type of project in both the NASA (Manned Space Station) and Defense (Military Orbital Development System) projections raises the question of the need for two development programs to furnish a basic facility and capability which could support many types of technical activity."[23] Summarized one analyst, "In the relatively exotic category of space stations, it did not seem likely that both the Department of Defense and NASA would each get to develop one. . . NASA and Air Force concepts for a space station were roughly equivalent. . . . One NASA engineer would later wonder if contractors had given the same study information to both NASA and the Air Force, but with differently colored covers" (emphasis in original).[24]

Nevertheless, the Air Force persevered with its Blue Gemini/MODS plans, and on 9 November 1962, Zuckert wrote McNamara with an official request for a $420-million increase in the Air Force space program for FY 64, of which $75 million was for MODS and $102 million for Blue Gemini.[25] As one Air Force historian said, by late 1962, "The Air Force attached great importance to the MODS."[26] However, given the fact that the entire DOD space budget was $1.55 billion in FY 63, a $420-million increase probably had limited, if any, appeal to the OSD.[27] McNamara stated he would favor MODS only "if it adds anything substantial to what we are already doing in X-20 [alternate nomenclature for Dynasoar] and the NASA Gemini and other programs."[28]

McNamara Responded

McNamara's response to the Air Force Blue Gemini and MODS proposals was to reject them but also to attempt to obtain a greater role for the DOD in NASA's Gemini program. Blue Gemini/MODS "never progressed beyond the proposal stage, partly because there was no unified position on it but also because other developments soon overshadowed it."[29] In January 1963, McNamara refused to include either Blue Gemini or MODS in the DOD's FY 64 budget request to Congress, apparently concluding the Air Force requests were duplicating Gemini.[30] A government report concerning MODS and Blue Gemini said that under "prevailing policies of restraint in space work, cost effectiveness, and precise program and requirements definition, the specific proposals did not survive."[31] The Air Force space proposals did, however, collectively "interest McNamara in exploring the possibility of a joint project with NASA."[32]

The November–December sequence of events in which McNamara proposed that DOD should assume management of the Gemini program survives only in oral-history recollections because apparently McNamara did not make a formal written offer, just a verbal proposal. The documents that do survive are NASA's pointed rebuttals to McNamara's position. Seamans recalled that he and Webb concluded by November 1962 that the DOD should be able to make greater use of the Gemini hardware NASA was developing: "If they didn't have their own program, at least shouldn't they have the opportunity to put experiments in our program and to run tests that would be useful to them?" So he and Webb went to discuss their idea with Deputy SECDEF Roswell Gilpatric, at which point McNamara happened to enter the room. McNamara said, "This is a really good idea. It's exactly what we would like to do—get the most we can out of these programs." But Seamans then said,

All of a sudden it seemed as though the thing [Gemini] was going to be grabbed hold of and almost taken away. You know, this came up several times, incidentally, that—wouldn't it be a good thing, not to have sort of two programs, but wouldn't it be a good idea to transfer the Gemini program over to the DOD? And McNamara made quite a strong case for this. Jerry Wiesner [Kennedy's science advisor and head of the Office of Science and Technology (OST) in the White House] made a strong case for this. McNamara and Gilpatric and Rubel said they were making the strongest case they could for this transfer, and we were making the strongest case we could for not transferring it. We were into the program. We had people trained. We said, "What are you going to do? Take over in Houston? How will you manage it?" It had some of these elements, again, of a sort of overcontrol by the Defense Department of our business. At least, we looked at it that way. But it finally shook down to a group that would review the experiments that were going to be carried out [on board Gemini], and some money was put in the DOD budget for experiments.[33]

Gemini's official history, based on interviews and correspondence with Seamans, contains a similar version, explaining that when Webb and Seamans made their offer to McNamara for a larger DOD role in the Gemini program, "His response to their offer was more than the two NASA spokesmen had bargained for; it took the Air Force by surprise as well. McNamara not only welcomed the idea of cooperation—he proposed merging the NASA Gemini program with the Air Force project and moving the combined effort to the Department of Defense."[34] Webb's biographer stated, "Webb saw the stakes as nothing short of NASA's independence as an agency." Webb explained that the Gemini incident was typical of McNamara's way of doing business, which was to "knock you down on the floor with a sledgehammer, and then, while you're down, ask you to sign off on a particular decision."[35]

Documentary evidence does verify the NASA officials' accounts of the role of Dr. Jerome Wiesner and his OST in the White House. One Kennedy administration insider explained Wiesner's close relationship to Kennedy, "President Kennedy turned to the Science Advisor, Dr. Wiesner, on many occasions on issues ranging from desalination of sea and brackish water to a whole series of defense issues related to research and development. What was new in this picture was the close personal relationship the President had with Dr. Wiesner. . . . The President saw a great deal of the Science Advisor. . . . The President had a very high regard for him and there was a very personal relationship between them."[36] Certainly Kennedy never took all the advice from a particular confidant. In fact, Wiesner was not keen on the overall lunar-landing program or on the specific NASA decision to pursue the LOR mission mode over the EOR method. Nevertheless, OST's and Wiesner's strong backing of a DOD takeover of Gemini shows the legitimate nature of the threat NASA faced.

Nicholas Golovin, by late 1962, working for OST after departing NASA on less than amicable terms, wrote Wiesner in December that since "the NASA program has been expanding at an extremely rapid rate," NASA's resources "will obviously be strained to an increasing degree in Apollo technical management." The administration had to decide "whether greater success in the national space program would be achieved by shifting part or all of the responsibility for Gemini from NASA to the DOD." Golovin supported such a transfer, citing its "direct management benefits," the fact that it would "enable more effective and rapid development of the military space program," and its "obvious domestic political advantages." DOD was the logical choice because "DOD resources and capabilities for technical space program management have been, and are likely to continue to be, substantially less strained." NASA could still use space-station equipment the DOD developed because "it is difficult to see any differences in the requirements for an engineering space laboratory between DOD and NASA—substantially the same technological problems involved in developing equipment suitable for extended operations in space will be met by both agencies." Therefore, "Only one Manned Space Station Program should be undertaken for meeting all national space needs. This program should be assigned for implementation to the DOD."[37]

Wiesner took Golovin's inputs, endorsed them, and incorporated them into a memo for Kennedy. Wiesner stated that earth orbit activities will become "an increasingly important and costly part of both the military and scientific space efforts, therefore we should make a major effort to unify them now before we become committed to two large programs." Wiesner recommended, "Arrangements be initiated for a major investment of the DOD, including funding, in the Gemini program and that the DynaSoar effort be collaterally reprogrammed to a small fraction of its current level. . . . It would seem advisable that the DOD be assigned responsibility for this development" of the Gemini and any follow-on space station.[38]

NASA Held Fast

Walter McDougall stated, "Webb exploded at this open assault on NASA."[39] NASA wasted no time in marshaling its forces. Boone would spearhead NASA's response.[40] NASA stated the primary reason it opposed transferring Gemini to the DOD was that "it is estimated that the Gemini schedule would slip at least one year, with a concurrent major increase in program cost." This would in turn delay the lunar-landing program by at least the same amount of time. NASA granted that further "national benefits could be derived through greater Air Force participation in NASA's Gemini program" but that a wholesale DOD management takeover was not required.[41]

Other internal NASA documents reveal that additional concerns buttressed NASA's opposition to a DOD Gemini transfer. For instance, many of the agreements NASA had forged with other countries to place NASA tracking stations on their territory were predicated on the notion that the facilities not be used for military purposes. Should DOD manage Gemini, NASA was likely to lose access to its stations in, at a minimum, Mexico, Nigeria, Spain, and Zanzibar.[42] Boone's summary memorandum stated, "The Gemini program should continue under the direction of NASA, with increased DOD (USAF) participation, on a not-to-delay basis, in order to further DOD objectives in space." However, management of the program had to remain in NASA because "dislocation and loss of continuity in the developmental effort, which would inevitably accompany a transfer of management, would result in a substantial delay and increased cost in the Apollo lunar landing program. The Apollo program, as currently planned, could not be accomplished without the experience to be gained from Gemini. . . . Any delay would reduce the chances that the United States will make a manned lunar landing before the Russians do."[43]

McNamara's first response to NASA's adamant refusal to consider transferring Gemini to the DOD was to propose joint management of the program. On 12 January 1963, he sent a presigned agreement to Webb (a common McNamara tactic) that proposed an eight-person Gemini Program Steering Board consisting of four representatives from each institution. It would control and manage the Gemini Program to ensure it was "planned, executed, and utilized in the overall national interest so as to avoid duplication of effort . . . and to insure maximum attainment of objectives of value to both the NASA and the Defense Department."[44] Webb responded, "I cannot agree that your proposed version of an agreement would set up management arrangements suitable to a national Gemini program. Nor do I consider its basic pattern one which can be made acceptable through a series of negotiated changes. . . . To join the DOD and NASA programs in a monolithic effort would inevitably cause the total program to be characterized as military with substantial loss of flexibility in our national posture." Webb counterproposed that the DOD submit experiments to NASA for inclusion on the Gemini manifest and that the DOD "participate in the development, pilot training, preflight check-out, launch operations and flight operations of the Gemini Program to the extent necessary to meet the DOD objectives." However, those concessions to increased DOD participation were "about as far as we in NASA feel we can go at this time."[45] Privately, Webb wrote Seamans, "I do not see how we can discharge our responsibilities and give him a veto. . . . We must not recede from this position except as we reach a settlement that all of us can live with."[46]

Three days later Webb and McNamara arrived at a settlement. It appears McNamara was not willing to push the situation any further because the 21 January 1963 NASA-DOD Gemini agreement incorporated primarily NASA's viewpoint on Gemini management, not OSD's. NASA would permit the DOD to include experiments on the Gemini flights, but the DOD would not assume an active role in managing the program. The experimental program, as well as the DOD support role in Gemini, would be implemented and supervised by a new five-person body called the Gemini Program and Planning Board (GPPB). The GPPB would report directly to Webb and McNamara, be chaired by the NASA associate administrator and the assistant secretary of the Air Force for R&D, and have two additional members from NASA and the DOD. The agreement made clear, "NASA will continue to manage the GEMINI project. It is, however, agreed that the DOD will participate in the development, pilot training, preflight check-out, launch operations and flight operations of the GEMINI Program to assist NASA and to meet the DOD objectives." DOD would contribute funds in accordance with the GPPB's determination. Probably the most important clause of the NASA-DOD Gemini agreement stated, "It is further agreed that the DOD and the NASA will initiate major new programs or projects in the field of manned space flight aimed chiefly at the attainment of experimental or other capabilities in near-earth orbit only by mutual agreement."[47]

The GPPB and the DOD

The GPPB was strictly advisory in nature and met 14 times between its inception and its final meeting on 12 April 1965. Its duties basically entailed: overseeing the planning and conducting of Gemini experiments to include establishing priorities; processing and disseminating the results from these experiments; and establishing the criteria for and then monitoring the process whereby the USAF Titan II was manrated, that is, made reliable enough to be used as a space booster that could carry humans. Overall, "The arrangements worked out very satisfactorily."[48] The specific list of DOD experiments to be incorporated into the Gemini Program was not finalized until 1964 and so will be discussed in chapter 9; they were closely linked to the missions of reconnaissance and satellite inspection. However, there were some indications in the last year of Kennedy's presidency that the relatively limited nature of DOD's participation in the Gemini Program, and its lack of any managerial input, was perceived as being an inadequate forum in which to conduct the necessary investigation into the usefulness of military officers in space. Therefore, the approval of a wholly DOD human-spaceflight program, the MOL, increased in likelihood over the course of 1963.

Lawrence Kavanau, the special assistant for space in the Office of the DDR&E, said in May 1963, "We are finding that, although there are many important and worthwhile things that can be done with GEMINI, due to the late stage of development, no significant DOD input can be made to the GEMINI design. GEMINI, while highly useful, could have been made even more so by joint participation earlier in the game."[49] DDR&E Brown amplified this sentiment the next month: "There is a disadvantage to entering a program that someone else is running which has been going on for some time. On the other hand, Defense does not stay a junior partner indefinitely in anything that it gets into."[50] As will be seen below, by June

1963 Brown was already seriously investigating the DOD's requirements for its own separate orbital platform, a concept that in six months would be approved as the MOL.

Dynasoar in 1961–62

The necessary preliminary to examining the MOL's emergence is understanding the progress of Dynasoar in the Kennedy administration and the close link that existed in OSD's thinking between Dynasoar and NASA's Gemini. The only time during the Kennedy administration in which the Dynasoar's future looked bright was in the first few months of his administration when Kennedy was dramatically increasing virtually all categories of defense spending (see chap. 5). As part of this upswing, Dynasoar's FY 62 budget was increased from the final Eisenhower figure of $76.5 million to $106.5 million. In April 1961, McNamara told the Senate, "This project is, of course, only a first step toward the development of a militarily useful vehicle and at the present time is conceived of strictly as a research effort. The additional $30 million requested would permit the work on this project to go forward at a more efficient rate."[51] Not only did Congress grant this increase, but also the House Appropriations Committee added another $85.8 million to the DOD request, an amount that McNamara declared he had no intention of committing to Dynasoar: "I doubt very much that we can expend that effectively and efficiently."[52] Therefore, there were limits to even the OSD's early support of Dynasoar. During this initial period, the Air Force still planned for a three-step Dynasoar development program. In step 1, preliminary suborbital R&D would be conducted. In step 2, a larger booster would lift manned and unmanned gliders to global range and orbital flight for tests of military equipment. In the final step, actual weapons systems would be studied and operational systems developed.[53] The OSD, however, had only approved step 1.[54] The first of six piloted flights was scheduled for May 1966, and program costs before completion in December 1967 were estimated at $921 million.[55]

Dual Reorientation

In late 1961 and early 1962, McNamara and the OSD reoriented Dynasoar in two senses. First, the suborbital phase of Dynasoar was dropped because McNamara had concluded the Titan III would have adequate capacity to boost the glider to orbital velocity without extensive testing in the suborbital realm. This elimination of the suborbital step in the Dynasoar program would, in turn, reduce overall R&D costs. Second, the OSD ordered the Air Force to drop all references to the potential and future military applications of the Dynasoar and to view it wholly as an orbital, not a suborbital, R&D project. This meant that the Dynasoar's research focus, in turn, shifted from exploring the intricacies of hypersonic flight, a topic in which NASA had great interest, to investigating the challenges of controlled and maneuverable atmospheric reentry and landing at a selected Air Force base, a topic in which NASA had little interest.

McNamara asked Congress in January 1962 for a $115-million FY 63 Dynasoar budget, despite Congress' desire to allocate $185 million. He said his figure was all the OSD believed was required and could be effectively utilized: "As you may know, last month we reoriented the entire program, eliminating the suborbital flight phase which would have involved the use of a modified Titan II booster. This intermediate step is no longer necessary inasmuch as we are now proposing very substantial investments in the Titan III booster program."[56] NASA noted that same month, "The Dynasoar was originally planned as a pilot controlled hypersonic Mach 16 glider. The project has recently been changed to the development of a pilot controlled earth orbital spacecraft suitable for winged reentry through the earth's atmosphere to an aerodynamically controlled earth landing."[57]

Simultaneously, the Air Force was instructed to play down the military applications of the space glider and emphasize that it was supposed to be only a hypersonic R&D or test vehicle. McNamara wrote Kennedy that he believed it proper to "reorient the program to solve the difficult technical problem involved in boosting a body of high-lift into orbit, sustaining man in it and recovering the vehicle at a designated place

352

rather than to press on with a full system development program" of military applications.[58] The edict went out: the USAF was no longer to actively explore, nor discuss, the potential military applications of the Dynasoar system such as reconnaissance. The office of the DDR&E suggested McNamara even give Dynasoar a new name in the tradition of the X-series of aircraft that represented purely research projects with no connotation of military operations or mission preparation whatsoever. Such a step would make the Dynasoar "more properly identifiable as an experimental development program (non-mission-oriented) with an appropriate research vehicle designation e.g., 'X-10'" and possibly give it more programmatic stability because in the past, "The DYNA SOAR program has alternately been considered for elimination, for stretch-out, for considerable acceleration, and for transfer to NASA."[59]

Accordingly, McNamara declared on 22 February 1962, "The principle of proceeding directly to orbital flight test is endorsed." He also ordered the program's name be redesignated to "an appropriate research designation (e.g., X-19) to indicate more specifically that this is an experimental program and to eliminate any further connotation of previous weapon system and military test system studies within the presently approved development effort."[60] After several months of wrangling, the new numerical designation for the Dynasoar program was announced on 26 June 1962 as "X-20."[61] The Air Force dutifully amended its Dynasoar development plan by deleting all references not only to suborbital flights but also to the development of military subsystems or applications.[62] By June 1962, DDR&E Brown testified to Congress that in the past the Dynasoar had been improperly presented as leading toward an operational system: "That has never been accepted as the purpose by the DOD and it is not now so accepted. What was accepted as a program was a vehicle which would serve to develop the technologies associated with manned space flight and some particular applications—not uses," such as short-notice deorbit and landing.[63]

An official Air Force history added that by mid-1962 the dual Dynasoar reorientation process was complete, and McNamara had approved a budget of $135 million for Dynasoar. However,

he also "instructed that technical confidence and data acquisition would have precedence over flight schedules. It was
quite clear that the X-20 Dyna-Soar program was exclusively
an experimental program which was directed towards demonstrating the ability of the Air Force to orbit the glider, reenter,
and land at a pre-selected site. . . . The X-20 program was not
directed towards developing a weapon system, nor even defining future military applications of the dynamic-soaring glider
[because] such references had been deleted."[64]

Although one might expect that the Air Force would have
cheered the reorientation of the Dynasoar toward an orbital
vehicle, this was not the case. Explained Houchin, the foremost Dynasoar scholar, this reorientation "placed Dynasoar in
a perilous position. Its mission competed with NASA's Mercury
and Gemini programs for the manned space mission and with
the NRO's unmanned satellites for the national reconnaissance mission. . . . While only a few officials within OSD knew
about them, NRO's highly classified unmanned reconnaissance satellites were fulfilling the military requirement to
gather information, even if they could not make conventional
landings. . . . Without knowing what type of reconnaissance
systems it had to compete with, Dyna-Soar's proponents
found it much harder to sell their system to OSD."[65] The threat
to Dynasoar's viability due to perceived duplication came not
only from NASA's Gemini, but also from NRO's robotic spacecraft, although the latter factor is difficult to directly assess
due to the continuing high level of secrecy and classification
pervading NRO's history.

It was simply a matter of time before OSD, and others,
would accuse Dynasoar of multifaceted duplication: "Knowing
the military capabilities of NRO's reconnaissance satellites,
the ability of NASA to place a man in orbit, and the burgeoning promise of NASA's Gemini program to perform military requirements in space, OSD officials began to question the need
for a separate Air Force–sponsored manned-spaceflight program."[66] Houchin also makes clear that "the Air Force faced a
'Catch-22.' How could it demonstrate a military need for man-
in-space before it placed one in space to prove his capabilities?
Ultimately, Dyna-Soar proponents would have to prove their

point by quantifying and qualifying Dyna-Soar against space systems they knew little, if anything, about."[67] Indeed, the chairman of the House Space Committee declared in September 1962 concerning the X-20's objectives of rapid launch, space maneuverability, flexible reentry, precision recovery, and conventional landing and reuse: "NASA is already conducting a program, Project Gemini, designed to accomplish all of these objectives several years sooner than will be possible with the X-20." In addition, Representative Miller pointed out that Gemini was much lighter than the X-20, much smaller, much less expensive, and could carry two men instead of one.[68] By December Golovin in the OST was urging that "the X-20 project be either canceled or drastically reduced, and DOD assigned all or a major part of the responsibility for development of the Gemini system" and that "the DOD space program be explicitly broadened to include early application of Gemini or Gemini-modified systems for reconnaissance and surveillance, and associated military operations, in near-earth space."[69] While the GPPB emerged to serve this role, the official Dynasoar cancellation did not take place for another year.

NASA's Interest in Dynasoar Waned

The official government position concerning Dynasoar was that it was a "manned test vehicle capable of maneuverable reentry from orbit to a conventional landing at an air base which can be selected by the pilot." There was no mention of potential military applications, only its scientific R&D components.[70] One might expect NASA to have cheered this aspect of the reorientation. This, however, was also not the case. In fact, another consequence of reorienting Dynasoar away from a suborbital vehicle to an orbital vehicle was the loss of genuine NASA interest in the project. As long as the vehicle was predicated on the notion of exploring the hypersonic-flight regime within the atmosphere, NASA had a legitimate interest in the glider. However, when Dynasoar's primary R&D objective became exploration of the orbital challenges of maneuverable reentry and landing at conventional air bases, NASA's interest waned. As NASA's chief for high-speed aerodynamics R&D, John Becker explained, "NASA's influential involvement with

Dyna-Soar came to an abrupt end in 1961." The OSD's dual reorientation of the Dynasoar described above only became known to NASA through William Lamar, the director of Dyna-soar engineering. Lamar "rather apologetically informed us during the fall of 1961 of the drastic redirection that was to be implemented in December 1961, without any participation or consultation with NASA." As a result of the elimination of Dyna-soar's suborbital hypersonic-research flights,

> As far as our NASA DS [Dynasoar] team was concerned, Dyna-Soar as a research airplane was dead. During the remaining two years of Dyna-Soar's existence NASA continued as a largely inactive nominal partner, completing the tests to which we were committed. It was now obvious that the USAF was interested in DS only as a prototype of an orbital system and not as a research vehicle. . . . As time went on it also became increasingly apparent that USAF did not have a clear believable vision of what their orbital system requirements really were, and thus doubts increased as to whether DS-1 was an appropriate development vehicle. . . . The Air Force has essentially eliminated NASA from policy decisions.[71]

In March 1962, the NASA Dyna Soar Coordinating Committee concluded, "The Dyna-Soar Project has changed in character from the X-15 type of hypersonic and reentry research and test system originally contemplated to a prototype for possible military space systems. Air Force emphasis is now being placed on exploring the potential of man to accomplish military functions in space, a mission that this system is poorly designed to accomplish. NASA was not represented in the technical management deliberations leading to these drastic changes and our subsequent objections have been largely overruled." The committee agreed that if the original, 14 November 1958, NASA-USAF MOU on Dyna-Soar—declaring Dynasoar a joint project (see chap. 4)—continued in force, "NASA will be held jointly responsible with the Air Force for the doubtful outcome of this project while in fact its destiny is being decided wholly by the Air Force." The committee felt there was enough research value left in the Dynasoar vehicle as a "highly maneuverable radiation-cooled manned-reentry vehicle" to warrant some continued NASA support. However, "This can be supplied in the traditional manner without the necessity for a joint project." Therefore, NASA should terminate the categorization of Dynasoar as a joint NASA-DOD project.[72]

NASA's institutional dissatisfaction with recent developments and concern over its future role in the Dynasoar program caused Webb to write Zuckert and explain, "A number of events have occurred which have prompted us to reexamine this project and our relationship to it." Webb said they should create a new memorandum of understanding with a "more accurate statement of NASA participation in the remainder of the program." Webb explained NASA still supported research on "the problems of highly maneuverable winged vehicles in the critical environment of the hypersonic flight corridor." However, "The additional uses of the glider by the Air Force as a space vehicle for exploring the potential of man to accomplish military functions in space are considered beyond the scope of NASA interests." While NASA was ready to "provide continuing technical support in the form of consultation and ground based testing," it no longer wished to be listed as a partner in the program.[73]

Accordingly, Webb and Zuckert signed a new MOU on 7 August 1962 that simply stated, "Dyna-Soar is an Air Force Program." The document reported that certain aspects of Dynasoar R&D such as exploring high maneuverability at hypersonic speeds with a conventional landing did interest NASA and therefore, "NASA endorses this objective as necessary to the national aerospace program." However, NASA's future role would be limited to "technical support (consulting and ground-facilities testing)" and "instrumentation and flight test support."[74] From this point forward, NASA's official role in the Dynasoar program would be distinctly circumscribed.

Dynasoar and Gemini

Unofficially, however, and in the minds of high-level OSD officials, NASA played a central role in Dynasoar's fate. The capabilities of its Gemini system, when augmented by a cylindrical laboratory, were eventually deemed a more capable, cheaper, and earlier available human-spaceflight alternative for the Air Force than Dynasoar.

McNamara Ordered a Comparative Review

One source calculated that by the end of FY 62 the Air Force had spent $240 million on the Dynasoar with only a full-scale mock-up to show for these expenditures (it should be remembered that the entire Mercury program would cost under $400 million) and that it would cost an estimated $1.3 billion to continue Dynasoar through its first piloted flight in 1966. Consequently, the X-20 program was coming under increased scrutiny in the fiscally minded OSD.[75] Only a few days before McNamara and Webb finally reached their NASA-DOD Gemini agreement on 21 January 1963, and virtually simultaneously with his rejection of the USAF's MODS and Blue Gemini proposals, McNamara informed DDR&E Brown: "I should like to review in detail the DYNASOAR program" both in Washington, DC, and at the main contractor facilities. McNamara explained, "In particular, I am interested in considering the relationship of DYNASOAR to GEMINI and the extent to which the former will provide us with a valuable military capability not provided by the latter." One day later he added, "I am interested in the extent to which the Gemini program as presently conceived by NASA will meet our military requirements."[76] McNamara openly pondered the X-20's fate before Congress in early 1963: "Do we meet a rather ill-defined military requirement better by proceeding down that track [spending $1 billion more on the X-20] or do we meet it better by modifying Gemini in some joint project with NASA?"[77] In less than one year, McNamara would become convinced that NASA's Gemini did (when attached to a laboratory cylinder), in fact, better meet the OSD's military requirements for a human-spaceflight program, focused on the reconnaissance mission, than did Dynasoar.

Webb recorded a conversation he had with McNamara in February 1963 concerning Dynasoar in which McNamara stated that "he was prepared to look carefully at the values that might be retained from the Dynasoar program, although he had serious doubts that there were any values in it worth the eight or nine hundred million dollars that it was costing."[78] Privately, Webb confided his personal views on Dynasoar "as an orbital vehicle it is going to be obsoleted [sic] by both Gemini and Apollo and that what we need now is careful, thoughtful work on hypersonic reentry."[79] Also in February 1963, Zuckert reported to McNamara that a congressman had asked him while testifying to the House Appropriations Committee about McNamara's opinions concerning the Dynasoar program: "I told him that I realized it was your disposition to cancel or substantially reorient the Dynasoar program, but that this matter had not finally been settled."[80] A final indication of McNamara's skepticism toward the X-20 program even before his formal review of it in March 1963 was his testimony to the House in February: "It appears to me that Gemini is advanced beyond the Dyna-Soar in technique and potential. There is no clear requirement, in my mind, at the present time for manned military operations in space. . . . But were we to require manned military operations in low earth orbit, it appears to me that the Gemini approach is a far more practical approach."[81] Even before McNamara's review trip, the trade press was speculating, "For all intents and purposes, the Dyna-Soar (X-20) program is dead. There will now be a family discussion on the best way to bury the body."[82]

In March McNamara embarked on an intense review of the X-20 program that included briefings not only in Washington, DC, but tours of the facilities across the country of the major contractors for the glider itself and its launch vehicle, such as the Martin Corporation and Boeing, as well as similar facilities associated with the Gemini program. Brockway McMillan, assistant SAF for R&D, provided the best synopsis of McNamara's tour:

> It was clear that the briefings on Dyna-Soar opened Mr. McNamara's mind in a way it had not been opened before on the point of Dyna-Soar as a space vehicle rather than as a research vehicle. . . . [However] Mr. McNamara several times said that he was concerned that in the Dyna-Soar project we were putting too great an emphasis on controlled reentry when we didn't even know what we were going to do in orbit. He felt the first emphasis should be on what missions can be performed in orbit and how to perform them, then worry about reentry at a later date. In other words, start looking at the problem from the end objective . . . and then worry about secondary problems like controlled reentry at a later time. It is not clear at this point that Mr. McNamara is willing to buy Dyna-Soar. In any event, he is not going to cancel it right away. He is clearly arguing with himself and several times raised the same questions. . . . It is clear that Mr. McNamara is concerned with the great cost of space flight and the great cost to the taxpayer of Gemini

and Dyna-Soar. It is also clear that he feels we will have to have some kind of test bed in space—presumably manned—in order to test out concepts related to manned space flight. . . . He suggested that we take as much as six months to study, what in the long run, would be the optimum test bed for military space. He thought it might be space stations serviced by a ferry vehicle.[83]

McNamara had, in effect, given Dynasoar a six-month lease on life. When he returned from his review trips he tasked the SAF with a detailed examination of the Dynasoar and Gemini programs and their relation to the four most likely DOD space missions: inspection and identification of hostile satellites; protection of our own satellites from destruction; the capability of carrying out reconnaissance missions from space; and the introduction of offensive weapons into near-earth orbit. McNamara alluded to the Dynasoar in his memo to Zuckert: "It appears to me that too much emphasis and too much money has been placed on the development of certain techniques such as controlled reentry and not enough attention has been directed to the specific military missions to be performed. In particular, I am interested in reviewing the contribution which the X-20 and GEMINI programs can make to each of the missions referred to above."[84] McNamara summarized his conclusions after his Dynasoar review to the House: "I seriously question whether our nation requires that both programs be completed. We have no clear military requirement for either."[85]

The Air Force Response

The USAF's response to McNamara's 15 March tasking order indicated that as an institution it was unwilling to strongly endorse either Dynasoar or Gemini as the best system for the four missions McNamara described. AFSC's Space Systems Division's (SSD) bottom line was, "Neither vehicle can, through modification, acquire all the characteristics desired of a military space system for routine operational use." SSD did present in detail the advantages and disadvantages of each system as they related to McNamara's four specified missions. But it made no firm recommendations as to how the SECDEF should proceed.[86] McMillan incorporated SSD's ambiguity in the memo that actually went to McNamara, reiterating, "Our analysis shows that neither the

X-20 program . . . nor the NASA Gemini program as presently defined will provide significant capabilities relative to the four missions. There is a very limited operational capability inherent in the two vehicles." McMillan passed on AFSC's analysis of the pros and cons of each system for the particular missions, but in the end concluded, "Neither the DOD X-20 nor the NASA Gemini program *as presently defined* will produce on-orbit operational capabilities of any military significance" (emphasis in original). Therefore, both programs should be continued.[87] Since the USAF appeared unwilling to decide between Dynasoar and Gemini, it would fall to the OSD, and particularly the Office of the DDR&E, to do so. Therefore, "if by July 1963 Dyna-Soar was not dead, its hold on life was at best tenuous."[88] Indeed, in July 1963 McNamara limited Dynasoar FY 64 funding to $125 million per year for the indefinite future, $10 million less than FY 63's level.[89]

The MOL Emerged as Dynasoar Expired

On 29 August 1963, Senator Clinton Anderson wrote Gilpatric, deputy SECDEF, to ask him what the situation was with Dynasoar and whether the DOD planned to continue the program. Gilpatric replied that the relative military usefulness of Gemini and Dynasoar "is the most difficult question facing me. In fact, neither Gemini nor Dynasoar, in their present form, can perform a genuine military mission. . . . The fundamental point is that no militarily useful mission to which these vehicles could contribute has been defined, although we have studied the problem intensively for several years. Should a mission be defined, it might favor one or the other approaches, but most likely would require the initiation of a third approach to circumvent the obvious limitations of the other two."[90] In fact, the OSD was already considering a third approach by September. This third approach was MOL, a Gemini capsule with an attached laboratory module. On 10 December 1963, McNamara officially sanctioned it and canceled Dynasoar.

Webb, McNamara, and Space Stations

One bone of contention resulting from the NASA-DOD Gemini agreement of January 1963 was at exactly what point in

NASA's exploratory space-station studies was NASA required to obtain "mutual agreement" with the DOD that it was not "initiat[ing] major new programs or projects in the field of manned space flight aimed chiefly at the attainment of experimental or other capabilities in near-earth orbit"?[91] McNamara took a restrictive view of this clause, believing that DOD should be involved in NASA space-station exploratory studies from an early point. Webb interpreted it more liberally, not wanting to sacrifice NASA's autonomy, and said that as long as NASA was engaging only in paper studies, either in-house or with contractors, and not actually building hardware, it was not required to consult with the DOD.

Webb wrote McNamara that there appeared to be a "lack of a meeting of the minds concerning the proper coordination between NASA and the DOD in the area of exploratory studies. . . . We feel here in NASA that we must constantly be looking well into the future in order that our progress will be such as to achieve and maintain a position of world leadership for the United States in [the] field of space sciences and technology." Concerning space-station exploratory studies, "In my view, such advanced exploratory studies do not fall within the purview of existing DOD-NASA agreements as they relate to the initiation of 'major new programs or projects.'"[92] McNamara simply included his response to Webb in a reply he made to Johnson when the vice president asked McNamara and Webb for their opinions on the five space-related questions Kennedy had asked Johnson to investigate in 1963 (see chap. 5). McNamara maintained, "It is essential that all major space programs be integrated with military requirements in the early stages of their development. . . . I am more concerned with the potential dangers in the divergence of our efforts in the study and planning of potential new large projects" such as the space station. Concerning a space station, McNamara declared, "While it is not yet clear that the project is justified, either on a military or nonmilitary basis, it is clear that it should be undertaken only as a national program, which meets the requirements of both NASA and DOD, and that it must be jointly planned from its inception. . . . Coordination and joint planning of our efforts must extend to all so-called 'advance studies.'"[93] More pointedly in a letter to Zuckert, McNamara admitted that he concurred in Zuckert's assessment that the DOD should be awarded the space-station mission: "I agree that this assignment, and the near-earth interests of the DOD, might be considered logical reasons for assigning to the DOD this new undertaking."[94]

Throughout May and June 1963, Webb and McNamara exchanged numerous letters, but the deadlock over what was and was not an acceptable level of coordination on space stations and exploratory studies of them continued. McNamara sent Webb presigned agreements that Webb would not sign. Webb did the same in return to McNamara.[95] In the midst of this, McNamara continued to try to move the Air Force toward some recommendation on the Dynasoar-Gemini situation. He wrote Zuckert late in June: "The Department of Defense will be faced with major new program decisions regarding manned space flight within the next year. Since space vehicle developments are so expensive it is necessary that we utilize every opportunity to minimize the number of separate developments." Therefore, he ordered Zuckert to submit "a plan for insuring the integration of the several study efforts now underway which may involve GEMINI and thus provide additional basis for comprehensive program decisions in the area of manned space flight as it relates to military missions."[96]

Later that summer Brown provided Zuckert additional OSD guidance on what OSD had in mind for an Air Force orbital platform. He authorized Zuckert to spend $1 million on the study McNamara had ordered and added, "Because of the national importance which could be attached to the outcome of this work, the Secretary of Defense and I will have a more detailed interest than usual in its progress." Brown then gave the USAF specific guidance:

> The immediate objective to which this study must be directed is the building of a space station to demonstrate and assess quantitatively the utility of man for military purposes in space. The space station so contemplated would be a military laboratory, and its characteristics must be established with some specific mission in mind if its function is to be a genuine military one. The principal missions to be considered are those that can be included in a broad interpretation of reconnaissance: surveillance, warning and detection can be considered in this context. Other missions such as those assuming the use of offensive

and defensive weapons shall not be considered unless it can be explained in detail how such missions might be done better from a space station than any other way. The successful conclusion of this study must provide answers to at least the following questions: What specific answers about what specific military capabilities will the space station answer? . . . What is the smallest kind of space station which will still provide a meaningful demonstration and measurement of man's utility?[97]

Clearly, within OSD, the need for some kind of orbital platform to finally test, once and for all, if military officers had any justifiable reason to operate in space had now been established. The MOL was edging closer to reality. However, even as such a long-duration orbital platform seemed more and more certain because OSD now wanted one to serve as a test bed, this meant that Dynasoar's chances for survival dimmed, given the drive for eliminating duplication and cost-efficiency inherent within PPBS and systems analysis.

While the situation was thus finally becoming clarified within the DOD concerning the need for some type of orbital testing of the military requirements of human spaceflight, the exact balance between NASA and the DOD concerning responsibilities in this area was not. McNamara and Webb could not agree on the degree and level of coordination required for their respective space-station exploratory studies. It finally took vice-presidential intervention to clarify the situation. At an NASC meeting on 17 July 1963, both Webb and McNamara expressed satisfaction over the progress of and level of coordination in the Gemini program since the promulgation of the January 1963 agreement. But when the discussion turned to space stations, Johnson asked if the various study contracts required mutual agreement. Predictably, "Webb answered the question about the need for agreement on studies in the negative and said, 'Not in my view.' He continued that NASA will furnish DOD outlines of its studies for comment and discussion but not for concurrence. He did not believe anyone else should have a veto over studies NASA proposed." McNamara then entered the discussion: "He differed with Webb by stating that he did feel that the other party should agree before a study is pursued, and that, if an agreement can't be reached, the matter should come to the Vice President or President to be settled."[98]

Shortly thereafter, Johnson sent each man an identical letter stating, "I was pleased to note both you and Secretary McNamara/Administrator Webb expressed satisfaction with the coordination existing in the Gemini program. . . . The situation regarding space stations was less clear, however, and I would like to get your best thinking as to what needs to be done." Therefore, each was to submit "a paper expressing the possible uses of space stations."[99] Webb's response recognized that any space station would not only be a major undertaking but also "a mandatory forerunner of any long-duration manned space operational system." Therefore, a single national program should be able to meet "the initial technological requirements of all interested parties." Concerning whether NASA or the DOD should manage the initial project, Webb simply said that after all study efforts were completed, the NASC should forward to the president a "recommendation as to management responsibility based on predominant interest and consideration of other pertinent factors, such as management competence, relation to other programs in progress, and international political implications."[100]

McNamara's response also foreshadowed his backing of the MOL and pending cancellation of the Dynasoar: "The real potential of manned space flight may not be understood until there has been the opportunity to conduct a program of long-duration multimanned orbital flights in a facility which permits men to move about and perform useful tasks." The Dynasoar, in its present configuration, did not permit officers to orbit for long periods, had only one person, and did not permit people to move about. Concerning the specific military uses of a space station McNamara postulated, "It may be that reconnaissance and surveillance techniques could be improved by human judgment and adaptability," and so a space station "may provide a platform for very sophisticated observation and surveillance."[101]

Apparently, the direct involvement of Johnson in the Webb-McNamara space-station dispute was sufficient not only to finally bring about a NASA-DOD accord, or at least an armistice, on space-station planning but also to increase the momentum for acceptance of the MOL within the OSD. The

NASA-DOD agreement covering a possible new manned earth-orbital R&D project of August/September 1963 stated that the two organizations' advanced exploratory studies on space stations and any follow-on actions "should be most carefully coordinated through the Aeronautics and Astronautics Coordinating Board. . . . Insofar as practicable all foreseeable future requirements of both agencies in this area should be encompassed in a single project." There followed an eight-step administrative procedure detailing exactly how NASA and the DOD would coordinate their continuing advanced studies through the AACB; and that the SECDEF and NASA administrator would jointly determine whether or not a space-station program should be started and then formulate a recommendation to the president as to managerial responsibility. If the president accepted their recommendation, then NASA and the DOD would create a joint board to formulate the specific objectives of the newly approved space-station program. However, the project would be under single-agency management, in accordance with the presidential decision.[102]

When McNamara finally signed this agreement, almost a month after Webb sent it to him, he offered several serious reservations to it centering on the fact that NASA continued to design space stations without DOD input but still insisted a single orbital platform would have to meet both agencies' needs. The core impression from McNamara's "acceptance" letter is that he seemed simply to have been fed up with the whole question, stating, "We have discussed this matter as much as is useful." He therefore signed it and hoped Webb would accept his reservations and instruct his staff to obtain DOD input on any space-station studies budgeted at greater than $100,000 in a single year.[103] When it was all said and done, the AACB's Manned Space Flight Panel formed a National Space Station Planning Subpanel to enforce this NASA-DOD space station agreement. However, this subpanel met only four times and "then lapsed into inactivity."[104] This whole infrastructure created to carefully coordinate NASA and the DOD space-station programs played absolutely no role in the MOL design and approval process because the senior leadership of both agencies "chose to regard MOL as something other than a space station, hence not covered by the September agreement."[105]

Approving the MOL/Canceling Dynasoar

By late October 1963, McNamara wanted to take another tour of the primary Dynasoar and Gemini facilities so he could conduct a second intensive review of the Dynasoar program, just as he had in March. The difference this time was that in the interim he had come to accept the need for some sort of a multimanned, large orbital test bed for military experiments in which more than one officer could live for an extended period and have the ability to move around. As a consequence, the Dynasoar's prospects looked bleak. While on this second tour, McNamara asked assorted questions that revealed his disposition favoring the concept of a laboratory module attached to the Gemini and against the Dynasoar.

The crucial briefing of McNamara's tour seems to have been on 23 October. Lamar, the Air Force's director of engineering for Dynasoar, recorded that McNamara's real interest was in getting answers to his basic questions of "a. What does the military want to do in space, and why? b. What is the relative cost-effectiveness of manned and unmanned space systems, and how do they compare with other means of doing the job?" Over the course of the discussion that day, Lamar said it became clear that McNamara "feels that a space system will be expensive and he does not understand why the Air Force wants to establish a mission by such an expensive method. He has asked these same questions a number of times over the past few years." Lamar added, "It was quite evident that Mr. McNamara felt considerable progress should have been made in obtaining answers to his questions. . . . He is not satisfied with the answers he received, and drastic consequences are likely if better answers are not forthcoming."[106]

In a separate memo, Lamar created a paraphrased transcript of the actual question-and-answer session on that day. In it McNamara is presented as remarking, "I want to know what is planned for the X-20 after maneuverable reentry has been demonstrated. I cannot justify the expenditure of $1 billion for a program that is dead-ended. I am not engaging in additional Dynasoar expenses until I have an understanding of what the space missions are. . . . It is imperative that a mission analysis be conducted in order to determine what has to be done. . . . The program will not have security until its purpose is fixed." Perhaps McNamara's attitude was best summarized by his question, "What does man do other than fly the vehicle?" This was quickly followed by an implied warning, "We are planning to spend a large amount of government resources when in fact we don't know why. In other words, we don't have a clear purpose in mind for follow-on use of the Dyna-Soar technology." When a Boeing official stated that the Air Force had repeatedly explained that reconnaissance was the primary justification for the Dynasoar, McNamara replied, "Agreed, but I can do it cheaper. . . . Is it worth $25 million per launch for the single orbit reconnaissance mission? I want to know what the military space missions are and how they get done."[107]

McNamara's critique of a supposed Air Force failure to elucidate the Dynasoar's mission seems not entirely fair. First, the Air Force did, in fact, frequently explain that Dynasoar would supply the ability to gather intelligence information over any portion of the globe on demand and in a short period. This was compared to the robotic reconnaissance satellites that were limited to covering the area directly beneath their orbital plane, although some limited adjustment to their coverage was possible in the early satellites. Second, McNamara seemed to have been searching for additional military applications that Dynasoar could perform. Yet this was the very role which McNamara had forbade the Air Force to explore in his dual reorientation of late 1961/early 1962. By late 1963 he was asking the Air Force to supply him with information resulting from investigations he had specifically prohibited it from performing for almost two years. Seamans accompanied McNamara not only to all the briefings during the October tour but spoke with him extensively during the hours of the aircraft flights. Seamans simply stated, "I could tell McNamara had made up his mind to cancel it [Dynasoar] and was looking for a good rationale. I could tell that whatever he saw in Houston [concerning NASA's Gemini program], he'd made up his mind he liked it. He was all exuberant about our Gemini program." While McNamara did not overtly state on the flight back to Washington, DC, that he had decided to terminate the X-20, "I knew damn well he had and justified it on the basis that we had the Gemini program."[108] As Houchin said concerning McNamara's demand that Air Force officers supply specific information relating to the Dynasoar's military applications: "For their answers to be useful, the secretary needed to be listening."[109] Third, two of the missions the Air Force had concluded Dynasoar could fulfill were as a delivery platform for nuclear weapons and as a satellite interceptor/inspector/neutralizer. However, with the adoption of UN General Assembly Resolution No. 1884 (see chap. 5) which led to the Declaration for the Legal Principles for the Use of Outer Space, which renounced the stationing of mass destruction in space, these two potential X-20 roles disappeared. While its third, specific possible mission—reconnaissance—was still viable in the USAF's opinion, the NRO already had operational versions of the robotic reconnaissance satellites providing valuable intelligence data to national policy makers. Once reconnaissance was the only remaining Dynasoar justification, this placed it "in direct conflict with the NRO and its highly classified 'black' reconnaissance satellites and their follow-on programs."[110] The Dynasoar's fate was almost certainly sealed by late October.[111]

What remained was for the DDR&E to determine the exact configuration of the Gemini-based MOL that would replace the Dynasoar and coordinate this with NASA. This took most of the month of November. In addition, the turmoil surrounding the assassination of President Kennedy also probably pushed the official announcement of Dynasoar's cancellation into the second week of December. NASA's top-level leadership offered no public support of Dynasoar and in private did not lament its potential death.[112] In his personal correspondence to Webb, Seamans noted, "We have not felt that the orbital operation capability inherent in the present X-20 configuration will significantly increase our knowledge over that already obtained from Mercury."[113] Gemini, of course, had even more capability than Mercury. When asked if NASA leaders concluded Dynasoar was not needed because NASA was developing similar capabilities in the Gemini program, Seamans replied to the author of this book, "Exactly." When asked if NASA leaders had any objection to the OSD decision to cancel Dynasoar, Seamans stated, "It didn't bother us. I can't remember any problem with that."[114] Finally, a memorandum from Kennedy's special assistant for national security affairs, McGeorge Bundy, to Kennedy preparing him for an upcoming session with Webb informed Kennedy that Webb "is quite cool about the use of Titan III and Dinosoar [sic] and would be glad to see them both canceled."[115] One may conclude that while the elements within NASA that had been closely working with the Dynasoar and had some direct interest in its continuation did support the program, NASA's top-level policy makers had no serious objections to its cancellation.

Brown, DDR&E, laid out his conclusions concerning the Dynasoar/Gemini/MOL programs on 14 November 1963. In one sense, it represented significant movement toward the Air Force's position that not everything the military needed to learn concerning military requirements in space could be learned by using NASA-developed systems or conducting "piggyback" experiments on NASA flights. Brown explained, "Although the NASA research and development will have broad applications toward any type of space program, it is not sufficiently attuned toward the needs of military missions to be commensurate with the cost which might be identified within the national budget as providing military support. There is a growing recognition that from the standpoint of economy as well as for other reasons, a directed military program would be preferable . . . for the assessment and measurement of the utility of man as a component in an operating military system." Brown added that in his analysis, "Principal attention was directed toward the tasks of surveillance, detection, and inspection," highlighting once again the central role of reconnaissance in the military space decision-making process.[116]

Brown then presented McNamara with a detailed analysis of six possible configurations for a DOD space station. He defined a space station as an earth-orbital platform that was designed for a relatively long orbital life, could be resupplied by other spacecraft, could have personnel ferried to and from it, and could maintain a comfortable, pressurized internal environment in which the officers could move around without wearing space suits. Such a station "will be in the nature of a military laboratory with adequate arrangements for military equipment and with provision for the crew to perform reasonable duplication of military missions in space." As with all OSD programs under PPBS/systems analysis, "The cheapest and most direct routes to this end will be considered. Extensive use will be made of other developments, principally those from the GEMINI and APOLLO programs."[117] Of the six alternatives he supplied for DOD space stations, Brown preferred two possibilities. The one that bore the closest resemblance (albeit much larger) to the MOL's ultimate configuration was a four-room, four-person, 2,140-cubic-foot station launched on a Titan IIIC with docking and storage capability, a living room, sleeping room, and laboratory. The Gemini capsule would serve as a ferry vehicle, and crews would be rotated every 30 days with resupply arriving every 120.[118]

In another sense, however, Brown's 14 November 1963 memo was mired in the past because it continued to maintain that as DOD built its space station, "good management would call for the transfer of GEMINI to the DOD" around September 1965. Given the OSD's experience with the proposed transfer of Gemini just one year earlier, it should have been clear that such a transfer was politically impossible. Be that as it may, the fundamental assumption in Brown's memo was that Dynasoar should be canceled: "Cancellation of the X-20 program and pooling of presently planned national funds related to manned earth-orbit programs would provide more than enough money in FY 1965. . . . A choice of this kind would provide the Air Force with a series of manned earth-orbital launches beginning 9 months earlier than it could expect from the X-20 program." Brown's summary recommendation to McNamara was that, "a military space station program be initiated, taking advantage of the GEMINI developments, based upon a package plan which cancels the X-20 program and assigns responsibility for GEMINI and the new space station program to the Air Force, the effective date for transfer of management responsibility for Gemini being October 1, 1965. . . . Something like the recommended program represents . . . the best way out of the NASA/DOD man-in-orbit problem."[119]

It should come as no surprise that NASA was not thrilled to learn that: a. the DOD was again recommending that Gemini should be transferred from NASA to the DOD; and b. the DOD was proposing that America's first space station be developed and managed under firm DOD control. Between this memo and a revised proposal Brown submitted on 30 November, there were two weeks of NASA-DOD negotiation from which no documentation apparently survives (except the resulting Brown 30 November memo) that one Air Force contemporary source described as "not fully known to persons other than the principals."[120]

In his 30 November memo to McNamara, Brown did mention that since his previous memo, NASA had offered "somewhat in the form of a counter-proposal" a request for the DOD to examine a "manned military program which would not extend quite as far as the establishment of a space station." NASA had suggested the DOD "develop a system consisting of the Gemini personnel carrier weighing 7,000 pounds attached to a pressurized and habitable military test module weighing approximately 15,000 pounds, the combination to be injected into orbit by a TITAN IIIC."[121] This was MOL in a nutshell. NASA supplied the idea for its basic configuration and proposed its creation as an alternative to the DDR&E's full-blown space-station proposals earlier that month, not the OSD. The Air Force was relegated to the role of a passive observer to the policy-making process. Brown relayed that DDR&E personnel's discussions with the NASA staff "have caused us to think it likely that they will advise Mr. Webb to agree, in principle, to a manned military space program which is separate from, but coordinated with, the NASA activity. They may not be prepared at this time, however, to agree to the assignment to the DOD of the responsibility for a space station."[122]

The crucial hair-splitting distinction was that by mutual agreement MOL was not to be considered a space station, but rather a military orbital test platform. The Webb-McNamara agreement on space stations signed in August and September respectively stated that it applied to spacecraft capable of prolonged spaceflight and larger and more sophisticated than Gemini and Apollo; both NASA and DOD could argue that MOL's projected 30-day occupancy was not prolonged nor was its overall configuration larger or more sophisticated than Gemini or Apollo. Brown reported that NASA leaders "have suggested that the DOD could fulfill its needs for an orbiting military laboratory by a system which does not involve the complications of personnel ferry, docking, and resupply." Brown said that the design he was submitting to McNamara "conforms to the NASA suggestion but which, at the same time, would continue as a design objective the preservation of an internal compatibility allowing it to be convertible with only minor additional development into a useful military space station."[123] Thus, while the OSD might agree with NASA that for purposes of strict bureaucratic definition and public relations that the MOL was not technically a space station, the OSD was also preserving the fundamental design characteristics that would enable the MOL to be easily converted into a fully functional space station.

The specifics of the MOL that Brown suggested involved the use of the Titan IIIC booster and the Gemini capsule modified so that it could join with and attach to a cylindrical, partially pressurized military test module of about 1,500 cubic feet. Two to four men could work and live there for 30 days. The laboratory module(s) would be equipped with "complete docking equipment" at both ends as well as a rudimentary propulsion system "so that two modules could be joined together" to form a space station of 3,000 cubic feet for up to eight people. Therefore, "Through a logical progression of development, a space station of any desired proportions could be achieved." One negative to adopting NASA's suggestion for a DOD MOL was that it would "have the effect of imposing a delay in arriving at a decision on the assignment of management responsibility for a space station, since their proposal [for the MOL] would not be defined as a station."[124] An incisive BOB analysis of the MOL proposal pointed out that the incorporation of future rendezvous and resupply features into the MOL "would result in a situation in which a space station project would most logically be an outgrowth of the present MOL project.

This would be a difficult situation for NASA to accept."[125] Still, the president's unclassified annual space report stated, "Rendezvous provisions will be designed into the MOL so that the laboratory could later be resupplied and reused if justified by progress made in defining man's military role in space."[126]

Nevertheless, McNamara quickly adopted Brown's supposed scaled-back recommendation as the OSD-preferred alternative and in 10 days announced the cancellation of Dynasoar and the beginning of the official study phase of the MOL. Before the 10 December announcement, the Air Force generated a flurry of memoranda to support the Dynasoar's existence, but to no avail. Near the end, the Air Force was proposing the Dynasoar be used as the ferry vehicle for any proposed space station, but it seemed extremely unlikely that the OSD would authorize a billion dollars for that purpose.[127] At the 10 December news briefing, McNamara explained OSD's calculations showed canceling Dynasoar and substituting MOL would save $100 million over the next 18 months; he maintained Dynasoar had cost $400 million so far, "but there are hundreds of millions left to be spent to achieve a very narrow objective." He elaborated that while the Dynasoar would have explored precisely controlled reentry techniques, "It was not intended to develop a capability for ferrying vehicles or personnel or equipment into orbit, nor was it intended that the Dynasoar would provide a capability for extended stay in orbit, nor was it intended that it would provide a capability for placing substantial payloads, useful payloads, in orbit, and hence, it had a very limited objective. It was very expensive."[128] Later he stated, "I think this is a good illustration of what happens when we start on a program with a poor definition of our end objective."[129]

When the Dynasoar cancellation was explained to the Congress the next month, McNamara said, "The X-20 was not contemplated as a weapon system or even as a prototype of a weapon system. Its distinguishing feature, as compared with MERCURY and GEMINI, was to be its substantial lifting maneuver capability. . . Yet, from the military point of view, the determination of man's ability to perform useful military missions in space is the more immediate problem, and for this purpose DYNASOAR was so limited as to make it a very poor choice. The maneuverability feature of DYNASOAR, while of great interest, is not needed now."[130] McNamara did not mention that it was he who had ordered the Air Force just two years earlier to stop studying the military applications of the Dynasoar and focus solely on its research potential. He also did not mention that it was he who just three months earlier had harshly criticized the Air Force for lacking the kind of information that would have resulted from the studies he prohibited it from making nor that it was he who nevertheless used this lack of information as a justification for canceling the program.

On 10 December McNamara also attempted to make clear his thinking about the MOL: "I have said many times in the past that the potential requirements for manned operations in space for military purposes are not clear. But that, despite the fact that they are not clear, we will undertake a carefully controlled and carefully scheduled program of developing the techniques which would be required were we to ever suddenly be confronted with a military mission in space."[131] The MOL was presented, at least for public consumption, as primarily a test bed to experiment with the functions of and evaluate the effectiveness of the military man in space. McNamara said the MOL was not created to perform a "precise, clearly defined, well-recognized military mission, but because we feel that we must develop certain of the technology that would be the foundation for manned military operations in space should the specific need for those ever become clear and apparent."[132] In other words, it was the building block approach. The press release distributed after McNamara's briefing described the MOL as "approximately the size of a small house trailer" that would "increase the Defense Department effort to determine the military usefulness of man in space." Its design would enable the two astronauts to move about freely without a space suit for up to a month. The first of six planned manned launches was expected in late 1967 or early 1968.[133]

The basic operational concept of the MOL was that the two astronauts would be positioned in the modified Gemini capsule that was itself attached to the laboratory module. This entire unit was placed on top of what would come to be called the

Titan IIIM and launched into orbit. Then, the astronauts would open the hatch between the Gemini capsule and the laboratory, enter the laboratory, and seal up the now inactive Gemini capsule. For the next 30 days they would perform the mandated experiments and observations. Then, they would reposition themselves into the Gemini capsule, separate from the laboratory module, reenter the earth's atmosphere, and land in the ocean just like a standard NASA Gemini reentry. Eventually, the laboratory module's orbit would decay, and it would burn up upon reentering the atmosphere. While McNamara's remarks cited above indicated a continuing skepticism about the role of military officers in space, his backing of the MOL was of some consolation to the Air Force in the context of losing Dynasoar: "Significantly, this was a departure from earlier Defense pronouncements that the military had no clearly defined mission for men in space. Now at least Secretary McNamara showed himself willing to investigate the subject seriously."[134]

Other documents cited above make clear that what the OSD had in mind was experimenting specifically with what role humans could play in gathering intelligence data via space-based reconnaissance. The DDR&E alluded to this when it described the MOL to the USAF and tasked the AFSC's SSD with responsibility for developing it, explaining that the MOL's goal was for "employing man in his most useful functions of discrimination, quality improvement and quick reaction through his ability to recognize information and transmit it back to the ground."[135] The core of MOL's mission was clear to perceptive analysts. The *New York Times* stated two days after McNamara's announcement, "The primary purpose of the Air Force's newly authorized orbiting laboratory will be to determine the effectiveness of manned space stations for photographic reconnaissance of the earth."[136] When asked about MOL's central mission, Seamans told this author, "Obviously that was going to be largely reconnaissance."[137]

NASA's Attitude Concerning MOL

Webb supported the MOL decision in public, stating, "The decisions announced by Secretary McNamara today . . . follow discussion with NASA and were fully coordinated with the programs of this agency. . . . The decisions announced by Secretary McNamara are based on the best use of resources to maximize our national capability in space and NASA fully supports them."[138] Privately, he was more concerned, especially over the fact that the media were pressing "for some statement as to why the terms of the agreement announced on October 17 to coordinate our approaches to a possible new program for manned orbital operations were not followed." Webb explained to Seamans,

> Some newsmen are taking the view that all this was bypassed and in a sense, I was forced by McNamara to go forward faster than this agreement calls for. My own view of what has happened is that in connection with our joint review of both the 1964 and 1965 budgets, it became clear that Dyna Soar could not hold up in the competition for funds and we have made an interim arrangement to use the Titan III booster and the Gemini spacecraft to accomplish a number of things the military need [sic] to do on an experimental basis.[139]

Webb then tasked Seamans with developing a NASA position paper detailing NASA's exact role in the development of MOL.

The resulting internal NASA document from Seamans made the following points that would in fact represent both NASA's and the DOD's long-term "party-line" position on MOL. It served as a guide for the next six years concerning public releases, congressional testimony, and speeches by leaders of both NASA and the DOD. It is therefore quite important because it represents virtually everything stated or written about the MOL in the public record and in unclassified documents from 1963–69.

- MOL is a single project with a specific goal within the overall U.S. space effort, not a broad space station program.

- MOL is being implemented in response to military requirements established solely by the DOD.

- NASA's technology, hardware, facilities, and operational know-how "will be made available to the DOD, and the DOD will take full advantage of these national assets. NASA will, in turn, take full advantage of the research and development opportunities presented by the MOL."

- MOL "should not be construed as the national space station" and does not fall under the Webb/McNamara agreement on manned orbital research and development systems larger than Gemini and

377

Apollo signed earlier that fall. "The MOL is, rather, a specific experimental test bed utilizing NASA's Gemini project and the Titan III for certain potential military space applications not within the scope of NASA's activities. NASA projects will be considered for test in the MOL on a noninterference basis."

- MOL was coordinated between the two agencies and concurred in by NASA. The DOD originally indicated its requirements for testing military equipment in space [Brown's 14 November memo] and then a system concept was evolved by NASA and DOD during the coordination phase [prior to Brown's 30 November memo] "and accepted in lieu of the original DOD concept for meeting these requirements."

- "NASA and DOD worked together in defining this project in the spirit of the Gemini agreement."

- "The DOD MOL, as a special-purpose experimental military project, does not conflict with the NASA unmanned and manned flight projects, and does not affect the high priority of the Nation's major close-range space goal of landing a man on the moon before the end of the decade."

- The timing of the MOL and Dynasoar decisions "were dictated by the urgency of the budget." Major savings will result from the cancellation of Dynasoar.[140]

This comprised the majority of information anyone but the most senior policy-making officials and Air Force personnel working on MOL had access to concerning MOL between its commencement in December 1963 and its cancellation in June 1969.

An Addendum: Reconnaissance Satellites and Space Policy in the Kennedy Administration

Within days of its beginning, the Kennedy administration tightened and extended Eisenhower's policies on releasing information concerning reconnaissance satellites in particular and military space launches in general. An OSD official explained to Kennedy that the information the DOD planned to release to the media on upcoming Samos launches "represents a severe reduction from what had previously been issued. Eliminated from former procedures are four pages comprising 22 questions and answers. Press briefings before and after launching have been eliminated." This assistant SECDEF for

Public Affairs stated, "Dr. Charyk has reviewed these changes and is satisfied that they meet all his security requirements and those of his SAMOS Project Director."[141] Traditionally the undersecretary of the Air Force served as the NRO director, a position Joseph Charyk held in the late Eisenhower and early Kennedy administrations. The assistant secretary summarized for the president, "This readjustment is a big step toward the gradual reduction of volunteering information on our intelligence acquisition systems which Mr. McNamara informed me is your desire."[142] Clearly, Kennedy offered no objections to the new policy, given the fact he apparently initiated it through McNamara.

After a year, the Kennedy administration in general and the OSD in particular concluded their new policy of withholding information on reconnaissance satellites was the proper policy and not only made it official but broadened it to include all military space launches. The OSD issued a directive, S-5200.13, *Security and Public Information Policy for Military Space Programs*, in March 1962 that stated, "Adequate protection of military space programs is vital to the security of the United States. This requires the capability to launch, control, and recover space vehicles without public knowledge of the timing of these actions or of the specific missions involved. It is impractical to selectively protect certain military space programs while continuing an open policy for others since to do so would emphasize sensitive projects." Therefore, in the future, *all* military space projects, vehicles, and launches would be identified only "by means of numerical or alphabetical designators selected and assigned at random," no nicknames could be used (emphasis in original). All public information releases had to be cleared through the OSD Public Affairs Office. All reports, plans, and other documents relating to all military space programs "will be severely limited and controlled." The number of people with access to information concerning military space programs was to be reduced.[143] In other words, the few people privy to information concerning the military space program could say or write virtually nothing about it. No US official would even formally admit the United States operated reconnaissance satellites until Pres. Jimmy Carter did so in 1978.

Apparently, the Kennedy administration's increasing the security surrounding reconnaissance satellites was an attempt to avoid provoking the USSR into threatening American reconnaissance satellites. Indeed, throughout 1961 and 1962, the Soviets waged a sort of diplomatic offensive in the UN and elsewhere against reconnaissance satellites. The United States denied satellite reconnaissance was espionage, but the Soviet campaign stopped only in the latter half of 1963 as the USSR perfected and began employing its own reconnaissance satellites.[144] America and the Soviet Union signed no accord concerning the legality of satellite reconnaissance; there simply emerged an unstated understanding that both countries conducted and accepted the practice.

The Kennedy administration did craft an official policy concerning satellite reconnaissance in 1962. Kennedy signed NSAM 156 (no title) on 26 May 1962. In it he explained,

> We are now engaged in several international negotiations on disarmament and peaceful uses of outer space. . . . They raise the problem of what constitutes legitimate use of outer space, and in particular the question of satellite reconnaissance. In view of the great national security importance of our satellite reconnaissance programs, I think it desirable that we carefully review these negotiations with a view to formulating a position which avoids the dangers of restricting ourselves, compromising highly classified programs, or providing assistance of significant military value to the Soviet Union and which at the same time permits us to continue to work for disarmament and international cooperation in space.[145]

One peek inside the resulting NSAM 156 Committee that was formed under U. Alexis Johnson, deputy undersecretary of state for Political Affairs, was provided by its executive secretary, Raymond Garthoff, in an article where he stated the fundamental purpose of the committee was to review the political aspects of United States' policy on satellite reconnaissance. The very existence of the committee, any reference to its function, and all of its work was considered Top Secret.[146] In addition, Garthoff related that after the committee submitted its report on 2 July, the NSC met on 10 July 1962 to discuss the report. After the meeting, Kennedy supported all 19 of its recommendations, except an arms control measure. Garthoff does not, however, provide specific information concerning the

380

nature of the 19 recommendations.[147] A military assistant to the SAF specializing in space explained that the NSC's passage and Kennedy's approval of the basics of the NSAM 156 Committee's report was translated into NSC 2454, which contained 18 points that formed "a firm foundation to space policy in this Government under President Kennedy's personal aegis. We all knew where we stood in space, what we would say at the United Nations, what we would say to the outside world, and what was absolutely not negotiable."[148]

The NSAM 156 Committee's report opened by stating, "The reconnaissance satellite program is extremely important to Free World security, and will continue to be necessary to provide crucial information about Soviet activities, capabilities, and targets."[149] After an extensive discussion of the international complexities of conducting a satellite reconnaissance program given the then-current Soviet diplomatic offensive against reconnaissance satellites, the report offered 19 recommendations. The recommendations directly relevant to reconnaissance satellites said the United States should maintain that international law applies to outer space in the same sense as it does to the high seas, and therefore, states are free to pursue defensive military pursuits in space; avoid declaring or implying that reconnaissance satellites are anything but a peaceful use of space; seek to gain acceptance of the principle of the legitimacy of space reconnaissance, even when confronted by specific Soviet pressure to outlaw satellite reconnaissance; conduct an R&D program into a completely clandestine reconnaissance satellite program in case circumstances should ever make it necessary; continue to refuse to "publicly disclose the status, extent, effectiveness or operational characteristics of its reconnaissance program; discreetly disclose to certain allies and neutrals selected information with regard to the US space reconnaissance program" with the goal of "impressing upon them its importance for the security of the Free World; in private disclosures emphasize the fact of our determination and ability to pursue such programs because of their great importance to our common security, despite any efforts to dissuade us;" and continue to study the role of space reconnaissance in disarmament inspection.[150]

The above recommendations were all unanimously agreed upon by NSAM 156 Committee members. It seems likely that they were included in NSC 2454, which was designed to take the report's recommendations and state them as official governmental policy. One document from August 1962 made clear the impact of the NSAM 156 Committee on Kennedy. In it the White House staff explained that Kennedy wanted American space policy to "be forcefully explained and defended" at forthcoming UN meetings, with an emphasis on three points. First, "to show that the distinction between peaceful and aggressive uses of outer space is not the same as the distinction between military and civilian uses, and that the U.S. aims to keep space free from aggressive use and offers cooperation in its peaceful exploitation for scientific and technological purposes." Second, "to build and sustain support for the legality and propriety of the use of space for reconnaissance." Third, "to demonstrate the precautionary character of the U.S. military program in space."[151] Clearly the NSAM 156 Committee's recommendations had been accepted by Kennedy and served as the core of his "marching orders" to the American diplomats at the United Nations. The NSAM 156 Committee's recommendations were the only official, written space-policy document to emerge from the Kennedy administration.

Finally, it is necessary to state that the NRO continued to serve as a kind of management overlay under which the USAF and the CIA continued to exhibit some degree of conflict in their administration of the nation's satellite reconnaissance program. Albert Wheelon was a participant in the Kennedy-era NRO. He became the CIA's first deputy director for Science and Technology in 1963. In this capacity, he was the chief architect of the CIA's space efforts and oversaw the Corona program during his tenure. He reported that McNamara believed the CIA's role in the NRO should be confined to defining requirements, doing some advanced research, and examining the film from the reconnaissance satellites. When McMillan became undersecretary of the Air Force and therefore NRO director, he tried to implement McNamara's desires by notifying the CIA he was transferring the CIA's responsibilities for Corona to the Air Force. For a year John McCone, director of the CIA, remained undecided as to how to respond to the DOD drive for sole control of the NRO. However, Wheelon finally convinced McCone that the CIA should continue to play a strong role in the NRO: "After a period of readjustment in the expectations of the Defense Department, the partnership between CIA and the Air Force on CORONA resumed and served the country well to the end of the program in 1972." However, Wheelon stated, "The debate between CIA and DOD then shifted in 1963 to whether [the] CIA ought to pursue new reconnaissance systems." OSD officials such as Eugene Fubini, assistant SECDEF, and McMillan "argued against each system that [the] CIA was developing." This debate continued until 1965 when Flax became the NRO's director. He "saw the CIA and the Air Force as valuable and complementary assets." Wheelon reported the OSD/USAF-CIA difficulties within the NRO faded from that point forward.[152]

Secondary accounts from this period of intra-NRO difficulties during the Kennedy administration, some based on interviews with the principals, seem to buttress Wheelon's account and even indicate the situation was quite heated. William Burrows concluded that McMillan was actually "determined to break the agency's [CIA's] hold on the design and procurement of reconnaissance systems through the NRO and, apparently, to wrest management of strategic reconnaissance away from the CIA in the process." This resulted in a collision course that "soon developed into a series of battles over turf that were so vituperative that they are still talked about by old hands."[153] Richelson says the situation was calmed only with the creation in 1965 by McNamara and McCone of the National Reconnaissance Executive Committee (NREC) to oversee the NRO's budget, structure, and R&D activities.[154]

Finally, the NRO's history office confirms the tensions that existed in the early 1960s. Its report (the research that surveyed applicable primary sources) stated that during the Kennedy administration, "the Air Force now moved to secure control over the entire reconnaissance effort." McMillan "recommended that the entire photosatellite program be turned over to the Air Force in order to streamline the command and achieve greater success. [For McMillan], the NRO was primarily an Air Force activity and the CIA was irrational and obstructionist. . . . The rivalry between the Air Force and the CIA intensified." In this battle McNamara "often sided with McCone against the Air Force in order to maintain his position as arbiter of DOD planning and resource allocations." The NRO account confirms that the situation finally got so bad that McCone and new deputy SECDEF Cyrus Vance formed an executive committee to make funding and other decisions for the national reconnaissance program. Finally, by 1965 the efforts of this three-person executive committee consisting of the director of the CIA, the assistant SECDEF, and the president's science advisor were able to establish the NRO as a separate agency within the DOD and designate the SECDEF as its primary executive agent. The new decision-making structure "worked well."[155]

In this chapter the intricate relationship between the specific programmatic efforts of NASA and the DOD has been examined. Neither the DOD's Dynasoar nor its MOL can be analyzed in isolation from NASA's Gemini. Under the imperatives of McNamara's systems analysis, the Air Force's human-spaceflight effort had to mesh with NASA's R&D, and it had to promise distinct and quantifiable advantages to national security. While these criteria doomed Dynasoar by December 1963, they were flexible enough to permit McNamara to authorize the creation of the MOL program that had as its avowed purpose the experimental evaluation and assessment of exactly what military officers could accomplish in space. The primary category of investigation would be the role humans could and should play in the gathering of reconnaissance information. During the Johnson administration, the delicate interplay between NASA and DOD's human-spaceflight efforts would not cease. If anything, the concerns over possible NASA-DOD duplication in this area became even more pronounced as NASA's budgets actually began to decline as a result of the financial demands of the Vietnam War and Johnson's Great Society programs. While MOL did manage to survive Johnson's tenure, Nixon would cancel it six months after his inauguration. NASA's proposed follow-on to the Apollo program appeared to be in little better shape during this Johnson era of mounting financial pressure. The next chapter will set the overall political and space-policy context as well as the NASA-DOD institutional stage for the three human-spaceflight projects (Gemini, Apollo, MOL) extant during the Johnson administration.

Notes

1. For a complete history of the Project Gemini, see Hacker and Grimwood, *On the Shoulders of Titans*. For a synopsis, see Ezell, *NASA Historical Data Book*, vol. 2, 149–70.

2. Boone, *NASA Office of Defense Affairs*, 83. The specific nature of this "deep" involvement was difficult to define from the beginning. For instance, during the LLVPG deliberations Nicholas Golovin recorded that OMSF director D. Brainerd Holmes "proposed having Air Force officers associated in all activities at STG [Space Task Group], NASA's organization at its Langley Research Center responsible for Mercury, and the early stages of Gemini, [before the Manned Spacecraft Center in Houston was created], but no organizational responsibilities." However, the LLVPG's deputy director, Lawrence Kavanau, of the office of the DDR&E "argued very strongly for direct Air Force participation at the STG level, suggesting that this participation should be at the Holmes level. Holmes opposed this concept strongly." Golovin, Chronological File entry, 1–5 December 1961, 1.

3. Rubel and Seamans to McNamara and Webb, memorandum, 7 December 1961, 1. The Air Force Agena vehicle would be launched on an Air Force Atlas rocket and serve as the target vehicle for the Gemini capsule's (launched on a modified Air Force Titan II) rendezvous exercises.

4. Ibid., 2.

5. Kennedy, "Message to the Congress," 31 January 1962, 9. In fact, the longest Gemini mission turned out to be *Gemini 7* in December 1965, which orbited the earth for over 13 days.

6. NASA, internal position paper on Project Gemini, 7.

7. Low to director of the Manned Spacecraft Center, memorandum with attachment, 7 February 1962, 1–2.

8. Office of the DDR&E to McNamara, report on Manned Military Space Programs, 20 February 1962, 2. Document is now declassified.

9. Berger, *The Air Force in Space*, 39.

10. McNamara to the SAF, memorandum, 22 February 1962, 2. Document is now declassified.

11. Berger, *Air Force in Space, FY 62*, 36.

12. "Blue" in this context refers mainly to the color of the uniforms worn by USAF personnel. "Blue Gemini" would thus refer to a program whereby the Air Force would, in some way, shape, or form, own and operate its own Gemini capsules separate from NASA.

13. Senate, Committee on Aeronautical and Space Sciences, Hearings, 24 February 1966, 33–34.

14. Hacker and Grimwood, *On the Shoulders of Titans*, 118.

15. This general Blue Gemini and MODS description is from Killebrew, USAF, *Military Man in Space*, 25. The official USAF development plan for Blue Gemini is, unfortunately, still classified at AFHRA and has been exempted from declassification IAW E.O. 12958. The term *Blue Gemini* thus becomes extremely confusing because of the uncertainty resulting from Air Force refusal to declassify its exact meaning. Various authors have speculated that it represented the program described above whereby Air Force astronauts would fly on NASA Gemini flights, a separate program of the Air Force acquiring Gemini capsules and independently launching them in an AF-only program, the MODS program itself, or the program that would eventually emerge in January 1963 whereby the DOD was permitted to include DOD experiments on NASA Gemini flights. The author was, however, able to obtain portions of the MODS development plan, and they are discussed below. Information extracted is unclassified.

16. Air Force, Partial Systems Package Plan for Military Orbital Development System (MODS) System Number 648C, 4 June 1962, iii, 1-1–1-2, 12-1. Document is now declassified.

17. Houchin, *The Rise and Fall of Dyna-Soar*, 226.

18. Gruen, *The Port Unknown*, 171.

19. "Air Force Space Plan," 474.

20. Houchin, "Interagency Rivalry?" 37. For the full story of the MORL, see Hansen, *Spaceflight Revolution*, 274–93.

21. Gruen, *The Port Unknown*, 157.

22. NASA, Long Range Plan, January 1962, 95.

23. BOB, Military Special Space Review, August 1962, V-1.

24. Gruen, *The Port Unknown*, 167.

25. The balance of the $420 million was $193 million for Midas, the system designed to provide early warning of ballistic missile attack, and $50 million for the unmanned Saint, satellite interceptor program. See Carter, "An Interpretive Study of the Formulation of the Air Force Space Program," 9; and Cantwell, *The Air Force in Space, FY 63*, 8. Document is now declassified.

26. Cantwell, *The Air Force in Space, FY 63*, 26. Document is now declassified.

27. NASA, *Aeronautics and Space Report of the President*, A-30.

28. McNamara, interview in *Missiles and Rockets*, 22 October 1962.

29. Cantwell, *The Air Force in Space, FY 63*, 30. Document is now declassified. One should note that there was also dissension within the Air Force concerning how far the Air Force could and should push cutting-edge space proposals like Blue Gemini and MODS without endangering the already-existing Dynasoar effort: "If Gemini were stretched beyond 1966 in any modified capsule version . . . then Gemini clearly would become competitive with Dyna-Soar" (ibid.).

30. See Houchin, *Rise and Fall of Dyna-Soar*, 238; Futrell, *Ideas, Concepts, Doctrine*, vol. 2, 143; and Richelson, *America's Secret Eyes in Space*, 83.

31. House, *Government Operations in Space*, 81. Flax stated that Blue Gemini and MODS "never received very serious consideration at the higher echelons of the Air Force and the Department of Defense at that time." Pealer, "Manned Orbiting Laboratory," pt. 2, 33.

32. Cantwell, *Air Force in Space, FY 63*, 26. Document is now declassified.

33. Seamans, oral history interview, 26 May 1966, 14.

34. Hacker and Grimwood, *On the Shoulders of Titans*, 34. Webb's recollection was simply, "Wiesner and McNamara were working very closely together, they were having lunch once a week. . . . They began to sort of mount a game, an effort to prevent us from moving independently." Webb, oral history interview, 11 April 1974, 34. Arnold S. Levine interviewed anonymous NASA insiders for his book: "According to one source, McNamara proposed that DOD take over all manned flight in Earth orbit; NASA all flights beyond Earth orbit. . . . NASA officials sensed that they could not accede to such a proposal and still retain control over their programs." Levine, *Managing NASA in the Apollo Era*, 230.

35. Lambright, *Powering Apollo*, 119. Webb's characterization of McNamara's negotiating style is from a Lambright interview with Webb on 8 January 1991.

36. Staats, oral history interview, 13 July 1964, 32–33.

37. Golovin to Jerome Wiesner, memorandum, 21 December 1962, 6–9. Document is now declassified.

38. Wiesner to the president, memorandum, 10 January 1963, reel 9, 1, 4.

39. McDougall, *The Heavens and the Earth*, 340.

40. In his memoirs, Boone confirmed Webb's and Seamans's account of the origins of the controversy: "McNamara, in joining a meeting late in 1962 at which Webb was present, had orally proposed that the NASA and the Air Force manned space flight programs be combined and the entire package placed under DOD management." Boone, *NASA Office of Defense Affairs*, 9.

41. NASA, internal position paper on Project Gemini, 7 January 1963, 21, 28.

42. Buckley to Boone, memorandum, 8 January 1963.

43. Boone to Webb, memorandum, 9 January 1963, 1–3.

44. McNamara, proposed Agreement between the DOD and NASA, Gemini Program Management, 12 January 1963, 1.

45. Webb to McNamara, letter, 16 January 1963, 338–39.

46. Webb to Seamans, letter, 18 January 1963, 339–41.

47. Webb and McNamara, Agreement between NASA and the DOD Concerning the Gemini Program, 21 January 1963, 341–42.

48. Boone, *NASA Office of Defense Affairs*, 84, 87.

49. Cited in "DOD Asks Space Station Role," *Space Daily*, 22 May 1963, 651.

50. Cited in "The Dyna Soar Paradox," *Space Daily*, 26 June 1963, 827.

51. McNamara, statement before the Senate Armed Services Committee, 4 April 1961, 26.

52. McNamara, press conference, 23 June 1961, 5. See also Hungerford, *Organization for Military Space*, 45; and Berger, *The Air Force in Space, FY 62*, 29.

53. Air Force, Advanced Systems: DYNASOAR, 1.

54. Berger, *The Air Force in Space, FY 62*, 26.

55. As of October 1961, Houchin, *Rise and Fall of Dyna-Soar*, 202.

56. McNamara, Statement before the Senate Armed Services Committee, 19 January 1962, 97. Document is now declassified.

57. NASA, Long Range Plan, January 1962, 87.

58. McNamara to Kennedy, memorandum, 7 December 1961, 438.

59. Brown to the SAF, memorandum, 20 February 1962, 1. Document is now declassified.

60. McNamara to the SAF, memorandum, 22 February 1962, 2–3.

61. Subsequently to this name change, the Dynasoar program was referred to interchangeably as either the X-20 program or as the Dynasoar program. This book will reflect that practice.

62. Houchin, *Rise and Fall of Dyna-Soar*, 208.

63. Brown was emphatic when he stated, "However, we [OSD] have not supported specific military uses for such a vehicle, be it destructive of other vehicles, be it maintenance and repair of satellites or whatever, because it is not possible to lay down military needs which would be fulfilled in an obviously useful way by such a vehicle." Brown to the Senate space committee, June 1962, as cited by Wilson, "Defense Denies Bid for NASA Programs," 34.

64. AFSC, *History of the Aeronautical Systems Division*, vol. 1, I–I-54. Document is now declassified.

65. Houchin, *Rise and Fall of Dyna-Soar*, 4, 175. On 23 October 1963, McNamara reportedly asked William E. Lamar, director of Dynasoar Engineering, "What can the X-20 do that SAMOS can't do?" Lamar replied, "I don't know. I'm not cleared for the program." To which McNamara could only respond, "Well, you should be." Ibid., 221.

66. Ibid., 218.

67. Ibid., 219–20.

68. Miller (*Congressional Record*, speech, 6 September 1962, 18673).

69. Golovin to Jerome Wiesner, memorandum, 21 December 1962, 11.

70. *U.S. Aeronautics and Space Activities, 1961*, 31 January 1962, 36.

71. Becker, "The Development of Winged Reentry Vehicles," 434, 437–38. Becker summarized that NASA had 55 personnel continuously working on Dynasoar support; through the end of 1961, they had devoted 3,900 hours of wind-tunnel time to Dynasoar R&D (ibid). By February 1963, NASA had devoted 6,135 hours of time in its various wind tunnels to Dynasoar R&D. NASA, *Chronology, NASA Participation in X-20 Project*, 13 March 1963, tab B, 3.

72. NASA, Minutes of the NASA Dyna Soar Coordinating Committee meeting, 30 March 1962, 2.

73. Webb to Zuckert, letter, 28 May 1962, 1–2.

74. Webb and Zuckert, memorandum of understanding, 7 August 1962, 1.

75. Elliott, *Finding an Appropriate Commitment*, 210.

76. McNamara to Brown, two memoranda, 18 and 19 January 1963, 1.

77. House, *Government Operations in Space*, McNamara testifying before the House Armed Services Committee, January–February 1963, 80.

78. Webb to Boone, letter, 13 February 1963, 1.

79. Webb to Dr. Arthur E. Raymond, letter, 13 February 1963, 2.

80. Zuckert to McNamara, memorandum, 26 February 1963, 1.

81. Cited in Laurence Barrett, "The Death of the Dyna-Soar Project," *New York Herald Tribune*, 26 December 1963.

82. Coughlin, "Eulogy to a Dyna-Soar," 50.

83. McMillan to Zuckert, memorandum, 15 March 1963, 1–2.

84. McNamara to the SAF, memorandum, 15 March 1963, 1. Document is now declassified.

85. Kolcum, "Defense May Ease Impact of X-20 Loss," 31. McNamara told *Missiles and Rockets* that "perhaps the Gemini project can be modified with relative slight effort to better meet the Air Force's needs for capabilities in manned spaceflight, it may be possible for the Air Force to cut back the Dyna-Soar project substantially." *Missiles and Rockets*, 1 April 1963, 46.

86. Air Force, response to Secretary McNamara's 15 March 1963 questions, 10 May 1963, 3. Document is now declassified.

87. McMillan to the SECDEF, memorandum, 5 June 1963, 3–4.

88. Cantwell, *Air Force in Space, FY 63*, 15. Document is now declassified.

89. Houchin, *Rise and Fall of Dyna-Soar*, 251.

90. Gilpatric to Anderson, letter, 27 September 1963, 2–3.

91. Webb and McNamara, *Agreement between NASA and the DOD Concerning the Gemini Program*, 21 January 1963.

92. Webb to McNamara, letter, 24 April 1963.

93. McNamara to Johnson, report, 3 May 1963, 342–47.

94. McNamara to the SAF, memorandum, 25 May 1963, 1. Document is now declassified.

95. Boone, *NASA Office of Defense Affairs*, 88, for the bureaucratic wrangling.

96. McNamara to the SAF, memorandum, 20 June 1963, 2.

97. Brown to the SAF, memorandum, 30 August 1963, 1–2.

98. Webb interjected that the issue was largely academic because men could stay up in the Apollo capsule for 2–3 months and so "we may not need a space station for some years in the future." NASC, Summary Minutes, 17 July 1963, 3–5.

99. Johnson to McNamara and Webb, letters, 22 July 1963, 1.

100. He added that NASA's interest in a first-generation station was in the fields of biomedical experiments, engineering R&D, and space science. Webb to the vice president, memorandum, 9 August 1963, 1–3.

101. McNamara did add that orbital bombardment "does not appear to be an effective technique at the moment." McNamara to the vice president, memorandum, 9 August 1963, 1–2.

102. There was also an attachment to the agreement entitled "Procedure for Coordination of Advanced Exploratory Studies by the DOD and the NASA in the Area of Manned Earth Orbital Flight under the Aegis of the Aeronautics and Astronautics Coordinating Board" which spelled out exactly how the AACB's Manned Space Flight Panel would coordinate NASA's and the DOD's space-station studies. Webb and McNamara, *Agreement Covering a Possible New Manned Earth Orbital Research and Development Project*, 17 August 1963 for Webb's signature; 14 September 1963 for McNamara's signature, 357–58.

103. McNamara to Webb, letter, 16 September 1963, 359–60.

104. Boone, *NASA Office of Defense Affairs*, 93. Likely explanations for the subpanel's lack of substance, after all the months of McNamara-Webb's dissension over space-station planning can be found in two factors. First, Webb had no desire for a large and capable, yet expensive, NASA-managed space-station program that would compete internally with the Apollo program for NASA budgetary priority. Second, once McNamara endorsed the MOL system in December 1963 the OSD also had no desire to seriously plan for any kind of a larger, more capable, next-generation space station until the results of the various experiments to be conducted on MOL could be performed and analyzed.

105. Levine, *Managing NASA in the Apollo Era*, 149.

106. Lamar, memorandum for record, n.d., 1–2. Numerous other accounts of the session corroborate Lamar's synopsis. See also Ruegg to Schriever, personal message, 23 October 1963; Ruegg to McNamara, status briefing, 29 October 1963; and Goldie to George Snyder, memorandum, 23 October 1963 (both men were senior officials at Boeing, the main contractor for the Dynasoar vehicle). Documents are now declassified.

107. Lamar to McNamara, paraphrased transcript of discussion, by Colonel Moore in Denver, 23 October 1963, 3, 4, 6. Document is now declassified.

108. Oral history interview of Seamans by the author, 5 July 1996. Col Walter L. Moore, who served as the Systems Program Office director for the X-20, the AF's top Dynasoar officer, explained that in January 1963 he attempted to brief Deputy DDR&E Rubel on exactly the topic McNamara would ask about 10 months later: the capabilities of the X-20 to test military equipment and man in space. Moore recalled, "Mr. Rubel strongly recommended that this kind of information be deleted from the presentation and indicated that such talk would jeopardize the program." When McNamara in October asked for just that type of information, Moore explained, "This interest was completely reversed from any direction or indications which had been received over the preceding years from the DOD level." Moore to SECDEF McNamara, memorandum, 23 October 1963. Document is now declassified.

109. Houchin, *Rise and Fall of Dyna-Soar*, 260.

110. Ibid., 264.

111. The best explanation of the Dynasoar's difficulties caused by the international situation late in 1963 and by the growing importance of the NRO's reconnaissance satellites is Houchin, *Rise and Fall of Dyna-Soar*, 254. He states, "When the United States and the Soviet Union accepted mutual satellite overflight in 1963, Dyna-Soar became a hindrance, threatening to unbalance international stability" which meant that the Kennedy administration soon "deemed the project a diplomatic liability." Houchin, "The Diplomatic Demise of Dyna-Soar," 274–80.

112. During the fall of 1963, as it became increasingly clear that Dynasoar would not survive, some lower-level officials within NASA did try to offer support to the beleaguered system. Most vocal was NASA's associate administrator for advanced research and technology Raymond L. Bisplinghoff, whose portion of NASA was responsible for working with the Air Force on Dynasoar. He wrote Seamans on 22 November 1963 that NASA should still support the X-20 because of its contributions to the "technologies of aerothermodynamics and high-temperature metallic structures applicable to maneuverable hypersonic vehicle systems." He maintained that vehicles with those characteristics "will become important components of the future national space program" and so, "The X-20 flights will therefore provide vital new technological data unobtainable from ground facilities." Bisplinghoff added, "Should the X-20 program be canceled, it is our belief that the time is so critical that action should be taken at once to develop a substitute program. The question is therefore one of considering whether the X-20 program can be completed at less cost than a substitute." Bisplinghoff to Seamans, memorandum, 22 November 1963, 1–2. Document is now declassified. One should note, however, that Bisplinghoff was at the fourth level of the NASA hierarchy. Above him were Associate Administrator Seamans, Deputy Administrator Dryden, and Administrator Webb.

113. Seamans to Dryden and Webb, memorandum, 11 September 1963, 2.

114. Oral history interview of Seamans by the author, 5 July 1996.

115. Bundy to Kennedy, memorandum, 18 September 1963, 2.

116. Brown to the SECDEF, memorandum, 14 November 1963, 1–2. Document is now declassified.

117. Ibid., 2–3.

118. Ibid., 6–9. Brown's other preferred alternative was to use NASA's Apollo's command and service modules converted into a three-person station with 3,400 cubic feet that would be launched on a Saturn IB and have capabilities at least equivalent to, if not in excess of, the previous configuration. Brown said this Apollo-Saturn alternative was the most useful but also the most expensive.

119. Ibid., 11.

120. Air Force, "Chronological Listing and Highlight Summary," December 1963, in *History of Aeronautical Systems Division, July–December 1963*, vol. 4, 2. Document is now declassified.

121. Brown to the SECDEF, memorandum, 30 November 1963, 1. Document is now declassified.

122. Ibid.

123. Ibid., 1–2.

124. Ibid., 2–4, 8.

125. Shapley to BOB director, memorandum, 6 December 1963, 1.

126. Johnson, *U.S. Aeronautics and Space Activities, 1963*, 27 January 1964, 41.

127. Flax to the SAF, memorandum, 4 December 1963, 1. Document is now declassified. Another memo representative of the AF's final attempts to save the X-20 was Maj Gen J. K. Hester, assistant vice-chief of staff of the Air Force, Memorandum to the CSAF, Approaches to a Manned Military Space Program, in which Hester said if some element of the space program had to be curtailed then "the cancellation of the GEMINI program should be considered since such action would result in considerable savings to the Nation which could be supplied toward other manned space flight efforts" (ibid., 9).

128. McNamara, transcript of news briefing, 10 December 1963, 1–2. It should be noted that in addition to canceling Dynasoar and starting MOL, McNamara also announced a program called Aerothermodynamic Structural Systems Environmental Test (ASSET) designed to use unmanned glide-type smaller vehicles launched on USAF Thor IRBMs to explore some of the same questions concerning the hypersonic-flight regime that Dynasoar was supposed to have investigated.

129. Cited in Futrell, *Ideas, Concepts, Doctrine*, vol. 2, 225.

130. McNamara, Statement before the House Armed Services Committee, 27 January 1964, 105.

131. McNamara, transcript of news briefing, 10 December 1963, 6.

132. Ibid.

133. DOD, Press Release No. 1556-63, 10 December 1963, 1.

134. House, *Government Operations in Space*, 9.

135. Office of the DDR&E to the assistant SAF for R&D, memorandum, 11 December 1963, attachment 1, 2.

136. John Finney, "Space Stations to be Tested for Reconnaissance and Command-Post Roles," *New York Times*, 12 December 1963.

137. Seamans, oral history interview, 5 July 1996. Seamans was intimately familiar with the MOL program because he defended it and its capabilities before and during the process as Nixon's secretary of the Air Force; Nixon canceled it in June 1969.

138. DOD, Press Release No. 1556-63, 10 December 1963.

139. Webb to Seamans, memorandum, 13 December 1963, 1.

140. Seamans, memorandum for record, 19 December 1963, 1. Only the portions of this indented section within quotation marks are direct citations; the entire section has been indented and uses bullet statements for purposes of organization and clarity and because that is the general format of the original document.

141. Sylvester to Kennedy, memorandum, 26 January 1961, 1.

142. Ibid.

143. DOD, Directive S-5200.13, 1–3.

144. The particulars of the administration's justification for tightening the policy of secrecy surrounding reconnaissance satellites as well as the USSR's campaign against them (and its cessation) are not germane to this study. For full details see Burrows, *Deep Black*, 105; Stares, *The Militarization of Space*, 66; Steinberg, *Satellite Reconnaissance*, 44; and Klass, *Secret Sentries in Space*, 75.

145. NSC, NSAM 156, 1. Document is now declassified.

146. Garthoff, "Banning the Bomb in Outer Space," 26.

147. Ibid., 27–28.

148. Worthman, oral history interview with SAF Zuckert, 25 July 1964, reel 12, 20–21.

149. Johnson, Report of the NSAM 156 Committee, 2 July 1962, 1. Document is now declassified.

150. Ibid., 7–9.

151. NSC, NSAM 183, 1. Document is now declassified.

152. Albert D. Wheelon, "Lifting the Veil on CORONA," *Space Policy*, 11 November 1995, 252–53.

153. Burrows, *Deep Black*, 199–200.

154. Richelson, *Secret Eyes*, 82. Richelson, like Burrows, also concluded, "McMillan wanted to seize control of the reconnaissance program for the Air Force. As Director of NRO he believed that he should be in full control of the satellite reconnaissance program and that the CIA should take orders from him, not be an equal partner." Ibid.

155. Haines, "The National Reconnaissance Office," 152–54.

Chapter 8

Johnson's Philosophy, Space Policy, and Institutional Continuity

This chapter covers two topics concerning the Johnson administration: exploration of the president's attitudes concerning the Cold War, and their impact on space policy as well as the race for prestige in space; and the institutional relationship between NASA and the DOD as expressed by the interacting components of support, coordination, and rivalry. There is sufficient continuity in these topics from the Kennedy into the Johnson administration so that one chapter should suffice. One historian summarized that "the anxiety raised by Sputnik did not end until Neil Armstrong and Buzz Aldrin took their historic steps in July 1969."[1] The lunar landing, which took place just months after Johnson left office, almost simultaneous with the cancellation of MOL, and serves as the chronological endpoint of this book. "Although Lyndon Johnson had remained committed to completing the Apollo program, the twin crisis of the conflict in Southeast Asia and urban unrest in the United States had not allowed him to allocate resources to any major post-Apollo space objectives. As the first lunar landing approached, the space program was clearly at a crossroads."[2]

Johnson, the Cold War, and Détente

During Johnson's term there was additional movement away from directly confronting the Soviet Union and a continued lessening of inflammatory Cold War rhetoric that had often crescendoed during the Kennedy administration. However, this budding détente was not enough to cause Lyndon Johnson to curtail the drive to ensure America was first to land on the moon. Nor was it enough to bring about a close rapprochement between the two countries, given the continuing presence of mitigating factors such as America's involvement in Southeast Asia and the Soviet invasion of Czechoslovakia in 1968. Therefore, while there was enough of a lessening of Cold War tensions during the Johnson era so that he did not feel impelled to extend the space race beyond Project Apollo, the détente was not pervasive enough to endanger Apollo's funding, momentum, or completion of Kennedy's May 1961 goal of landing a man on and safely returning him from the moon.

Continued Quest for Peace within Containment

Throughout Johnson's five years in office, he regularly spoke words of reconciliation. In his first month as president he said, "One of my first concerns has been to make it clear to the Soviet Union, and to Mr. Khrushchev personally, that the United States will go its part of the way in every effort to make peace more secure." Of course he also added, "On strength and the need for fully effective defenses I yield to no one. . . . We have to live on the same planet with the Soviet Union, but we do not have to accept Communist subversion."[3] Just as Eisenhower and Kennedy shared the trait of vigorously pursuing the containment policy while searching for verifiable disarmament measures and other means of lowering Cold War tensions, so did Johnson's Cold War policy incorporate these dual approaches. There are seemingly infinite examples, though the few below will suffice, of declarations throughout his presidency that at first seem contradictory but upon closer reflection fit the Eisenhower-Kennedy pattern described above.

For example, Johnson declared on 20 April 1964, "Communists, using force and intrigue, seek to bring about a Communist-dominated world. Our convictions, our interests, our life as a nation, demand that we resolutely oppose, with all of our might, that effort to dominate the world. This, and this alone, is the cause of the cold war between us."[4] Yet five days later he said, "We are constantly searching for any agreements that can be effected that will ease tensions and promote our national interest and promote better relations. . . . I do hope always for better relations. I am searching for them. I am doing everything I can to promote them."[5] Johnson summarized, "Our guard is up, but our hand is out."[6] These sentiments of containment and national defense on the one hand, coupled with a desire for lessening tensions on the other, characterized the Cold War rhetoric of senior administration officials from Johnson on down.[7]

In private Johnson revealed a certain strain resulting from balancing these two impulses, especially as they came together in Southeast Asia. He told his biographer concerning the Vietnam imbroglio,

> I knew from the start that I was bound to be crucified either way I moved. If I left the woman I really loved—the Great Society—in order to get involved with that [b----] of a war on the other side of the world, then I would lose everything at home. All my programs. . . . But if I left that war and let the Communists take over South Vietnam, then I would be seen as a coward and my nation would be seen as an appeaser and we would find it impossible to accomplish anything for anybody anywhere on the entire globe. . . . I knew that if we let Communist aggression succeed in taking over South Vietnam, there would follow in this country an endless national debate—a mean and destructive debate—that would shatter my Presidency, kill my administration, and damage our democracy.[8]

Little wonder that a radical reorientation of American space policy was not at the top of Johnson's priorities. He was inclined to support the lunar-landing goal, do what McNamara felt necessary in space for national security purposes, but not authorize any large next-generation space endeavors.

By early 1965, McNamara spoke for the administration when he explained the "gradual relaxation of the previously rigid bipolarization of world power. . . . Long frozen positions are beginning to thaw and in the shifting currents of international affairs there will be new opportunities for us to enhance the security of the Free World and thereby our own security." He added that while America's involvement in places such as Vietnam was worrisome and difficult, "We do ourselves a grave disservice if we permit them to obscure the more fundamental and far reaching changes in our position in the world vis-à-vis the Soviet Union."[9] NSAM 352 of July 1966 was entitled *Bridge Building* and stated, "The President has instructed that . . . we actively develop areas of peaceful cooperation with the nations of Eastern Europe and the Soviet Union. . . . These actions will be designed to help create an environment in which peaceful settlement of the division of Germany and of Europe will become possible."[10] By early 1967, Johnson openly declared, "Our objective is not to continue the cold war, but to end it"[11] and that "there is abundant evidence that our mutual antagonism is beginning to ease."[12] In June, Johnson met with Soviet Premier Alexei Kosygin in Glassboro, New Jersey, and while they reached no breakthroughs, Johnson felt comfortable enough by the end of 1967 to summarize, "We don't think that things are as tense, or as serious, or as dangerous as they were when the Berlin Wall went up, in the Cuban missile crisis, or following Mr. Kennedy's visit with Mr. Khrushchev at Vienna."[13]

The thaw, or at least the perception of one, between the two countries was sufficient for the Johnson administration to build upon the limited but concrete agreements Kennedy had forged with the Soviets near the end of his term such as the Limited Test Ban Treaty, the Washington-Moscow "hot line," and sales of surplus American wheat to the Soviets (see chap. 5). The tangible results from the Johnson administration included a Civil Air Agreement resuming US-Soviet air service; a Consular Convention to establish diplomatic posts throughout each country; and assorted accords on East-West trade and cultural exchanges. Johnson called the Nuclear Nonproliferation Treaty designed to halt the spread of nuclear weapons components and technology "the most important international agreement in the field of disarmament since the nuclear age began."[14] Johnson expressed hope that the United States and USSR could "enter in the nearest future into discussions on the limitation and the reduction of both offensive strategic nuclear weapons delivery systems and systems of defense and ballistic missiles."[15] Johnson later summarized, "We all had a long way to go, but slowly the Cold War glacier seemed to be melting."[16] Probably most important from the space historian's perspective was the Outer Space Treaty (see below).

One must maintain a sense of balance, however. After the USSR invaded Czechoslovakia in August 1968 to crush the movement toward loosening Communist party control, Johnson emphasized, "The events in Eastern Europe make it clear—and make it clear with the force of steel—that we are still a long way— a long way—from the peaceful world that we Americans all wish to see. The message out of Czechoslovakia is plain: The independence of nations and the liberty of men are today still under challenge. The free parts of the world will survive only if they are capable of maintaining their strength. . . . Peace remains our objective. But we shall never achieve it by wishful thinking, nor by disunity, nor by weakness."[17] Simultaneous with all the agreements of the previous paragraph, Johnson also steadily increased the American military presence in Southeast Asia from 35,000 in 1965 to over 500,000 in 1968[18] because "a Communist military takeover in South Vietnam would lead to developments that could imperil the security of the American people for generations to come. . . . If we had not drawn the line against aggression in Vietnam . . . some American President someday would have to draw the line somewhere else."[19] The Soviet leaders, on the other hand, made clear their position on Vietnam: "The Soviet Union will not remain unconcerned about the fate of a fraternal socialist state; she will be ready to render it all needed help."[20]

George Herring aptly concluded that the quest for peace and the Cold War dynamic coexisted in a sort of transition period during Johnson's tenure, "The Johnson years thus marked a time of adjustment between the unqualified globalism and militant anticommunism of the early Kennedy years and the détente and retrenchment of Richard Nixon and Henry Kissinger. . . . The cold war underwent significant modification during the Johnson years. The international system was changing from the bipolar structure of the immediate post–World War II years to a 'polycentric' system with multiple centers of power."[21]

Johnson, International Cooperation in Space, and the Outer Space Treaty

The Outer Space Treaty was perhaps the most heralded of the agreements directly relevant to the space arena indicative of some closing of the gap between the USSR and the United States. It was one of two developments in the international cooperation in space field during the Johnson administration, both of which were extensions of initiatives that began during Kennedy's term. The other was the fact that the Dryden-Blagonravov talks and initiatives resulting from them continued. However, neither the talks nor the resulting actions led to any significant level of US-Soviet cooperation in space. The assessment of those who participated in the Dryden-Blagonravov experiments during the Kennedy administration remained the same during the Johnson administration: "The performance of the Soviet participants on these projects for many years is best described as indifferent."[22] It will be recalled from chapter 5 that in 1962–63 the United States and USSR signed agreements on coordinating their efforts in certain aspects of communications satellite experiments, meteorology satellites, worldwide geomagnetic surveying, and exchange of experimental data pertaining to bioastronautics and space medicine. According to one analysis, "The Soviet performance was disappointing. By the end of 1972, only the communications project had been completed." For instance, while the two countries agreed to exchange information on bioastronautics and space medicine in October 1965, the Soviets did not submit any research data until January 1970.[23] In a general sense the Soviets regularly failed to respond to the frequent and wide-ranging American offers for cooperation in space, exchange of information, visits to each other's facilities, observation of each other's launches, and so forth.[24] Early in Johnson's presidency Webb wrote Johnson, "No new high-level U.S. initiative is recommended until the Soviet Union has a further opportunity to discharge its current obligations under the existing NASA-USSR Academy agreement."[25] Since the Soviets made little effort to 'discharge its current obligations' under the initial Dryden-Blagonravov agreements, the situation progressed very little over the course of Johnson's tenure.

Webb's summary to Johnson on this issue late in 1964 succinctly encapsulates the US-Soviet cooperation in space situation until the end of Johnson's presidency:

> Our experience since June suggests that the Soviets are willing to cooperate in a generalized and limited way, but that they remain relatively inflexible with respect to commitments in negotiation and are laggard in execution. Their performance does not seriously reflect the assurances . . . that the Soviet Union is receptive to expanded cooperation in space research. . . . For the immediate future, it might be useful to convey to the top Soviet leadership . . . our dissatisfaction with the painfully slow and limited progress to date, as well [as] with Soviet reluctance to enter into reasonable arrangements for implementing agreements.[26]

The foremost scholarly analysis of the US-USSR cooperative effort summarized, "As 1968 faded into 1969 and a new Administration prepared to take over in Washington, the watchword for space in both the United States and the Soviet Union was success in ongoing competition, not greater cooperation."[27] Given the lack of genuine Soviet interest, there is simply very little more to report concerning direct US-Soviet cooperation until the Apollo-Soyuz Test Project in 1975 (which was "the result, not the cause, of political détente"[28]), a topic well beyond the scope of this book.

However, the second prong of the international cooperation in space effort during the Johnson presidency involved the United States and USSR within the forum of the United Nations and the resulting Outer Space Treaty of 1967. This treaty essentially codified the principles enunciated in the two UN resolutions in the fall of 1963 (numbers 1884 and 1962), which banned the orbiting of weapons of mass destruction and reserved space generally for peaceful purposes only, respectively. The fact that the US-USSR Cold War relationship had progressed at least to the point where they could work together in the UN, plus the mutual tacit acceptance of overhead satellite reconnaissance, meant that the resolutions could evolve, albeit very slowly, into a treaty between 1963 and 1967.

On 7 May 1966, Johnson publicly called for a treaty that would make official the UN resolutions from almost three years earlier.[29] Events moved quickly from there. Both the United States and the USSR introduced draft treaties into the United Nations in June, and by December the two main spacefaring nations worked together within the COPUOS to draft a full-treaty text. The UN opened it for signatures on 27 January 1967, and more than 60 nations, including the United States and the USSR, quickly signed. The US Senate ratified the treaty 88 to zero on 25 April 1967. McNamara assured the Senate that the United States could verify its provisions "through our space observation and other technical surveillance systems."[30] The treaty entered into force on 10 October 1967. In essence, it made official the resolutions of four years earlier: it was forbidden to place weapons of mass destruction in outer space or on celestial bodies; it restricted military activities on celestial bodies; it barred claims of sovereignty and national appropriation; and it generally reserved space for peaceful uses only.[31] As McDougall has pointed out, however, the treaty "denuclearized outer space and demilitarized the moon. But it did not demilitarize outer space."[32] Both the United States and the USSR were free to continue their military activities in space such as reconnaissance, navigation, communications, early warning, and so forth, so long as they avoided deploying offensive weapons of mass destruction in space. While the process of simply codifying principles promulgated four years earlier is by no means a major diplomatic breakthrough, it can perhaps at least be considered both noteworthy and indicative of some small thawing in the previously universally frigid US-USSR relationship. Further, it is one admittedly small indication that Johnson did not want to extend the competitive dynamic in space beyond Apollo and the quest to be the first to land on the moon.

Johnson, Space Policy, Prestige, and Budgets

Within the Cold War dynamic described earlier, Johnson's specific space policy had two main thrusts. First, he did maintain enough of a commitment to the space-for-prestige principle to ensure that Apollo was adequately funded and stayed on schedule to land Americans on the moon by the end of the decade. Second, however, was the fact that within a fiscal environment increasingly constrained by the Vietnam War and exploding social-welfare spending, his commitment to competing in space was not great enough to impel him to approve any large, ambitious, and expensive next-generation follow-on space projects. In fact, the next major commitment to a large space system after Kennedy's lunar-landing speech in May 1961 did not come until January 1972, when Nixon approved construction of the space shuttle. The perceived lessening of Cold War tensions with the Soviets also strengthened this lack of desire to extend space competition beyond Apollo. Related to these two general principles was the fact that concerning military space Johnson continued to rely, as had Kennedy, on the conclusions of McNamara concerning the DOD's space requirements. As long as McNamara continued to see some value in the MOL, it continued. But by mid-1969, when both McNamara and Johnson had left their positions, Nixon terminated it.

Space and Prestige

There are similarities between Johnson's pronouncements on space policy and his declarations on the Cold War. Just as he could call for a continued strong military effort in support of the containment policy while also supporting détente, so could he also call for continuing the Apollo competitive effort while not extending the competitive ethos beyond it. Perhaps the primary factor in Johnson's desire to limit the space-for-prestige competitive dynamic to the Apollo program was economic considerations. One of Johnson's first acts as president was to make clear that all agencies would hold the FY 65 budget "to the barest minimum consistent with the efficient discharge of our domestic and foreign responsibilities"; therefore, each departmental head must "submit to me promptly a . . . statement of the steps which you propose to take in the next year to tighten your operations and effect savings."[33]

These economic measures impacted NASA as hard as, if not harder than, other agencies. In December 1963, Johnson told Webb concerning the FY 65 budget, "I've just got to get some kind of a tax bill through, and Harry Byrd [powerful Democratic senator from West Virginia] will not support it unless I guarantee I will hold expenditures of NASA under $5 billion and I want you to do that." It will be recalled that NASA's FY 64 budget had been $5.1 billion. Webb later admitted that once Johnson "became president, he had a different set of problems than he had had before. He was not quite as free to press those areas that he had a particular interest in; he had to look at the total."[34] Johnson's only mention of space in his first State of the Union address mixed both the competitive and the cooperative dynamic: "We must assure our preeminence in the peaceful exploration of outer space, focusing on an expedition to the moon in this decade—in cooperation with others if possible, alone if necessary."[35]

There is no shortage in the historical record of Johnson's statements that are firmly in the space-for-prestige/competitive camp. In January 1964 Johnson said, "If the goal of being first in space is to be achieved and maintained, there can be no slackening of effort and no dampening of enthusiasm for space achievements."[36] He wrote for a popular magazine, "The fate of free society—and the human values it upholds—is inalterably tied to what happens in outer space, as humankind's ultimate dimension."[37] Later that spring in a speech Johnson averred, "For the United States has nothing to fear from peaceful competition. We welcome it and we will win it."[38] Khrushchev seemed to agree, as he stated in June 1964, "And in the not too distant future we plan to fly to the moon. Not to live there, but to see what is going on there. And we shall reach the moon."[39] However, in January 1964 the State Department concluded, "The Soviet Union and the United States have backed into a race for the moon for psychological and prestige reasons. . . . Whether the Soviet Union regards itself as engaged in a 'race' with the United States for a moon landing has not yet been proven."[40] The CIA reported in May 1964, "It has been almost a year since the Soviets orbited a manned satellite."[41] In March 1965 Dryden wrote Johnson, "There is no evidence that they [Soviets] are building a booster as large as Saturn V," the type and size required to go to the moon. Dryden continued, "At present there is no indication of effort peculiar to a manned lunar landing effort as, for example, reentry tests at speeds equivalent to lunar return."[42] At a minimum, there were elements within the executive branch wondering if a race really did exist. While such questions at the highest policy-making levels probably could not imperil the progress of Apollo toward the moon, they would make it difficult for any follow-on competitive effort to gain momentum.

There seemed to be a growing perception in Johnson's mind during his presidency that the United States had, in fact, become *the* (not *a*) leader in space. In August 1964 after a successful American lunar probe he declared, "We started behind in space. . . . We know this morning that the United States has achieved fully the leadership we have sought for free men."[43] In February 1965 he even seemed to back off a bit from the basic space-for-prestige idea: "Our purpose is not, and I think all of you realize never will be, just national prestige. Our purpose remains firmly fixed on the fixed objective of peace. The frontier of space is a frontier that we believe all mankind can and should explore together for peaceful purposes."[44] The next month he told the press, "It was really a mistake to regard space exploration as a contest which can be tallied on any box score. . . . Now the progress of our own program is very satisfactory to me in every respect. . . . And while the Soviet Union is ahead of us in some aspects of space, U.S. leadership is clear and decisive and we are ahead of them in other realms on which we have particularly concentrated."[45] While Johnson would not sacrifice this leadership by slowing Apollo, he was also unlikely to spend billions on some Apollo follow-on such as a human flight to Mars if he had concluded that the United States continued to lead the Soviets in overall space capability.

All the above examples illustrate the dual thrust of Johnson's space thinking—maintaining Kennedy's commitment to competing with Apollo but demonstrating little willingness or desire to extend competition beyond the lunar landing—were taken from the early stage of Johnson's presidency. However, the same dynamic could be traced with a plethora of documents and citations from mid-1965 and on, but the fundamental point would remain unchanged. As he summarized in his memoirs, "Early in my Presidency I reaffirmed the national policy that I had helped to forge. 'Our plan to place a man on the moon in a decade remains unchanged,' I said in my first budget message. I restated that plan often enough to insure that there was no mistaking our purposes. . . . Throughout my time in office I supported the program to the limit of my ability."[46] What changed during his own term as president was the increasing financial demands upon Johnson stemming from the Great Society and America's escalating involvement in Southeast Asia.

Budgetary Slide

The real squeeze began in the fall of 1965 as the FY 67 budget process began. For reference and overview purposes, see table 8 for the last of Kennedy's and all of the Johnson's NASA and military space budgets.

Table 8. NASA and military space budgets for the years shown (in billions of dollars)

Fiscal Year	1964	1965	1966	1967	1968	1969
NASA	5.1	5.25	5.175	4.966	4.587	3.991
DOD	1.599	1.574	1.689	1.664	1.922	2.013

Adapted from NASA, *Aeronautics and Space Report of the President, Fiscal Year 1995 Activities* (Washington, DC: Government Printing Office [GPO], 1996), A-30.

As a general trend over Johnson's full term, the NASA budget declined over a billion dollars, greater than 20 percent. The DOD's space budget increased some $400 million or almost 25 percent, due mostly to increasing the MOL expenditures before its cancellation in FY 70. Similarly, total NASA employment (see table 9 below) including civil service positions as well as contractor jobs, peaked in 1965.

Table 9. NASA employment (includes civil service positions as well as contractor jobs)

1965	1966	1967	1968	1969
411,000	396,000	309,100	246,200	218,000

Adapted from Jane Van Nimmen, Leonard C. Bruno, and Robert Rosholt, *NASA Historical Data Book,* vol. 1, *NASA Resources 1958–1968*, NASA SP-4012 (Washington, DC: GPO, 1988), figure 1-4, 14; and Arnold S. Levine, *Managing NASA in the Apollo Era* (Washington, DC: GPO, 1982), 1969 figure, 107.

The timing of NASA's budget slide starting in 1965 was unfortunate, as one analyst explains, because Apollo was in full stride and reaching its highest financial requirements and because, "The heavy NASA spending coincided with the far-larger sums that were suddenly needed by the escalation of the Vietnam War in 1965." While Johnson did permit BOB director Charles Schultze to reduce NASA's FY 67 budget to an even $5 billion, he did protect it from further BOB-desired cuts because it was agreed such cuts would mean delaying the lunar landing until the 1970s.[47] The increase in spending for the Vietnam War was from $4.6 billion in FY 66 to $10.3 billion in FY 67.[48]

Webb told Congress in February 1966 that Johnson's $5 billion NASA budget figure for FY 67 "reflects the President's determination to hold open for another year the major decisions on future programs—decisions on whether to make use of the space operational systems, space know-how, and facilities we have worked so hard to build up, or to begin their liquidation."[49] In private vice president and chairman of the NASC, Humphrey, tried to explain, "It is my firm belief that these cuts in no sense reflect any decreased interest in or evaluation of the importance of the national space program. Rather, such cuts reflect realities—military, political, and economic—of the war in Vietnam."[50] During the FY 67 budget battle in January 1966, for the first time since Sputnik, a president did not mention space in the State of the Union address. An internal NASA history simply summarized, "The emphasis in 1966 was on carrying out 'Great Society' programs."[51] Congressmen also commented on the linkage between FY 67 NASA budget cuts and Vietnam. Rep. Olin Teague, D-TX, chairman of the Manned Space Flight Subcommittee of the House space committee said on 3 May 1966, "The war in Vietnam has already forced a substantial reduction in the NASA budget for the coming year."[52] Finally, Humphrey explained to the NASC in November 1966, "The President has a lot of problems to solve, with the requirements of the war in Viet Nam carrying heavy priority."[53]

A Case Study of the 1968 Budget

The FY 68 budget negotiations over the course of the second half of 1966 and most of 1967 were even worse for NASA, resulting in a budget cut of almost half a billion dollars. Webb fought the good fight, maintaining 1968 was a "year of decision" because NASA would require $6 billion in FY 68 "to stay in business with what we have, but that $7 billion would be required to really move forward with things."[54] In the end he would get just over $4.5 billion. When it became clear that Johnson was not prepared to ask for seven, or six, but closer to five billion dollars, Webb wrote him: "I have done my best to obtain support in Congress for the reductions you have had to make and to minimize any political risk to your administration from the fact that we are operating substantially under what would be the most efficient program." Webb again stated that FY 68's budget would likely be "a major turning point with indicated requirements on the order of $6 billion of new obligational authority."[55]

In August BOB director Charles Schultze told Webb he should count on only $5.15 billion for FY 68: "In view of the above-normal expenditures in Southeast Asia, and the threat of inflationary pressures on the economy, it is not feasible to plan on the program extensions and program levels" Webb desired. Schultze continued, "In fact, in the light of our review of budget totals it is quite likely that we shall have to go below this figure in the final budget."[56] Webb characterized this figure as disastrous and that such a budget would cause the "liquidation of some of the capabilities which we have built up." Webb spoke quite frankly and seemed to question Johnson's commitment to the space program: "There has not been a single important new space project started since you became President. Under the 1968 guidelines very little looking to the future can be done next year. Struggle as I have to try to put myself in your place and see this from your point of view, I cannot avoid a strong feeling that this is not in the best interests of the country. . . . We cannot deliver the kind of successes we have had with the thin budgetary margins of the past three years."[57]

Schultze replied for Johnson that "the space program is not a WPA [Work Progress Administration]" and given the fact that the federal budget for secondary and elementary education was only $2 billion and that for the war on poverty only $1.8 billion, "I don't believe that in the context of continued fighting in Vietnam we can afford *another* $600 million to $1 billion in the space program in 1968" (emphasis in original).[58] In December 1966 Johnson sided with Schultze, recommending a NASA budget for FY 68 of just over $5 billion.[59] Johnson's fundamental mind-set can be seen in his remark in March 1966, "We haven't wiped out all the deficiencies in our program yet, but we have caught up and we are pulling ahead."[60] Therefore, there seemed little reason for Johnson to fight for any increases in the NASA budget, or to strongly resist slight yearly reductions as long as they did not imperil the lunar-landing goal.

An author who has carefully examined tapes of internal NASA meetings related that during the FY 68 budget process Webb spoke of LBJ: "We are not dealing with the guy who said, 'I am your champion, I will go out there and fight your battles, I will get Kennedy and his Congress to give you the money.' He is saying, 'By God, I have got problems and you fellows are not cooperating with me. You could have reduced your expenditures last year and helped me out, you didn't do it.' " Webb lamented that the operative principle in the BOB was *cost-effectiveness*: "It is a byword over there. . . . I must say that all I get is a cold, stony demand that we act like the Post Office when I go over there."[61] Johnson publicly stated, "We are not doing everything in space that we are technologically capable of doing. Rather, we are choosing those projects that give us the greatest return on our investment."[62] An internal government report concluded concerning the US space effort of 1966, "The United States, which as recently as two years ago was on the defensive with respect to the Soviets, now commands a clear cut lead. In the eyes of world opinion, the United States was exhibiting a virtuosity and capability that the Soviets were not matching and which evidenced leadership in space."[63] Again, there seems to be little reason for Johnson to have felt compelled to extend the competition for prestige beyond Apollo, to increase NASA's budget, or to oppose its gradual decline.

Webb may have thought his troubles with the FY 68 budget were over when the process moved from the White House to Congress in 1967 but the situation only became bleaker from the NASA administrator's perspective. Infinitely worse than financial concerns was the tragic fire on 27 January 1967 as Apollo-Saturn 204 was undergoing a series of simulation tests on the launch pad at Cape Kennedy, Florida. A fire broke out in the pure-oxygen atmosphere of the capsule and killed Virgil "Gus" Grissom, Roger Chaffee, and Edward White. This horrific accident came on the eve of Congress beginning its deliberations over the FY 68 NASA budget. When it entered the serious stages of budgetary negotiations in the summer of 1967, "Congress seemed out to punish NASA—and Webb. It was in a cutting mood."[64] In August the House Appropriations Committee recommended a cut of half a billion dollars in the $5.1 billion administration request. To Webb's consternation, Johnson did not oppose this. One scholar explained, "Johnson felt he had to show Congress he would cut space to get his new tax bill (a 10 percent tax increase). Senator [Margaret Chase] Smith was furious and charged that Johnson had 'literally pulled the rug out from under those who direct the space program.'"[65]

Between Johnson's recommendation for a $5.1 billion FY 68 NASA budget late in 1966 and his acceptance of a $4.5 billion level in August 1967, several things had changed. McNamara informed Johnson in November that the true cost of the Vietnam War each year was going to be more in the vicinity of $20 billion, not the $10–12 billion he had previously estimated. In addition, the federal budget deficit skyrocketed from Johnson's announced figure in January 1966 of $1.8 billion to an all-time high by the end of the year of $9 billion.[66] As a result of the disastrous situation that developed in 1966, austerity was the goal for 1967. As Humphrey told the NASC in June 1967, "I know there are going to be problems this year with the budget, not so much because of the Apollo accident as because of the other major budgetary strains, particularly from the Vietnam War."[67] In addition, indications continued that perhaps America was indeed ahead in the space race and that an all-out crash effort was no longer necessary. A CIA estimate in March 1967 concluded, "Two years ago, we estimated that the Soviet manned lunar landing program was probably not intended to be competitive with the Apollo program as then projected, i.e., aimed at the 1968–69 time period. We believe this is still the case. . . . We believe that the most likely date [for a Soviet lunar-landing attempt] is sometime in the 1970–71 time period."[68]

Johnson himself explained when he signed the reduced NASA FY 68 appropriation in August 1967, "Under other circumstances I would have opposed such a cut. However, conditions have greatly changed since I submitted my January budget request." He detailed the "economic and fiscal realities now facing the Nation": increased expenditures and reduced revenues; a threatened deficit as high as $29 billion; and a 10 percent tax surcharge he has asked the American people to bear. Therefore, as every federal dollar is scrutinized, "In the process some hard choices must be made. The test is to distinguish between the necessary and the desirable. Our task is to pare the desirable. The administration and the Congress must face up to these changes in the space program." Johnson said he knew the reductions in NASA's budget would "require the deferral and reduction of some desirable space projects. Yet, in the face of the present circumstances, I join with the Congress and accept this reduction." Johnson closed by emphasizing the cuts did not indicate a lack of confidence in NASA or the space program. However, "Because the times have placed more urgent demands upon our resources, we must now moderate our efforts in certain space projects."[69] Clearly in Johnson's mind by 1967, the space program above and beyond Apollo was desirable but not necessary. Privately, Johnson could simply relate to Webb that he did not "choose or prefer to take one dime from my [NASA] budget for space appropriations this year and agreed to do so only because [House Committee on] Ways and Means in effect forced me to agree to effect some reductions or lose the tax bill."[70] Within the general gloom, however, Apollo's budget within NASA was "left virtually intact at about $2.5 billion."[71] The Apollo program director explained that the cuts within NASA were highly selective, "And, with relatively few exceptions, the Apollo program budget has been appropriated at approximately the required level I have stated."[72]

Indicative of Johnson's mind-set was a remark two months later at a ceremony for the Outer Space Treaty: "The first decade of the space age has witnessed a kind of contest. We have been engaged in competitive spacemanship. We have accomplished much, but we have also wasted much energy and resources in duplicated or overlapping effort."[73] There remained in Johnson, however, enough of a commitment to the space program, particularly Project Apollo, for him to disapprove any and all proposals from Congress or the BOB for reductions that would endanger the lunar landing or its accomplishment within the 1960s. His rhetoric could still heat up when he spoke at NASA facilities, as he did in December 1967:

"If we think second, and if we look third, then we are going to wind up not being first. . . . We may not always proceed at the pace we desire. I regret—I deeply regret—that there have been reductions and there will be more." However, "We will not surrender our station. We will not abandon our dream. We will never evacuate the frontiers of space to any other nation."[74] Nevertheless, by January 1968 Seamans (NASA's second-ranking official) had resigned and by the end of the year so would Webb.

Said one scholarly team concerning the difficulties for NASA created by rising social welfare spending, along with the Vietnam War's costs, "There was little support in the Johnson administration or Congress to increase NASA's budget; indeed, Great Society programs and the Vietnam war were pushing in the opposite direction."[75] Johnson also tied NASA's budgetary difficulties at least in part to the Great Society: "One of my regrets is that because of the demands, of the cities, and the poor, and the hungry, and the educational and health needs, that we found it necessary in the last few budgets of the Space Administration to trim our sails, and to make reductions that the Administrators did not think wise."[76] Seamans told the author that NASA leaders never "really understood the pressure that Johnson was under. . . . Johnson had an agenda. His number one priority was his social agenda, the Great Society. And then he was saddled with Southeast Asia. So there were real pressures on Johnson and what had been near and dear to his heart, namely the space program, was looming extremely insignificant."[77]

Levine postulated, "NASA was not a closed system; one cannot entirely discount the budgetary impact of the Vietnam War and Johnson's policy of . . . continued social service spending."[78] A NASA document comparing three categories of federal expenditures for actual FY 67 budgets and expected amounts for FY 68 and 69 (in billions of dollars) illustrated the fundamental reality of the impact of the Vietnam War and Great Society programs on Johnson's space policy (see table 10). In his final budget message Johnson stated that his "efforts to widen the opportunities for the disadvantaged" meant that "outlays for major social programs have risen by $37.4 billion, more than doubling since

Table 10. Federal expenditures for actual FY 67 budgets and expected amounts for FY 68 and FY 69 (in billions of dollars)

Fiscal Year	National Defense	Space Research and Technology	Health/Labor/Welfare
1967	74.2	5.4	30.0
1968	80.3	4.8	45.3
1969	83.9	4.5	50.4

Adapted from NASA, "FY 69 Budget Briefing," 29 January 1968, folder: Webb Budget Briefing, Webb subseries, Administrators series, NHDRC, 1.

1964. This is twice the rate of increase of outlays for any other category of Government program."[79] One scholar concluded, "The Great Society and the Vietnam war diverted attention from the challenges of spectacular technology as Americans were humbled by rural guerrillas or by the persistence of urban poverty and pretechnology prejudice."[80] By the end of Johnson's tenure, "The social agenda and the war spawned large demonstrations and engendered deep feelings that made NASA seem increasingly irrelevant."[81]

At the press conference in September 1968 announcing his retirement,[82] Webb was forthright, "I am not satisfied with the program. I am not satisfied that we as a nation have not been able to go forward to achieve a first position in space. What this really means is we are going to be in a second position for some time to come." When asked if the need to spend money elsewhere, such as for Vietnam and antipoverty programs, had taken the urgency out of the space program, Webb replied, "I think that is right. . . . I think a good many people have tended to use the space program as a sort of whipping boy. . . . In essence if it were not for the fiscal problems faced by the President and the Director of the Budget I would believe that the program would have been supported in the Congress and the country at a higher level than it has been."[83] A lengthy BOB review a month later designed to inform the next administration of the NASA and general space-program situation opened with what was by October 1968 an accepted fact: "The resource requirements of the Viet Nam war and of pressing domestic needs, coupled with an apparent acceptance of the Soviet presence in space, have tended to push the civil space program down the scale of national priorities. As funding requirements for on-going programs have declined, it has been very difficult to obtain funds for new starts." The BOB actually turned the competition for prestige argument on its head when it suggested, "An alternative to the policy of competition would be a policy of cooperation with [the] U.S.S.R. in large manned flight endeavors. Reasons for proceeding other than competition including enhancing the national prestige, advancing the general technology, or simply faith that manned space flight will ultimately return benefits to mankind in ways now unknown and unforeseen."[84]

This detailed case study of the FY 68 budget could be repeated with the same level of detail for the FY 69 process, whereby NASA's budget dropped to just under $4 billion or the FY 70 process that cut NASA's funding to $3.75 billion. But the fundamental conclusions would remain the same. As Apollo approached its climactic moment of the July 1969 lunar landing, NASA's presidential, public, and congressional support was eroding. NASA was, and would be for several years, unable to forge either an internal consensus on what the next steps in space beyond Apollo should be, or an external coalition to support any such future goals. NASA seemed adrift and Johnson appeared unwilling to prescribe a course of action beyond ensuring that the lunar landing took place on time.

To the Moon

One of the most visible symbols of Project Apollo was the giant Saturn V rocket blasting off from Cape Kennedy, Florida. Few realize that one consequence of NASA's budgetary restrictions was that NASA suspended production of the Saturn V in 1967 and officially discontinued it in 1970.[85] However, despite any criticisms that might come his way for reducing NASA's budget, one fundamental fact remained: Johnson did maintain sufficient momentum and financing for Project Apollo to enable Americans to land on the moon on 20 July 1969, six months after he left the White House. Neil Armstrong and Edwin "Buzz" Aldrin planted the American flag on the lunar surface five and one-half months before the deadline Kennedy had established eight years earlier. While presidential programmatic implementers such as Johnson often receive less attention and credit than presidential programmatic originators, one must give Lyndon Johnson due credit for shepherding NASA and Project Apollo through the tumultuous 1960s in a manner that enabled the organization and the program to fulfill a high-visibility pledge made by a previous president. Johnson himself explained: "People frequently refer to our program to reach the moon during the 1960s as a national commitment. It was not. There was no commitment on succeeding Congresses to supply funds on a continuing basis. The program had to be justified, and money appropriated year after year. This support was not always easy to obtain."[86]

On the other hand, "The space program's grip on the public imagination had begun to fade even before the first moon landing. . . . What had been imagined as a natural process of growth in manned space travel had by 1970 come to be seen as a technological exercise that wasn't worth the effort. In the political arena, the opposition to manned space flight was not just a matter of indifference, but of growing hostility. . . . A new all-purpose political truism entered the language: 'If this nation can put a man on the moon, then it should be able to.'"[87] Exploration of that development is beyond the scope of this work. However, if Lyndon Johnson is given a large measure of credit for the success of Project Apollo, he must also be seen as chiefly responsible for the fact that "much of the prestige America hoped to gain on the surface of the moon had already been lost in the jungles of Southeast Asia by the summer of 1969."[88]

Continuity in the Air Force and OSD Perspectives

This chapter now turns to the institutional climate that existed between NASA and the DOD. As with the realm of space policy discussed above, the organizational relationship during the Johnson era also had significant continuity with the

Kennedy period. The Air Force continued to desire a more rigorous investigation of the military applications of humans in space. The OSD continued to demand quantitative justification for new space-based systems. As the 1960s progressed, however, one can see the Air Force beginning to embrace the idea that operations in space should be done only if they offered a cost or an operational advantage over ground-based means of accomplishing a particular mission. Once work on the MOL began, most of the OSD-USAF tension centered on exactly what it would be designed to do and how fast work should proceed, and so these questions will be discussed in the next chapter.

The Air Force and Space

Project Forecast was an Air Force–organized effort late in the Kennedy administration to "reassess Air Force missions and weapons systems in light of current policy and the most likely developments in the period extending to 1975. Emphasis was placed on a study of the technological requirements involved."[89] The Air Force appeared to be concerned over its inability to secure OSD approval for space and other systems, as well as the cancellation of Dynasoar, Skybolt, the nuclear airplane, and other cutting-edge technological ventures. Project Forecast, headed by General Schriever and his AFSC, was designed to chart a reasonable and attainable future course for the Air Force. Its space-related sections revealed the continuity in Air Force thinking with previous declarations. Nodding to the nation's space-for-peace policy, the Air Force nevertheless emphasized, "At the same time, we must take such steps as are necessary to defend ourselves and our allies. We should develop and apply space competence to enhance our ability to cope with any military challenge in outer space, to keep the peace and to deter aggression." At times the USAF even seemed to echo the OSD's building-block rhetoric: "Within the national space program, present military efforts toward manned space missions should be to establish the necessary technological base and experience upon which to expand, with the shortest possible time lag, in the event firm military manned space requirements are established in the future."[90]

On the other hand, the Air Force remained firmly committed to the principle that humans in space would be an integral component of any long-term military presence in space: "Manned space flight is not only desirable but necessary to significantly improve current military space capabilities." The USAF admitted that "space flight today is where aviation was at Kitty Hawk." Despite the fact that "today, the only seriously considered missions for spacecraft are the message carrying and ground surveillance roles once considered the useful limits of aircraft," the Air Force believed that just as in the case of early aircraft "the ingenuity and flexibility of man as an operator made many military functions possible, and with his increasing experience these functions contributed significantly to national defense. It seems inevitable that this process will occur with space systems as well. . . . It is certain that the full military potential of space will be obtained only through the development of manned space systems."[91] The Air Force remained firmly wedded to the concept that officers in space would be required to maximize the use of space for national defense. Therefore, the MOL was key.

Springing from Project Forecast was a new set of "military space capabilities which are the goals of the US Air Force through the 1970 time period" which CSAF LeMay issued on 20 April 1964. LeMay listed two general categories. First was "early space operational objectives required and attainable in the 1960s." Included were seven systems: a satellite system capable of collecting systematically or on request pre- and poststrike intelligence data on the Sino-Soviet area; a "credible and operationally effective" early warning system against ballistic missile attack; a nonorbital satellite interception and negation system; an orbital system for inspection and negation of uncooperative satellites; an enhanced communications satellite; a next-generation weather satellite; and a recoverable satellite system "able to effect co-orbital rendezvous and docking for the purposes of conducting space rescue and logistic support operations." The second overall category included those "objectives which must await extensive and important technological advances." This consisted of three systems: one to perform interception of ballistic missiles; one capable of quick reaction and economic launching of varied mission modules into orbit; and a "large-scale manned-maneuverable vehicle system containing elements of defense, strike, reconnaissance, and command control, located and operating in relatively permanent orbit."[92] While ambitious, at least this 1964 set of objectives recognized that there would be limits in the short term to what the Air Force could expect to accomplish, and how they could prioritize these objectives.

The role of the MOL would be to experiment with the feasibility of the reconnaissance-related missions, the highest priority category of all. As an internal Air Force document stated late in 1964, the MOL "has as its immediate objective the assessment of man's utility in performing military functions related to reconnaissance, surveillance, inspection, detection, and tracking mission areas."[93] The MOL was at the heart of the Air Force's program. As the office in the Pentagon responsible for monitoring its progress said, "MOL is our entree to manned space capabilities. . . . MOL is the focus of our man in space efforts and is, therefore, the key program to the development of future military missions in space."[94] Headquarters Air Force MOL personnel regularly stated,

> The Air Force believes that man is the key to the future in space, and that certain military tasks and systems [reconnaissance] will become feasible only through the discriminatory intelligence of man. . . . We consider the MOL to be a bridge from R&D experiments, techniques, and embryonic operational experience to our being able to conduct the more classical military missions and roles in space if and when they are needed. . . . History indicates that throughout time new technologies and new regions have been thoroughly exploited for military advantage. The USAF exploration of space is aimed at preventing a mid-twentieth century "Trojan Horse" from being built 160 miles overhead of our Nation. An exploration program such as the MOL appears to be the best insurance which can be provided for the Nation's complete defense posture.[95]

By 1965, near the end of his Air Force career, Schriever was no longer delivering speeches describing how he felt "inhibited" or "shackled" by the nation's space-for-peace policy. That complaint faded from standard Air Force–space rhetoric. It was replaced with Schriever maintaining that MOL was simply "one part of a large and varied space effort. The MOL does not exist in isolation from other military developments in space, and it certainly does not exist in isolation from the programs of the National Aeronautics and Space Administration. We have worked closely with NASA in defining the program." On the other hand, he still vigorously protested what he considered artificial divisions in the US space program: "I think it is high time for people to stop trying to divide the national space effort into a series of airtight little compartments, each of which can be neatly labeled as 'peaceful' or 'nonpeaceful.'. . . In actual fact, all of our space programs serve peace."[96] Schriever's contribution to a book stated, "Both NASA and the Department of Defense have valid and distinctive roles in the national space program. . . . Preparation for national defense in space is not inconsistent with the national policy that space be used for peaceful purposes."[97]

By the end of the Johnson administration, the Air Force's philosophy on space had evolved to the point where pragmatic considerations ruled, and there was a much closer congruence between Air Force declarations and those of the OSD's of many years earlier. For instance, a 1968 version of the USAF Planning Concepts stated the Air Force would develop space capabilities only when space afforded the sole reasonable means to perform an essential military task.[98] General Ferguson, who took over from Schriever as AFSC commander in 1966, stated in 1968, "We have to prove that space projects can pay their way—that our space program can earn its keep. . . . Military space systems must show distinct promise of directly enhancing national security. Further, those space programs must represent either the <u>only</u> way to get the job done or the most cost-effective way of doing it." Ferguson hastened to add that the MOL was justified because it "will provide an operational test bed for the development of equipment for use in both manned and unmanned military space projects; additionally, it will provide empirical 'cost-effectiveness' and technical data on the ability of man to perform militarily useful tasks in space" (emphasis in original).[99] Virtually gone from rhetoric was the old "high-ground" idea of occupying space because if the United States did not, the Soviets were sure to. Of greater concern by the end of the 1960s was justifying space R&D in accordance with the edicts of the PPBS and systems analysis. Space was indisputably a place in which particular missions might be performed, not a mission in and of itself. Given the fact that DDR&E Harold Brown, McNamara's foremost space expert, became SAF in October 1965, this came as no great surprise. In addition, the Outer Space Treaty of 1967 made it virtually certain that the military would not emplace or maintain an offensive presence in space.

The OSD as Continuing Watchdog

The OSD continued to insist that Air Force space programs meet two criteria: the systems had to mesh with NASA efforts, and they had to hold the promise of enhancing military power and effectiveness. As Brown explained, "The Secretary of Defense continues to insist that, as a fundamental criterion, the Department of Defense space program must be coordinated closely with that of the National Aeronautics and Space Administration in all important areas and that DOD and NASA programs taken together constitute an integrated national program."[100] At the beginning of the Johnson administration McNamara explained, "Space technology is still very new and its implications, especially for the military mission, cannot be fully foreseen at this time. This is particularly true with regard to the potentials of a 'man-in-space.'. . . The time has come when, in our judgment, these efforts should be more sharply focused on areas which hold the greatest promise of military utility" and so the DOD had embarked on the MOL program as a military experimental orbital platform.[101] Albert G. Hall became deputy DDR&E for Space in the Johnson administration and stated,

> Sober consideration of military potential in space has not yet developed a decisive case for <u>manned</u> space supremacy as a primary constituent of military supremacy. . . . While we are not yet able to define a specific military mission for man in space, we believe we should purchase insurance against the possibility that a manned operational system may be required in the middle 1970s. This insurance will take the form of a flight test system to determine man's effectiveness in performing useful military functions in space. . . . The MOL program will be directed specifically to fulfilling the need for an early, effective determination of man's utility in performing military functions in space. . . . Despite several years of thinking about the subject, there is no clear,

common agreement on the ultimate military significance of manned space technology. Perhaps there is a mission for military man in space. Perhaps not.[102] (emphasis in original)

On the one hand, the Air Force had moved toward the OSD position that space had to "pay its own way" in the sense of justifying its costs when compared to similar ground-based systems. On the other hand, the OSD at least allowed for an investigation of the potential utility of military officers in space. Since each party had made some concessions to the other's viewpoint during the late Kennedy administration and throughout the Johnson presidency, the level of tension decreased, but did not disappear, between the OSD and the USAF.

One must not form the impression, however, that McNamara and the OSD were sudden converts to the military man-in-space cause. As Brown stated late in 1964, "The problems of manned military space flights are, and generally will continue to be, more complex and more difficult and expensive to solve. I want strongly to emphasize that as of this time even the requirement for manned military operations is still in question."[103] As will be seen in the next chapter, McNamara's granting official approval to the Air Force in December 1963 to study the MOL for possible construction was only the first of many steps the Air Force had to take in justifying to the OSD that the MOL should actually be built. The OSD had not been convinced by the end of 1963 that the MOL should actually be fabricated; Johnson would make that decision in August 1965. Rather, the OSD was simply willing to let the Air Force officially investigate this possibility throughout 1964 and early to mid-1965. However, these studies required money, and the FY 66 DOD military-space budget was $1.67 billion (20 percent of all DOD R&D funding), or $124 million more than FY 65 and double that of FY 61.[104]

Continuity in NASA-DOD Relations I: Overview and Coordination

Just as there was some lessening of tensions between the policy-making levels of the Air Force and the OSD, so too did the tension and rivalry between NASA and the DOD described in chapter 6 begin to abate, though not disappear, in the Johnson administration. Only two primary areas of direct NASA-DOD conflict continued to play themselves out during the Johnson era, and they both involved the question of exactly how much support the Air Force would continue to render to NASA, not whether or not such support would continue to be forthcoming. These two areas of conflict were NASA reimbursement of DOD support expenses (mostly at ETR) and how many military officers would continue to be transferred to NASA.

An Overview of NASA-DOD Relations in the Johnson Era

Webb summarized in 1964, "I am happy to report that during the past six years there has been a steady strengthening of understanding, coordination, and mutual support between the Air Force and NASA. . . . [We] are cooperating effectively in many ways which benefit both agencies and which serve the best interests of the nation. . . . The rapid rate of progress in the NASA part of the national space program over the past six years would have been impossible without the launch vehicles and related technology derived from Air Force missile programs." He detailed some of the extensive coordinating and supporting aspects of the relationship in the national launch vehicle program, space medicine, operations support, cross-use of facilities, astronauts (three of seven Mercury and 13 of 29 Gemini astronauts were Air Force officers), management personnel like Gen Samuel Phillips, improved liaison, and the GPPB.[105] President Johnson himself said, "I doubt that we have spent but very few hours resolving disagreements between the Administrator of the Space Agency and the Secretary of Defense, and yet I have seen hundreds of reasons why we could have had serious disagreements and had the Government divided among itself."[106]

Shortly before he retired, Schriever seemed to have reconciled himself to NASA's existence: "I get impatient with allegations that the two agencies are in some kind of wasteful competition. Where there is competition, it is productive, not wasteful. The NASA and Air Force programs are complementary, not duplicating."[107] Schriever's successor at AFSC, James Ferguson, declared, "In our space program, it is hard to tell today which area of national effort—the civilian or the military—has contributed most to the exploration and use of space for our benefit here on earth. <u>And it doesn't really matter</u>. The close relationships between National Aeronautics and Space Administration and the Department of Defense have always been very evident to those of us engaged in the NASA-DOD partnership" (emphasis in original).[108] While there was likely some residual resentment within portions of the Air Force that it had been superseded in space by NASA, by the end of the 1960s the leaders of that portion of the Air Force that worked most closely with NASA apparently harbored little animosity toward NASA and pledged continued cooperation and support.

NASA-DOD Coordination

In essence, the comprehensive coordination network of boards, panels, subpanels, groups, and committees that originated in the Eisenhower administration and grew deeper and more extensive in the Kennedy administration continued to function as expected during the Johnson years. An internal NASA report of April 1969 called the overall coordinating mechanisms "generally adequate," with the AACB and its six panels remaining the most important component. However, the report did state that as with any complex and multifaceted phenomenon involving two large bureaucracies "the effectiveness with which these organizational entities are being utilized could be increased." Its main suggestion seemed to allude to the Webb-McNamara difficulties discussed in chapter 6 because it stated (even though both men had recently departed their positions), "The absence of a close working relationship at the top renders it much more difficult to overcome the divisive tendencies that are bound to be latently present where two dynamic agencies have responsibilities and aspirations in a common field of activity."[109]

The assorted coordinating groups continued to add to the ever-growing body of NASA-DOD official agreements. As has been discussed in chapters 3 and 6, a government report from 1965 listed 88 separate "major" NASA-DOD agreements,[110] and a comprehensive NASA accounting from 1967 described 176 NASA-DOD accords out of a total NASA inventory of 302 interagency agreements.[111] A government accounting in 1965 determined that NASA, at the headquarters level alone, was involved in 203 interagency coordination and advisory bodies.[112] Obviously this chapter is not the place for a description of each one. What is important, however, is the degree to which almost every possible facet of the NASA-DOD relationship was legalistically and contractually spelled out.[113] During the Johnson administration, some major coordination agreements included: operation of the instrumentation ships and aircraft collecting data from space vehicles; coordination of the space medicine-bioastronautics design, development, and test program; separate agreements for the coordination of the geodetic, communication, navigation and weather satellite programs; reimbursement to the US Navy for recovery operations; and coordination of the respective space-science programs. In addition, of course, these formal agreements were supplemented by many informal understandings and working arrangements at lower levels within the agencies that contributed to the meshing of the programs into a single, national effort.

In addition the Launch Vehicle Panel of the AACB conducted multiple and extensive studies designed to achieve closer integration of the nation's family of launch vehicles. However, the coordination effort in the field of space boosters was one which continued to show relatively little progress compared to other aspects of the coordination process. The detailed case study of the LLVPG in chapter 6 explained the general pattern that emerged for these launch vehicle coordination efforts that, in fact, continued to exist during the Johnson administration. Neither NASA nor the DOD had any great desire to rely on the other organization to provide it with a critical member of its space launch vehicle fleet, thereby ceding control over a vital aspect of its overall space program. The fundamental conclusion of these launch vehicle studies continued to be: "No financial gain would accrue from either reducing the numbers of different launch vehicles in the national inventory or from substituting vehicles in existing programs."[114] A November 1968 study explained the reasons why, for the past 10 years, such attempts to closely integrate NASA's and the DOD's launch vehicle fleets had not succeeded. First, "The lack of future manned mission requirements prevents focusing of the vehicle studies" because neither the DOD nor NASA knew exactly what it expected to accomplish with human spaceflight well into the future. Second, "A relative comparison of the costs of the candidate vehicles is not possible because they are not based on equivalent studies and have not been developed on common ground rules." Therefore, this study could only recommend that "studies be continued by both agencies as required."[115]

The simple fact was that "most of the studies involving AACB panels were technical and noncontroversial."[116] Their goal was to ensure there was as little duplication as possible between the NASA and DOD space programs. While the coordination effort was not always 100 percent effective, such as in the launch vehicle field, for the most part it was a good-faith attempt at ensuring the American taxpayer did not pay twice for a particular space capability. A congressional report concisely summarized, "Because of this cooperative NASA-DOD effort a more aggressive and meaningful space program is being pursued."[117] An Air Force history described the extensive 1968–69 study effort concerning injecting greater economy and efficiency into the NASA and the DOD space programs and ensuring the nation's space program was not wasting money due to duplication. After over a year of effort, the institutions concluded the space programs were not wasteful or duplicative: "Conclusions drawn from the study effort attested to the effectiveness of DOD-NASA cooperation and indicated that significant economies were not possible unless specific projects were curtailed or canceled."[118] The NASA-DOD coordination effort was not a perfect one, but it did seem to be functioning well by the end of the 1960s.

Continuity in NASA-DOD Relations II: Support and Tension/Rivalry

The report describing America's 1966 space activities mentioned there were over 400 separately identifiable activities in which the DOD was supporting NASA at an annual cost of at least $500 million. These activities included those with which the reader will be familiar from past chapters: national launch ranges and host-base support; launch vehicles; recovery operations; use of aircraft and ships; and construction by the Army Corps of Engineers were only some of the categories with higher-dollar totals mentioned.[119] McNamara regularly pointed this out in his testimony while emphasizing that NASA reimbursed only 80 percent of the DOD's costs.[120] This figure of half a billion dollars of annual DOD support to NASA held relatively steady, though by early 1968 the DDR&E stated it had declined to $407 million; however, NASA's reimbursement level had dropped to 62 percent.[121] By the next year, this figure had dropped to $225 million.[122] According to one calculation, the USAF had supplied the launch vehicles and launch crews for 67 percent of American space launches through June 1968. In addition, the Air Force provided 95 percent of the United States' space tracking and control capability.[123] This sampling of facts illustrates two points. First, DOD's support for NASA was at a significant level throughout the 1960s but was declining near the end of the decade as NASA completed its first 10 years of existence and began to enjoy a greater institutional autonomy and independence from the DOD due to the development of its own capabilities and facilities. Second, the OSD believed NASA should reimburse a higher percentage of this support, even if the overall support level was declining.

Specific Support for Gemini

A very basic outline of DOD support for NASA's three human-spaceflight projects revealed the following. For Mercury the DOD provided astronauts, launch facilities, launch vehicles, range support, and recovery operations. For Gemini the DOD supplied most of the astronauts; participated in the training, launching, and launch operations; developed the man-rated Titan II; conducted assorted checkout and operational procedures; provided range support and recovery forces; and provided some of the onboard experiments. For Apollo the DOD's role was limited to providing most of the astronauts, range support, and recovery forces.[124]

By the end of the 1960s, one assessment of the specific support the DOD rendered to NASA Gemini and Apollo missions

426

concluded, "It is now routine to gather support forces around a manned space flight with little confusion, duplication and wasted motion, dismiss these forces and repeat the process in a similar manner for the next mission." Any problems were on the order of minor aggravations.[125] A more specific listing of the functions the DOD (and particularly the Air Force) performed to support Gemini would include supplying and launching the Agena target vehicle and its Atlas booster for rendezvous and docking exercises; supplying and man-rating the Titan II launch vehicle for the actual Gemini capsule; providing the actual launch facilities in Florida and much of the network, tracking, data acquisition, range, recovery, and medical functions associated with space launches; and supplying many of the services supporting space operations such as communications, security, transportation, photography, and public affairs personnel. As one Air Force document pointed out, "Support of Gemini operations is in many instances an added task to be performed by resources originally fully programmed for other purposes."[126]

One of the more difficult challenges the Air Force faced in supporting Gemini was modifying the Titan II so that it could be considered safely capable of launching humans. In addition to retrofitting the vehicle with redundant systems for electrical power and flight control, replacing the inertial guidance with radio guidance, and installing a malfunction detection system, the Air Force confronted several technical problems. The Titan II's first-stage engines had a tendency to oscillate longitudinally in what observers called "the pogo effect" in a manner severe enough to endanger human life. This problem cost $3.3 million to fix. There were also problems of combustion instability in the second-stage engine chambers that cost $11.3 million to fix. Finally, the Air Force spent $1.7 million flight-testing the vehicle to verify its fixes. Therefore, total Air Force expenses just to ensure the Titan II was ready for delivery to NASA were $16.3 million.[127] Whereas a Titan in its ICBM configuration cost $4–5 million, one modified as a Gemini booster cost $19 million. A NASA document explained, "Necessary and stringent requirements were established by NASA. The re-

427

sponse to these requirements by the Air Force and by its contractors was usually prompt and vigorous."[128]

The first Gemini mission was in April 1964 and the last in November 1966. Over the course of flights ranging up to almost 14 days, NASA perfected the necessary lunar-prerequisite techniques of rendezvous, docking, personnel transfer, and EVA. For any single mission, DOD's contribution could include up to 11,301 personnel, 134 aircraft, 27 ships, and 13 worldwide tracking stations. At the beginning of the program, Gemini's estimated total cost was $531 million; it actually cost $1.147 billion.[129] The NASA deputy administrator wrote the SAF after Gemini's completion, "Jim Webb and I are very conscious of our debt to the Air Force officers and men who have played a major role in this program. Titan certainly performed magnificently throughout Gemini, and has earned our complete confidence and respect."[130] There was also debt in the literal sense of continuing unreimbursed expenses for Gemini support, as seen in table 11. According to the DOD's accounting in millions of dollars:[131] Two facts, therefore, stood out. First, the DOD's support of over half a billion dollars was the equivalent of one-half of Gemini's overall $1.1 billion NASA budget. Second, the DOD continued to absorb unreimbursed expenses on the order of 16 percent. As seen in table 11, these nonreimbursed expenses continued as one of the few major points of contention between NASA and the DOD.

Table 11. Unreimbursed expenses for Gemini support (in millions of dollars)

Service	Reimbursed	Nonreimbursed	Total
US Air Force	435.564	54.455	490.019
US Navy	18.311	31.090	49.401
US Army	1.246	.214	1.459
Total	455.140	85.772	540.912

Adapted from Lt Col Alfred C. Barree, Policy and Plans Group, Directorate of Space, deputy chief of staff for R&D, USAF, memorandum for record, subject: Summary Report—DOD Support of Project Gemini, 17 April 1967, IRIS 1003006, AFHSO, 1.

Specific Support for Apollo

The Air Force submitted to the OSD on 12 May 1966 its official plan for rendering support to Project Apollo. McNamara approved it on 28 July 1966. One source related, "The plan called for essentially the same kind of support provided Gemini, employing the identical service units."[132] One must quickly add, however, that Air Force support to Apollo did not include providing launch vehicles because Apollo was launched on the Saturn family of boosters developed and procured by NASA (albeit largely by the ABMA team which NASA inherited from the DOD in 1960). Therefore, DOD assistance continued in areas of ETR support, network operations, recovery, communications, meteorology, medical personnel and supplies, public affairs, and so forth described above for Gemini. For instance, starting in 1965, 85 percent of the Air Force's tracking equipment was modified so it could support Apollo requirements; this eventually cost $50 million.[133]

One noteworthy aspect of the DOD's support for Apollo was the use of DOD reconnaissance-related resources such as cameras and map-making facilities to survey the moon. In 1965 alone, "The DOD is currently engaged in 88 man-years of work in support of Project APOLLO for NASA in the form of lunar maps, charts and other materials."[134] The agreement for this function said the Air Force would provide technical assistance to NASA by developing and providing lunar mapping and survey flight equipment. Given the role of this equipment in the NRO specifically and satellite reconnaissance in general, the agreement delicately stated, "DOD security classifications and procedures, as prescribed by the Air Force for application to mapping and survey equipments furnished under this agreement, will be observed by both agencies."[135] For instance, NASA's Lunar Orbiter that photographed the surface of the moon in preparation for the lunar landings featured a camera system that was developed, in NASA's words, "in a DOD project with classified aspects" with the Eastman Kodak Corporation. Since NASA wanted to deal with that company for Orbiter, "Arrangements were made with the appropriate element of DOD for the contractor to propose to NASA, under DOD supervision, a suitable unclassified camera system. NASA had no access at any time to the classified equipment. This procedure has proven to be very satisfactory and assures that any classified technology is appropriately protected."[136]

Secondary sources have recently plainly stated that Orbiter's photographic system used a "high-resolution camera system [which] was a derivative of a spy satellite photo system created specifically for earth reconnaissance missions specified by the DOD." This source added that its two lenses worked automatically and "with the precision of a Swiss watch" to take pictures of the lunar surface from 28 miles above it with one-meter resolution. However, "Few NASA people were ever privy to many of the details of how the 'black box' actually worked, because they did not have 'the need to know.' " All five of the Lunar Orbiter missions "worked extraordinarily well," generating a total of 1,654 photographs.[137] The trade press reported that the astronauts aboard *Apollo 7* had taken 700 photographs of the earth's surface using "very high-resolution film developed for Air Force reconnaissance satellites." Therefore, the DOD had "for the first time demanded seats on the NASA board selecting photographs for release. . . . NASA was permitted to release only 13 pictures," and officials doubted any more would ever be cleared.[138] The reality of omnipresent secrecy concerning the NRO and American reconnaissance satellites pervaded even the relevant aspects of the DOD's support to NASA human-spaceflight projects.

Since this study does not examine the entire Apollo program but only its portion up to *Apollo 11*'s lunar landing in July 1969, complete figures for DOD support will not be presented. However, representative are figures for the *Apollo 11* mission for which the DOD provided 6,927 people, 54 aircraft, and nine ships.[139] After this first lunar landing, new NASA administrator Thomas Paine wrote new SECDEF Melvin Laird to express NASA's "deep sense of gratitude" to the DOD for its many contributions to NASA human spaceflight over the years: "Without the assistance and cooperation of the Defense establishment, the nation would not have been able to achieve this goal." Paine pointed to the many "truly outstanding officers" such as General Phillips "who turned in a magnificent performance as Director of the Apollo Program. In these and many other ways, the Department of Defense has been one of the principal essential members of the Apollo team."[140] NASA's Office of Defense Affairs concurred that the lunar landing "could not have been accomplished without the vast amount of assistance and support received from the Department of Defense."[141] Even allowing for the standard inflated rhetoric in these bureaucratic exchanges, it nevertheless was undeniable that the DOD played a vital role in assuring *Apollo 11* landed on the moon and safely returned its three astronauts to earth.

Support Could Lead to Tension I: Personnel

Chapters 3 and 6 have described the important role that the approximately 300 military officers played in providing NASA with valuable managerial talent and expertise during the late 1950s and 1960s. The highest-ranking, and probably most important, figure was General Phillips, who served as Apollo program director. It will be recalled that his experience in the Air Force ICBM development program led him and NASA associate administrator George Mueller (whose systems management expertise also came from working with the Air Force ICBM program, although on the civilian side) to reorient the Apollo test program from a lengthy stage-by-stage, system-by-system approach to the Air Force "all up" procedure. This meant that "NASA could with reasonable confidence test the entire stack of stages in flight from the beginning, at great savings to budget and schedule."[142] Wernher von Braun, whose normal methodical testing procedures were overruled in favor of all up testing later stated, "In retrospect it is clear that without all-up testing the first manned lunar landing could not have taken place as early as 1969."[143] Phillips's enduring reputation within NASA was such that after the space shuttle *Challenger* exploded in January 1986 he was asked to head up a review of NASA management and procedures.

However, not all was well in the NASA-DOD personnel arena. Some within the Air Force felt the procedure was unbalanced in that NASA received all the benefits and the Air Force provided all the personnel. Instead of requesting certain talented high-level managers such as Phillips, or those at the colonel level, NASA began asking for large blocks of more junior officers. For instance, in April 1964 NASA deputy administrator Dryden wrote SAF Zuckert to request 55 USAF officers be transferred to NASA to perform not management functions but the regular lower-level, day-to-day duties of operating consoles, manning tracking stations, and so forth.[144] Zuckert replied that he would like to avoid supplying officers just to alleviate NASA's manpower shortages and would prefer "to assign experienced officers of exceptional ability who . . . indicate an intent to return to the Air Force upon completion of their NASA tour." After all, "There is a limit to the numbers of such people who can be assigned to NASA [and so] we think that they should be placed in key and middle-management level positions." Therefore, he suggested "A joint review of the total program in light of our collective experience would provide a sound basis for responding to your recent request, and would establish guidance for the continuing management of the program."[145] Dryden agreed and Phillips was placed in charge of this review of DOD personnel transfer procedures.[146]

This review under Phillips eventually validated 42 of the 55 NASA requests and forged new guidelines for future personnel transfers from the DOD to NASA. The new 15 September 1964 NASA-DOD memorandum of understanding required NASA to first deplete civilian sources for filling its vacancies before turning to the Air Force. It also restricted future NASA requests for AF personnel to positions within the fields of engineering, physical/life sciences, and technical program management (not equipment operators) that required the specific education, experience, or skills developed by that officer. By-name requests could be made only for colonels and generals.[147] Even within these new guidelines, however, the Air Force remained flexible and went out of its way to meet NASA's requests. For instance, in 1965 Dryden and Zuckert worked out an agreement for a wholesale transfer of no less than 128 USAF officers (84 lieutenants, 38 captains, and six majors) to NASA to do exactly the kinds of day-to-day operational duties that the September 1964 memorandum of understanding said they should not perform. It appears this transfer was feasible because the phasing out of several Atlas and Titan ICBM units within the Air Force created a condition of surplus officers with the type of operational skills that NASA needed in its burgeoning Gemini and Apollo programs.[148]

However, the Air Force made one requirement crystal clear concerning the transfer of these 128 officers: "Under no circumstances should this action be connected with the proposed MOL program" in any public discussions or releases, even though the avowed purpose of their going to NASA was "to receive on-the-job training and experience in the operational control of manned space flights."[149] In fact, the vice director of the MOL program wrote that the 128 officers were "to receive training in the skills required in the operational control of manned spacecraft for subsequent application to Air Force programs, e.g., MOL."[150]

An overall evaluation of the usefulness of the personnel-transfer program to the Air Force in late 1965 revealed USAF reservations. The report concluded, "Benefits to the Air Force accruing from the assignment of nearly 200 officers do not appear to be commensurate with the potential" that existed when the program began. Overall, the results of the program of assigning Air Force personnel to NASA "have not been very encouraging."[151] Almost a year later the MOL's vice director stated, "The Air Force has acceded to many requests from NASA in the past for officers with qualifications critically short within the Air Force. It is questionable whether the Air Force has received a sufficient return on these investments. In many cases the officers so assigned, for one reason or another, have not returned to the Air Force."[152] NASA seemed to sense the growing dissatisfaction within the Air Force. Seamans wrote the Air Force's top personnel officer early in 1967, "We have been very pleased with the USAF-NASA detailee program, and believe we could effectively continue it at its present level of activity. However, we recognize that your manpower requirements are not static and have been greatly impacted by the Vietnam situation. With this in mind, we are working toward a reduction in our future requirements for Air Force officers to be assigned to NASA."[153] It will be recalled from statistics presented in chapters 3 and 6 that NASA translated this pledge into action: after peaking at 323 in 1966, the number of military detailees to NASA decreased to 318 in 1967, 317 in 1968, and 268 in 1969 with further decreases thereafter.[154] Additional data from 1967 indicated that only a limited number of NASA personnel were assigned to DOD.[155]

It is difficult to argue with the assessment of the NASA official who monitored NASA-DOD relations concerning Air Force personnel serving in NASA: "The military detailee program was eminently successful" because the officers were of "inestimable value" to NASA projects.[156] As was true concerning DOD material support to NASA's efforts, DOD personnel provided NASA with managerial talent not available anywhere else but desperately needed during NASA's first decade of existence. Military officers serving in NASA furnished it the time required for it to develop internally its own managers and technical experts as part of its overall move toward greater institutional independence.

Support Could Lead to Tension II: Reimbursement

The roots of DOD dissatisfaction with its unreimbursed NASA support expenses went back to the Kennedy administration (see chap. 6 for the specific details of cost accounting, etc.). This area of discontent only continued and culminated during the Johnson administration. It was the one NASA-DOD disagreement that was so stubborn that it had to be referred above the SECDEF/NASA administrator level for arbitration. Both agencies turned to BOB director Schultze for resolution of the seemingly eternal reimbursement question in 1967.

This dispute tended to focus on the Eastern Test Range in Florida, extending into the Atlantic Ocean. NASA's general position was that "each agency should be responsible for the management direction and technical operation of its own facilities, and for budgeting and funding for such operations." Reimbursement should be avoided as much as possible. Therefore, since the DOD had been assigned responsibility for the national ranges, including the ETR, it should budget for and fund their annual operating costs, while NASA should be held responsible only for those additional range costs directly attributable to its activities: "It merely adds to administrative and other overhead costs to seek reimbursement. . . . Reimbursement should be restricted to those areas where one agency performs unique or unexpected services for the other, the nature of which precludes normal planning and budgeting for by the supporting agency."[157] On the other hand, McNamara and the OSD's position was that some type of cost-accounting system could be created that would determine exactly what portions of the ETR's resources and time were used by NASA and how much by the DOD; from that point, each agency would be billed accordingly. NASA's reply to this was that any attempt to prorate general operations and maintenance costs "would require a major and costly expansion of Air Force and NASA accounting and auditing groups."[158]

On 25 August 1966, McNamara promulgated a revised DOD Directive 5030.18, which made official the policy he had unofficially been trying to implement since at least 1963. It directed across-the-board reimbursement for all NASA support: "It is in the national interest for the Department of Defense, to the extent compatible with its primary mission, to make its resources available on a reimbursable basis, as appropriate, to NASA."[159] However, "This reversal of reimbursement policy was not accepted by NASA and a DOD/NASA management group was established to resolve the conflicts."[160] Boone added that McNamara issued this directive "without any prior discussions or coordination with NASA."[161] DDR&E John S. Foster explained the crux of the problem to McNamara as he urged McNamara to hold firm to his August directive: "The Eastern Test Range is the most complicated and highest-cost example of DOD support to NASA. Although NASA programs received about one-half of the total FY 66 range support, NASA reimbursed only $27 million of total range costs of about $250 million. . . . Lack of clear association of non-reimbursed costs with the NASA mission to some extent lessens NASA motivation to minimize requirements for DOD support." Foster added, "NASA maintains that if significant additional reimbursement is requested, NASA must enter into the general management of the range. It is not clear whether NASA actually believes that this would be necessary, or whether they are using this threat to coerce the Defense Department into abandoning plans for increased NASA funding contribution."[162]

Whatever the reason for NASA's refusal to pay its way with the DOD, McNamara held firm in his demand that it do so. He wrote Webb on 2 November 1966, "I propose to initiate a policy of full recovery of DOD costs of support to NASA . . . to the maximum extent possible in FY 68 [at the ETR], and that all DOD support to NASA, and NASA support to DOD will be on this basis by FY 69."[163] NASA's response continued to be, "If NASA is to significantly increase its contribution to funding of the ETR development, operations, and maintenance costs, then NASA should have a commensurate voice in ETR management; and that under the present 'National Range Concept' and present reimbursement policies, NASA does not have a responsibility for, nor a significant voice in developing and justifying [Eastern Test] Range planning and funding."[164] A NASA-DOD special working group under Boone had in fact been discussing this very issue since 1965 without success. They were ordered anew to forge a compromise concerning this Gordian knot of accounting, but to no avail. Boone summarized that he and the DDR&E representative could not even achieve a meeting of the minds on a report summarizing their differences: "It appeared at this point that we could not even agree as to how we should report that we disagreed."[165]

Accordingly, in April 1967 Seamans and Foster signed a joint letter referring the whole ETR/reimbursement problem to the BOB for arbitration because "fundamental differences in the views of the two agencies will continue to retard our progress toward agreement."[166] Schultze accepted the task of arbitration and in February 1968 decided: that the DOD would continue to provide management functions at the ETR without reimbursement from user agencies such as NASA; that support functions would continue according to present practices in that NASA would pay for only the direct costs it incurred for equipment and at facilities with no charges for overhead or administration imposed; but that there would be, however, a prorated division of costs related to operations at the ETR and that NASA should pay 40 percent of these costs, except for the Apollo range instrumentation aircraft, for which NASA would pay 85 percent. Therefore, whereas under the old system in effect for FY 68 NASA had paid only $25 million for its use of the ETR, in FY 69 it would have to pay $51 million; DOD ETR FY 69 costs would be $209 million.[167]

Schultze also stated that this entire scheme was only an interim arrangement until a financial management system could be installed at the ETR that would fully identify costs based on valid accounting procedures. He also emphasized that these guidelines for the ETR did not apply to other areas of cross support (e.g., cost of Communications Satellite Corporation launches, Apollo recovery forces, and support services at the Western Test Range and the White Sands Missile Range); discussions concerning these other areas had to continue. Boone's interpretation was that "for FY 1969, at least, the Director of the Bureau of the Budget had accepted essentially the NASA position on ETR funding."[168] Nevertheless, an internal NASA memorandum stated, "As a result, a review of all NASA requirements is under way to reduce these requirements and insure full utilization of NASA facilities wherever possible."[169]

When NASA and the DOD asked for yet another round of arbitration for FY 70, the BOB simply extended the FY 69 guidelines for another year. However, an Air Force historian explains that in 1970, "Air Force officials soon discovered that the reimbursement issue could cut both ways. Hence, it would cost far less for the Air Force to participate in NASA's Space Shuttle development under an additive cost arrangement," which was NASA's interpretation of the reimbursement issue (paying for only the direct costs one agency added to another agency's program by its participation in the program but not for any overhead or administrative costs). Therefore, the reimbursement issue began to fade. A second factor lessening the issue's importance was that as NASA's Apollo program passed its peak funding requirements stage, it required a lower level of Air Force support, and so NASA generated a lessened amount of unreimbursed expenses. See table 12 for the specific amounts. In other words, the Air Force was absorbing less than $25 million in 1970 compared to almost $80 million in 1967 while simultaneously looking at a potentially expensive involvement in the space shuttle program if it, the Air Force, had to pay for its participation on a strictly reim-

Table 12. Apollo support (in millions of dollars)

Year	Total Support	Reimbursed	Nonreimbursed
1967	244.7	164.7	79.9
1968	160.9	116.2	44.6
1969	175.7	128.2	47.4
1970	125.2	101.7	23.4

Adapted from Jacob Neufeld, *The Air Force in Space, Fiscal Years 1969–1970* (Washington, DC: Office of Air Force History, July 1972), 13–14.

bursable basis. Given the fact that NASA had already agreed to pay a greater percentage of the operations-related expenses at the ETR by 1970, "the full cost issue became a moot point. At least for the moment, the reimbursement level was closed."[170]

When all was said and done, Levine probably best summarized the complicated NASA-DOD reimbursement controversy when he characterized the whole situation as "hopeless, but not serious." The whole imbroglio suggested more than anything else that the NASA-DOD relationship had matured to the point where it could survive the strain, even when confronted with a problem that was "not amenable to any simple or permanent solution."[171]

The Reciprocal: NASA's Contribution to National Security

The respective NASA and OSD perspectives concerning NASA's direct contribution to national security continued to follow the trends outlined in chapter 6. Senior NASA officials regularly averred that the facilities, experience, and technology generated by NASA's human-spaceflight program were a direct contribution to national defense because they were a national capability available to all. Webb wrote the chairman of the JCS at the beginning of Johnson's presidency, "The entire Gemini program and more than seventy-five percent of the Apollo program are devoted to developing a national capability to conduct extended operations in near earth orbit, including the development of operational techniques for rendezvous, docking, and maneuvers in space." Webb said these capabilities were "applicable to all regimes of manned space flight, and therefore should be included in any study of the contribution which the NASA program makes to military objectives in space."[172] Webb and the rest of the NASA leadership took care not to clarify this conclusion or give it much publicity because of its potential international implications on NASA's worldwide tracking stations. For instance, while NASA was negotiating with Madagascar to augment its facilities there, Pres. Philibert Tsiranan "revealed sensitivity to any possible military implications of the station. He expressed a desire to avoid publicity abroad concerning the station."[173] Therefore, NASA officials had a delicate balancing act in which they tried to partially justify NASA's expenditures because of their presumed military relevance, yet had to deny this relevance to secure or maintain access to some foreign countries. This meant Webb and the others rarely clarified exactly what they had in mind when they stated NASA's experience and hardware were applicable to national security; they simply maintained NASA's capabilities were relevant.

Thus, it tended to be only in military forums that Webb declared, "The future of man in space cannot yet be distinguished from his possible military value there. Even purely scientific inquiries into the nature of the space environment will be necessary for the employment of any military systems in space. . . . We have no choice but to acquire a broadly-based total capability in space—a capability that can enable us to insure that protection of our national security while we actively seek cooperative peaceful development of the scientific and practical resources of space."[174] In internal reports NASA stated its entire complex of unique facilities constructed by the Army Corps of Engineers and the Navy Bureau of Yards and Docks at a cost of a billion dollars "is available to the Department of Defense to meet whatever requirements it may have in the future for manned space systems"; the facilities could support "defense measures in space if required" and served collectively as "a national resource of substantial value to the military." Therefore, in a general sense, NASA made available "an expanding flow of new scientific knowledge and more advanced technology for use in the development of weapon systems of ever-increasing effectiveness . . . which will give our nation the capability to insure that space cannot be used by a hostile power to gain a military advantage over us."[175]

As NASA's budgets began to decline from 1965 on, Webb did not hesitate to regularly point out to Congress the military relevance of NASA's R&D. In April 1966 he maintained, "Every airplane in Vietnam today is a better airplane because of the work in NASA. . . . The missiles that we have as a major part of our deterrent force all have benefited, including the largest one in the military service, from the work which comes from our research and development program."[176] Webb wrote to Senator Everett Dirksen in August 1966 to oppose the proposed cuts in NASA's FY 67 budget, stating they would make it difficult for NASA to develop the space technology necessary "to make sure we do not wake up some day and find others in possession of the power to deny us the use of space."[177] As mentioned earlier, however, Webb did not clarify these general statements with specific examples of exactly to what he was referring.

It will be recalled from chapter 6 that McNamara was not overly impressed with NASA's claims of the direct relevance of its work to national security; at one point he determined that only $600–675 million of NASA's requested FY 64 budget of $5.7 billion was in fact the kind of R&D activity the DOD would undertake if NASA did not. This general OSD assessment continued throughout the Johnson administration; there is no need to belabor it here. What was true during Kennedy's tenure, moreover, remained true during Johnson's: the DOD evidently did not feel strongly enough about the matter to expend much time or energy in publicly refuting Webb's claims or those of other NASA leaders concerning the potential military utility of NASA's R&D, facilities, and experience. DDR&E Brown testified to the Senate space committee in 1965 that if NASA was not conducting its $5 billion annual program, then the DOD might have to spend "hundreds of millions a year" or perhaps even "$1 billion a year to develop that technology." But McNamara remarked in 1965 that the Apollo program had "no direct military worth."[178]

A four-volume, 600-page April 1964 Air Force assessment of the lunar-landing program concluded,

> With the exception of Gemini . . . no system or subsystem in the National Lunar Program is directly applicable to established military requirements. There are a number of techniques being explored and experience being acquired in the National Lunar Program, which are applicable to military requirements and interests in both the midrange and long-range time periods. Unique military needs, however, are not covered in the NASA program. The most significant military benefit of the National Lunar Program is the overall contribution to the advancement of space technology.[179]

An Air Force document from mid-1967 stated, "The value of past and current USAF/NASA association is most difficult to quantify. The primary value is new technology, which will not have an impact on operations for a considerable time. . . . At present, there is no way to extrapolate from current NASA efforts to determine future value to the Air Force."[180] At best, the military community was ambivalent about the direct relevance of most of NASA's work.

NASA and Its Vietnam Support Effort

One area in which NASA did make a direct and concretely identifiable support effort for the DOD was in adapting and originating technology for the DOD's use in Vietnam and Southeast Asia. In December 1965, NASA created a special Limited Warfare Committee "to coordinate the overall NASA effort to support the Military Services in Southeast Asia."[181] By the end of 1966, Webb wrote the CSAF: "We have had a modest effort underway for a year now, aimed at applying space derived technology or techniques to the solution of some of these problems [in Vietnam], and we have two or three projects which are about ready to be turned over to the Services at this time."[182] Seamans reported to DDR&E Foster that "we are most pleased to have the opportunity to assist in these difficult matters [concerning the] application of NASA's competence, capabilities, and resources to the problems you are facing in Southeast Asia." Seamans reported that NASA was currently working on numerous projects for eventual DOD use such as: reflector satellite, quiet aircraft, target marker, countermortar system, ambush detection system, passive communications satellite, and new battery systems.[183] Boone reported NASA's efforts by the end of 1966 were budgeted at about $4 million a year and occupied 65 scientists and engineers.[184]

By the middle of 1967, Seamans considered two of NASA's projects in this field as major: an effort to improve the use of white phosphorous as a target marker and an attempt to develop an acoustic-mortar locator. Seamans estimated NASA's FY 68 expenses for its Southeast Asian support effort at $3.7 million. This figure included not only the two major projects but also exploration into many other possibilities such as: tunnel destruction, ambush detection, and a napalm cannon.[185] In December 1967, however, NASA's support for the DOD's war effort in Vietnam leaked to the press. The *Washington Post* reported NASA's Office of Advanced Research and Technology was spending between four and five million dollars per year directing 100 scientists and engineers on tasks "vital to the Vietnam war." The *Post* quoted an unnamed NASA official: "I don't think anybody is so naive that he might feel an agency spending $4 billion a year on technology shouldn't spend some of it trying to win a war we're fighting."[186]

It is doubtful NASA welcomed this type of publicity, but its effort to support the war in Vietnam continued. Boone stated that NASA personnel eventually considered 89 specific problem areas for the DOD relating to the Vietnam War.[187] For instance, a NASA document from 1969 revealed, "This agency is studying the development of a surveillance system for helicopter patrol of urban areas" in Southeast Asia.[188] Another 1969 NASA document listed numerous contributions to the US forces in Vietnam, to include: $3 million worth of computer technology, highlighted by the sound-location system for detection of mortars; electronics such as a small device to locate a survival radio when separated from a downed pilot; fire-suppressant foam to reduce airplane hazards; and a target-marker rocket.[189] Given the lack of documentation surviving from this effort after 1969, however, Levine's conclusion that it was phased out in 1969 seems correct.[190] Given the fact that a $4 million annual effort in a budget of $5 billion represents less than one-tenth of 1 percent of NASA's total funding, its Vietnam War effort was not a major factor in the NASA-DOD relationship. However, it is one of the few concrete areas that existed in which NASA tangibly supported the DOD.

Summary

Space policy as well as the NASA-DOD relationship during the Johnson presidency was marked by elements of continuity with his predecessor. Johnson remained committed to the competitive dynamic within the Apollo drive for the moon but not so committed that he could be persuaded to endorse any large space projects to follow it. As one Johnson scholar summarized, "Johnson never abandoned his determination to beat the Russians to the moon, but the course of events, especially the Vietnam War, forced him to impose some very real limits on the American effort in space."[191] Logsdon concurred: "Lyndon Johnson may have believed in the importance of space leadership, but he found himself unable to allocate to the space program the resources required to sustain that leadership once America reached the Moon. His support for space is unlikely to be recorded as one of the highlights of Lyndon Johnson's years in the White House." After all, by the time Johnson left office, NASA's budget had declined from its peak of $5.2 billion to less than $4 billion.[192] Levine added, "Once NASA began to lose the support of the White House and Congress—roughly from 1967—the difficulty of running the agency became greater and NASA began to resemble any other large government organization which redoubles its efforts as it forgets its aim."[193]

Support, coordination, and rivalry continued to characterize the relationship between NASA and the DOD although NASA's institutional independence continued to grow throughout the 1960s, and any tension remaining between the two organizations seemed to be confined to questions of exactly how much the DOD would support NASA in the sense of transferring personnel and receiving financial reimbursement for services rendered. As an Air Force source noted in mid-1968, the first 10 years of NASA's existence "has been a build-up phase acquiring talent and facilities needed to support their activities. This build-up has been essentially completed, and they have an impressive array of engineering and scientific manpower, facilities, and experience in space development and operations." In addition to the smoothly functioning formal relationship between the two organizations, "a fine informal relationship exists between the agencies."[194] It was only natural that NASA's dependence on the DOD had waned simultaneously with the growth of its own internal capabilities.

Shortly before he resigned as SECDEF, McNamara remarked, "A whole network of formal and informal channels has been established with the National Aeronautics and Space Administration . . . to ensure the maximum interchange of men, ideas, technology and hardware. . . . In every case, I have insisted that the space projects undertaken by the Defense Department must hold the distinct promise of enhancing our military power and effectiveness, and that they mesh in all vital areas with those undertaken by NASA, so that, together, they constitute a single, fully integrated national program."[195] Given the close supervision McNamara exercised over Air Force space proposals, there is little likelihood the fundamental balance of support, coordination, and rivalry existing between NASA and the DOD could have been significantly altered, even if the Air Force had wanted to do so. As former DDR&E and then-SAF Harold Brown emphasized concerning the Air Force in space, "These programs must be mature and well thought out. We should not be doing things just to be doing them. Rather, they must have a direct relation to established military needs. Space is not a mission, but a place to perform a mission. When a mission can best be performed from space, the Air Force will perform it from there."[196] The next chapter will largely focus on how the Air Force attempted to justify the MOL within that framework and the impact of NASA's Apollo Applications Program on that process.

Notes

1. Robert A. Divine, *The Sputnik Challenge*, vii.
2. Logsdon et al., *Exploring the Unknown*, vol. 1, 495.

3. Johnson, *Public Papers of the Presidents*, no. 26, 5 December 1963, 28.

4. Johnson, *Public Papers of the Presidents*, no. 272, 20 April 1964, 495.

5. Johnson, *Public Papers of the Presidents*, no. 294, 25 April 1964, 554.

6. Johnson, *Public Papers of the Presidents*, no. 272, 20 April 1964, 496.

7. As Johnson said when he referred to the old days after Khrushchev's removal from office in late 1964, "Our relations with the Soviet Union have come a long way since shoes were banged on desks here in New York and a summit meeting collapsed in Paris." Johnson, *Public Papers of the Presidents*, no. 662, 14 October 1964, 1330. And yet four days later: "We must never forget that the men in the Kremlin remain dedicated, dangerous Communists." Johnson, *Public Papers of the Presidents*, no. 686, 18 October 1964, 1377.

8. Kearns, *Lyndon Johnson and the American Dream*, 251–52.

9. McNamara, Statement before the House Armed Services Committee on the 1966 Defense Budget, 18 February 1965, 4–5. The CIA also signaled the opportunities for reducing tensions in its intelligence estimates: "Mutual disarmament will probably be conceptually attractive to some of the Soviet leadership as a means for reducing the economic burden of their defense establishment. . . . Any progress toward international arms limitation agreements will probably be slow. But we think that the Soviets probably will continue to seek ways to curtail the arms race in a moderate degree by mutual example." CIA, NIE 11-4-65, 207.

10. NSC, NSAM 352, 1.

11. Johnson, *Public Papers of the Presidents*, no. 3, 10 January 1967, 10.

12. Johnson, *Public Papers of the Presidents*, no. 59, 17 February 1967, 207.

13. Johnson, *Public Papers of the Presidents*, no. 554, 19 December 1967, 1163.

14. Johnson, *Public Papers of the Presidents*, no. 308, 12 June 1968, 713.

15. Johnson, *Public Papers of the Presidents*, no. 349, 1 July 1968, 764. However, the Strategic Arms Limitation Treaty (SALT I) and the Antiballistic Missile (ABM) Treaty would have to wait until well into the Nixon administration. One pact directly applicable to the space arena was an Agreement for the Rescue and Return of Astronauts and Space Objects that mandated countries render assistance to astronauts in distress as well as the return of space objects and components to the country that launched them. Johnson called it "one more link in a growing chain of international cooperation, which helps protect the peace of this planet." Johnson, *Public Papers of the Presidents*, no. 387, 15 July 1968, 810.

16. Lyndon B. Johnson, *The Vantage Point*, 473.

17. Johnson, *Public Papers of the Presidents*, no. 472, 10 September 1968, 937.

18. Hammond, *Cold War and Détente*, 231.

19. Ibid., 939.

20. *Pravda*, 1 February 1965, cited in Adam B. Ulam, *The Rivals*, 360.

21. Herring, *Lyndon B. Johnson National Security Files*, v. In the words of another scholar, "As the Vietnam War illustrated, the pursuit of détente did

not end Cold War assumptions and behavior." Frank Costigliola, "Lyndon B. Johnson, Germany, and 'the End of the Cold War,'" 207.

22. Newell, *Beyond the Atmosphere*, 313. Newell was NASA's associate administrator for space science and applications during the 1960s. Another NASA insider concurred, "With regard to substantive matters, the Soviet participation, like water, tended to seek lower levels." Frutkin, *International Cooperation in Space*, 103. Frutkin was NASA's director of international affairs in the 1960s.

23. Portree, *Thirty Years Together*, 7.

24. Hirsch and Trento, *The National Aeronautics and Space Administration*, 151.

25. Webb to Johnson, letter, 28 January 1964, 2. Document is declassified.

26. Webb to Johnson, memorandum, 18 December 1964, 3. The next year Webb echoed his earlier sentiments, "The plain fact is that the Soviets have been competitors in this field and not cooperators." Dryden added, "I would describe the situation as a form of limited coordination of programs and exchange of information rather than a true cooperation. . . . They have not responded to any proposals which would involve an intimate association and exposure of their hardware to our view or anything in the nature of a joint group working together." House, *Independent Offices Appropriations for 1966*, Hearings, pt. 2, 1006–07. Nor could Webb report any change in 1966 in the Soviet attitude toward cooperating in space: "We have looked for evidence that they are interested and found none. In fact, I would say the evidence has been the other way. . . . They show no evidence of any kind of giving us a key or even a partial key, that might unlock the door to cooperation." Portree, *Thirty Years Together*, 7. The stalemate continued into 1967 with Webb commenting, "We have made repeated efforts to persuade the Soviets to enter new projects, but our initiatives have not been accepted. . . . We regret that the Soviets have not been prepared to move more rapidly and broadly. . . . It has been made plain again and again that we stand ready to explore any and all possibilities for meaningful cooperation." Webb, US-USSR Space Cooperation, report, 30 January 1967, 2–3.

27. Harvey and Ciccoritti, *U.S.-Soviet Cooperation in Space*, 187.

28. McDougall, "Technocracy and Statecraft in the Space Age," 1022.

29. Johnson, *Public Papers of the Presidents*, no. 209, 7 May 1966, 487.

30. Garthoff, "Banning the Bomb in Outer Space," 37.

31. The full text is available in DOS, *United States Treaties and Other International Agreements*, vol. 18, pt. 3, 2412–18.

32. McDougall, *The Heavens and the Earth*, 417.

33. Johnson to the Heads of Departments and Agencies, memorandum, 30 November 1963, 1–2.

34. Both men cited in Nathan C. Goldman, *Space Policy*, 12.

35. Johnson, *Public Papers of the Presidents*, no. 91, 8 January 1964, 117.

36. Johnson, *Introduction to U.S. Aeronautics and Space Activities, 1963*, 1.

37. Johnson, "The Politics of the Space Age," 22.

38. Johnson, *Public Papers of the Presidents*, no. 272, 20 April 1964, 495. Johnson stated in the fall of 1964: "I recognize that we cannot be the leader of

the world and the follower in space. . . . We cannot be second in space and first in the world. . . . As long as I'm permitted to lead this country I will never accept a place second to any other nation in this field." Johnson, *Public Papers of the Presidents*, no. 572, 15 September 1964, 1071.

39. In *Pravda*, 20 June 1964, cited in Sheldon, *A Comparison of the United States and Soviet Space Programs*, 34.

40. DOS, Planning Implications for National Security of Outer Space in the 1970s, 30 January 1964, 15.

41. CIA, Forecast of Soviet Space Spectaculars in Balance of 1964, memorandum, 30 May 1964, 1.

42. Dryden to Johnson on the Soviet space program, report, March 1965, 5.

43. Johnson, *Public Papers of the Presidents*, no. 493, 1 August 1964, 922.

44. Johnson, *Public Papers of the Presidents*, no. 81, 25 February 1965, 215.

45. Johnson, *Public Papers of the Presidents*, no. 117, 20 March 1965, 306. Less than a week later he emphasized that the US space program had "but one purpose—the purpose of exploring space for the service of peace and the benefit of all mankind. We are not concerned with stunts and spectaculars, but we are concerned with sure and with steady progress." Johnson, *Public Papers of the Presidents*, no. 134, 26 March 1965, 330. By mid-1965 Johnson went so far as to proclaim, "But the need of man—the need of these times—is not for arms races or moon races, not for races into space or races to the bottom of the sea. If competition there must be, we are ready and we are willing always to take up the challenge and to commit our country to its tasks. But this is a moment when the opportunity is open and beckoning for men of all nations to come and to take a walk together toward peace." Johnson, *Public Papers of the Presidents*, no. 302, 6 June 1965, 644.

46. Johnson, *Vantage Point*, 283.

47. Divine, "Lyndon B. Johnson and the Politics of Space," in *The Johnson Years*, vol. 2, 237–39.

48. Redford and White, *What Manned Space Program After Reaching the Moon?* 177.

49. Lambright, *Powering Apollo*, 139.

50. Hubert Humphrey, Opening Statement by the Vice President, 2.

51. NASA, *Preliminary History of NASA*, II-14. Testifying to Congress, Seamans more delicately stated that the FY 67 NASA budget cut of $163 million "reflects the constraints upon the total national budget imposed by the needs to balance our commitments overseas and our needs at home. . . . The budget constraints do not permit the initiation of major new projects." Seamans (now NASA's deputy administrator after Dryden's death in late 1965), Statement before the Subcommittee on Manned Space Flight, 18 February 1966, 3.

52. Hechler, *The Endless Space*, 185.

53. Humphrey, Introductory Statement by the Vice President, 1. McNamara later outlined for Johnson the incremental cost of the Vietnam War "over and above the normal costs of the Defense establishment": FY 66, $9.4 billion; FY

67, $19.7 billion; FY 68, projected $22.4 billion. McNamara to the President, memorandum, 26 October 1966, reel 9, 1.

54. NASC, Summary Minutes, 15 June 1966, 3.

55. Webb to Johnson, letter, 16 May 1966, 1.

56. Schultze to Webb, letter, 13 August 1966, 1. Document is now de-classified.

57. Webb to Johnson, letter, 26 August 1966, 2–3.

58. Schultze to Johnson, memorandum, 20 September 1966, 859.

59. Divine, "Johnson and the Politics of Space," in *The Johnson Years*, vol. 2, 240–42.

60. Cited in Evert Clark, "President Reaffirms Goal of Moon Landing by 1970," *New York Times*, 17 March 1966, 1.

61. Levine, *Managing NASA in the Apollo Era*, 127.

62. Johnson, *Public Papers of the Presidents*, no. 13, 24 January 1967, 50.

63. "Johnson Sees Major Soviet Space Efforts Upcoming," *Space Business Daily*, 2 February 1967, 180. United States Information Agency report on the worldwide perception of US space activities in 1966.

64. Lambright, *Powering Apollo*, 184.

65. Ibid., 185.

66. Shapley, *Pride and Power*, 375.

67. NASC, Summary Minutes, 22 June 1967, 1.

68. CIA, NIE 11-1-67, 2.

69. Johnson, *Public Papers of the Presidents*, no. 362, 21 August 1967, 800–01.

70. Johnson to Webb, memorandum, 29 September 1967, 1.

71. Joshua Lederberg, "Demands of Vietnam Hobble Our Steps to Outer World," *Washington Post*, 2 March 1968, 1.

72. Mayhall and Appleton, "Military Applications of Space Are Inevitable, 17.

73. Johnson, *Public Papers of the Presidents*, no. 425, 10 October 1967, 920.

74. Johnson, *Public Papers of the Presidents*, no. 533, 12 December 1967, 1123–24.

75. Compton and Benson, *Living and Working in Space*, 20.

76. Cronkite, oral history interview of Johnson, 5 July 1969, 24.

77. Oral history interview of Seamans, 5 July 1996, by the author.

78. Levine, *Managing NASA in the Apollo Era*, 207.

79. Johnson, *Public Papers of the Presidents*, no. 678, 15 January 1969, 1281.

80. Maier, "Science, Politics, and Defense in the Eisenhower Era," xvi.

81. Goldman, *Space Policy*, 14.

82. There were probably both positive and negative factors that led Webb to decide to resign in the fall of 1968. He had completed seven years at the helm of one of the government's largest bureaucracies and was likely weary from the constant budgetary battles he had waged since 1965. However, "Webb had grown increasingly concerned about the presidential transition,

worried that some last-minute interference from the new administration would wreck everything. While he had done all in his power to give his team a fighting chance to succeed, he did not believe that he would be with them at the finish line. . . . NASA had to be depoliticized, in fact and in appearance." Accordingly, Webb and Johnson felt it would be best if Webb resigned so that Deputy Administrator Thomas Paine "would have to succeed Webb sooner rather than later so he could build a record of technical success. To depoliticize the transition at NASA, the change should take place before the November election." While the actual announcement of Webb's resignation on 16 September 1968 may have been a bit hastier than Webb would have preferred, it seems both men agreed it was time for Webb to resign. Lambright, *Powering Apollo*, 200–201.

83. Webb, press conference, transcript, 16 September 1968, 4–6.

84. BOB, NASA, report, 30 October 1968, 496–97.

85. Levine, *Managing NASA in the Apollo Era*, 226.

86. Cited in Bornet, *The Presidency of Lyndon B. Johnson*, 215.

87. Murray and Cox, *Apollo*, 447–48.

88. Koppes, "The Militarization of the American Space Program," 15.

89. Air Force, *Project Forecast*, R-1. Document is now declassified.

90. Ibid., V-10.

91. Ibid., VIII-10, G-8.

92. Air Force, Memorandum on Space Objectives, 1–2. Document is now declassified.

93. Schultz, Inputs on MOL to the Chief of Staff's Posture.

94. Air Force, Supporting Presentations, 22 December 1964, 1.

95. Air Force, Unclassified Supporting Witness Statement, Manned Space Programs, 9 March 1965, 1, 8.

96. Schriever (speech to the American Institute of Aeronautics and Astronautics, 12 October 1965, 1–2).

97. Schriever, "Does the Military Have a Role in Space?" 62–63.

98. Cantwell, *The Air Force in Space*, pt. 1, October 1970, 1.

99. Ferguson (speech on Bioastronautics and Orbiting Space Stations, 25 June 1968, 6).

100. House, *DOD Appropriations for 1966*, hearings, pt. 5, 14.

101. McNamara, Statement before the House Armed Services Committee on the 1965 Defense Budget, 27 January 1964, 104.

102. Hall (speech on The Objectives of the Military Space Program, 5 February 1964, 1, 6–7, 11).

103. House, *DOD Appropriations for 1965*, hearings, pt. 5, 12. McNamara clearly shared this sentiment, stating in March 1965, "The orbital laboratory might be manned or unmanned. . . . The important point is not whether the man is there. The important point is that there may be a military requirement and we should meet it." House, *DOD Appropriations for 1966*, hearings, pt. 3, 170.

104. McNamara, Statement before the House Armed Services Committee on the 1966 Defense Budget, 18 February 1965, 137.

105. Webb, "NASA and the USAF, 7. Johnson's report on 1964's space activities said during the year there developed "a much improved degree of cooperation and coordinating action as between the major agencies engaged in the national space program. Not only was there improvement in the exchange of information between such agencies, but there also was a useful interagency assignment of experienced personnel." Johnson, *US Aeronautics and Space Activities, 1964*, 7.

106. Johnson, *Public Papers of the Presidents*, no. 81, 25 February 1965, 216.

107. Schriever (speech to the Aviation Writers Association, May 1966, A3274).

108. Ferguson (speech on Science in a Synergistic Society, 20 May 1968, 6).

109. Boone to Dr. Eggers, memorandum, 8 April 1969, 3, 5.

110. House, *Government Operations in Space*, 123–32.

111. NASA, "Inventory of NASA Interagency Relationships," 13 October 1967.

112. House, *Government Operations in Space*, 101. Some, but not a significant proportion, of these would have been with other agencies besides the DOD.

113. See chap. 6 for some examples.

114. Cantwell, *The Air Force in Space, FY 65*, 4. Specifically, Cantwell was summarizing the July 1964 conclusions of an AACB Launch Vehicle Panel study.

115. AACB, *Intermediate-Class Launch Vehicles for Future DOD/NASA Manned Missions*, November 1968, 15.

116. Levine, *Managing NASA in the Apollo Era*, 229.

117. House, *The NASA-DOD Relationship*, 6.

118. Neufeld, *The Air Force in Space*, 10.

119. Johnson, *US Aeronautics and Space Activities, 1966*, 1 January 1967, 72.

120. McNamara, Statement on the FY 68 Defense Budget, 23 January 1967, 129. Document is now declassified.

121. Foster, Department of Defense Activities in Space, 1967, 26 March 1968, I-30.

122. According to the testimony of DDR&E Foster to the Senate space committee in June 1969 and summarized in "DOD/NASA Support Costs Revealed," 31. Johnson, *Aeronautics and Space Report of the President, 1969*, January 1970, 42, reported a similar figure for FY 69 of $235 million.

123. Kelly, "Ten Years in the Outer Realm," 22.

124. House, *The NASA-DOD Relationship*, 1964.

125. Clements, *The Coordination of Manned Spaceflight Operations between DOD and NASA*, 49.

126. DOD, Department of Defense Support for Project Gemini Operations, 21 October 1964, II-1.

127. Gemini Program and Planning Board, "Minutes of Meeting," 1 February 1965, 2.

128. Rosen to Admiral Boone, memorandum, 8 October 1965, 4.

129. Hacker and Grimwood, *On the Shoulders of Titans*, 387, 596.

130. Seamans to Harold Brown, letter, 22 November 1966, 1.

131. Barree, memorandum for the record, 17 April 1967, 1. Numerous other sources support this accounting. See, for instance, NASA's official Gemini history: Hacker and Grimwood, *On the Shoulders of Titans*, 595.

132. Cantwell, *The Air Force in Space, Fiscal Year 1966*, 12.

133. Benson and Faherty, *Moonport*, 470.

134. Johnson, *US Aeronautical and Space Activities, 1965*, 21 January 1966, 68.

135. McMillan, "DOD/NASA Agreement on the NASA Manned Lunar Mapping and Survey Program," 20 April 1964, 1.

136. Spriggs, memorandum for the record, 16 February 1967, 1.

137. Hansen, *Spaceflight Revolution*, 328, 338, 346. "Need to know" refers to one of the two conditions that must be met for a person to be granted access to DOD classified material. First, the person must have the appropriate level of clearance: Confidential, Secret, Top Secret, etc. Second, even with the proper clearance, a person must demonstrate a specific need to know the information for purposes of officially conducting one's assigned duties. Without a need to know, even a properly cleared person will be denied access to a requested piece of specific information.

138. "Washington Roundup," 4 November 1968, 228.

139. Smith, *Department of Defense Support: Apollo 11*, July 1969, 5.

140. Paine to Melvin Laird, letter, 11 August 1969, 1.

141. Boone, *NASA Office of Defense Affairs*, iii.

142. Bilstein, *Orders of Magnitude*, 78.

143. Cited in Murray and Cox, *Apollo*, 162.

144. Dryden to Zuckert, letter, 1 April 1964, 1.

145. Zuckert to Dryden, letter, 5 May 1964, 1.

146. Another problem in the background of the personnel issue is that the Air Force was getting back less than a quarter of the officers it sent to NASA. An accounting at the end of 1964 showed that the Air Force had sent 174 people to NASA, and of the 80 who were no longer working there, only 18 had returned to the Air Force because 46 separated from the service (many to continue working in NASA) and 16 retired (many to continue working with NASA). Anderson to Schultz, memorandum, 18 December 1964, 1.

147. Phillips, Report of the Joint Air Force–NASA Military Requirements Review Group, September 1964, 13.

148. See the document collection *Air Force Support of Project Gemini, Inputs from the Major Commands*, 1967, for the actual 22 July 1965 Dryden-Zuckert MOU on the 128 officers. The background is provided by Cantwell, *Air Force in Space, FY 65*, 13; and Boone, *NASA Office of Defense Affairs*, 52.

149. Air Force, SAF, message to all subordinate commanders and offices, 13 August 1965, 1.

150. Evans to the deputy chief of staff for R&D, memorandum, 6 December 1965, 1.

151. Miller to the directorate of NASA Program Support, report, 9 December 1965, 1, 3.

152. Evans to the Air Force Personnel director, memorandum, 10 September 1966, 1.

153. Seamans to Lt Gen H. M. Wade, letter, 4 January 1967, 1.

154. Van Nimmen et al., *NASA Historical Data Book*, vol. 1, 80; and 1969 figure from Gawdiak and Fodor, *NASA Historical Data Book*, vol. 4, 68. The downward trend of military detailees continued in the 1970s: in 1970 (231), 1971 (172), 1972 (119), 1973 (78), 1975 (61), and 1975 (45): Ibid., 68.

155. AACB, *A Survey of Information Exchange between NASA and DOD*, 16 November 1967, 8. This document mentioned only two NASA personnel at DOD: Dr. Michael Yarymovych serving as the MOL technical director and Mr. Duncan Collins assigned to the MOL Office of Systems Engineering.

156. Boone, *NASA Office of Defense Affairs*, 59.

157. Ginter to deputy associate administrator for Defense Affairs, memorandum, 10 December 1964, 1, plus attach.

158. Seamans to Schriever, letter, 12 April 1965, 1.

159. DOD, Directive 5030.18.

160. Barree, memorandum for the record, 17 April 1967, 1.

161. Boone, *NASA Office of Defense Affairs*, 141.

162. J. Foster to the SECDEF, memorandum, 6 October 1966, 1, 2, and attach. 6.

163. McNamara to Webb, letter, 2 November 1966, 1.

164. Bryant to Boone, memorandum, 9 November 1966, 1.

165. Boone, *NASA Office of Defense Affairs*, 143.

166. Foster and Seamans to Charles Schultze, joint letter, 4 April 1967, 1.

167. Boone, *NASA Office of Defense Affairs*, 149–50.

168. Ibid., 150.

169. Ginter to the deputy associate administrator for advanced R&D, memorandum, 4 October 1968, 1.

170. Neufeld, *The Air Force in Space, Fiscal Years 1969–1970*, 13–14.

171. Levine, *Managing NASA in the Apollo Era*, 213, 237.

172. Webb to Taylor, letter, 16 January 1964, 1. As NASA's Office of Defense Affairs often stated, "NASA is acutely conscious of the need to render maximum support to the Department of Defense. It is a primary policy of NASA to assist the DOD in every way possible to meet its needs in the use of space for national security. . . . We estimate that 75 percent of the cost of the Apollo program will be devoted to the development of a capability for conducting near earth orbit operations, an essential basis for the development of any manned space weapon system." NASA, "The Use of the National Space Capability in Military Affairs," 3 July 1964, 312–13.

173. Callaghan to Moyers, memorandum, 12 May 1964, 1.

174. Webb (speech to the US Naval Academy, 14–15).

175. NASA, "NASA's Contributions to National Security," 1, 16–17, 19.

176. Webb added that one of NASA's priority areas was developing the "technology and operational capability from the surface of the Earth out-

ward through the air and outward in the immediate environment of the Earth. This capability has a direct bearing on our military capability." House, *Independent Offices Appropriations for 1967*, hearings, 1417–18.

177. Webb to Dirksen, letter, 9 August 1966, 492.

178. Both men cited in Nieburg, *In the Name of Science*, 31.

179. Air Force, *Study Pertaining to the National Lunar Program*, vol. 1, 6. Document is now declassified.

180. Air Force, *History of the Directorate of Space: January–June 1967*, July 1967, 13.

181. Boone to Hilburn, memorandum, 7 December 1965, 1.

182. Webb to McConnell, letter, 29 December 1966, 1.

183. Seamans to Foster, letter, 22 November 1966, 1.

184. Boone, *NASA Office of Defense Affairs*, 250.

185. Seamans to Foster, letter, 17 July 1967, 1–3.

186. Thomas O'Toole, "NASA's Role in War Grows," *Washington Post*, 4 December 1967, sec. F, 4.

187. Boone, *NASA Office of Defense Affairs*, 251.

188. Ginter to commanding general, memorandum, 24 April 1969, 1.

189. Ginter to NASA assistant administrator for international affairs, memorandum, 24 June 1969, 1, plus attach.

190. Levine, *Managing NASA in the Apollo Era*, 230.

191. Divine, "The Politics of Space," 233.

192. Logsdon, "National Leadership and Presidential Power," 209.

193. Levine, "Management of Large-Scale Technology," 48.

194. Air Force, *History of the Directorate of Space*, July 1968, 8.

195. McNamara, Statement on the 1969 Defense Program, 22 January 1968, 161.

196. "Air Force Secretary Brown," October 1966, 69.

Chapter 9

Apollo and the MOL

This chapter examines the development and ultimate cancellation of the MOL, the USAF's last, best hope for an independent human-spaceflight program. The fate of the MOL cannot be analyzed in isolation because its cancellation was closely tied to factors both within the DOD and external to it. One of the factors was a perception that the MOL largely duplicated the intelligence-gathering capabilities of the NRO's robotic-reconnaissance satellites; therefore, one must briefly look at the question of reconnaissance satellites in the Johnson administration. Another factor in the MOL's cancellation was the conclusion by some that the MOL duplicated NASA's Apollo Applications Program (AAP) because they were both basically early versions of space stations; therefore, one must also look at the relevant portions of the AAP story insofar as they impacted the MOL. It is hoped that this strategy of tying in all the relevant inputs to the MOL's cancellation will complete the picture of the NASA-DOD relationship in the 1960s.

Preliminary and Background Information

Chapter 7 described how, after his failed attempt to gain management control over NASA's Gemini program, McNamara signed a NASA-DOD agreement in January 1963 that included provisions for the DOD to conduct experiments on NASA Gemini flights. These experiments serve as a sort of introduction to the main MOL story because they highlight the underlying reason the OSD and the Air Force felt the military needed to determine if officers had a role in space—reconnaissance. While public discussion and rhetoric continued to characterize the MOL as an experimental test bed, the reality was that throughout the Gemini experimental program, the quest for data on exactly what, if anything, humans could contribute to the process of gathering overhead reconnaissance was paramount in the military space program. As assistant secretary of the Air Force for R&D, Flax wrote DDR&E Brown two months after McNamara approved the concept of the MOL, "It is believed that the experience gained in the Gemini experiments will be of considerable value for the MOL program."[1]

Chapters 3 and 6 briefly touched on the situation with reconnaissance satellites and the NRO during the Eisenhower and Kennedy administrations. Discussion of this topic during the Johnson era must rely almost wholly on secondary sources, due to the continuing secrecy surrounding specific space reconnaissance methods and systems. The relevant point is simply that as automated reconnaissance satellites increased in capabilities, developed a performance track record, and became key players in arms control and disarmament verification, the MOL's justification as another reconnaissance platform (this one manned and thus, dramatically more expensive) became increasingly difficult to maintain. In the end, it appears that the purported capabilities of the MOL, above and beyond those of robotic satellites, were not sufficient to convince high-level space policy makers that the MOL was worth its cost.

DOD Experiments on Gemini

By the spring of 1964, NASA and the DOD had jointly selected a total of 23 experiments for the Gemini flights, 10 of which were reserved for the DOD. NASA was very careful in its description of some of the more sensitive DOD experiments. For instance, one was titled "Visual Definition of Objects" that NASA defined as the "exploration of the technical problem areas associated with man's use of visual and optical equipment during spaceflight. Commercially available photo/optical equipment will be integrated into the Gemini spacecraft in a manner allowing the astronaut to view and photograph selected objects." Another DOD experiment was "Visual Definition of Terrestrial Features," which NASA said was for the "optical and photographic observation of terrain features to compare what man says he can see to that verified by photographs." A third was "Astronaut Visibility" to "precisely determine man's capability to see Earth's objects clearly. Calculations which can be made need to be checked before man's visual discrimination capabilities can be ascertained. A simple optical system will be operated by the astronaut in making visual observations." Of course not every single DOD experiment aboard Gemini dealt with reconnaissance. Several were oriented toward the satellite inspection mission such as "Visual Definition of Objects in Near Proximity in Space" designed to "demonstrate human proficiency and functional capability in space while maneuvering. The astronaut will maneuver his spacecraft so as to visually observe nearby objects in space from various aspects."[2]

When the trade press translated NASA's generic descriptions of the DOD experiments, it was much clearer: "The Air Force has restricted its experiments to those it considers vital to the Manned Orbiting Laboratory prelude where a military crew will be charged with the responsibility of spying and inspecting from his space platform. . . . The DOD experiments are obvious—the determination of the feasibility of operating a reconnaissance and spying system from a manned platform in space." It should also be noted that among the strictly NASA Gemini experiments there were also several that seemed to support the DOD's desire for gathering information on the role of humans in collecting intelligence from space. For instance, three of NASA's experiments were titled "Visual Definition of Terrestrial Features," "Synoptic Terrain Photography," and "Visual Acuity in Space," and carried descriptions similar to the DOD reconnaissance-related experiments.[3]

Internal Air Force documents also summarized, "Experiments have been chosen to make maximum contribution toward the objectives of satellite inspection and observation from spacecraft" with an emphasis on "investigat[ing] man's ability to discriminate, acquire, track and photograph preselected orbital and terrestrial objects from Gemini."[4] Flax told Secretary of the Air Force Eugene Zuckert: "The Air Force experiments have been chosen to provide maximum contribution toward objectives of satellite inspection and for observation. These are rudimentary experiments which will contribute to later programs such as the Manned Orbiting Laboratory. It is reasonable for this interpretation to be drawn from unclassified test descriptions. Reconnaissance and inspection, however, have not been publicized" (emphasis in original).[5] Al-

457

though one must mention the DOD's Gemini experiments program, one must also not make too much of it. In the words of the Air Force, "Due to physical space limitations, our Gemini experiments must be of a rather basic nature." As such, the Gemini experiments were "a first minimal effort toward the development of a reservoir of manned military space experience. However, our participation in this limited way in NASA's Gemini program falls far short of satisfying our requirements. We cannot gain the experience which we require to build a firm foundation for a manned military space program by looking over the shoulders of the people who are designing, conducting, and managing space programs." The MOL therefore remained "our calculated program which offers the best promise of military preparedness for any space eventuality."[6]

The apex of the DOD Gemini experiments program was probably on the *Gemini V* flight, 20–29 August 1965, on which the following DOD experiments were conducted: basic-object photography; celestial radiometry/space-object radiometry; and surface photography. Astronaut-visibility nearby-object photography was also scheduled but had to be canceled because it could not rendezvous with the target vehicle. NASA also conducted its experiments of synoptic-terrain photography and visual acuity. NASA's official report concerning basic-object photography for this flight indicated that "acquiring, tracking, and photographing celestial bodies present no problems." The radiometric experiments, designed to detect and measure energy emitted from various nonnatural sources such as ICBMs, were successful: "Visual observation of the rocket plumes was possible in all cases." The surface-photography experiment used enhanced but commercially available cameras to photograph the earth from space, and "results obtained indicate that visual acquisition with visual tracking can be successfully applied to obtain photographs of a preselected terrestrial object." NASA commented on its own synoptic-terrain photography, "Ground resolution is remarkably high; many small roads, canals, pipelines, and similar features are clearly visible."[7] The astronauts conducted the photographic experiments on *Gemini V* using a modified Hasselblad camera and telephoto lenses with 10-inch and 48-inch focal lengths. Photographs of downtown Dallas "clearly showed the individual runways, taxi-strips, and buildings of Love Field."[8] Perhaps it was entirely coincidental that in August 1965 President Johnson also gave his official approval for the Air Force to proceed with actual construction of the MOL—perhaps not.

Overall NASA reported encouraging results from its and the DOD's experiments relating to the human ability to conduct photoreconnaissance from space. This must have been heartening to supporters of the MOL program. However, when the press openly discussed these reconnaissance-related experiments, vice president and NASC chairman Hubert Humphrey was not pleased.[9] He wrote Webb, "I was disturbed and concerned about the attached news story [cited above]. . . . I was under the impression that all of this reconnaissance activity was top secret. If I am in error, I would like to be so informed. You may want to look into this, and I do hope so."[10] While this author discovered no reply from Webb to the vice president, there was a trailing off of publicly released information about the DOD experiments aboard Gemini flights until the program's termination a little over a year later.

The Air Force evaluation of the Gemini program was also positive.[11] One of its reconnaissance- and observation-related conclusions stated, "Astronaut capability to acquire, track and photograph predetermined objects in space was confirmed." Concerning the photographic definition of terrestrial features, "The ability of an astronaut to acquire, track and photograph predetermined ground targets with equipment having a narrow field of view was clearly demonstrated. Information was developed on requirements and procedures for accomplishing manned spacecraft photography." The Air Force's astronaut visibility experiment on Gemini "confirmed techniques for predicting the capability of astronauts to discriminate small objects on the surface of the earth in daylight." The Air Force was troubled by its inability to classify its reconnaissance-related activities in the Gemini program, however, due to NASA's insistence on an open-information policy: "With complete exposure of the DOD experiments certain aspects such as photography, low light level television, and radiometric measurements which inherently convey implications of intelligence objectives, became especially trouble-

459

some. The concern for public impression on these subjects eventually caused curtailment of activities in these areas and resulted in limitation of experiment technical product." In the end, however, the Air Force concluded the Gemini experiment effort "has been worthwhile. Valuable technical information and experience has been acquired at relatively low Air Force cost . . . which will be valuable in obtaining information and support for the MOL program."[12] The overall cost of the experiments' program was $28.5 million.[13]

There was no equivalent follow-on program from the Gemini experiments aboard Apollo, due primarily to the fact that the Air Force had the MOL program to conduct investigations of the military applications of human spaceflight and therefore did not need to piggyback on further NASA missions. Except for Apollo 7 and 9, there was little time spent in earth orbit during Apollo missions. After an exchange of correspondence over 1964, the DOD and NASA did agree in March 1965 to continue onboard Apollo some of the nonreconnaissance-related work that the DOD had done on Gemini in the areas of radiation measurements, manual-autonomous navigation, and carbon dioxide reduction.[14] However, after the fire tragedy in January 1967 delayed Apollo's flight-test schedule, Seamans recounted that NASA's leadership decided to include only those experiments relating directly to the lunar landing in Apollo earth-orbital flights. This left the DOD's Apollo experiments without a spacecraft assignment, and in June 1967 all three DOD experiments were officially deleted from Apollo flights.[15]

Reconnaissance Satellites in the Mid- to Late-1960s

One of the most famous declarations concerning space reconnaissance is from the Johnson administration. It survives only because Johnson believed he was speaking off the record to a group of educators and government officials in Nashville, Tennessee, but apparently was not. He said in March 1967,

> I wouldn't want to be quoted on this but we've spent 35 or 40 billion dollars on the space program. And if nothing else had come of it except the knowledge we've gained from space photography, it would be worth 10 times what the whole program cost. Because tonight we know how

> many missiles the enemy has and, it turned out, our guesses were way off. We were doing things we didn't need to do. We were building things we didn't need to build. We were harboring fears we didn't need to harbor.[16]

This enthusiastic presidential endorsement of space reconnaissance—and indirectly, the unmanned satellites of the NRO—gives some indication of the importance of these space assets by the end of the 1960s. While the MOL was designed to be part of the general family of reconnaissance-gathering systems, it would encounter difficulties in cost-effectively adding any capabilities to what the robotic satellites already offered.

The only unclassified primary-source document readily available concerning the NRO in the Johnson administration is DOD Directive 5105.23, *National Reconnaissance Office*, 27 March 1964. It was apparently the end product of the intra-NRO squabbling between the Air Force and the CIA outlined in chapter 7. This directive stated the NRO was "an operating agency of the Department of Defense, under the direction and supervision of the Secretary of Defense." It was responsible for "consolidation of all Department of Defense satellite and air vehicle overflight projects for intelligence into a single program . . . and for the complete management and conduct of this Program in accordance with policy guidance and decisions of the Secretary of Defense." By 1964 the blackout of information on the satellite reconnaissance program was complete: "All communications pertaining to matters under the National Reconnaissance Program will be subject to special systems of security control. . . . with the single exception of this directive, no mention will be made of the . . . National Reconnaissance Program [or] National Reconnaissance Office. Where absolutely necessary to refer to the National Reconnaissance Program in communications not under the prescribed special security systems, such reference will be made by use of the terminology: 'Matters under the purview of DOD TS-5105.23.'"[17] Beyond this single document, all other statements concerning the NRO and reconnaissance satellites from the Johnson era are from secondary sources and thus, by definition, have an element of speculation and conjecture.

461

Perhaps the only facet of the NRO and reconnaissance satellites as breathtaking as the security procedures surrounding them were the claims concerning the satellites' capabilities by the end of the 1960s or at least those under development in the late 1960s that debuted in the early 1970s. These capabilities were of course the ones against which the MOL was indirectly competing as the Air Force struggled to justify its continued funding at that time. In 1970 Philip Klass claimed that "current designs have cloud-cover sensors to prevent them from wasting film on targets obscured by weather, a valuable feature not found on the first photographic satellites. Still more advanced designs in the future are expected to provide real-time photographic and electromagnetic reconnaissance."[18] Two years later Klass described the nation's newest reconnaissance satellites, often referred to as KH-9 or "Big Bird" as "nearing full operational status," delivering photographs with "fantastic resolution" with "resolution approximately twice that of previous designs, provid[ing] discrimination of individual persons from an altitude of more than 100 miles. Big Bird is designed to perform both the search-and-find and the close-look type missions that have required two different spacecraft." Klass stated the first Big Bird was orbited on 15 June 1971, so the system clearly would have been in development during the mid- to late-1960s, simultaneously with the MOL, and presented the MOL with a formidable competitor in the space-reconnaissance-collection mission field.[19]

Richelson concluded that the KH-9 was initially developed as a backup to the MOL and did, in fact, become the nation's primary system when the MOL was canceled in June 1969. The KH-9 satellite supposedly weighed 30,000 pounds, measured 50 by 10 feet, and featured not only conventional photographic cameras, but also infrared and other multispectral systems.[20] Richelson calculated that the KH-9 had two cameras with 60-inch lenses that produced 24-inch resolution over an 80-by-360-mile swath of territory and carried four film canisters instead of two.[21] Another analyst stated that by 1972, "military reconnaissance satellites in the Keyhole series had resolutions on the order of three inches."[22] If those descriptions were even generally true, the MOL faced a formidable competitor in the NRO, especially considering the extra weight and expense of the MOL generated by the life-support equipment necessary to support humans in orbit.

The exact nature of the competition between the MOL and the NRO's robotic satellites, and how this rivalry may have contributed to the MOL's ultimate cancellation, will not be known until the NRO declassifies its historical documents. A more general point about the fundamental importance of reconnaissance satellites to national security and geopolitical stability seems certain. It may be an exaggeration to declare, "In simplest terms, there is strong reason for believing that observation from space is *the* most significant development in man's experience" (emphasis in original).[23] Nonetheless, even the most sober assessments make clear that "the NRO produced, according to some estimates, nearly 90 percent of all intelligence data on the Soviet Union" since its creation in 1961. The NRO's satellite systems "established, with considerable accuracy, the actual military capability and preparedness of the Soviet Union. Cost was rarely a question asked. The NRO mission held the highest priority. . . . There is little doubt that the NRO played a major role in the U.S. 'victory' in the Cold War."[24]

The Concept of the MOL

The Douglas Aircraft Corporation's final MOL configuration was a cylinder 42 feet long and 10 feet wide, weighing 30,000 pounds. The Gemini capsule sitting atop the MOL cylinder would add another 13 feet.[25] A mission module added to the payload meant the entire stack on top of the Titan IIIM would be 72 feet high.[26] The pressurized portion of the cylinder in which the two officers would live would reportedly be 14 feet long; the rest of the laboratory would have been unpressurized.[27] The volume of the pressurized portion was to have been approximately 1,300 cubic feet; the MOL would maintain an orbit between 125 and 250 nautical miles above the earth's surface.[28] One historian concluded the MOL's camera would have a lens six feet wide that offered six- to nine-inch resolution.[29] Another said MOL's massive camera would "provide near real-time reconnaissance of the earth [with] ground resolution of four inches."[30] If these conjectures are even marginally accurate, such a capability would be very expensive: initial MOL cost estimates were around $900 million[31] but quickly jumped to $1.5 billion[32] and continued to climb from there to as high as $3 billion upon program termination in 1969.

The MOL as a Reconnaissance Platform

It is important to understand that the MOL was conceived of, designed, and evaluated as a reconnaissance-gathering system. The difficulty with this concept is that it was not publicly discussed as such. Open testimony and unclassified documents of the time consistently described it as a system that "will be able to test and evaluate experimental equipment and determine man's ability to use the equipment in the discrimination, evaluation, filtering, and disposal of data."[33] Observers knowledgeable of space affairs could extrapolate that such generic descriptions were referring to Sino-Soviet bloc reconnaissance, but the DOD did not publicly acknowledge this, and so it was not entirely clear that the MOL was in a sort of dual competition: one to justify its publicly declared functions as an experimental test bed when compared to NASA's Apollo-derived earth-orbital capabilities; and another, more clandestine competition against the NRO's robotic-reconnaissance satellites. Therefore, the true nature of the MOL's planned mission bears some description.

The gathering of intelligence information from space is usually considered to include both the creation of photographic images as well as the collection of electronic emanations from ground-based military systems such as radars and assorted communications devices. The MOL was apparently designed to accumulate both types of information, optical and electronic. By May 1964, AFSC petitioned Headquarters USAF for, and received approval of, a manned electromagnetic-signal detection experiment aboard the MOL for the "detection of electromagnetic signal radiation, with its included elements of reception, demodulation, processing, display, measurement, and recording."[34] This type of intelligence gathering is generally referred to as SIGINT or signals intelligence (or occasionally as ELINT or electronics intelligence), to distinguish it from the gathering of photographic images of the earth's surface.

In internal documents, the DOD could be slightly more forthright concerning what it envisioned the MOL doing. In July 1964, it stated the MOL's basic purpose was to investigate and assess the utility of humans for military missions in space. "The tasks will be derived primarily for the reconnaissance, surveillance, inspection, detection, and tracking mission areas."[35] Internally, the Air Force was even clearer when it described what the MOL astronauts would do:

> The 2-man crew will discriminate, detect, point, track, evaluate, re-program and command as appropriate in missions of reconnaissance, fly-by inspection, co-orbital inspection . . . and perform support tasks such as navigation, re-entry, etc. The reconnaissance mission tasks seem to be well conceived. . . . They [the crew] examine the area photographs and look for targets and then program themselves on a suitable orbit to take high resolution photographs of targets of interest. High resolution photos are then taken of these targets.[36]

Also in August 1964, Flax described to George Mueller, NASA associate administrator for Manned Space Flight, the primary military objectives of the MOL. One of the most important was "Acquisition and Tracking of Ground Targets," to evaluate human performance in acquiring ground targets and tracking them "to an accuracy of better than 0.2%." This would involve direct viewing of the targets through a pointing and tracking scope controlled by a computer and connected to a camera. Desired targets were military airfields, operational missile sites, ships, submarines, and "various targets of opportunity." A second MOL objective was "multi-band spectral observation," to evaluate man's ability to operate specialized radiometric equipment for the "acquisition and tracking of orbiting objects and/or ballistic missiles during their boost, mid-course and re-entry phases." The third intelligence-related objective of the MOL was "electromagnetic signal detection," to evaluate man's ability "for making semi-analytical decisions and control adjustments to optimize the orbital collection of intercept data from advanced electromagnetic emitters." Finally, the Air Force did hope the MOL could also be active in other areas such as: acquisition and tracking of space targets, autonomous navigation, and geodetic survey.[37] The Air Force considered these descriptions of the MOL's true reconnaissance-related purposes to be highly sensitive. In the fall of

465

1964, the Air Force did not concur in a proposal to brief the British on the MOL because "the discussion of surveillance and reconnaissance experiments is most inappropriate for a foreign audience and, for that matter, for any audience which does not have a very real need to know."[38] Throughout the MOL's life, the Air Force restricted information concerning the actual reconnaissance missions of the MOL almost as zealously as the NRO protected its unmanned satellites.

The above explanation of the MOL as a reconnaissance platform has deliberately been confined to the first year after McNamara sanctioned official study of the MOL concept in December 1963. The reason is that in December 1964, McNamara "reoriented and expanded the MOL program, essentially changing it from a research to a developmental and operational program." The MOL's budget increased from $10 million in FY 64, to $38 million in FY 65, and a projected $150 million for FY 66. After McNamara's reorientation, the NASC said the MOL's primary objectives were "a. Development of technology to improve capabilities for manned and unmanned operations of military significance, e.g. reconnaissance, surveillance, inspection. This includes the necessary steps toward operational systems. b. Development and demonstration of unmanned extravehicular assembly and service of large structures in orbit with potential military applications, e.g., large antennas."[39] Large antennae are one of the requirements for collecting the electromagnetic emanations necessary for signals intelligence. A contemporary BOB document confirms that a group consisting of the BOB, McNamara, Webb, and Hornig as presidential science advisor in December 1964 did expressly state that the MOL's two primary purposes were "a. Development of technology contributing to improved military observational capability for manned or unmanned operation. This may include intermediate steps toward operational systems. Examples are side-looking radars, optical cameras of high resolution and large size, etc. b. Development and demonstration of manned assembly and service of large structures in orbit with potential military applications such as a telescope or radio antenna. This will interact strongly with a."[40] From January 1965 until its cancellation in June 1969, the

MOL was even more directly engineered to be a reconnaissance and intelligence-gathering platform. One consequence of this would be the release of less and less information about its progress.

The most valuable piece of evidence that highlights the centrality of the MOL's reconnaissance missions is a 428-page detailed description of 12 of the MOL's 14 primary experiments as of March 1965. A quick synopsis of this document described the MOL's experiments as follows:

- P-1: Acquisition and Tracking of Ground Targets: "Measures man's ability to acquire and track pre-assigned ground targets under varying conditions."

- P-2: Acquisition and Tracking of Space Targets: "Measures man's ability to acquire and track satellite targets under varying conditions."

- P-3: Direct Viewing for Ground Targets: "Measures man's ability to detect surface targets of opportunity and to make cursory intelligence assessments."

- P-4: Electromagnetic Signal Detection: "Measures man's ability to make semi-analytical decisions and adjustments based on information from electromagnetic emitters."

- P-5: In-Space Maintenance: "Measures the crew member's ability to perform in-space maintenance as applied to present and future manned space missions."

- P-6: Extravehicular Activity: "Determines what functions man can perform outside the spacecraft and what tools he will require for these functions."

- P-7: Remote Maneuvering Unit: "Measures crew member's ability to control a maneuvering unit by remote control."

- P-8: Autonomous Navigation and Geodesy: "Measures man's ability to navigate in space and to perform geodetic survey of uncooperative targets."

- P-9: Deleted.

- P-10: Multiband Spectral Observations: "Determines the crew member's ability to operate radiometric and related equipment in the completion of military and scientific activities."

- P-11: General Human Performance in Space: "Measures the day-to-day general performance capabilities of the crew members."

- P-12: Biomedical and Physiological Evaluation: "Measures the physiological and biomedical factors of the crew members under conditions of long-term orbit and weightlessness."

- P-13: Ocean Surveillance: "Evaluates the capability of man to control, coordinate, and use a system consisting of various sensors and subsystems to detect, track, classify, and catalogue sea targets."[41]

This document then described each experiment in intricate detail. The Air Force was clearly planning to experiment with cutting-edge reconnaissance techniques.[42] Unfortunately, two of the most important primary MOL experiments, P-14 and P-15, were so highly classified that they were only briefly alluded to in this internally classified Air Force document. A footnote on page 164 said P-14 was an "Antenna Experiment" and that one would have to "See Experiment P-14 Data Book" for more information. A similar footnote referred to an "Optics Experiment" and stated "See Experiment P-15 Data Book."[43] In essence the MOL's role had clearly "changed from a test-bed program to an operational manned reconnaissance program, with Air Force officials now seeing an open-ended program that would not be limited by six launches as demonstrators of military missions in orbit."[44] It is hard to believe that in the fiscally constrained environment of the mid- to late-1960s, with NASA's budget being pared every year, that Johnson and the OSD would approve over a billion dollars for a military test bed in space. The Air Force almost certainly had to sell the MOL to McNamara as an operational intelligence-gathering system or risk its immediate cancellation.

This reorientation necessarily resulted in the transfer of planned MOL launches from Florida's ETR to Vandenberg AFB on the California coast north of Santa Barbara. Reconnaissance satellites must be in a polar orbit because orbiting over the north and south poles eventually brings all points on the earth's surface beneath the satellite's orbit, given the rotation of the earth. In contrast, an equatorial orbit, or one slightly inclined off the equator—the type usually achieved by a launch from Florida—meant that a significant portion of the northern and southern hemispheres would never pass beneath the satellite, making such nonpolar orbits unsuitable for reconnaissance of the Sino-Soviet bloc. Launches for polar orbits need to take place from Vandenberg because the earth's rotation ensures that if the rocket were to fail, the satellite would fall into the ocean, and the rocket debris would not impact populated areas. Neither of these conditions can be guaranteed in a launch from Cape Canaveral. This meant that when the MOL was reoriented from a test bed to an operational-reconnaissance system, it had to be launched from California, not Florida. The Florida political delegation complained about this transfer, and Senator Spessard Holland even called for hearings. Deputy SECDEF Vance had to explain to President Johnson in February 1966 that using Vandenberg for the MOL's polar launches "has been firm since the primary intelligence mission for the program was approved last year. Prior to that time, when the MOL was being considered for a variety of other experiments, equatorial launches would have been a possibility."[45]

In the Senate hearings regarding the MOL's move from the east coast to the west, the DOD steadfastly refused to explain why a polar orbit was essential to program requirements. Senator Holland pressed DDR&E Foster on the point, but Foster simply repeated again and again, "I am sorry, I can only say that it [a polar orbit] is a requirement of the program. . . . To fulfill the purpose of the program, these inclinations [90° relative to the equator] are required." Holland stated angrily, "I have asked what I think is an answerable question, and an intelligent question, and you haven't answered yet." Howard W. Cannon of Nevada tried to defuse the situation by asking Foster, "Can the same areas be overflown with an equatorial flight that can be overflown with a polar flight?" Foster replied, "No, of course not." Cannon: "Does the objective of this flight require that areas be overflown in a polar orbit that cannot be overflown in an equatorial orbit?" Foster: "That is correct," because the polar orbit is the only one that "goes over all regions of the Northern Hemisphere or the Southern Hemisphere."[46] The political controversy over moving the MOL's launch location faded, and the MOL's FY 67 budget jumped to $228.4 million, from FY 66's $150 million.[47] Projections for FY 68 were $430 million, even though the Air Force said $510 million was required to keep the program on track.[48] Overall estimated program costs for the MOL continued to rise from $1.5 billion near the end of 1964, to $2.2 billion by the spring of 1967, to $3 billion by 1969. The Titan IIIM necessary to launch was expected to cost another $2 billion to develop.[49]

As seen above, the OSD deemed protecting the reconnaissance-related nature of the MOL's true mission worth the risk of angering certain members of the Senate space committee. For public consumption, the party line on the MOL as an experimental, not an operational, system remained the same. DDR&E Foster told the Senate space committee in March 1968, "MOL's objectives remain unchanged. The system is designed to: develop technology and equipment for the advancement of manned and unmanned space flights; perform meaningful military experiments; and, improve our knowledge of man's capability in space to support defense objectives."[50] The reality, according to some analysts of the US reconnaissance satellite effort, was that "MOL was now part of the KEYHOLE program. Its camera was given the designation KH-10 and the program to use the MOL for reconnaissance was codenamed DORIAN."[51] Logic would seem to dictate that since the MOL was regarded as an operational, reconnaissance-gathering system after late 1964, it would have had to be analyzed, justified, and funded in the context of the NRO's unmanned satellites designed to do the same job.

The MOL's Progress through August 1965

December 1963 through August 1965 was a period during which the Air Force struggled to justify the MOL's actual physical construction to McNamara and the rest of the OSD and ultimately the president. The USAF finally achieved success in the summer of 1965 when Johnson approved the MOL's fabrication, but only after it was reoriented from an experimental test bed to an operational reconnaissance-gathering platform, as described above. The official term for this almost two-year period during which the Air Force attempted to find a suitable justification for the MOL was the *Preprogram Definition Phase*. One Air Force office described this as a study and analysis period "oriented to the definition of the optimum concept for accomplishing the development criteria and requirement." This would, in theory, be followed by Phase I, Project Definition Phase, and Phase II, Systems Acquisition Phase.[52] One analyst quipped that the reality of the Preprogram Definition Phase from 1963 to 1965 was "spending a year or so to decide whether you want to do something after you've announced you're going to do it."[53] Dr. Albert Hall was deputy DDR&E for Space, and in a February 1964 speech he explained, "At this point in time what we are calling MOL is a concept rather than a specific piece of hardware." Therefore, the immediate task was "to detail, by thorough study, what is to go into the program and what we expect to get out of it."[54] As described above, the eventual conclusion was that the MOL had to serve as an operational reconnaissance system.

By 1964 the Air Force had gained several years of experience with McNamara's PPBS and systems-analysis procedures and knew that generalized statements about the MOL's purported value and acceptance would not be adequate to ensure its approval. Hard data and quantifiable facts were required. As Undersecretary of the Air Force, McMillan wrote to the USAF Headquarters office responsible for the MOL, "Orbital tests will be conducted only when it is determined, from all necessary studies and tests short of orbital, that it is both desirable and necessary to perform tests in space." It was assumed that if the Air Force recommended such orbital tests with humans, then "the determination must be supported by results of a substantive comparison of man's capabilities helped by automatic equipment against purely automatic equipment" because that was the exact comparison the SECDEF would make. Therefore, "It is implicit that a clear analysis and summary of 'man's contribution' must be defined, taking into account the relative performance of man versus unmanned systems, the worth of 'man's contribution,' relative costs, confidence of success, comparative risks, and the probability that most of the penalties of the life support system are borne by other experiments."[55] The relatively low MOL budgets of $10 million for FY 64 and $38 million for FY 65 are thus explained by the fact that little was taking place besides studies and analyses. These inquires did enable the Air Force to publish the 400-page document in March 1965 outlining the primary experiments for the MOL discussed above.

The USAF's conclusion, as expressed in September 1964 by Gen James Ferguson, deputy chief of staff for R&D, was that "we have studied these testing requirements, and we have concluded that a manned military test station in space provides the only reasonable solution to the problems of testing equipment designed for use in space."[56] McNamara was equally clear in December:

> My principle is a very simple one. I believe we are a military organization, we are not interested in space except insofar as it bears directly on our military mission. If there is anything that NASA can do, that we can in effect hire them to do as our agent, I am 100% in favor of doing so. . . . I want to be sure that the MOL program, the details of it are fully analyzed by NASA and fully taken into account when NASA establishes any portion of its Apollo program not directly related to the lunar program. . . . I want to be certain that the Air Force, when it establishes the MOL, takes account of what NASA is required to do as part of the Apollo program that in turn is directly related to the lunar program. And, in turn, I want to be certain that NASA, to the extent that it expands Apollo beyond the limits required for the lunar program, takes into account whatever we must do in the MOL to meet bona fide and military objectives.[57]

The same principle that caused the Dynasoar difficulties in justifying its existence in the context of the capabilities offered by NASA's Gemini was going to play itself out again as the MOL faced the fact that NASA was also planning an extensive follow-on program to the Apollo lunar landing, called the Apollo Applications Program (AAP) that had a component calling for continued exploration of earth-orbital applications.[58] There was little that the USAF could do besides emphasizing the reconnaissance capabilities of the MOL, because these would clearly not compete with NASA's mission (although it would throw the MOL into competition with the NRO's systems). Refashioning the MOL as an operational-reconnaissance vehicle did enable the Air Force to fulfill OSD's stringent criteria for justifying a space project. It also meant that in the long term the MOL had to be justified when compared to the KH-series of NRO reconnaissance satellites. As Johnson's science advisor, Hornig declared, "One shouldn't risk the life of man on things you can do with instrumented things."[59]

The preceding section described the late-1964/early-1965 process in which the MOL was reoriented into an operational-reconnaissance system. This proved to be the necessary step required to gain McNamara's, and eventually presidential, approval to begin building the MOL. The final sprint toward presidential approval began with an NASC meeting on 9 July 1965. McNamara simply stated that regarding unmanned versus manned systems for intelligence gathering, "He had concluded that you could get a better result using a manned system. However, they also had worked out the cost effectiveness and even though the manned program development costs would be two-to-one compared to the unmanned," the manned would turn out to be cheaper "for the same or better information, since fewer launches would be made. This, then, would offset the larger initial cost of manned launches versus unmanned launches." McNamara added that "NASA can't perform the reconnaissance mission and that the details of such a mission could not be discussed publicly." He stated in closing that the main reason to proceed with the MOL "was to obtain information quickly and on a selective basis . . . this would require the manned system. . . . Secretary McNamara indicated that the DOD was prepared at a later date in the program to go either way, manned or unmanned; and in fact they were recommending to go both ways initially."[60] Therefore, even on the brink of approving MOL construction, McNamara did not seem overly enthusiastic about the system and clearly was preserving the option of continuing with an entirely unmanned family of reconnaissance satellites.[61]

On 25 August 1965, the White House released a statement from Johnson announcing, "I am today instructing the Department of Defense to proceed with the development of a Manned Orbiting Laboratory." He estimated costs at $1.5 billion and the first manned flights in late 1968.[62] At a press conference that same day Johnson said, "This program will bring us new knowledge about what man is able to do in space." There was no direct discussion of exactly what the MOL would do, much less its central reconnaissance mission.[63] A briefing by an individual the press was instructed to refer to only as a "defense official" included the following exchange: "Question:

What's the purpose of a polar orbit that you plan? (Laughter) Defense Official: I didn't say we were planning polar orbits."[64] Again, fade to black.

Nevertheless, speculation concerning the MOL's true purpose was rampant in the press. The *Washington Post* speculated, "Large and powerful segments of the Johnson administration were sold on the Air Force's Manned Orbiting Laboratory simply as an intelligence tool. But it was the added possibility in the arms control field that brought unanimity. As one key official put it: 'If this does what we think it will do, MOL will be the greatest boon to arms control yet.' . . . The primary mission of MOL . . . is without doubt to have men supplement the machine as a shutterbug spy in the sky."[65] If nothing else, "The President's agreement to proceed with MOL meant the end of a ten-year struggle by the Air Force to gain a role in manned space flight."[66] The remaining open question was whether the MOL would ever literally get off the ground.

NASA and the MOL

NASA leaders never publicly questioned the need for the MOL nor did they ever concur publicly with any assessments that it duplicated NASA's general R&D program or the AAP specifically. However, outside commentators did not hesitate to write, "By the time MOL is placed in orbit, NASA's capabilities will exceed those envisaged for MOL. . . . When one examines the 'complex tasks' envisaged for military astronauts, one finds it difficult to locate a single function that NASA has not already performed or is planning to perform, or is capable of doing."[67] This illustrates the price the Air Force and the DOD paid for maintaining strict secrecy concerning the MOL. Since many analysts, and probably congressmen, had no official knowledge or confirmation that the crucial justification for the MOL's existence was to collect intelligence, the MOL could therefore be compared to NASA's AAP and declared redundant. In fact, its distinguishing characteristic of reconnaissance made it radically different from the AAP's R&D activities in low-earth orbit, but defense officials would not or could not point this out due to security restrictions.

NASA's public declarations continued to support the MOL in accordance with the position paper drafted internally by Seamans in late-1963/early-1964 and discussed in chapter 7. Shortly after the December 1963 preliminary MOL approval, Seamans wrote DDR&E Brown, "Since it is evident that NASA can support the MOL project in several ways, we are planning on providing such support in a manner that will give the greatest assurance of MOL success and at the same time maintain the momentum of the current NASA space program."[68] Throughout the MOL's history, NASA regularly expressed its willingness to support the MOL with its Gemini hardware and facilities as much as possible. It also continued to maintain that the MOL and the AAP were closely coordinated and not duplicative. Finally, NASA said it would be happy to take advantage of the MOL's experimental capabilities. Also in January 1964, Mueller, associate administrator for Manned Space Flight, declared "NASA will have a requirement for experiments which can be accomplished by the MOL system."[69]

Webb regularly testified to Congress that "the Gemini-B/MOL program was needed by the DOD to make an early determination of the utility of a man in space. The DOD will be able to move ahead rapidly with plans to make this determination within the desired time frame by virtue of the fact that the necessary technology and capacity to provide the hardware and to conduct such an operation have been developed by NASA and are available."[70] This is representative of NASA public statements concerning its use of the MOL for experiments and the DOD's use of NASA resources (Gemini capsules, tracking and data acquisition facilities, communications, command and control equipment, etc.) for the MOL during the MOL's 1963–69 existence. The MOL would, in fact, incorporate numerous components from the Gemini and Apollo systems such as power supplies, environmental control systems, guidance and navigation equipment, and reaction control systems.

As the minutes of a NASA meeting from July 1965 plainly stated, "We were in agreement that NASA cannot oppose a manned laboratory when the DOD puts it in the terms of a national security requirement. There is the implied priority of any program which is related to national defense." NASA was, however, honest with itself: "If one excludes the highly classi-fied military mission, there is a high degree of commonality in the experiments which NASA and DOD need or would like to perform in space."[71] In the NASC meeting of July 1965, "Ad-ministrator Webb indicated that he also supported the MOL program. He asserted that it was no different than many oth-ers where originally DOD's missiles were used as building blocks by NASA. Now the Department of Defense can use NASA's manned space flight experience for its purposes."[72]

NASA and the DOD coordinated the initial stages of the MOL through the regular AACB channel of the Manned Space Flight Panel. For instance, they signed an agreement in January 1965 concerning DOD use of NASA control centers and track-ing network stations that stated, "In general, NASA facilities will be made available to support the Air Force GEMINI B/MOL program, having due regard for national priorities and to the extent that such use is compatible with international agreements covering tracking stations on foreign territory."[73] Concerning the use of the MOL for NASA scientific experi-ments, however, NASA did not sound very optimistic. In an-swer to a vice-presidential question, NASA said, "It should be recognized that the security requirements of the DOD MOL program will impose limitations on such participation." In ad-dition, since military objectives would have first priority for MOL launches, "NASA would have difficulty in maintaining a high level of interest among the scientific community in ex-perimental efforts which, although meaningful, could only be flown on a space available basis."[74] NASA still maintained, however, after the MOL received final approval in August 1965 that the system could use NASA facilities: "We stand ready to plan with you for the maximum practicable utilization by the DOD of the NASA developed hardware and technology, our production, testing, checkout, simulation, training, mission control, and data acquisition and processing facilities, and our management and operational experience."[75]

Evidently the two organizations foresaw enough systemic in-teraction to merit creation of a separate committee outside the AACB parameters to concern it with NASA-DOD human-spaceflight issues. In January 1966 they created the Manned

Space Flight Policy Committee (MSFPC) as a "means of expediting coordination at a policy level the manned space flight programs of the two agencies."[76] Chaired by the DDR&E and NASA's deputy administrator, the MSFPC took the place of the old GPPB, which was then disbanded.[77] In March 1966 NASA and the DOD created yet another body, the Manned Space Flight Experiments Board (MSFEB), to recommend approval or disapproval of experiments to be conducted under NASA- and DOD-manned spaceflight programs (i.e., MOL and Apollo); recommend assignment of experiments to specific flights; and recommend relative priorities of experiments to be implemented and periodically review the numbers of experiments scheduled for specific missions.[78] The importance of bodies such as the MSFPC and MSFEB is not so much found in any specific decisions they might have made but rather that they indicated the continued good-faith attempt by NASA and the DOD to coordinate their human-spaceflight programs as closely as possible to avoid duplication and waste. By March 1966, NASA had not yet indicated the desire to conduct any experiment on the MOL.[79] Seamans told this author that he had no recollection of NASA ever designing any experiments to be flown on the MOL.[80] A secondary source, however, suggested that NASA did have one experiment prepared for MOL deployment—a carbon dioxide sensor.[81]

Nevertheless, Mueller reported that by mid-1966 NASA had transferred to the DOD's Gemini B/MOL effort some $20–25 million worth of equipment, to include two Gemini capsules, an environmental control system, an attitude control system, a communication system, pressure suits, and a fuel-cell power supply.[82] In March 1968, Webb told Congress that NASA had turned over more than $100-million worth of hardware and support equipment to the Air Force for use with the MOL.[83] There are some fleeting indications, however, that not all was completely smooth in the NASA-DOD human-spaceflight coordination arena.

Perhaps some tension in the NASA-DOD/AAP-MOL field was inevitable. After all, in the words of one government report, the MOL "ran full tilt into competition with NASA plans for near-earth orbiting laboratories and stations."[84] Webb's biographer concluded that when the MOL was approved, "NASA was not happy about this. However, there was strong support in Congress for a military manned space program of some sort."[85] Levine reported his NASA sources concluded that "considerably less than 1 percent of the data obtained by MOL would be superior to what would be obtainable from available systems. . . . In effect, some NASA technical managers suspected that MOL really did duplicate NASA programs, at a time when influential Congressmen were demanding less duplication and more standardization."[86]

Some primary source evidence does exist pointing to a certain level of negative feelings within NASA toward the MOL. Levine examined tapes of NASA meetings and presented a transcript from one in September 1966 involving NASA center directors and Webb. One director stated, "MOL is a rather poor program at best and they [DOD] have never justified it properly. Now, you [Webb] haven't wanted to attack them . . . because I don't think McNamara is a nice guy to attack, he is rough. Webb: Well, hell, he has attacked MOL worse than I have. Official: Well, my point is that MOL is a very poor program. At one time it would have been a halfway decent program but it is way out of date now. . . . I say it right now that MOL is no good. They are always too late."[87] One of Webb's personal consultants wrote him through Seamans, "The rub is, however, that a MOL operation such as is now planned by the Air Force would go far beyond what is necessary for direct military purposes—will in fact lead to a second and strictly military national space program. In this lies a serious danger not alone for NASA and its assigned mission, but for the basic philosophy underlying the whole U.S. approach to space exploration and utilization. The Air Force plan calls for a complete space system, one that would parallel NASA practically every step of the way." This consultant said that with a second strictly military space program, "NASA would obviously lose ground and in a variety of ways from the development." He recommended, "Accelerate AES [Apollo Extended System] [AAP] so that it may be operational by roughly 1970 in order that it may obviate any need for an extension of the MOL program beyond its original purposes."[88]

Golovin of LLVPG fame, who had subsequently moved to the Office of Science and Technology, now under Hornig, continued to criticize the space program. He concluded, "If NASA and DOD are left to themselves, they will not internally generate either the necessary will or the administrative means for effective technical coordination in orbital manned space flight. If so, the total national costs for these activities will turn out to be greater than they need be and, more importantly, the rate of progress . . . might turn out to be slower than it otherwise could be."[89] Despite these disparate grumblings concerning the NASA-DOD human-spaceflight interface, the NASA leadership at the Webb-Seamans level remained officially and seemingly supportive of the MOL. A Seamans letter to Webb perhaps expressed the NASA-MOL situation best:

> I feel that concern for peaceful versus military "image" is often overrated in importance, and that this consideration is not the basic rationale for the Space Act of 1958. I believe the fundamental issue is how best to make effective use of aeronautical and space exploration both nationally and internationally. Certain activities must be kept classified for reasons of national security, and I believe this is generally recognized and accepted internationally. . . . There is no basis for our questioning the primary objectives stated by the DOD for their MOL. These objectives are peculiar to stated military operations at this time. . . . If the Gemini B-MOL-Titan III is implemented, we should consider its use along with Apollo-Saturn to meet national aeronautical and space objectives under NASA control.

Seamans closed by reiterating that if the MOL program were implemented, "NASA should support its development" (emphasis in original).[90] For the most part NASA did support the MOL as much as possible during the developmental phase and seemed prepared to continue to do so if it ever became operational. On the other hand, there seemed to be little effort on NASA's part to develop scientific experiments for MOL deployment.

The Concept of the AAP

There are a number of allusions above to NASA's Apollo Application Program. The discussion now turns to a brief survey of its origins and early evolution.

479

This treatment need not be as comprehensive as that of the MOL for two reasons. First, the AAP's ultimate programmatic execution came in the 1970s, primarily with the three Skylab missions and is, therefore, outside the scope of this study. Second, this author is convinced that the role played by the perceived duplication with the AAP in the MOL's demise was only one of three main factors, along with financial considerations and perceived duplication with the NRO's reconnaissance satellites. The first factor in looking at the AAP is to examine exactly why Webb was not enthusiastic about mapping out a specific path toward NASA's future, of which the AAP was one part.

NASA's Reluctance to Plan for the Future, Including the AAP

On 30 January 1964, Johnson asked Webb to review NASA's future space exploration plans to relate hardware and development programs to prospective missions. Webb's preliminary 28-page reply on 20 May 1964 was completely lacking in specifics and was simply a laundry list of past accomplishments, studies currently being conducted, and the numerous possibilities for the future. On the final page Webb concluded, "An extensive analysis of each mission is being made to determine its requirements in manpower, facilities, and other resources and to balance these against the value of potential returns in the form of new national space capabilities, new knowledge, new civilian and military applications and new industrial capabilities."[91] NASA's final report in response to Johnson's asking about its future did not come until January 1965, a full year after the presidential tasking. NASC staffer Charles Sheldon commented on a draft version, "NASA defensively points out all the reasons why no one should rock the boat at this time. I agree they cannot set new goals without the building of a national consensus, but I think we are entitled to more leadership in this regard than has been illustrated." Sheldon characterized the report as "so safe and so sane that it does not really make anyone feel a new sense of purpose or enthusiasm, or that the country is going to strike out boldly and achieve a true position of leadership. There is more a plea that we wait to see what others will do, and then will try hard to match them in some way." He concluded that the "galloping conservatism" of NASA's future planning constituted a "continual shying away from new missions [which] plays right into the hands of those who would end advanced work on the grounds of lack of requirements."[92]

NASA's final report to Johnson in January 1965 bore out Sheldon's preliminary conclusions. Its 61 pages did not in essence progress beyond the statement, "Unless an urgent National need arises, large new mission commitments can, better than in previous periods, be deferred for further study and analysis based heavily on ongoing advanced technical developments and flight experience." NASA's position was that the main requirement of future programs was simply, "First, apply available resources to every aspect required for success in the ongoing programs, especially the Apollo program, and to bring these to fruition as quickly and efficiently as possible." Second, NASA should define an "intermediate group of missions and work toward them using the capability being created in the on-going programs." The final step would be to "continue long range planning of missions that might be initiated late in this decade or early in the 1970s." The remainder of the report outlined present capabilities NASA had built up, the types of intermediate and long-term capabilities that could be created, the general categories of experiments that NASA might undertake, and the various potential configurations of possible systems such as a large, manned orbiting research laboratory, a lunar base, or a manned planetary exploration mission. There was, however, nothing in the way of preferred specific missions or concrete recommendations in the report.[93]

As former NASA historian Launius has noted, "Webb was quite reluctant to commit NASA to specific goals and priorities in advance of any expression of political support, preferring instead to list a range of possible tasks and to ask top policymakers to choose the options they wished to pursue. This was the approach taken in this January 1965 report by NASA to President Johnson."[94] In addition, that approach marked NASA planning throughout Webb's tenure and assured the AAP would not be crafted as a large and ambitious project, nor

481

would it move quickly to fruition along with Project Apollo's lunar landing because Webb wished to avoid any possible internal NASA competition for funds between projects. According to the minutes of an NASC meeting in March 1966, "Mr. Webb doubted the wisdom of setting new goals without some reason to expect Congressional support, and that we should preserve our options. Most definitely we should not tell the public about future plans until the President had made a decision." When Seamans suggested that there might be Saturn-class vehicles left over after completion of Project Apollo and thus available for something like the AAP, Webb interjected, "but they could not be released without the danger of failing to meet the lunar landing commitment."[95] NASA's George Mueller recalled, "It was rather clear that Jim Webb did not want a plan." When Mueller's office floated the idea of a Mars expedition in 1966 or 1967, Webb replied, "Absolutely not. We don't want to have a plan like that. First we've got to do the moon before we begin to put into effect a longer-range plan" because for every person that would support it, there would be 10 to shoot it down. Mueller said Webb was unable to "find any overall national consensus that said we had to have a plan past the moon. . . . after the [January 1967] fire it became even more obvious that we ought to be sticking to our knitting and not producing what he would call grandiose plans for the future."[96]

In all fairness to Webb's perspective, he was not alone in his reluctance to embrace any next-generation space goals. A State Department report in October 1966 stated, "From the standpoint of our foreign policy interests, we see no compelling reasons for early, major commitments to such goals, or for pursuing them at the forced pace that has characterized the race to the moon. Moreover, if we can deemphasize or stretch out additional costly programs aimed at the moon and beyond, resources may to some extent be released for other objectives . . . which might serve more immediate, higher priority U.S. interests." The State Department's bottom line was, "Instead of indefinitely extending the space race, it would be preferable to work toward a twofold objective: 'De-fusing' the space race between the U.S. and Soviets [and] . . . Bridging the gap between the space powers and others."[97] There was, therefore, no groundswell of support anywhere within the executive branch to boldly forge new space initiatives. As the State Department speculated, "after the U.S. and Soviets have achieved manned lunar landings, it is likely that international interest in the space race as such will subside. Excitement concerning specific space spectaculars may also diminish."[98] Therefore, the United States should "seek to move away from an extension of the space race and toward more orderly and internationally responsible ways of doing business in space."[99]

Webb's biographer also tried to put the director's refusal to set NASA on a particular course for the future into its proper context. Lambright said Webb was fully aware of four factors that made it impossible to forge a consensus on a future space program and therefore did not push any such effort because to do so only would have created additional political, budgetary, and bureaucratic difficulties for NASA. "First, NASA's own success made a difference," in the sense that America by the mid-1960s was doing well in space while the Soviets appeared to be losing momentum. Lambright continued, "Second, the nation and President Johnson were increasingly distracted from space by two other larger efforts: the Great Society and the Vietnam War. Third, at a time when gaining support for post-Apollo was most critical, Apollo was yet to be completed. Fourth, the overall space budget was suffering cutbacks in a period of general financial stringency, and the NASA priority had to be to spend its diminishing resources to maintain Apollo rather than to establish post-Apollo efforts." Therefore, "It is doubtful that anyone could have sold a post-Apollo program in the environment of the late 1960s."[100] However, to understand the difficult circumstances with which Webb had to deal in crafting NASA's future is not to deny the ultimate consequences of his failing to do so.

Lambright fully admits that the cost of Webb's focus on completing the lunar-landing mission to the exclusion of future planning "was retrenchment in the early efforts to launch a post-Apollo program. . . . The decision *not* to sustain the momentum of Apollo through an equally large-scale, follow-on effort had been made incrementally, year by year. . . . The consequence was drift and frustration" (emphasis in original). The MOL also figured into the equation. Its approval "showed NASA that it could not take Johnson's support for granted. If NASA did not move quickly enough, the other agencies would."[101] Other scholars take these points even further. Levine concluded, "One of the major reasons for the decline in the NASA budget was the agency's failure to plan effectively for the long term."[102]

The AAP's Origins and Early Evolution through Late 1965

The ambiguity surrounding NASA's future plans meant that the AAP got off to a rocky start and had significant difficulty being defined. NASA Associate Administrator for Space Science and Applications Homer Newell wrote, "During the muddy period of planning for an Apollo Applications Program that was not going to sell, Webb often stated to his colleagues in NASA that he did not sense on the Hill or in the administration the support that would be needed to undertake another large space project." He explained, "Webb preferred to hold back and listen to what the country might want to tell the agency. It was his wish to get a national debate started on what the future of the space program ought to be, with the hope that out of such a debate NASA might derive a new mandate for its future beyond Apollo. But no such debate ensued. In a country preoccupied with Vietnam and other issues, the space program no longer commanded much attention."[103] In fact, the primary decision concerning the AAP's basic configuration as a fully equipped space laboratory (later renamed Skylab) launched on a Saturn V was not made until late July 1969 by new NASA administrator Paine, after both the MOL's cancellation and the accomplishment of the lunar landing and is thus largely outside the scope of this work. Nevertheless, the AAP does have an important role in the history of space in the 1960s in general and in the MOL's fate in particular.

NASA's planning for large space stations continued after the process described for the Kennedy administration in chapter 7. Edward Z. Gray was the director of Advanced Studies in NASA's Office of Manned Space Flight. In a January 1964 interview, he suggested NASA's space station would be more so-phisticated than the MOL. He also described more than a dozen study projects NASA had underway, that when com-pleted, would allow NASA to appraise its requirements and pursue the best approach to developing a space station.[104] Throughout 1964 the AACB's Manned Space Flight Panel's National Space Station Planning Subpanel (NSSPS) met four times to coordinate NASA-DOD space-station studies but "then lapsed into inactivity."[105] The simple fact was that once the OSD decided to support the MOL it had little desire to con-sider a larger, more capable, and even more expensive space station (even though the MOL was technically not defined as a space station) until the MOL's experiments could be con-ducted and analyzed. For example, in FY 64 NASA had at least 11 separate space-station studies with a total budget of $22.1 million examining concepts ranging from a modified Apollo system weighing 15 tons for four to six humans, to a 100-ton giant for 18–24 people. DOD had just one study at $1 million, which was scheduled for termination after FY 64 due to the MOL approval in December 1963.[106]

NASA was, in essence, left alone to study possible configu-rations and purposes for a large space station. In the words of the Air Force officials assigned to monitor NASA's space-station work, "Taken in total, it is evident that the NASA space-station study program encompasses the entire spectrum of space-station capabilities from the small, with limited capability and orbital lifetime, to the large, with extensive capability and life-time."[107] As explained above, Webb was not keen on any future space-station effort progressing beyond the study stage. If NASA were to have an earth-orbital presence outside of that inher-ent in Project Apollo, it would clearly have to somehow modify the available Apollo-Saturn hardware for additional and ex-tended earth-orbital experimentation. Thus was born the idea of the Apollo Extended System/Extended Apollo System or Apollo X, more commonly referred to as the AAP.

This discussion of the AAP focuses on one particular ele-ment of the concept that was most relevant to the MOL—the idea of and planning for an earth-orbital workshop. The broad concept of the AAP, however, in its planning stages included

485

using Apollo-Saturn hardware for many different types of missions. The workshop element was the only one that survived. An internal NASA document from October 1965 stated, "Basically, the objective is to acquire data and experience in earth orbit, in lunar orbit and on the lunar surface, by the early 1970's." It was hoped that the AAP would lead to space stations in earth orbit, lunar observatories, and manned planetary exploration in the 1970s and 1980s. Of the 254 experiments considered for the AAP inclusion in this document, 20 were categorized as "lunar-orbital survey" and 36 as "lunar-surface exploration."[108]

Similarly, in March 1966 Robert Gilruth, the director of NASA's Manned Spacecraft Center, mentioned two major clusters of experiments within the AAP. First, there would be "follow-on lunar missions for exploration, mapping, and scientific studies" in which elements of the Apollo system would be modified. The LEM would be changed so it could "be used as a shelter on the lunar surface." Second, there would be "earth orbital operations with remote sensors to observe surface phenomena and with optical and radio telescopes for outward observations, in addition to conducting experiments of medical or other scientific interest." In this case, Apollo items such as the Apollo command module would be updated to increase its earth-orbital capabilities from 14 to 45 days.[109] While the orbital-workshop concept—the one most relevant to the MOL—turned out to be the only part of the AAP to survive (as Skylab), before the late 1960s, it was only one of many concepts under the general rubric of the AAP.

In November 1963, North American Aviation, the main NASA contractor for the Apollo capsule, issued a final study on modifying the spacecraft for extended earth-orbital missions to experiment with unknowns such as prolonged exposure to weightlessness. The study explained how Apollo systems could be modified to meet the requirements of extended missions.[110] NASA's official Skylab history stated that the agency began plans in 1964 to fly an extended Apollo as its first space lab, designed to lead to an intermediate-space-logistics system and then finally a sophisticated space station.[111] By August 1964 NASA's Manned Spacecraft Center in Houston proposed an

Apollo X consisting of a modified Apollo lunar spacecraft to be used in earth orbit for biomedical and scientific missions. In the first phase, two humans would orbit for up to 45 days, but by the fourth phase, three men would orbit for 120 days.[112]

At this point McNamara apparently sensed enough NASA interest in earth-orbital systems that he deemed it prudent to make another attempt at joint planning. He wrote Webb on 25 September 1964:

> It is my understanding that your staff have [*sic*] been studying a configuration called APOLLO X planned as a possible forerunner to a National Space Station. I am also informed that NASA may tentatively plan to devote an appreciable amount to studies having a bearing on this matter in FY 1965, continuing an effort of approximately $12 million committed to space station studies and related programs in FY 1964. . . . In view of the very large expenditures which would be involved in a National Space Station, its possible significance to national security as well as its importance to the country as a predecessor to manned planetary exploration, it seems to me that it may be timely to consider how we might jointly manage separate large programs.[113]

McNamara specifically proposed a management plan based upon several principles. First, he and Webb would agree that the MOL "is the flight forerunner to the definition of a scientific or militarily operational space station." Second, NASA would accept managerial responsibility for a program of scientific experiments to be flown on the MOL, though the Air Force would continue as overall MOL program manager. Third, "Following flight results from the MOL, a determination will be made on (1) the necessity of a new large military operational or scientific space station, (2) the extent to which both scientific and defense needs might be met by a single operational program, and (3) the agency of the government that should carry the development responsibility."[114]

Webb would have none of it. He admitted that NASA studies had revealed that the Apollo spacecraft plus the Saturn IB and Saturn V rockets would permit up to 100-day orbits without resupply or personnel transfer. He felt, however, that these studies had already been properly coordinated with the DOD in accordance with their space-station agreement of the previous fall and the AACB channels specified therein. Webb believed the AACB coordination bodies "represent sound and adequate measures to insure the most effective and economical action in the area of manned space flight." Therefore,

> it seems to me that we should not attempt rigidly to interpret or classify current programs in terms of possible undertakings in the future. . . . I view Gemini, Apollo, and the DOD MOL all as important contributors to the ultimate justification and definition of a national space station. All are forerunners and precursors in this sense. While it is inevitable that there will be some duplicative capacity for experimentation in these three projects, each has its essential role in the national space program. . . . I believe that the predominant mission and objectives of a national space station, if and when justified, will in turn indicate which agency of the government should be designated to carry the primary responsibility for development and management.[115]

Webb was arguing for the status quo. DOD would continue with the MOL, and NASA would continue with its low-level AAP studies. That is exactly what transpired. NASA ceded no managerial responsibility to the DOD.

Seamans stated in late October 1964 that NASA planned to initiate program definition studies of an *Apollo X* spacecraft in FY 65, but that a long-duration space-station program would not receive funding for actual hardware development until the 1970s.[116] Associate Administrator Mueller told Congress early in 1965 that "Apollo capabilities now under development will enable us to produce space hardware and fly it for future missions at a small fraction of the original development cost. This is the basic concept in the Apollo Extension System (AES) now under consideration. . . . This program would follow the basic Apollo manned lunar landing program and would represent an intermediate step between this important national goal and future manned space flight systems."[117]

On 6 August 1965 NASA established an official Saturn-Apollo Applications Program office at its headquarters within OMSF and under the direction of Maj Gen David Jones, USAF, one of the many senior-ranking officer managers on loan to NASA. By the end of the month, "Designers at MSFC [NASA's Marshall Space Flight Center in Huntsville, Alabama] began seriously to investigate the concept of a Saturn IVB-stage orbital workshop." On 10 September 1965, the AES was formally renamed the AAP.[118] The workshop concept involved the conversion of a spent Saturn IVB-stage to a shelter suitable for extended stay and use by humans; in these early years it was thought this conversion would take place in orbit.[119] One team of historians said NASA's early presentations of the AAP concept to Congress "found no enthusiasm for the program" because "the straightforward extension of Apollo's capability smacked too much of busywork—of 'boring holes in the sky.'"[120]

NASA pressed on and in November General Jones solicited the views of the chief executives of America's major aerospace companies on NASA's proposed goals for the AAP,[121] which at the time were simply described as using the Apollo and Saturn hardware for extended earth-orbital experimentation "to develop operational equipment and techniques; to obtain direct benefits to man; and to conduct further scientific exploration in space."[122] NASA's Mueller characterized such generic descriptions of the AAP as, "This is suicide. You just can't get anybody interest[ed] in it. It's everything to everyone, but nothing that really grabs anyone." To which Seamans replied, "That's right."[123] Nevertheless, in the fall of 1965 NASA's budget submissions for the first time included a separate line item for the AAP, only a few weeks after Johnson gave final approval to the MOL.[124] The figure projected for FY 66 was $13.8 million, with $83 million requested for FY 67 and $122 million for FY 68.[125] As 1965 ended, the AAP's status continued to be nebulous, as it was still in the conceptual study stage, akin to the MOL's evolution between December 1963 and August 1965.

The Execution of the MOL and the AAP

The final section of this study brings to a close the discussion of the MOL and the AAP programs for the decade of the 1960s by examining four issues. First, were the MOL and the AAP seen as duplicative by anyone, and, if so, did this influence the progress of either? Second, and closely related, were there any attempts by the DOD or NASA at commonality concerning the use of either Apollo-Saturn hardware for the MOL or the MOL hardware for the AAP, and if not, why not? The final two sections attempt to trace the respective histories of the AAP, through the summer of 1969 and the MOL through its cancellation during the same period, just a month before the lunar landing that July.

Were the MOL and the AAP Duplicative?

Senator Anderson, chairman of the Senate space committee, thought so. He wrote BOB director Kermit Gordon in November 1964, "Unless the MOL is changed to some degree, the Air Force will spend a billion dollars on it and then have no place to go. We think the NASA and the DOD programs can be prepared to save a substantial part of this money, and that either the basic MOL or Apollo can be oriented into the first generation space platform."[126] The BOB's resulting report concluded that two orbital systems were not required, noting, "Proceeding with the MOL does not now appear justified on the basis of the originally stated need for an experimental testing of the potential capabilities of manned space flight for high priority military purposes. The need for proceeding with the MOL is now very questionable in view of the diminished justification and the possibility of conducting experiments on most, if not all, of the problems of interest in due course with an extended Apollo system." If the MOL continued, "It should be 'nationalized' and oriented to serve as a test vehicle for technical and scientific experiments of both military and general interest. In this role, it should have full and tangible support of both NASA and the Department of Defense." In the BOB's assessment, another acceptable alternative would be, "Transfer the entire MOL program to NASA . . . with Defense to provide experiments of military interest in accordance with the present Gemini pattern."[127]

The BOB was clear that the AAP "should not be justified on the general grounds of continuing the utilization of Apollo-Saturn capabilities beyond those being procured for the MLLP." Instead, it "should be justified on a technical or other mission requirement basis in competition with other possibilities in the overall national space program and other demands on the Federal budget." If the MOL proceeded, then the AAP should not duplicate any of its experiments but make use of the "special capabilities" of the Apollo-Saturn system. If the MOL were canceled, the AAP program should be reoriented to provide "on a national basis the entire range of technological and scientific experiments of both military and general interest, on a schedule and plan that does not interfere with the MLLP."[128] The BOB's bottom line was clear: the MOL or the AAP should be "nationalized" so it could meet both NASA's and the DOD's needs because the United States did not need two separate programs.[129] By the end of the month, the trade press was declaring, "A merger of the Air Force's Manned Orbiting Laboratory with the civilian agency's counterpart Apollo-based project is in the works."[130]

A month after his letter to the BOB, Senator Anderson wrote President Johnson with the same suggestion: cancellation of the MOL would save the United States $1 billion over five years; in turn this money should be used to support NASA's AAP R&D.[131] As noted above, in December 1964 the MOL program was officially reoriented to an operational, reconnaissance-gathering platform. Possibly Anderson was briefed about this change and perhaps his campaign to merge the MOL with the AAP played a part in hastening the official designation of the MOL as an operational system and not an R&D test bed. Whatever the case, by the end of December, Anderson announced that the DOD and NASA had worked out an agreement that mollified his concerns. "The Department of Defense and NASA have gone a long way toward answering the questions I raised several weeks ago," he stated. "I have been told that the Air Force and NASA will take advantage of each other's technology and hardware development, with all efforts directed at achievement of a true space laboratory as an end goal."[132]

When asked about Anderson's desire to cancel the MOL, McNamara responded, "I think Senator Anderson was simply emphasizing the absolute essentiality of fully coordinating the NASA and the Defense Department programs. With that, I agree 100 percent."[133] The immediate controversy caused by the late 1964 Anderson letters and BOB analysis seemed to fade from that point.

The next major flashpoint in the MOL-AAP duplication discussion did not come until March 1966. The House Committee on Government Operations considered the MOL and the AAP and concluded,

> The greatest potential for cost savings in this program . . . would come from NASA participation in the MOL program. Both agencies have talked about the possibility of accommodating NASA experiments on a noninterference basis on the MOL, but to date little has been done to achieve this goal. Instead, NASA is proceeding with its plans for a similar near-earth manned space project which also will explore the effects on man of long duration space flights. . . . Despite the fact that Apollo Applications is not considered an approved program, there is the danger that both agencies soon will reach a point of no return where separate and largely duplicating programs cannot be avoided. Inasmuch as both programs are still research and development projects without definitive operational missions, there is reason to expect that with earnest efforts both agencies could get together on a joint program incorporating both unique and similar experiments of each agency. . . . Such a step would without question save billions of dollars.

This particular committee concluded the MOL could fulfill both NASA and DOD requirements: "A soundly conceived MOL with carefully devised experiments can serve both military and civil space requirements." NASA's merging of its earth-orbital requirements into MOL should be "effected within the existing scale of priorities which accords to the military experiments greater urgency."[134]

The BOB dutifully requested NASA prepare a study to see if the AAP could be designed around the MOL-Titan IIIM system. As Levine related, "Predictably, NASA concluded that it could not, since its current programs were adequately supported by existing vehicles."[135] General Phillips (still detailed to NASA as Apollo director) attempted to explain the difference between the MOL and the AAP and indirectly showed the difficulty of doing so because of his inability to touch upon the MOL's reconnaissance central mission, "The MOL objective is to develop manned orbital capabilities for accomplishing uniquely military tasks in narrowly constrained, low altitude earth orbits. . . . AAP, on the other hand, is planned to extend the technology and experience of the Gemini and Apollo programs by conducting experiments not only in a wide range of low earth orbits but also in earth-synchronous and lunar orbits and on the lunar surface."[136] Many policy makers were likely to ask if that distinction merited hundreds of millions of dollars and perhaps several billion in additional expenditures. Nevertheless, by the summer of 1966 Humphrey declared, "There al-

492

ready exists a high degree of cooperation between the Air Force and NASA in the MOL program, and I expect it will continue. I have no reason to predict an actual merger of the MOL research and development effort with any of NASA's manned projects."[137] When asked in the fall if there were any possibility that the MOL would be merged with any competitive NASA programs, Undersecretary of the Air Force Normal S. Paul replied, "Not the slightest possibility."[138]

The undersecretary's assessment seemed to be correct, though for unstated reasons. The MOL was conceived, designed, and its experiments focused on the particular challenges of humans gathering intelligence information from space. It was, therefore, part of the nation's Top Secret but high-priority overhead reconnaissance program. Given the likely delay that would result from merging it with the AAP or having to incorporate the AAP experiments and hardware into the MOL, there was little realistic chance either system would be merged with the other. While the MOL's reconnaissance mission may have given it a degree of "immunity" from merger into the AAP, the unfortunate flipside was that this mission could not be publicly discussed, nor could it even be revealed beyond a close circle of top-level national policy makers. This meant charges of duplication continued to be raised by various parties who were probably unaware of the national security imperatives behind the MOL's reconnaissance tasking.

In fact, Vice President and NASC chairman Hubert Humphrey asked Deputy SECDEF Paul Nitze in a November 1967 NASC meeting, "How do we explain to the public the difference between the MOL and the AAP? Nitze responded that the MOL is a military experiment and people understand fairly well that we do not talk publicly about military experiments." Another DOD official added that since the first MOL launch had slipped until 1971, "We have until then to decide how much and what to tell the public."[139] In other words, the US government should continue to tell the public virtually nothing about the MOL, as it had for most of its existence already. The price the DOD and the Air Force would pay for continued silence would be additional erosion of support for the MOL in

Congress and elsewhere because they could not or would not explain its central justification of reconnaissance.

NASA tried to urge groups such as PSAC, which had recommended closer MOL-AAP integration, to compare the actual designs of the MOL and the AAP in an attempt to objectively determine if one could do the mission of the other. NASA believed if the critics charging duplication did so, these charges would wane. NASA explained in January 1968 that the PSAC's recommendation that the AAP's objectives be merged into the MOL "does not appear to be a sensible approach. In the first place, we have given serious and repeated study to the use of MOL for NASA manned-earth-orbital missions as an alternative to the first set of AAP missions. In this context, the MOL fell far short of accommodating the minimum required experiments and goals that had been planned for AAP. In addition, the use of MOL for the AAP missions would be more expensive. To consider MOL as a follow-on to the first round of AAP just doesn't make sense."[140] Webb also tried to make this point without mentioning the reconnaissance-oriented nature of the MOL:

> The thing that the Manned Orbiting Laboratory is attempting to find out is whether a man will contribute more military information up there than an instrument, and do the job better across the board. If they find he won't, there won't be any more of that, I am sure. However, in contrast with the Apollo Applications Program, the MOL's going into different orbits. They have different missions and look down on different areas. The military mission for the Apollo Applications Program is about zero. For the MOL it is about 100 percent. I think you can note the absence of duplication for military purposes. One of the things the AAP does is look out into space for astronomy purposes, not looking down at earth all the time. . . . There are some 87 experiments on board the Apollo Applications Program, none of which duplicate the experiments that are in the MOL program.[141]

These were difficult distinctions to make but apparently NASA made them well enough to ensure that its AAP was not folded into the MOL. At times Webb faced incredulous congressmen. In February 1968 he appeared before the House space committee and was asked by Rep. William Fitts Ryan if there had been any "serious study of whether the two programs should be combined in an effort to avoid duplication." Webb replied, "There is no duplication that is not important in the development of the space capabilities of this Nation, between the Manned Orbiting Laboratory and the Apollo Applications Program. . . . There is no meaningful comparison between a Saturn V launched workshop and a Titan III launched Manned Orbiting Laboratory." When Ryan continued to maintain there was obvious duplication, Webb reiterated, "The fact a man may be orbiting in a military spacecraft and another man orbiting in a NASA spacecraft in my view is not duplication of a kind that should be considered unwise." Chairman Miller then interrupted, "I agree with you, and I don't think we will go into it any further."[142] DOD officials did not offer a great deal of comment or testimony on the question of the MOL versus the AAP. Perhaps they were confident of continued MOL survival and autonomy because of its intelligence-gathering raison d'être.

The administration's position remained clear. NASC executive secretary Welsh wrote the vice president in March 1968 after researching the MOL versus the AAP question in the Pentagon and NASA headquarters and explained to Humphrey, "I was assured in both instances that there is no program for merging the AAP and the MOL and there is no program for joint operation of a manned workshop."[143] NASC supporting material tried to highlight some of the differences between the MOL and the earth-orbital workshop portion of the AAP: the AAP would fly 87 experiments, but the MOL only a very few due to its military mission; the AAP was expected to have the ability to accommodate nine men to the MOL's two, have a 22-foot diameter to the MOL's 10-foot, have 10 times the MOL's internal volume, and sustain humans for 90 days to the MOL's 30; and the MOL's single aim was to advance specialized military missions "which almost completely absorb its capacity" while the AAP was aimed at the broad development of human spaceflight for the 1970s.[144] As Mueller told the Senate space committee on 19 April 1967, "The programs are not directly related."[145]

Nevertheless, the question remained open in the minds of some, the BOB in particular. Its October 1968 briefing on the nation's space program designed for the incoming administration stated, "A major policy problem concerns the future of earth orbital manned space flight in which DOD now has the Manned Orbiting Laboratory and NASA has the Apollo Applications Program. In [the] future, should we plan on two manned programs, a single program jointly run, or should a single agency be assigned responsibility for all manned space flight activities?"[146] Charges of duplication continued to flow from some congressional quarters. Rep. James Fulton stated in 1968 that the MOL and the AAP should be merged and placed under NASA's aegis, and it was reported he was prepared to reissue this call in 1969. Fulton believed that a single program should be created due to reasons of "prudent management and good judgment." He was characterized as "annoyed" by the secrecy surrounding the MOL, saying, "It's so super secret, members of the committee don't know what's going on." Welsh could only reply, "To combine these projects would be more expensive, not less; and less efficient, not more."[147] NASA's official response to Fulton was, "The NASA Apollo Applications Program and Air Force MOL Program have different objectives, require different orbits, and the equipment and supporting facilities of each are designed to meet the separate purposes of each program. We do not believe it is practical or prudent to merge the two programs. The end result would result in a compromise[d] spacecraft unable to satisfactorily meet the primary objectives of either program. . . . cost savings could not be achieved by merger or more extensive use of joint elements in the MOL and the AAP programs."[148]

Use Apollo-Saturn Hardware for the MOL or Vice Versa?

A closely related question asked why the DOD could not use NASA's preexisting Apollo-Saturn hardware for its MOL. Or, conversely, why could not NASA use the hardware that DOD was developing with the Gemini B-MOL-Titan III combination to conduct the experiments it wanted to do in the AAP? This question differs from the duplication question in that the duplication issue focused on whether or not one entire system should be merged with the other. The hardware question presumed both systems would continue to exist but asked Why could one system not make greater use of the other's equipment? The answer is that numerous studies, investigations, and queries were made into this question, but neither DOD nor NASA ever took any substantial action. Both agencies offered justification for proceeding with their separate programs with distinctive hardware configurations for each.

One NASA official involved with early studies on potential MOL-AAP hardware exchange reported in August 1964, "The Air Force is not interested in the Apollo for the MOL because they believe the program will slip and also they cannot count on the availability of hardware for their program."[149] Given the national priority accorded the lunar-landing program, and then the Apollo fire in January 1967, there was some legitimacy to these concerns. While the Air Force may not have desired to step back from its Gemini B-MOL-Titan IIIM configuration, there are hints of numerous studies throughout the mid- and late-1960s investigating the possibility of using Apollo-Saturn hardware for MOL objectives. For instance, in 1964 the Aerospace Corporation, a company that conducted various types of future studies as well as systems analysis for the USAF, conducted an MOL-AAP study that described four basic Apollo configurations, three of which would accomplish the MOL mission. It added, "However, it is doubtful that NASA would agree to the Air Force use of Apollo except on a strict noninterference basis." The potential for delaying the MOL program by redesigning it to use Apollo-Saturn hardware also concerned the Air Force because North American Aviation, the prime Apollo contractor estimated the earliest delivery date for an Apollo-based MOL would be 36 months after go-ahead, and AFSC's Space Systems Division "considers this to be optimistic since a new production line would have to be established and would not interfere with Apollo production."[150]

The Air Force regularly pledged that it would continue to "assess the Gemini B/Laboratory Module/Titan IIIC configuration and configurations of the Apollo system to determine which would satisfy the objectives in the more efficient, less costly, and more timely fashion."[151] But each time the status quo won out, the MOL continued to be defined as in its original design. The conclusions of numerous and continuing studies on using the MOL equipment for AAP's objective were much the same—the MOL could accomplish most of the AAP's

497

missions. But in response NASA would justify continuing to use the Apollo-Saturn equipment and not reverting to Gemini and a DOD-developed laboratory cylinder.

Even when the DOD and NASA got together within the AACB to study the entire launch vehicle fleet including every conceivable combination of not only Titans and Saturns but also the Thor, Delta, Atlas, Agena, and Centaur vehicles, no radical changes were recommended. After an extensive study in 1964, the AACB's Launch Vehicle Panel in December stated, "There is not a decisive difference between the total costs of the launch vehicle options considered in this study for either the maximum or minimum values of the mission model. . . . Cancellation at this time of entire Atlas, Titan or Saturn launch vehicle families . . . would not result in cost savings significant within the accuracy of the study. . . . The potential cost advantages to be obtained from substituting one booster for another, either entirely or in specific programs, may sometimes be illusory." In fact, the AACB study concluded, "The most striking result is the fact that so little difference in cost exists between options. . . . There is not a significant difference (less than 1%) between the total costs of the launch vehicle options considered."[152] Given that its primary coordination mechanism did not exert pressure to pare the launch vehicle fleet, it comes as little surprise that the Air Force had no desire to use the Saturn for the MOL, nor that NASA wanted to avoid incorporating the Titan III into the AAP.

Nevertheless, the studies continued. Webb and McNamara jointly pledged in January 1965, "DOD, with assistance from NASA, will compare configurations of Apollo which may be suitable for military experiments with the Gemini B-MOL configuration to determine the complete system that can meet the primary military objectives in a more efficient, less costly, or more timely fashion."[153] Each time, the Air Force would reply that it had investigated

> adapting Apollo to accomplish the MOL objectives [but] our preliminary studies have shown that the development of such a laboratory program would cost more than the Gemini B/MOL and that a laboratory would not be available any sooner than the Gemini B/MOL. Even if NASA were to build an Apollo laboratory and to agree to perform DOD experiments on a priority with theirs, the arrangement would not satisfy

all military objectives for a laboratory program. The military must develop and test its systems; we cannot gain operational experience in space by watching over the shoulder of the people who are planning, developing, directing, and conducting space programs.[154]

All in all, the Air Force concluded, "The Gemini/MOL configuration offers a more advantageous route to a space laboratory than does the modified Apollo"[155] because "this redevelopment of the Apollo lunar hardware would probably cost more, would not produce a laboratory faster, and the resultant laboratory would offer no functional advantages over the Gemini/Titan III/MOL."[156] Therefore, the idea of conducting a radical redesign of the MOL so its objectives could be met using Apollo-Saturn hardware never progressed beyond the study stage within the Air Force or OSD.

NASA's studies of using Gemini B-MOL hardware to meet the AAP's objectives followed much the same dynamic as did the USAF's studies of using Apollo-Saturn hardware to meet the MOL's objectives. NASA's basic conclusion was:

> We certainly support the fundamental concept that NASA should make the maximum effective use of all available technology in carrying out the U.S. objectives for manned space flight. However, the practical problems of NASA conducting manned space flights with two separate booster-spacecraft-ground support systems would require significant increases in NASA resources, particularly in manpower. NASA's goals will require the use of Apollo-Saturn systems, regardless of MOL availability, for missions in: a. Earth synchronous orbits; b. Low earth orbits; c. Lunar exploration. The MOL system will not have adequate performance for such missions.

NASA explained that the MOL could conduct only five of NASA's 12 planned earth-orbital missions and none of the lunar and planetary type of operations.[157] Therefore, proposals for using MOL equipment for the AAP missions made little headway within the NASA hierarchy. As Seamans explained to Albert Hall in the office of the DDR&E, "The Saturn-Apollo system designed for lunar exploration has inherent capabilities beyond those required for the MOL" and so NASA could not adapt the MOL for AAP use. Conversely, "These capabilities make an Apollo-based system more expensive than a Gemini-based system for the program now contemplated" and so DOD would probably not want to refashion Apollo-Saturn equipment for MOL use.[158]

One AAP history stated that NASA found "good reasons for not conducting its AAP program aboard the Air Force laboratory. The basic MOL configuration was inadequate to meet the AAP goals, while a DOD proposal for a larger MOL would take four years to develop and cost an additional $480 million in facility modifications. Even then, OMSF calculated, to achieve the same results, an uprated MOL program would cost more annually than the Saturn IB and Apollo." Costs to integrate the MOL and the AAP systems as of 1966 were estimated at $250 million and would require three and one-half years; therefore, 17 launches would be required just to pay back the conversion costs.[159] NASA's position throughout the late 1960s remained firm: "Introduction of either the Titan IIIM launch vehicle or the Titan IIIM/MOL systems into the post-Apollo manned space flight program is neither technically desirable nor cost effective. Such action could jeopardize the possible U.S. position in space by delaying for almost three years the low earth orbital application of proven U.S. space technology. Thus, continuation of the Saturn I-Apollo system for the AAP missions is in the best national interest."[160] NASA explained it had extensively studied the issue of using the MOL system for the AAP experiments, but its studies "have indicated that using the Titan III/MOL would cost over $500 million more during the next five years than using the Saturn I/Apollo combination." In addition, the Apollo-Saturn system would be available "several years earlier" than a modified MOL.[161]

NASA stuck by these conclusions through the remainder of the MOL's existence, despite studies from its prime contractor that the MOL could, with relatively minor modifications, be easily adapted to perform all of NASA's biomedical and behavioral assessments of humans in space for up to a year at a time and 85 percent of the engineering and scientific experiments. Douglas Aircraft's Missile and Space Systems Division stated, "It is concluded that use of MOL-derived hardware is conceptually feasible and cost effective in accomplishing early NASA objectives" and that "Existing MOL facilities and equipment for manufacturing and subassembly of the NASA space station could be made available without interference to the Air Force MOL program."[162] NASA was not swayed. It continued to maintain that switching from an Apollo-Saturn AAP to one employing Gemini-MOL-Titan III equipment would be much more expensive, would entail at least a three-year delay in the program, and would not be as capable as the original AAP design.[163]

As yet another in the seemingly interminable series of MOL-AAP studies was taking place in late 1968, the NASA and DOD representatives probably summarized the entire process best: "NASA and DOD have, over the years, made a number of studies relating to the use of APOLLO derivative hardware for MOL missions or MOL hardware for AAP missions. These studies have generally been nonconclusive or have reached negative conclusions for reasons of schedule or costs or unable to meet technical requirements." However, "Under today's conditions of drastically reduced funding for both programs and the resulting slipped schedules, it is desirable to reexamine the possible utilization of hardware from one or the other of these programs to meet the goals of both."[164] Given the fact that this very question had been studied since at least 1964 and that within 10 months the MOL would be canceled, this latest study effort had little impact. In fact, none of these studies led to any appreciable progress toward common NASA-DOD use of the AAP-MOL hardware before the MOL's June 1969 cancellation.

The AAP's Progress through 1969

The AAP's path from 1966 through 1969 was dictated by budgetary stringency. By the end of this book's time frame, it had progressed only to the point where the new NASA administrator had approved a final design. Construction had not begun and no launches would in fact take place until 1973. Therefore, the story of the AAP through 1969 is essentially one of financial struggle and clarifying design work. In February 1966 Seamans outlined to Congress that the basic thrust of the AAP was to extend earth-orbital stay times to 45 days or more through minor modifications of the present Apollo system but that "we cannot today look toward a permanent manned space station, or a lunar base, or projects for manned planetary exploration until our operational, scientific and technical experience with major manned systems already in hand has further matured."[165] Whereas NASA requested $250 million for the AAP in FY 67, it received only $50 million.[166] Given these limitations, Seamans met with NASA's program directors and outlined the three cardinal AAP tenets in March 1966. First, the lunar landing remained NASA's top priority and must not be compromised by any AAP activity. All changes to Apollo hardware for the AAP had to be approved by Webb or Seamans; so did any AAP procurement actions. Finally, any AAP experiment submitted had to have a "clear and defensible rationale."[167] This was not a recipe for a vibrant and flourishing program.

The first official NASA-AAP schedule was also released in March 1966 and was surprisingly ambitious. It envisioned 26 Saturn IB and 19 Saturn V AAP launches with the first launch scheduled for April 1968.[168] These would include three orbital workshops based on conversion of spent Saturn IVB stages, three Saturn V orbital labs and four Apollo Telescope Mounts (ATM), which was a human-tended astronomical observatory designed to study the sun and other celestial bodies. By January 1968 this formidable schedule had already been scaled back to three Saturn IB launches, three Saturn V launches, one Saturn IB orbital workshop, one Saturn V orbital laboratory, and one ATM, with the first launch scheduled for April 1970.[169]

During 1966 and until mid-1969, it remained undecided if these workshops would be "wet" or "dry." Wet workshops would be created from the spent stages of launched rockets, with which astronauts would rendezvous, dock, and outfit as an orbital laboratory. Dry workshops would be fully constructed and equipped on the ground, launched into space, and then receive the astronauts via a separate launch. This dry methodology ultimately characterized Skylab operations.

Whatever the workshop's configuration, one scholar stated, "All were subjected to very sharp criticism from NASA officials, from Congress, from the Bureau of the Budget, and from various scientific advisory groups." In June 1966, Newell pointed out "The lack of a substantial, visible end product to serve as a focus for the effort. After four or five years of activity, NASA will have spent many billions of dollars and have relatively little to show for it. . . . [AAP] as now configured just doesn't seem to justify such high costs for an extended period."[170] NASA did its best to succinctly define the AAP's goals, but its attempts paled when compared to the goal Kennedy tasked Apollo with in May 1961: "Before this decade is out, of landing a man on the moon and returning him safely." For instance, "The basic purposes of the Apollo Applications Program are to continue without hiatus an active and productive post-Apollo program of manned space flight, to exploit the capabilities of the Saturn Apollo system for useful purposes and to effect a progressive development of these capabilities as a stepping stone to whatever programs lie in the future."[171]

Some of the consultants Webb retained outside of the NASA framework tried to convince him of the inadequacy of such definitions, and indeed, the underlying AAP philosophy they reflected. Said one, "There is no valid requirement for 'applying' certain technologies just because they were developed by Apollo. Each space project has to justify itself on the basis of its merit—whether it is related to Apollo or not. . . . The immediate problem is to extricate NASA from the pitfall of trying to present an unconvincing concept."[172] The AAP's budgetary history suggests, however, that this never happened. As Lambright noted, "Under the impact of budget cuts, redesign became a way of life for the AAP."[173]

Webb entered the FY 69 budget cycle requesting that the BOB and President Johnson authorize $652 million for the AAP. The BOB approved a request of $454 million, but Congress initially appropriated only $253 million and subsequently reduced this to $150 million.[174] The result was the downsizing and schedule slippage described above. In February 1968, Webb declared concerning the AAP, "Our progress to date has been limited by the need to hold down expenditures in FY 1968 and those projected for FY 1969. . . . The amounts provided . . . will barely keep the program alive."[175] Accordingly, Webb implemented an AAP holding plan for the remainder of FY 68 "in order to maintain a reasonable balance in program content while avoiding major cuts to work in progress. This action became necessary because of funding restraints imposed on AAP."[176] When finally calculated in terms of direct NASA obligations, AAP's funding history was as follows: 1966, $13.8 million; 1967, $83.4 million; 1968, $122.2 million; 1969, $150 million.[177]

The final AAP development relevant to this chapter's time frame was NASA administrator Paine's decision on 18 July 1969 to officially approve the shift from a wet to a dry orbital-workshop concept for the AAP. The AAP would feature one space laboratory constructed on the ground, equipped with one ATM, and launched by one Saturn V. Three-person crews would subsequently be orbited by a Saturn IB and dock with the workshop for the rest of their tour and then return in the Apollo capsule upon completion. This meant the program consisted of four launches: one Saturn V to launch the workshop and ATM and three Saturn IBs to launch the three separate sets of three astronauts each that would inhabit it. The first launch had now slipped to July 1972.[178] The AAP was officially renamed Skylab on 17 February 1970.[179] It was not, in one NASA historian's opinion, "the versatile and long-lasting station that NASA had planned since the 1950s. Designed to satisfy the institutional need to do something after Apollo and to keep the NASA team together long enough to finish the lunar-landing missions, Skylab was makeshift and temporary. NASA's space-station engineers, in fact, deliberately built the station without the thrusters necessary to keep it in orbit for any significant amount of time" because "they hoped to ensure the construction of a more permanent and sophisticated station."[180]

It is impossible to know with confidence if the reason Skylab survived was the MOL's cancellation. Whatever the case, the $2.6 billion Skylab program was the closest the United States would get to a space station until the turn of the century.[181] The 100-ton workshop was launched into orbit on 14 May 1973 and marked the last time the giant Saturn V was used. The first crew of three astronauts joined it on 25 May and, after repairing some damage caused to the laboratory during its launch, stayed in orbit for 392 hours before returning on 22 June 1973.[182] The second set of astronauts occupied Skylab for 59 days starting 28 July 1973 and the third for 84 days, starting 16 November 1973.[183] In total the crews occupied Skylab for

171 days and 13 hours while conducting almost 300 scientific and technical experiments.[184] Probably the most media coverage the workshop received was when it reentered the earth's atmosphere. "In 1979, Skylab did fall to earth and made more news as a burning hunk of metal than it ever did as an operating space laboratory."[185] As will be seen below, the MOL never received even that transitory amount of publicity.

The MOL from Presidential Approval to Presidential Termination

Less than a month after Johnson officially approved MOL construction in August 1965, the press was already speculating about the competition it faced, not from NASA, but from NRO's reconnaissance satellites. *Newsweek* said the Air Force was concurrently developing a 10-ton unmanned reconnaissance satellite "stuffed with cameras, sensors and detectors, and possibly capable of maneuvering in orbit. . . . Such a surveillance system could conceivably give MOL stiff competition."[186] It would take almost four years, but shortly after Nixon assumed office, a combination of three factors led to the MOL's demise: perceived duplication with NASA's AAP earth-orbital programs; perceived duplication with NRO's reconnaissance satellites (both described above); and continued government-wide financial pressures resulting from persistent Vietnam War and social welfare expenditures. The MOL program would expire before its first launch.

Starting in 1966, however, a "massive expansion" of the MOL program began. For instance, the Air Force acquired a 15,000-acre ranch adjacent to Vandenberg AFB deemed necessary to ensure the safety and security of the burgeoning MOL facilities. The AFSC's Space Systems Division's deputy commander for Manned Systems, Brig Gen Joseph S. Bleymaier, predicted that within five years Vandenberg would have continuous manned operations involving 40 or more launches annually.[187] However, there were signs that perhaps the OSD would not permit development of the MOL and its facilities as quickly as the Air Force desired. McNamara permitted no increase in the MOL's FY 67 budget above the $150 million it had been allowed in FY 66.[188] *Missiles and Rockets* magazine said this amount was "far below what early DOD estimates had called for in FY '67. . . . It is widely held by Pentagon observers that program stretchout is indeed taking place" despite denials.[189] McNamara's only comment was, "Manned Orbiting Laboratory development should proceed on a deliberate and orderly schedule."[190]

Some within the defense community, albeit at a low level, even questioned if the MOL's mission as a reconnaissance platform merited the projected level of expenditures. An advisor to AFSC commander, General Schriever, forwarded a 50-page paper on the MOL. Duncan Macdonald's basic conclusion was, "There is no valid role for man in the acquisition loop of a high resolution operation. The present combination of high resolution operation and the MOL format compromises both reconnaissance and the MOL and, therefore, the reconnaissance program should be placed under NRO and USAF should restudy its MOL concepts." Macdonald elaborated,

> It is my opinion that the current program is not directed toward exploring a sufficiently broad range of military missions, but instead, is concentrating too narrowly on an evaluation and test of the reconnaissance mission. It should be clear that this selected mission as now constituted, provides, at best, a marginal role for man and certainly not a continuing role. As a consequence, U.S.A.F. may well be denying itself the opportunity to explore and establish timely programs for longer range and continuing roles and missions in space. . . . I urge prompt and frank recognition of the fact that the present program is not an MOL program, that it logically belongs as an (unmanned) NRO program and that the U.S.A.F. should redo its MOL concepts.[191]

There is no evidence that the OSD or the USAF ever seriously reconsidered refashioning the MOL into anything other than a reconnaissance platform. However, the Macdonald report does illustrate the fact that, whether intended or not, the MOL was being evaluated in the context of the capabilities and innovations of the NRO's reconnaissance satellites.

The only launch even indirectly associated with the MOL took place on 3 November 1966 when a Titan IIIC lifted off from Cape Canaveral with a modified, but unmanned, Gemini capsule. Its heat shield had been reconfigured to include the hatch the MOL astronauts would use to pass between the capsule and the laboratory. The goal was to determine if the heat shield's integrity would remain intact and protect the capsule. The capsule endured a 33-minute suborbital flight and then plunged into the atmosphere at 17,500 mph, generating temperatures approaching 3,000 degrees Fahrenheit before being recovered 5,500 miles downrange. The heat shield's integrity was proven, and the boilerplate capsule was undamaged.[192] For FY 68 the MOL's budget started to rise appreciably from FY 67's $150 million because the MOL began to enter its peak period of facilities construction and its initial phase of hardware procurement. McNamara asked Congress in January 1967 for $431 million (an amount which Congress later approved) for the MOL for FY 68, as part of the DOD's overall $1.99-billion space program. He admitted, however, that the MOL's first manned flight date had slipped to late 1969 from original projections of late 1968.[193]

Later that spring, the increasing weight of the MOL meant the Air Force had to redesign the Titan IIIM with either a larger central-core liquid-fueled engine or with larger solid-fuel strap-on rockets. Either option further delayed the MOL's first manned flight until 1970 at the earliest and increased overall program costs from $1.5 to $2.2 billion.[194] Nevertheless, the government's official space report in January 1968 declared, "Development of all the major components of the MOL system was initiated and progressed on schedule during the past year" to include the first stage of the Titan IIIM. Mock-up and structural assemblies of the laboratory and experiment modules were completed, procurement of the actual system components began, construction was initiated on the Vandenberg launch complex, and 16 MOL astronauts were in training.[195]

The MOL's budget for FY 69 was $515 million out of DOD's overall $2.22-billion space budget, and the MOL's projected FY 70 spending was $578 million.[196] The government's report of space activities in January 1969 stated the MOL was "approaching a point of peak activity" as structural test assemblies of major system components were fabricated; subsystem components were being manufactured; demonstration firings of the Titan IIIM's first stage commenced; construction of the MOL's launch complex at Vandenberg neared completion; and the training of astronauts continued.[197]

However, as costs climbed and its initial operational date slipped, doubt seemed to creep into McNamara's thinking about the MOL. At the end of 1967, he wrote SAF Brown, "I am concerned at the amount by which the cost of MOL (approximately $2.7 billion) exceeds the original estimate, on the basis of which the President approved the program ($1.5 billion). Even at $1.5 billion, some consider the program marginal. Should we not reexamine the role of the man and develop a plan for completing MOL, at least in the first phase, without a man? I believe that such a program could be financed in FY 69 at $400 million."[198] Lew Allen was an Air Force officer with a PhD in physics who later became the service's top ranking general, its chief of staff. He indirectly confirmed McNamara's interest in exploring the conversion of the MOL into an unmanned platform because he, Allen, was the one responsible for conducting the studies of this concept. Allen recalled,

> I was assigned the task of developing the technologies to operate the Manned Orbiting Laboratory unmanned, which was a very strange contradiction in terms and really a fascinating perversion of the whole intent. By this time, one had gone full circle; that is, one had decided to have a Manned Orbiting Laboratory without knowing the purpose of it. One had then decided that since you are going to have a Manned Orbiting Laboratory, the only thing to do in it was a particular sensor approach [reconnaissance] which was otherwise going to be done unmanned. Then having decided to do that manned and doing all of the designs for it to be manned, then one came back and said, "Well, could you automate the things you had already decided for the man to do?" We went through the technology studies of that and concluded that you could do the functions which had now been attributed to the man unmanned. . . . I think the advisory committees ended up perverting the whole process so much that cancellation was inevitable.[199]

The unstated but nevertheless intense competition between the MOL and the NRO's reconnaissance satellites was therefore clearly present as McNamara departed the DOD in April 1968 to become president of the World Bank.

Air Force Magazine in March 1968 reported in a general sense on the growing capabilities of American reconnaissance satellites. It explained that real-time reconnaissance (that which transmits images directly to ground stations electronically without the delay required to recover and develop film dropped from orbit) was now possible due to multispectral sensors, microelectronics resulting in massive but lightweight computer processing capacity, and high-volume communications enabled by efficient optical lasers. Thus, "There are strong reasons for believing that orbital cameras now have sufficient resolution to show objects the size of garbage-can tops. Progress in improving resolution has been steady and most of the experts believe it will continue." In addition, the next generation of cameras and film "should be able to photograph objects less than one foot in diameter from an altitude of 150 miles." Up to nine cameras could be put on a single, unmanned reconnaissance satellite, each with film sensitive to a separate wavelength region of the light spectrum. Similar technological progress in creating images through radar was also reported.[200] As Seamans related, "As time went on, after McNamara left, the interest in the project [MOL] became less, and partly because of dollars, but partly because other technologies had moved ahead more rapidly than expected" in robotic-reconnaissance techniques and systems.[201]

Given these detailed observations resulting from the NRO's robotic-reconnaissance satellites and McNamara's instructions to study the redesign of the MOL into an unmanned configuration, it seems likely that he was beginning to wonder at the end of his tenure if perhaps any additional capabilities the MOL could offer would ever merit total program expenditures of over $2 billion. Such thoughts indisputably arose upon the advent of the Nixon administration. Nixon's first order to heads of all executive branch departments and agencies revealed the stringent financial environment he initially established. "As we set the course of the new administration, a careful and thorough review of the budget must be the first order of business. The American people have a right to expect that their tax dollars will be properly and prudently used. They also have a right to expect that fiscal policy will help to restrain the present excessive rate of price inflation in our economy." Therefore, each agency would review the outgoing administration's budget to "identify activities of low priority which can be reduced or phased down and perhaps, over time, eliminated completely."[202]

The submissions received from executive branch agencies did not please the new president. He wrote his BOB director Robert P. Mayo, "I expected that review to result in a sizable reduction in the total Federal spending budgeted by the outgoing administration for the fiscal year 1970." However, "The report you have given me based on the responses of the department and agency heads is very disappointing. . . . Several billions of dollars <u>more</u> [emphasis in original] must be saved. The inflationary environment in which we find ourselves, our continuing commitment in Southeast Asia" and other factors "all demand decisive and substantial action to reduce the size of the budget and to keep Federal spending under strict control." He ordered Mayo to develop and recommend a revised 1970 budget "which will be <u>significantly below</u> [emphasis in original] the $195.3 billion forecast in the Johnson budget. In some cases our Administration will have to propose and fight strongly for legislation and appropriation reductions that will be unpopular in many quarters."[203]

Apparently new SECDEF Melvin Laird was able to preserve the MOL through the initial $1 billion cut from the DOD's $79-billion budget by reducing total MOL launches from seven to six, which meant manned launches decreased from five to four.[204] In April 1969 the trade press reported that in addition to reducing the MOL's scheduled launches by one, the MOL's FY 70 budget had been cut by $51 million from Johnson's original request and now stood at $525 million.[205] Seamans had moved from deputy administrator at NASA under Johnson to SAF under Nixon. He recalled that as the idea for canceling the MOL outright gained momentum during the early Nixon administration, he fought hard to preserve the system. When Seamans initially asked Laird and deputy SECDEF David Packard if the MOL was to be terminated, they replied, "Well, not really. There's just a little discussion going on." But he kept hearing such rumors and finally went to Laird and said he'd like to "have a day in court with the President before the cancellation takes place so that maybe I can convince him not to cancel it." He was in fact granted 30–45 minutes with Nixon, but the MOL was still canceled. "It really came down to the fact when you're putting in substantial sums for F-15's [the Air Force's next generation fighter] and B-1's [their next generation bomber] and satellite warning systems and so forth, it's pretty hard to justify more <u>speculative</u> and <u>large</u> developments" (emphasis in oral history transcript).[206]

Seamans elaborated to this author that it was Mayo's assistant budget director James Schlesinger (who later became SECDEF) who approached him "right out of the blue" and said, "You guys don't need that MOL, why don't we take that out of your budget? It's just an enigma. It's from the past. Why don't we clean things up and get rid of it?" At this point Seamans approached Laird and asked for the appointment with Nixon. It was granted and also in attendance—along with Mayo, Laird, Seamans, and an Air Force general—was National Security Advisor Henry Kissinger. Seamans recalled he made the case that higher and higher resolution was important to the DOD and that the MOL would provide it. He described to Nixon how the MOL would offer real-time intelligence and what a man in the loop could contribute to this process. His briefing took approximately half an hour, and Seamans said the president was "obviously interested" and took a lot of notes. Kissinger later told Seamans it was a good presentation.[207] Nevertheless, within a week, on 10 June 1969, Deputy SECDEF Packard announced the MOL's cancellation.

Seamans' overall conclusion was that, while the financial issue did play a part in the program's demise, the real crux of the matter was that the NRO had "reason to believe they were going to be able to send back, from satellites, really clear, real-time photographs. They were able to use the very cameras Eastman-Kodak was developing for the MOL" and simply placed them on the NRO's unmanned robotic satellites. Therefore, "With that capability coming along, I have to say, looking at it in 20/20 hindsight, the decision was correct. Technology had superseded it [MOL]."[208] Schriever's evaluation of the MOL's demise was, "I know that the NRO people were shooting at the MOL, saying we can do it without the man in the loop. So that had something to do with it. But I think the primary reason was the budget. And I think it's really marginal in terms of what the man in the loop would have provided to intelligence."[209]

Packard's official announcement on 10 June 1969 said the MOL was being canceled "because of the continuing need to reduce Federal defense spending and the advances made in automated techniques for unmanned satellite systems."[210] The DOD press release stated that since both Houses of Congress "are searching for ways of reducing expenditures. . . . The MOL cancellation will be a major step in reducing the budget." The DOD estimated the MOL's cancellation would ultimately save $1.5 billion. The press release stated it "was necessary to cut back drastically on numerous small programs or to terminate one of the larger, most costly R&D undertakings. We have concluded that the potential value of possible future applications of the MOL were not as valuable as the aggregate of other DOD programs that would need to be curtailed to achieve equal reductions." The DOD stated the MOL was unlikely to be ready before mid-1972, and it was these delays that "were largely responsible for the increase in estimated total cost from approximately two to three billion dollars, of which about $1.3 billion has been spent to date."[211] As Richelson explained, "Military programs not related to the [Vietnam] war effort were reduced or simply canceled to provide more money for the war. The MOL, the largest nonwar item in the Air Force research and development budget, made an inviting target. . . . Vietnam proved to be a budgetary black hole, absorbing funds without anything coming back."[212]

An Air Force document also emphasized the two primary reasons of cost reduction and the capability of unmanned satellites to do the MOL's job as the main factors in the MOL's cancellation. "First, it was determined that most essential DOD space missions could be accomplished with lower cost unmanned spacecraft. Second, the potential worth of possible future applications of the experimental equipment being developed for the MOL, plus the information expected from flights on man's utility in space for military purposes, while worthwhile, did not equate in immediate value to other DOD programs."[213] As the *New York Times* made clear, "Not mentioned by the Pentagon were the rapid strides that have been made in using satellites for detailed photographic reconnaissance."[214] Another secondary source later explained that by the time of the MOL's termination, "The design of a fourth generation of unmanned reconnaissance satellite was far enough along to indicate that it probably could perform most of the functions planned for the MOL and do so at lower total cost. While human judgment was extremely useful in space reconnaissance, the cost in terms of spacecraft payload to maintain human astronauts in the hostile environment of space resulted in a questionable trade-off."[215] In other words, while the MOL did offer reconnaissance capabilities beyond what the KH-9/Big Bird could, policy makers decided the price was too high. Even the trade press admitted that the MOL had been "so stretched out by funding cuts and low keyed management that its technology has become obsolete and its costs astronomical."[216]

The third factor, beyond cost control and redundancy with reconnaissance satellites, which this author has stated had some relevance in the MOL's cancellation, was the idea of perceived redundancy with NASA programs, in particular the AAP. This did not seem to be a significant factor within the calculations of the Nixon administration, but one can point to it as a reason why there was little protest from Congress after the MOL's cancellation. Perhaps enough members were persuaded that the two stated reasons for the MOL's cancellation—when combined with the idea many already had that the MOL was duplicative with the AAP—created a strong enough case that most representatives and senators accepted the MOL's demise with few second thoughts.

It is unlikely that many within NASA mourned the MOL's demise. One who did was Michael Yarymovych, on loan from NASA to the Air Force to serve as MOL technical director. He recalled, "When MOL was canceled there was cheering in the aisles of the AAP people. I was not cheering."[217] Whatever were the particulars within Congress and NASA, the fact remained that "the cancellation ended the Air Force's hopes for manned spaceflight and brought to a close a decade of political competition."[218] The MOL's death "served as another painful lesson to the Air Force and the military that their preferred military space doctrines and programs would not come to fruition."[219] The MOL's passing "signaled the death knell of Air Force efforts to make manned-military spaceflight the center of a space-oriented service. . . . The utility of military man-in-space activities remained untested."[220] All in all, "The history of the Air Force efforts to get a man in space is a spectacularly frustrating one. For various reasons, the Air Force was long denied the opportunity to try. When permission was finally begrudgingly given, delays and quibbles and fund-withholdings still were encountered" before, ultimately, the Air Force's future human presence in space was completely indeterminate.[221]

By the summer of 1969, a number of events had played themselves out and formed a logical stopping point for this work. Americans had reached the moon and returned safely. The USAF's MOL was terminated and with it the last attempt the Air Force would make for its own independent human-spaceflight program. The AAP's final configuration had finally been decided upon; it would be a dry workshop with an ATM launched on a Saturn V, and three subsequent crews of three astronauts would visit it, launched on Saturn IBs. As discussed in previous chapters, détente seemed to be growing, and the SALT process would soon commence. Therefore, there seemed little hope for extending the idea any further of competing in space for prestige.

Two main space policy tasks remained. First, the remainder of the Apollo lunar landings took place. Second, Nixon had to decide what his administration's space policy would be. It would not be until January 1972 that Nixon decided on the main thrust of America's next-generation space endeavors: the space shuttle. A distinct and fascinating NASA-DOD relationship during the shuttle era developed in turn. A full account of the remaining Apollo flights can and has been written.[222] The history of the NASA-DOD relationship from 1970 on has not been and stands as one of the many intriguing research tasks remaining to be accomplished in the political and organizational history of the space age.

Notes

1. Flax to Harold Brown, memorandum, 18 January 1964, 3.
2. NASA, News Release no. 64-78, 6–9.

3. "Gemini NASA/DOD Piggy-Back Experiments," *Space Daily*, 17 April 1964, 102. This list of experiments is also contained in Gregory, "DOD, NASA Agree on Gemini Experiments," 42.

4. Adair, *History of the Assistant for Manned Orbiting Laboratory*, 5.

5. Flax to Zuckert, memorandum, 17 August 1964, 1.

6. Air Force, Unclassified Supporting Witness Statement, 3–4.

7. NASA, *Gemini Program Mission Support: Gemini V*, 8-2, 8-5, 8-7, 8-10, 8-41.

8. Houchin, "Interagency Rivalry?" 38.

9. See Harold M. Schmeck, "Man's Role in Space Underlined as Astronauts Sight Missile Shot," *New York Times*, 25 August 1965, 1; and William Hines, "Gemini Flight Slated to Carry Sky-Spy Gear," *Washington Evening Star*, 15 August 1965.

10. Humphrey to Webb, memorandum, 21 August 1965, 1.

11. The final tally of DOD experiments scheduled for Gemini was 15; overall, 52 experiments were conducted on Gemini missions. The experiments, and their official results, were basic-object photography, successful; nearby-object photography, not performed; mass determination, once not performed, once successful; celestial radiometry, twice successful; star-occultation navigation, once not performed, once successful; surface photography, successful; space object radiometry, twice successful; radiation in spacecraft, twice successful; simple navigation, once not performed, once successful; ion-sensing attitude control, twice successful; astronaut-maneuvering unit, experiment only partially performed; astronaut visibility, once not performed, once successful; UHF-VHF polarization, once not performed, once only partially performed; night-image intensification, once not performed, once successful; and power-tool evaluation, not performed. In addition, NASA Gemini experiments included synoptic-terrain photography, successfully performed seven times; synoptic-weather photography, successfully performed seven times; and visual acuity, successfully performed twice. Ezell, *NASA Historical Data Book*, vol. 2, 169–70.

12. Air Force, *Preliminary Evaluation—Program 631A*, 13 September 1966, 14, 17, 20, 23, app. X.

13. Air Force, *History of the Directorate of Space, January–June 1967*, July 1967, 10.

14. Flax to the DDR&E, memorandum, 29 March 1965, 1.

15. Cantwell, *The Air Force in Space*, pt. 1, 22.

16. Evert Clark, "Satellite Spying Cited by Johnson," *New York Times*, 17 March 1967, 13. Also in the *Washington Post*, 18 March 1967. Official, on-the-record acknowledgment of US reconnaissance satellites did not come until Pres. James E. Carter declared at the Kennedy Space Center on 1 October 1978, "Photographic reconnaissance satellites have become an important stabilizing factor in world affairs in the monitoring of arms control agreements. They make an immense contribution to the security of all nations. We shall continue to develop them." Carter, *Public Papers of the Presidents, 1978*, vol. 2, 1686.

17. DOD Directive 5105.23, 373–75.

18. Klass, "Military Satellites Gain Vital Data," 55. *Real time* is a term meant to describe a process whereby the reconnaissance images are transmitted to ground stations virtually simultaneously, or with very minimal delay, and are shortly thereafter made available to national policy makers. In 1970 the traditional method of data return continued in operation—dropping the film inside canisters back to the surface of the earth to be recovered, processed, and manually delivered to decision makers.

19. Klass, "Big Bird Nears Full Operational Status," 17.

20. Richelson, "The Keyhole Satellite Program," 135. Probably the first author to postulate that the KH-9/Big Bird was begun as a backup to the MOL was Curtis Peebles. Peebles, "The Guardians," 381.

21. Richelson, *America's Secret Eyes in Space*, 106.

22. Burrows, "A Study of Space Reconnaissance," 227.

23. Butz, "Under the Spaceborne Eye: No Place to Hide," 93.

24. Haines, "The National Reconnaissance Office (NRO)," 154–55.

25. Pealer, "Manned Orbiting Laboratory, Part II," 28. See also "MOL Cancelled," *Space Daily*, 11 June 1969, 179.

26. Pealer, "Manned Orbiting Laboratory, Part III," 16.

27. Payne, "After Apollo Blasts Off, What Next in Space?" 11.

28. DOD, *Space Program Data Sheet on Manned Orbiting Laboratory*, 1.

29. Spires, *Beyond Horizons*, 130. Another source said the lens on MOL's telescopic camera was 90 inches wide. Stares, *The Militarization of Space*, 98.

30. Day, "Invitation to Struggle," 262. It was also stated that a six-foot-wide lens would make possible four-inch resolution. Burrows, *Deep Black*, 228.

31. Anderson, memorandum for record, 20 August 1964, 1.

32. Davis, memorandum for record, 8 October 1964, 1.

33. Schriever's description of MOL in an address to the National Space Club, 20 May 1964, cited in Lillian Levy, *Space: Its Impact on Man and Society* (New York: W.W. Norton & Co., 1965), 197.

34. Air Force to Headquarters, memorandum, 28 May 1964, 1.

35. DOD, *Space Program Data Sheet on Manned Orbiting Laboratory*, 23 July 1964, 1.

36. Anderson, memorandum for record, 20 August 1964, 1–2.

37. Flax to George Mueller, letter, 28 August 1964, 1–3. Geodesy involves the precise measurement and survey of the surface of the earth and the earth's magnetic fields from space. Among its military uses, geodetic information is absolutely vital to the accurate targeting of ballistic missiles.

38. Schultz, letter of nonconcurrence, 29 September 1964, 1.

39. Konecci to Welsh, memorandum, 17 June 1965, 1.

40. Gordon, memorandum for record, 10 December 1964, 1. NASA and the DOD confirmed this decision in a generically worded press release of 25 January 1965. NASA, News Release, 25 January 1965, 1–3.

41. Air Force, *Primary Experiments Data for the Manned Orbiting Laboratory System (MOL) Program*, March 1965, v–vi. P-9 was listed as "Deleted." The MOL program also had an extensive list of secondary or S-series experiments built into the program.

42. For instance P-1 "will require the capability to obtain very high-resolution photographs for technical intelligence. The high resolution photographs can be obtained if a sufficiently large optical system is provided and if precise image motion compensation can be accomplished." Air Force, *Primary Experiments Data for the Manned Orbiting Laboratory System (MOL) Program*, March 1965, 21. Or, under P-10, it proposed a low-light-level television that could provide "a capability for viewing targets in the visual spectrum at illumination levels ranging from dusk to quarter moonlight conditions" (ibid., 307).

43. One MOL analyst stated, "In order to bolster its case for a separate MOL program based on Gemini-Titan III hardware, the Air Force proposed two additional primary or P-experiments: P-14, essentially the assembly of large structures (radar antenna) and P-15, a large optical system for military space use." Pealer, "Manned Orbiting Laboratory, Part I," 13.

44. Ibid.

45. Vance to the president, memorandum, 12 February 1966, 1.

46. Senate, *Manned Orbiting Laboratory*, hearings, 24 February 1966, 35–36. In another hearing, NASA's Seamans explained that the MOL's polar orbit was "a requirement inasmuch as they [DOD] want to have world coverage, and they only get world coverage by going to a polar orbit and it is very difficult to go in a polar orbit from Cape Kennedy because of the overflight problems that ensue." House, *Independent Offices Appropriations for 1967*, hearings, pt. 2, 1266.

47. Pealer, "Manned Orbiting Laboratory, Part II," 34.

48. Pealer, "Manned Orbiting Laboratory, Part III," 20.

49. Prados, *The Soviet Estimate*, 174.

50. Foster, Statement to the Senate Committee on Aeronautical and Space Sciences, 26 March 1968, I-6.

51. Richelson, *America's Secret Eyes in Space*, 85.

52. Air Force, Memorandum to Headquarters, 28 May 1964, attach. 2, 1.

53. Gruen, *The Port Unknown*, 269.

54. Hall (speech, The Objectives of the Military Space Program, 5 February 1964, 7).

55. McMillan to Colonel Schultz, memorandum, 2 July 1964, 1, 3.

56. Cited in Schultz, Congressional Preparation Instruction Letter no. 1, September 1964, 2.

57. Cited in "Secretary McNamara Seeks NASA-USAF Cooperation on Manned Orbiting Laboratory," 21.

58. Before the term *Apollo Applications Program* (AAP) began to be universally employed, one could see the general concept of continuing to use Apollo-Saturn hardware for purposes of earth-orbital R&D referred to as Apollo X, as Apollo Extended System (AES), and as Extended Apollo System (EAS). For purposes of clarity, this chapter will employ AAP consistently.

59. Cited in Wilson, "President Will Outline U.S. Space Goals," 12.

60. NASC, Summary Minutes, 9 July 1965, 2–4. Document is now declassified.

61. Some within Congress were clearly eager for the OSD to finally make a decision on the MOL. One report concluded that the "Air Force should be commissioned, without further delay, to execute a full-scale MOL project." This would of course be done "without prejudice to NASA's future requirements for manned space stations." House, *Government Operations in Space*, H. Rep. 445, 4 June 1965, 11, 17.

62. White House, Statement of the President, 25 August 1965, 1.

63. Johnson, *Public Papers of the Presidents*, 25 August 1965, 917.

64. DOD, Background Briefing on the Manned Orbiting Laboratory, 25 August 1965, 11.

65. Howard Simons and Chalmers Roberts, "Role in Arms Control Clinched MOL Victory," *Washington Post*, 5 September 1965, 1. Similarly, an article in *Aviation Week and Space Technology* concluded the MOL "is now conceived primarily as a reconnaissance/surveillance payload." It would use both electronic and photographic sensors to relay data to ground stations by digital-data transmission. The pictures would be of adequate resolution so that ejection of photographic capsules was not necessary. Other sensors such as low-light-level television, zoom lenses, high-resolution radar, and a variety of electronic-ferret devices would also be available. Fink, "CIA Control Bid Slowed Decision on MOL," 21.

66. Kennan and Harvey, *Mission to the Moon*, 200.

67. Schwartz, "Manned Orbiting Laboratory—For War or Peace?" 56.

68. Seamans to Brown, letter, 6 January 1964, 1.

69. Mueller to the NASA associate administrator, memorandum, 15 January 1964, 1.

70. More specifically, Webb stated, "The DOD MOL program will be accomplished using many component systems and operational techniques which have been developed and proven by NASA. Necessary supporting facilities established by NASA will be made available and fully utilized. . . . At the same time, NASA will take full advantage of the opportunities presented by the MOL to further its research and development effort." Webb, Statement before the Senate Committee on Aeronautical and Space Sciences, 4 March 1964, 34–35.

71. Hilburn to associate administrator, memorandum, 1 July 1965, 1.

72. NASC, Summary Minutes, 9 July 1965, 4.

73. NASA, Agreement with the Air Force, 28 January 1965, 1.

74. NASA, "Answers to Vice President Humprey's 21 Questions," 28–29. Document is now declassified.

75. Seamans to Brown, letter, 27 September 1965, 1.

76. Johnson, *US Aeronautics and Space Activities*, 1966, 71.

77. For full details see Webb and McNamara, memorandum of understanding, 14 January 1966.

78. Seamans and Foster, memorandum of agreement, 21 March 1966, 1–2.

79. Flax, *DOD Authorizations for Fiscal Year 1967*, hearings, 528.

80. Seamans, oral history interview by the author, 5 July 1996.

81. Kennan and Harvey, *Mission to the Moon*, 204.

82. Mueller to Boone, memorandum, 3 May 1966, 1, 4.

83. Senate, *Independent Offices Appropriations, Fiscal Year 1969*, hearings, 1091.

84. House, *Government Operations in Space*, 89.

85. Lambright, "James Webb and the Uses of Administrative Power," 190. In another text Lambright added that when the Air Force was granted the MOL, "NASA was injured by the decision, and its post-Apollo planning with respect to possible laboratories in space was constrained. If it was to include a[n] MOL-type concept in its own plans, it would have to do so in spite of the Air Force program." Lambright, *Presidential Management of Science and Technology*, 68.

86. Levine, *Managing NASA in the Apollo Era*, 234.

87. Ibid.

88. Harvey to Seamans, memorandum, 2 July 1965, 1–3. Document is now declassified.

89. Golovin to Hornig, memorandum, 5 July 1967, 3. Document is now declassified.

90. Seamans to Webb, memorandum, 7 July 1965, 2–3.

91. Webb to Johnson, letter, 20 May 1964, 28.

92. Sheldon to Welsh, memorandum, 24 November 1964, 1, 3. Document is now declassified.

93. NASA, *Summary Report: Future Programs Task Group*, January 1965, 4–6.

94. Launius, *NASA*, 189.

95. NASC, Summary Minutes, 15 June 1966, 5.

96. Mueller, oral history interview, 8 November 1988, 182–83.

97. DOS, Space Goals after the Lunar Landing, report, October 1966, i–ii.

98. Ibid., 2.

99. Ibid., 19.

100. Lambright, "James Webb and the Uses of Administrative Power," 194–95, 200.

101. Lambright, *Presidential Management of Science and Technology*, 141, 143.

102. Levine, *Managing NASA in the Apollo Era*, 207. Another scholar concurred, "The fact that NASA leaders in the mid-1960s did not propose post-Apollo goals and defend them in the budget process virtually ensured a situation in which there was no clear future for the civilian program." Schichtle, *The National Space Program from the Eighties to the Fifties*, 28–29.

103. Newell, *Beyond the Atmosphere*, 285.

104. Newkirk et al., *Skylab: A Chronology*, 29.

105. Boone, *NASA Office of Defense Affairs*, 93.

106. Yarymovych and Schultz to the SAF, memorandum, 1.

107. Coulter and Loret, "Manned Orbiting Space Stations," 37.

108. NASA, *Manned Space Science and Advanced Manned Missions*, 7 October 1965, 35–37.

109. Gilruth to Mueller, letter, 25 March 1966, 1–2. For a detailed explanation of AAP's progress throughout the 1960s see Compton and Benson *Living and Working in Space*, and Newkirk et al., *Skylab: A Chronology*.

110. Newkirk et al., *Skylab: A Chronology*, 28.

111. Compton and Benson, *Living and Working in Space*, 14.

112. Newkirk et al., *Skylab: A Chronology*, 35.

113. McNamara to Webb, letter, 25 September 1964, 1–2.

114. Ibid.

115. Webb to McNamara, letter, 14 October 1964, 1–3.

116. Seamans, *Missiles and Rockets*, 26 October 1964, 14. An Air Force officer attended an AAP briefing the next month and reported that NASA had prepared over 200 separate charts as part of a 23-volume study describing the specific modifications required of the Apollo capsule to extend its orbital life. He also said NASA's proposed schedule included the first of nine eventual launches in late 1968. Anderson, memorandum for record, 10 November 1964, 1–2.

117. Mueller, testimony to the House space committee, 18 February 1965, in Newkirk et al., *Skylab: A Chronology*, 38.

118. Ezell, *NASA Historical Data Book*, vol. 3, 98.

119. Newkirk et al., *Skylab: A Chronology*, 47.

120. Compton and Benson, *Living and Working in Space*, 20.

121. Newkirk et al., *Skylab: A Chronology*, 55.

122. Johnson, *US Aeronautical and Space Activities, 1965*, 31 January 1966, 17.

123. Seamans, series of oral history interview, 3 June 1968, 152. Seamans recalled Mueller's characterization and his response.

124. Levine, *Managing NASA in the Apollo Era*, 173.

125. Nimmen et al., *NASA Historical Data Book*, vol. 1, 148.

126. Anderson to Kermit Gordon, letter, 9 November 1964, 1.

127. BOB, discussion paper, FY 1966 Budget Policy Considerations, 7–8.

128. Ibid., 8–9.

129. Ibid., 10–11. The White House's Office of Science and Technology largely concurred: "It is important that there be either a single national orbital system capable of generating the data and experience required by all consumers or, if for various compelling reasons there need to be two such systems, their orbital capabilities should be complementary rather than largely overlapping." Therefore, "It is not reasonable that both the MOL and the EAS systems (as currently defined) be approved for development and funded for FY 1966." Golovin to Donald Hornig, memorandum, 27 November 1964, 4–5. Document is now declassified.

130. "Washington Roundup," 17.

131. Fink, "Senate Space Head Pushes MOL Merger," 7 December 1964, 16; and Newkirk et al., *Skylab: A Chronology*, 37.

132. Anderson, Press Release, 20 December 1964, 1.

133. McNamara, Transcript of News Conference, 12 December 1964, 2.

134. House, *Missile and Space Ground Support Operations*, 46–47.

135. Levine, *Managing NASA in the Apollo Era*, 255.

136. "Interview of Samuel Phillips," 11. As mentioned above, lunar-orbital and lunar-surface missions were studied during the early stages of AAP planning. However, by mid-1969 at the latest, these efforts died as it became clear that political support for and funding of AAP would be extremely limited and that AAP would consist mostly of an earth-orbital workshop. Mayhall and Appleton, "Military Applications of Space Are Inevitable," 17–19.

137. "Humphrey: 'Space Program is Here to Stay,'" 5 September 1966, 13.

138. NASA, *Astronautics and Aeronautics, 1966*, 323.

139. NASC, Summary Minutes, 14 November 1967, 4.

140. NASA, Position Paper on the Report of the Space Panel, 26 January 1968, 6. NASC executive secretary Welsh helped in the effort to clarify the MOL versus AAP question. He told the House that the NASC had examined both systems in detail "to see if they had the same experiments on board both. We found they do not have the same experiments on board" though neither was yet operational. "Also," Welsh continued, "they are not designed to develop the same information." Welsh, testimony to the House Appropriations Committee, 5 February 1968, 50.

141. Webb, testimony to the House Appropriations Committee, 5 February 1968, 50.

142. House, *NASA Authorization, 1969*, 97–98.

143. Welsh to the vice president, memorandum, 4 March 1968, 1.

144. Mrozinski to William Moore, memorandum, 1 April 1968, 1.

145. Cited in Kennan and Harvey, *Mission to the Moon*, 204.

146. BOB, NASA: Highlight Summary, report, 30 October 1968, 499.

147. "Air Force Orbit Plan is Attacked," *Baltimore Sun*, 30 January 1969, 8A.

148. Allnutt to Fulton, letter, 1 April 1969, 3–4.

149. Gray to Mueller, letter, 31 August 1964, 1.

150. Davis, memorandum for record, 8 October 1964, 2. Excerpts from a report by a Mr. Strible from the Aerospace Corporation.

151. Schultz, Congressional Preparation Instruction no. 7, 27 November 1964, 10.

152. AACB, *Study of DOD and NASA Launch Vehicle Requirements*, 2, 4–5, 78–79.

153. McNamara and Webb, "Decisions on MOL and Related Matters," 25 January 1965, 3.

154. Air Force, Congressional Preparation Instruction no. 14, 26 January 1965, 2–3.

155. Air Force, Unclassified Supporting Witness Statement, 9 March 1965, 7.

156. Air Force, Unclassified Supporting Witness Statement, April 1965, 1.

157. Gray to Boone, memorandum, 21 July 1965, cover letter p.1, and separate page answer to NASC question no. 5.

158. Seamans to Hall, memorandum, 5 August 1965, 1. Document is now declassified.

159. Compton and Benson, *Living and Working in Space*, 47–48.

160. NASA, NASA Position Summary on Saturn I/Apollo vs. Titan III/MOL, 6 January 1967, 4.

161. Callaghan to Wolff, letter, 29 March 1967, 1.

162. Douglas Aircraft Corp., DAC-58060, vol. 1, October 1967, 1–2, 33. This summary volume was part of a 10-volume comprehensive study Douglas did on the question of NASA using MOL for early AAP missions. In addition, Douglas completed another multivolume study in July 1968 on the ability of its MOL to meet NASA's intermediate and long-term AAP goals and concluded: "MOL can accomplish intermediate objectives for the NASA. Extended durations up to 90 days are feasible without major redesign. A NASA MOL could be available in 1972." Douglas Aircraft Corp., *NASA Use of MOL for Extended Orbital Missions*, July 1968.

163. NASA, Complementary Nature of AAP and MOL Programs? 10 June 1968.

164. Foster and Newell, Terms of Reference, 6 August 1968, 1.

165. Seamans, testimony to the House space committee, 18 February 1966, in Newkirk et al., *Skylab: A Chronology*, 66.

166. Hale to Welsh, memorandum, 3 March 1966, 2.

167. Newkirk et al., *Skylab: A Chronology*, 68.

168. As explained above, these large numbers were partially explained by the fact that in 1966 AAP still included lunar-orbital and lunar-surface missions as well as earth-orbital ones. By 1969 AAP had become a single-mission project, an earth-orbital workshop.

169. Ezell, *NASA Historical Data Book*, vol. 3, 98. See also Newkirk et al., *Skylab: A Chronology*, 126.

170. Levine, *Managing NASA in the Apollo Era*, 248.

171. NASA, Budget Submittal to the BOB, 1 November 1966, 189.

172. Cabell, Memorandum for Presentation of A.A.P. Objectives, 21 December 1966, 1–2.

173. Lambright, *Powering Apollo*, 194.

174. Bilstein, *Orders of Magnitude*, 81. See also Levine, *Managing NASA in the Apollo Era*, 258; and Lambright, *Presidential Management of Science and Technology*, 148.

175. Webb, statement to Senate Subcommittee on Independent Offices, 17 May 1968, 10.

176. Newkirk et al., *Skylab: A Chronology*, 136.

177. Van Nimmen et al., *NASA Historical Data Book*, vol. 1, 156. See also Ezell, *NASA Historical Data Book*, vol. 3, 63.

178. Paine to Anderson, letter, 22 July 1969, 1. See also Newkirk et al., *Skylab: A Chronology*, 167.

179. Ezell, *NASA Historical Data Book*, vol. 3, 100.

180. Hansen, *Spaceflight Revolution*, 270.

181. Ezell, *NASA Historical Data Book*, vol. 3, 121.

182. Launius, *NASA: A History of the U.S. Civil Space Program*, 98–99.

183. Compton and Benson, *Living and Working in Space*, appendix A, 374.

184. Launius, *A History of the U.S. Civil Space Program*, 99.

185. Hansen, *Spaceflight Revolution*, 270.

186. "For $1.5 Billion: A New Air Force Eye in the Sky," 47.

187. NASA, *Astronautics and Aeronautics, 1966*, 11.

188. Ibid., 24.

189. "Missiles Gain, Space Suffers at DOD," 31 January 1966, 15.

190. NASA, *Astronautics and Aeronautics, 1966*, 87.

191. Macdonald to O'Brien, report for transmission to General Schriever, 27 May 1966, i, 1, 5. This AFHRA call number (see full citation in the Bibliography) represents its collection of Schriever's personal papers and demonstrates that the report did, in fact, reach Schriever.

192. Air Force, *Space and Missile Systems Organization: A Chronology, 1954–1970*, 180. See also Pealer, "Manned Orbiting Laboratory, Part III," 19.

193. McNamara, Statement on Department of Defense Appropriations on the FY 1968 Defense Budget, 23 January 1967, 129–131. Document is now declassified.

194. Pealer, "Manned Orbiting Laboratory, Part III," 20.

195. Johnson, *U.S. Aeronautics and Space Activities, 1967*, 44.

196. DOD, *Space Program Budget for Fiscal Years 1968–1970*, 31 December 1968, 1–2. Document is now declassified.

197. Johnson, *US Aeronautics and Space Activities, 1968*, 2, 36.

198. McNamara to the SAF, memorandum, 9 December 1967, 1–2. Document is now declassified.

199. Allen, series of oral history interviews, 30 September 1985–10 January 1986, 32–33. After retiring from the Air Force, Allen was named director of NASA's Jet Propulsion Laboratory.

200. Butz, "New Vistas in Reconnaissance from Space," 46–56.

201. Seamans, oral history interview, 15 December 1988, 395.

202. Nixon to heads of executive departments and agencies, memorandum, 25 January 1969, 1.

203. Nixon to Mayo, memorandum, 24 March 1969, 1–2.

204. "Laird Cuts $1.1 Billion; Battle Goes On," *Philadelphia Bulletin*, 9 May 1969, 8.

205. "MOL Delayed by Funding Cut," 21 April 1969, 17.

206. Seamans, oral history interviews, September 1973–March 1974, 521–22. Document is now declassified. New assistant secretary of the Air Force for R&D, Grant L. Hansen, stated, "Our aircraft fleet has gotten so behind the times that we have to have a great concentration of effort in that area to be able to get a modern fighter and bomber and airborne early warning system and combat air support aircraft. One of the things we are sacrificing in order to be able to afford to do those things . . . is the further ex-

ploitation of capabilities in space for the things in the future." Futrell, *Ideas, Concepts, Doctrine*, vol. 2, 683.

207. Seamans, oral history interview by the author, 5 July 1996.

208. Ibid.

209. Schriever, oral history interview by the author, 2 July 1996.

210. Air Force, *Space and Missile Systems Organization*, 198.

211. DOD, News Release No. 491-69, 10 June 1969, 1–2.

212. Richelson, *Secret Eyes*, 101–2. By the time of its cancellation MOL's weight had grown from 25,000 to 30,000 pounds, necessitating continued redesign of the Titan IIIM. Its first manned launch date had slipped to at least early 1971, and its final program cost estimates were $3 billion. Klass, *Secret Sentries in Space*, 169.

213. Shaughnessy, Statement re: MOL Cancellation.

214. John W. Finney, "Pentagon Drops Air Force Plans for Orbiting Lab," *New York Times*, 11 June 1969, 10.

215. Klass, *Secret Sentries*, 169.

216. Hotz, "No Tears for MOL," 11.

217. Yarymovych, oral history interview, 2 February 1976, 13.

218. Compton and Benson, *Living and Working in Space*, 109.

219. Hayes, *Struggling Towards Space Doctrine*, 232.

220. Spires, *Beyond Horizons*, 133.

221. Cabell to Jaffee, memorandum, 21 June 1968, 4.

222. Compton, *Where No Man Has Gone Before*.

Chapter 10

Conclusion

The director of NASA's Manned Spacecraft Center has stated, "The last time we flew to the moon, NASA had to pay the bill to put it on national television for the landing, because the networks wouldn't cover it."[1] Apparently by the time of that Apollo 17 mission in December 1972, the American public had become accustomed to American preeminence in space—it was now old news. The Nixon administration canceled the last three planned Apollo missions.[2] Most estimates of Project Apollo's final cost cite a figure of approximately $25 billion. To this one may add $2.6 billion for Skylab and $250 million for the Apollo-Soyuz Test Project (ASTP).[3] The ASTP was the final use of the Apollo-Saturn hardware. This first international spaceflight took place in July 1975 when détente was in full bloom. Launius explained the ASTP's purpose was to test if American and Soviet spacecraft could successfully rendezvous and dock in space and also "to open the way for international space rescue as well as future joint manned flights." The actual 15–24 July flight involved the docking of the two spacecraft and two days of experiments. It clearly demonstrated the fading of the competitive dynamic in space policy. As Launius summarized, "The flight was more a symbol of the lessening of tensions between the two superpowers than a significant scientific endeavor—taking 180 degrees the competition for international prestige that had fueled much of the space activities of both nations since the late 1950s."[4] By the summer of 1975, only six years after the first lunar landing, the Apollo era had ended.

In 1960 the portion of the federal budget devoted to the civilian space program was 0.5 percent. By 1965, after Kennedy's lunar-landing decision, this had risen ninefold to 4.5 percent; however, by 1970 it had decreased to 2 percent and by 1985 to 0.6 percent, almost where it was a quarter of a century earlier.[5] It appears highly unlikely that America will ever again devote a figure comparable to that of the mid-1960s to the civilian exploration of space, because after the dissolution of the Soviet Union space exploration will almost certainly never again be regarded as a vital-geopolitical instrument. Without the status conferred upon it as an integral component of national strategy, space exploration is highly unlikely to receive an increased proportion of a shrinking or stable federal pie. Individuals who forlornly lament the fact that America no longer has a civilian space-exploration program commensurate to that of the Apollo heyday seem either unable or unwilling to accept the fact that Apollo received such a high level of support only because Kennedy and Johnson saw it as vital to America's waging of the Cold War. Once that factor faded with détente, civilian space expenditures eventually settled back to a level roughly equivalent to where Eisenhower had pegged them decades earlier as part of his conscious decision not to employ space exploration as a centerpiece of Cold War competition. As one history of NASA correctly summarized, "In a way, the Soviet Union is responsible for creating NASA. It may well require a military turn in space to force action on the U.S. Congress for a full-speed-ahead program once again, but that is not likely as things stand."[6]

When Tsiolkovsky, Goddard, and Oberth speculated as to why humans should penetrate the realm of space, most of their thoughts focused on the potential scientific and possibly commercial benefits. Oberth even foresaw military applications. None of the three, however, postulated that nations would compete for spectacular accomplishments in space as part of a geopolitical struggle. Nevertheless, in the post-World War II environment, the quest for prestige and the search for intelligence information on potential adversaries became two of the primary motivating factors in humankind's struggle to escape gravity.

As early as 1946, Air Force contractors pointed out the potential utility of satellites for spaceborne reconnaissance. However, the practical application of this technology depended upon much greater advancement in the art of ballistic missile design and construction so that the satellites would have a launcher to put them into orbit. It was not until Eisenhower vastly accelerated America's drive for an operational ballistic missile that the likelihood of satellite reconnaissance passed from the theoretical into the probable. In 1955 both the TCP's report and America's first official space-policy document, NSC 5520, highlighted the importance of establishing a legal right for reconnaissance satellites in American space policy. This principle was one of the consistent themes of American space policy in the 1950s and 1960s.

Less constancy was found in presidential conceptions of using space, and particularly human spaceflight, for augmenting international prestige. After the almost Pearl Harbor–like impact of Sputnik in October 1957, one might have expected Eisenhower to accede to the pervasive demands for a crash American space program that would accomplish something, almost anything, before the Soviets. Eisenhower's deeply held beliefs about the danger of excessive government spending and his conception of the Cold War as a long-term struggle meant that he followed a more measured course. His creation of NASA ensured America would pursue a significant program of civilian space exploration, although one designed to guarantee America was simply *a* leader in space, not *the* leader. His creation of an overarching OSD space hierarchy of ARPA and the DDR&E also ensured the USAF's calls for R&D into virtually every facet of using space for the purposes of national defense would be tempered.

Eisenhower's approach to human spaceflight started to become clear in August 1958, when he assigned that mission to NASA instead of the military services. He and his coterie of civilian scientific advisors concluded that human spaceflight most likely held little potential relevance to America's deterrent strength. The search for a legal regime in which to operate reconnaissance satellites was almost certainly the most important factor in America's space policy under Eisenhower. Thus, human-spaceflight R&D in the form of Project Mercury would proceed at a deliberate and measured pace and would not be conceived of as a component of any type of a race for prestige. Eisenhower did nod to the importance of space in the quest for prestige, as evidenced by his strong support of the Saturn booster designed to lift heavy payloads into orbit. Still, both Mercury and the Air Force's sole hope in the human-spaceflight arena, Dynasoar, were funded at relatively low levels throughout the Eisenhower administration.

Another historical trend emerged during the Eisenhower administration—the basic support-coordination-rivalry structure of the NASA-DOD relationship. During its first few years, NASA was heavily dependent on the DOD, and particularly the Air Force, for launch vehicles, top-level managers, national ranges and tracking stations, and expertise in the initiation and administration of large aerospace systems. While NASA would slowly but surely forge its own capabilities in these and other areas and lessen its dependence on the DOD, the complexities of NASA-DOD interaction continued to involve supporting each other (but mostly the DOD supporting NASA), coordinating numerous programs so as to avoid waste and duplication, and occasionally resolving the conflicts and tension that inevitably arose when two large bureaucracies had to operate programs in the same basic area—in this case, space exploration.

The avoidance of space for prestige, characteristic of Eisenhower's terms, was a fundamental American space-policy principle that changed radically under Kennedy. Like Eisenhower, he believed in the ultimate Cold War goal of containing the Soviet Union while pursuing arms control and other measures designed to lessen tension. Unlike Eisenhower, he believed that one of the means the United States should employ to achieve the end of Cold War "victory" was a space race. Accordingly, in May 1961 he set America on its way to the moon when he officially authorized Project Apollo and began the process of quintupling NASA's budget. While Kennedy did order extensive reviews of the space program and his lunar-landing goal in 1962 and 1963, the available evidence suggests that his commitment to the goal of landing on the moon before the end of the 1960s held firm until his death in November 1963.

The NASA-DOD relationship in the Kennedy era also saw a flurry of activity. At first many accused the Air Force of waging a "campaign" to take over NASA. While it is possible that some Air Force officials not at the top policy-making levels harbored such desires, the relationship between chief of staff General White and NASA administrator Glennan was cordial, and their correspondence revealed determined efforts aimed at allaying congressional fears of Air Force hegemony. Kennedy's establishment of the lunar-landing goal soon rendered the point moot, as NASA's budget skyrocketed and its congressional patrons were strengthened as well in their support of NASA's interests. Another factor that quickly reined in any nascent Air Force desires for a larger role in the US space program was SECDEF McNamara and his new OSD managerial philosophy. Under systems analysis and PPBS, not only did the Air Force have to offer convincing and quantifiable proof that a space project added to America's national security, it also had to make clear that its space efforts did not conflict in any way with NASA's R&D; if they did, the USAF was extremely unlikely to win approval for a space proposal.

It appeared that shortly after they began their service in the Kennedy administration, Webb and McNamara had a sort of falling out and were unable to deal with each other on a face-to-face basis; further communication took place through subordinates such as Seamans and Rubel. Simultaneously, McNamara became convinced that the capabilities Dynasoar was designed to offer could largely be provided more cheaply and more quickly by NASA's Gemini. Therefore, McNamara made a bid in late 1962 for managerial control over Gemini. While his attempt failed, it did lead to a January 1963 NASA-DOD Gemini agreement that increased DOD's role in Gemini and permitted the DOD to place experiments aboard NASA's Gemini flights. Quite possibly, Dynasoar's days were numbered from this point. Meanwhile, the multifaceted DOD support for NASA continued, to include General Phillips as Apollo program director. At times, however, this very support could lead to tension and rivalry as illustrated by NASA-DOD differences over the questions of reimbursement and personnel transfers that emerged late in the Kennedy administration and increased under Johnson.

Over the course of 1963 McNamara became convinced that the DOD did not require Dynasoar. While he cited the fact that the Air Force had not provided him with specific military missions for the glider, the reality was that McNamara had prohibited the Air Force from investigating that very topic when he reoriented the Dynasoar into a strictly research vehicle called the X-20 and deemed it an orbital system designed to explore maneuverable reentry, not a suborbital vehicle searching for information on hypersonic flight. This McNamara-decreed reorientation also resulted in NASA's distancing itself from the project. By December 1963, it fell to new president Johnson to approve McNamara's recommendation that Dynasoar be canceled and replaced by an experimental space laboratory, the MOL. The final significant development of 1963 was the fact that the Soviet Union implicitly and quietly accepted the reality and legality of satellite reconnaissance, thereby fulfilling one of the primary American space-policy goals since 1955. As a result, the NRO and the American reconnaissance satellite program went even deeper into the "black" in an attempt to stabilize a situation in which satellite reconnaissance was no longer a diplomatic football but rather a tacitly recognized international fact.

Under President Johnson there was a great deal of continuity in both the approach to using space as a competitive Cold War tool and in the NASA-DOD relationship. While it was true that NASA's budget leveled off and then began its long-term decline under Johnson, he always ensured that Project Apollo had enough budgetary and political protection to stay on course and on schedule. The DOD's support of NASA continued and so did the two agencies' close coordination of plans and projects through such entities as the AACB. Tension and rivalry between the two was largely confined to the questions of NASA reimbursing the DOD for services rendered and how many military officers the DOD would provide NASA. Neither area of disagreement altered the fact that over the course of the mid- to late-1960s, NASA's institutional capabilities in all areas—from managerial personnel to launch facilities—continued to mature and resulted in less dependence on the DOD.

The most important feature of the NASA-DOD relationship under Johnson was the approval and then, shortly after he left office, the cancellation of the MOL. Even before McNamara left the DOD and Nixon became president, McNamara ordered the Air Force to explore the idea of reconfiguring the MOL into an unmanned system. As the MOL's budget continued to climb and its initial operation date regularly slipped, it was thrown into competition with the NRO robotic satellites that were more and more capable, such as the fourth-generation KH-9 that eventually debuted in mid-1971. Within six months of Nixon's inauguration and subsequent imposition of government-wide financial constraints, the MOL was terminated. It was a victim not only of budget cuts and indirect competition with the NRO's satellites but also a more obvious form of competition with NASA's plan to use Apollo-Saturn hardware for earth-orbital R&D. While both NASA and DOD regularly claimed NASA and the MOL were not duplicative, enough suspicion lingered in Congress that they were to prevent any significant protest over the MOL's demise.

Levine is the only analyst who has previously examined in detail the NASA-DOD relationship. He correctly explained that the DOD was "the one Federal agency with which NASA had to come to terms in order to carry out its mission at all. The essence of their relationship had far more to do with mutual need than with philosophical arguments concerning the existence or the desirability of one space program or two."[7] It is difficult to dispute this interpretation of the essentially pragmatic nature of the NASA-DOD relationship. NASA needed certain items (e.g., launch vehicles) and services (e.g., managerial expertise) from the DOD. The DOD honored NASA's requests, and NASA carried out its general mission of civilian space exploration and its particular tasking of landing an American on the moon before the end of the 1960s. Levine elaborated that within this relatively straightforward supportive relationship—which in turn gave rise to intricate coordination mechanisms so as to minimize duplication between the two multibillion-dollar programs—"Where the two agencies could not agree was in the sphere where program philosophy and program management overlapped, particularly in the cases of Gemini and Manned Orbiting Laboratory."[8]

McNamara's failed attempt to seize managerial control of Gemini did lead to a January 1963 NASA-DOD Gemini agreement that increased the DOD's level of participation in the program. This increased DOD involvement led in December 1963 to the OSD's conclusion that Dynasoar should be canceled and the MOL initiated. By the end of the Kennedy administration, "NASA succeeded in freeing itself from overt DOD control

531

by 1963."[9] The MOL matured as a reconnaissance-gathering platform over the course of the next five years, but it too expired as a new president concluded that the nation's finances could not support it and that its purported capabilities were, by 1969, largely superseded by the NRO's reconnaissance satellites.

While it is not entirely accurate to declare, "It remains imperative to have NASA keep its status as the decorous front parlor of the space age in order to reap public support for all space projects and give Defense Department space efforts an effective 'cover,' "; there was nevertheless some small element of truth in that scholarly team's assessment.[10] This central truth was that NASA did in a sense present a convenient focus for the publicity concerning America's highly visible civilian space exploration program while at the same time America's military uses of space, in particular the commencement and perfection of satellite reconnaissance, proceeded under a deepening cloak of secrecy. By the fall of 1963, the Soviets accepted the necessity of spaceborne reconnaissance for a stable and mutual deterrence and ceased their diplomatic campaign to outlaw it. American space policy thus continued to highlight NASA and its activities, not only because it was busy racing the USSR to the moon in a presidentially mandated quest for prestige, but also because spotlighting NASA diverted attention away from the military uses of space and thus was unlikely to upset the delicate diplomatic consensus that tacitly sanctioned reconnaissance satellites. Walter McDougall correctly summarized, "The principal concern of American [space] policy was always the protection of spy satellites." The resulting American space strategy encompassed a dual thrust that featured the "establishment of a legal regime in space that complemented the American propaganda line of openness and cooperation in space and held out hope of agreements to 'put a lid on the arms race,' and at the same time preserved American freedom to pursue such military missions in space as were needed to protect and perfect the nuclear deterrent."[11]

A former NASA historian explained that World War II made possible the exploration of space because it forced nations to focus on technical progress in rocketry, though obviously for purposes of weapons development. Then as the Cold War intensified after World War II, space technology was pursued largely for its military potential and its prestige-related aspects—"The security role of the Department of Defense and the function of NASA as a civilian space agency have been inextricably related ever since."[12] Another space historian explained that from the earliest days of Tsiolkovsky, Goddard, and Oberth, pioneering thinkers envisioned the increase in scientific knowledge and the possible practical benefits to humankind that could be generated through space exploration. Then after World War II, "Although as policy goals they [scientific exploration and commercial use of space] remain essentially unaltered, public clamor in the wake of Sputniks 1 and 2 introduced a third goal: ensuring national pride and international prestige. . . . What had begun as an evenly if slowly paced research and development effort would be spurred forward at a gallop."[13] Kennedy's contribution was to focus on human spaceflight as the primary prestige-gathering tool and accelerate its pace from a gallop into a full sprint. Johnson maintained the primacy of the lunar-landing goal but refused to extend it to any follow-on-space projects.

It seems beyond dispute that civilian space exploration was a child of Cold War politics: "As it had in the past and would in the future, international politics more than international dreams advanced the development of space technology."[14] Launius correctly summarized that, "The history of space and rocketry during the twenty years after World War II was almost entirely propelled by the rivalry between the United States and the Soviet Union, as the two great superpowers engaged in a 'cold war' over the ideologies and allegiances of the non-aligned nations of the world." This intense US-USSR competition "ensured that they would dedicate significant resources to the effort" of exploring space and in the end, "It was this rivalry that prompted the development of a formal U.S. civil space program."[15] Again and again space historians have emphasized, "The initial driving force for a strong American space program was not scientific, economic, or romantic, but political—the pursuit of national prestige and power by a new means and in a new frontier. This no doubt accelerated the development of spaceflight capabilities and the attainment of high-visibility goals." In this sense, "The astronauts were our modern Cold War equivalents of the medieval knights who stepped forward to engage in single-man combat with the enemy."[16] This book's examination of the specific component of the NASA-DOD relationship within the broader fields of space history and policy supports the correctness of the fundamental thesis linking the Cold War, prestige, and space exploration.

Did NASA, the lunar-landing goal, and Project Apollo merit the national priority and approximately $25 billion accorded them? Any answer to that question must admit, "In the final analysis, it is difficult to think of a way to identify and measure the independent contribution to U.S. international prestige of being perceived as a leader in space. There is no equation linking prestige with influence, power, and control over events and choices."[17] One's answer necessarily reveals more about one's opinions concerning NASA, space exploration, and the wisdom of human spaceflight than it does about any objective evaluation of facts, figures, and geopolitical consequences.[18] Having said that, it should come as no surprise that opinions occupy all points along the spectrum.

Alex Roland is a pointed critic of human spaceflight in general and Project Apollo in particular. He accuses NASA of being "locked into a climate of opinion bred of Sputnik, Gagarin, and Apollo. It is intent upon extending the romantic era of spaceflight—indeed upon building our whole future in space around a program of barnstorming. We are in a state of suspended adolescence, deferring mature exploitation of space in a childish infatuation with circus." He believes the origins of this "anachronism" are found in the Apollo program. He posits that Apollo established a long-term NASA focus on human spaceflight and that the problem with this "is that it is driven by romance not practicality. There are many worthwhile things to do in space; sending people there is one of the most expensive and least productive. . . . It costs ten times as much to conduct a space mission with people as it does with automated spacecraft." Further, in Roland's opinion, "Any specific mission we can identify to conduct in space we can build a machine to do. And we can do it more quickly, more safely, and at a fraction of the cost of sending people up to do it." Roland did grant that "Apollo returned to the U.S. just what it went after—the international prestige of being the best in space." But he questions if this was worth $25 billion and, even if it were, disputes the subsequent centering of most of NASA's space effort around human spacecraft and facilities in space.[19]

In another format Roland declared, "The space race, however, has no payoff beyond prestige. A victory in one heat achieves nothing unless you also win the next one. . . . The prize is in the prestige; the purse is filled with Tang." He likened the Cold War space programs to historical antecedents such as the Great Pyramids of Egypt that served as "awe-inspiring monuments to the power of the state and its ability to waste incredible resources on otherwise pointless enterprises. . . . The space program, and manned spaceflight in particular, surely fits this mold: an enormous, expensive, inspiring technological artifact whose cost in labor, lives, and treasure exceeds its practical utility. . . . No national consumption in the last thirty years has been more conspicuous than manned spaceflight."[20] Other analysts who reach negative conclusions concerning America's space program in the 1950s and 1960s express sentiments similar to Roland's, though perhaps not as eloquently.

This author, however, subscribes to another set of conclusions offered by what appears to be a wider range of scholars. At a minimum it seems likely that "A Soviet first landing on the moon in the 1960s would undoubtedly have been interpreted throughout the world as a humiliation and a grave reverse for the West."[21] The alarmed reactions to Sputnik in October 1957 and the Gagarin flight in April 1961 support this conclusion. As Webb's biographer stated, "In the broad sweep of history Apollo was a critical victory in the Cold War technological competition between the United States and the Soviet Union. Certainly, subsequent history between the two nations would have differed greatly had Russians walked the moon, rather than Americans."[22] It is possible that the space race acted as a sort of relief valve for the Soviet-American rivalry:

> Had we not had the peaceful space rivalry of the 1960s, the Soviet Union and the United States might have been forced into military demonstrations of their technological prowess. . . . Without the space race, there might have been more incidents like the Cuban missile crisis. . . . Apollo relieved some of that pressure. It permitted the United

States to prove that it had the technology to deliver military warheads anywhere it wanted. . . . Ironically, America's civilian space program made credible the military capabilities of the Department of Defense.[23]

As is the case with many facets of space history and policy, the dean of scholarly studies in this field ably tied together the important trends. Logsdon summarized:

> Given the context of 1961, the Apollo *decision* was an appropriate choice of a symbol to serve the national interest at the time; however, given the drastically changed social and political context of the late 1960's and early 1970's, the culmination of the Apollo program was a rather inappropriate manifestation of what was receiving priority in this country. . . . Kennedy cannot be faulted for not anticipating the domestic and international upheavals of the sixties; few people did. . . . Thus it is possible to conclude that starting Apollo was a "good" decision, while still having a mixed evaluation of itself. . . . The message communicated by Apollo in 1969 and later was not what Kennedy intended when he started the project in 1961 (emphasis in original).[24]

Logsdon explained that when Kennedy started Apollo, he viewed it as a "remedial action" designed to respond to "a variety of political and psychological needs which were present in the nation at the time the decision was made." In the ultimate evaluation,

> Apollo *did* serve the short-term objectives of Kennedy rather well. While it is extremely difficult to isolate the influence of deciding to compete with the Soviet Union from the influence of other actions and decisions of the period which led to the end of Cold War hostilities, there may well have been such an influence. If the United States had not opted to compete, the Soviet Union would have continued to reap political benefit from its space successes. By entering the race with such a visible and dramatic commitment, the United States effectively undercut the Soviet monopoly of space spectaculars, without doing anything except announcing its intention to compete (emphasis in original).[25]

Logsdon added, "Without having done Apollo first, the decision to commit to the shuttle and to the level of space activity in the 1980s that is implied by that commitment would not have occurred in 1972. . . . The fact that the United States began, and completed, Apollo created the context within which the focus of the space program could be turned to earth-oriented activities. . . . Without having first accomplished Project Apollo, almost everything else which has been done in space since would have been much more difficult to initiate."[26]

Logsdon was not alone in his characterization of Apollo as "an important victory"[27] and "a substantial success."[28] After all, given the turmoil of the late 1960s, if not for the space program, "there would be little positive for Americans to remember from that time."[29]

Lambright concurred: "With Vietnam a disaster and Johnson's Great Society falling apart, space remained the one positive legacy from the Kennedy-Johnson years."[30] Whether hard-core human-spaceflight enthusiasts realized it or not, once America reached the moon, "The extraordinary flurry of technological activity to get humans off the planet and on their way to other worlds far, far away was over—at least for the time being, until external circumstances would once again come together to spur the inner disquiet that launches such space odysseys. . . . If Apollo was about leaving, the period after Apollo was about staying home."[31]

Such a repeat of the requisite external circumstances has not yet transpired, and it appears extremely unlikely that it will in the foreseeable future. Until another president calculates, as did Kennedy and to some degree Johnson, that pre-eminence in space is an important element of US power, NASA's program of civilian space exploration will likely remain focused on earth-orbital activities. The unmanned robotic exploration of Mars could certainly accelerate speculation concerning the possibility of microscopic life there, but a human expedition to Earth's nearest planetary neighbor would seem to be, at a minimum, several decades away. One scholar noted, as the Cold War was ending, "Space exploration has been intimately tied to the Cold War that followed the hostilities of the World War. As the Cold War ends, so, I assert, does much of the energy and momentum that propelled us to do some wonderful things in space exploration. Without that drive, and with increasing competition for public funds, it is apt to ask whether space exploration can survive the end of the Cold War."[32] Now, more than a decade after the end of the Cold War and well into a new century and millennium, there seems little evidence of a renewal of the requisite "energy and momentum" for increased space exploration. However, one conclusion is firm: the multifaceted relationship between NASA and the DOD involving support, coordination, and rivalry formed an important component of America's first decade in space and its Cold War strategy. The NASA-DOD relationship does and will continue to play a vital role in determining the nature, pace, and international posture of America's presence in space.

Notes

1. Gilruth, oral history interview, 2 May 1987, 326.
2. Compton, *Where No Man Has Gone Before*, chap. 12, http://www.hq.nasa.gov/office/pao/History/SP-4214. Therefore, the final numbering system for the Apollo missions is Apollo 11 through 17. Since Apollo 13's accident prevented its landing on the moon, six Apollo missions and 12 astronauts actually reached the surface of the moon and returned.
3. Ezell, *NASA Historical Data Book*, vol. 2, 121.
4. Launius, *NASA: A History of the U.S. Civil Space Program*, 100. For the full story of the ASTP, see Ezell and Ezell, *The Partnership*.
5. Fries, "Introduction," to *A Spacefaring Nation*, n. 1, 7.
6. Hirsch and Trento, *The National Aeronautics and Space Administration*, 217.
7. Levine, *Managing NASA in the Apollo Era*, 211.
8. Ibid., 229.
9. Ibid., 236.
10. Kennan and Harvey, *Mission to the Moon*, 217.
11. McDougall, *The Heavens and the Earth*, 178, 187.
12. Launius, "Early U.S. Civil Space Policy," 64–65.
13. R. Hall, "Thirty Years Into the Mission," 134.
14. Goldman, *Space Policy*, 7.
15. Launius, *A History of the U.S. Civil Space Program*, 17.
16. Wilford, "A Spacefaring People," 70.
17. Logsdon, "National Leadership and Presidential Power," 216.
18. The author fully admits to a generalized support of space exploration and NASA's civilian pursuit of it via human spaceflight with the caveat that America must first provide for its defense-oriented needs in space.
19. Roland, "Barnstorming in Space," 39, 42, 46–47.
20. Roland, "The Lonely Race to Mars," 41–43.
21. Levine, *The Missile and Space Race*, 206.
22. Lambright, *Powering Apollo*, 214.
23. Ezell, "The Apollo Program," 30.
24. Logsdon, "The Apollo Decision in Historical Perspective," 5–9.
25. Ibid.
26. Ibid.

27. Logsdon and Dupas, "Was the Race to the Moon Real?" 23.
28. Logsdon, "Evaluating Apollo," 19 July 1989, 3.
29. Ibid., 4.
30. Lambright, *Powering Apollo*, 189.
31. Hansen, *Spaceflight Revolution*, 428.
32. Murray, "Can Space Exploration Survive the End of the Cold War?" 23.

List of Abbreviations

AACB	Aeronautics and Astronautics Coordinating Board
AAF	Army Air Forces
AAP	Apollo Applications Program
ABM	antiballistic missile
ABMA	Army Ballistic Missile Agency
AEC	Atomic Energy Commission
AFB	Air Force Base
AFBMD	Air Force Ballistic Missile Division
AFHRA	Air Force Historical Research Agency (Maxwell AFB, AL)
AFHSO	Air Force History Support Office (Bolling AFB, DC)
AFSC	Air Force Systems Command
AMC	Air Materiel Command
AMR	Atlantic Missile Range
ARDC	Air Research and Development Command
ARPA	Advanced Research Projects Agency
ARS	American Rocket Society
ASTP	Apollo-Soyuz Test Project
ATM	Apollo Telescope Mounts
BMD	Ballistic Missile Division
BOB	Bureau of the Budget
Bomi	Bomber-Missile
BuAer	Bureau of Aeronautics (US Navy)
CIA	Central Intelligence Agency
CMLC	Civilian-Military Liaison Committee
COPUOS	Committee on the Peaceful Uses of Outer Space
Corona	joint USAF-CIA program, part of WS-117L extracted and renamed

CSAF	chief of staff of the Air Force
DDEL	Dwight D. Eisenhower Library
DDR&E	director of Defense Research and Engineering
Discoverer	unclassified cover name for Corona program
DOD	Department of Defense
DOS	Department of State
Dynasoar	Dynamic Soarer (boost-glide research airplane)
EAS	Extended Apollo System
ELINT	electronic intelligence
EO	Executive Order
EOR	earth orbit rendezvous
ETR	Eastern Tracking Range
EVA	extravehicular activity
FRUS	*Foreign Relations of the United States*
FY	fiscal year
GALCIT	Guggenheim Aeronautical Laboratory, California Institute of Technology
GNP	gross national product
GPO	Government Printing Office
GPPB	Gemini Program and Planning Board
HDLO	Historical Division Liaison Office
HSTL	Harry S. Truman Library
ICBM	intercontinental ballistic missile
IGY	International Geophysical Year
IRBM	intermediate range ballistic missile
IRIS	Inferential Retrieval Indexing System (AFHRA's automated system)
JCS	Joint Chiefs of Staff
JPL	Jet Propulsion Laboratory
lbs.	pounds
LEM	Lunar Excursion Module
LLVPG	Large Launch Vehicle Planning Group

LTBT	Limited Test Ban Treaty
LOR	lunar orbit rendezvous
MA	Mercury-Atlas
Memcon	memorandum of conference
MILA	Merritt Island Launch Area
MLLP	Manned Lunar Landing Program
MODS	Military Orbital Development System
MOL	Manned Orbiting Laboratory
MORL	Manned Orbital Research Laboratory
MOSS	Manned Orbiting Space Station
MOU	memorandum of understanding
mph	miles per hour
MR	Mercury-Redstone
MSFC	Marshall Space Flight Center
MSFEB	Manned Space Flight Experiments Board
MSFPC	Manned Space Flight Policy Committee
MTSS	Military Test Space Station
NACA	National Advisory Committee for Aeronautics
NAS	National Academy of Sciences
NASA	National Aeronautics and Space Administration
NASC	National Aeronautics and Space Council
NASM	National Air and Space Museum
NHDRC	NASA Historical Data Reference Collection (Washington, DC)
NIE	national intelligence estimate
NREC	National Reconnaissance Executive Committee
NRL	Naval Research Laboratory
NRO	National Reconnaissance Office
NRP	National Reconnaissance Program
NSAM	National Security Actions Memorandum

NSA MUS	National Security Archive, Military Uses of Space
NSA PD	National Security Archive, Presidential Directives
NSC	National Security Council
NSF	National Science Foundation
OCB	Operations Coordinating Board
ODA	Office of Defense Affairs
OMB	Office of Management and Budget
OMSF	Office of Manned Space Flight
OMSS	Office of Missile and Satellite Systems
ONR	Office of Naval Research
OSANSA	Office of the Special Assistant for National Security Affairs (White House)
OSAST	Office of the Special Assistant for Science and Technology (White House)
OSD	Office of the Secretary of Defense
OST	Office of Science and Technology (White House)
PARD	Pilotless Aircraft Research Division
PBCFIA	President's Board of Consultants on Foreign Intelligence Activities
PL	Public law
PPBS	Planning, Programming, and Budgeting System
PSAC	President's Science Advisory Committee
R&D	research and development
RAND	Research and Development Corporation (Air Force think tank)
RG	Record Group
Robo	Rocket-Bomber
SAC	Science Advisory Committee
SAF	secretary of the Air Force
SALT	Strategic Arms Limitation Treaty

SAMOS	Satellite and Missile Observation System See WS-117L
SASC	Senate Armed Services Committee
SCORE	Signal Communication Orbit Relay Experiment
SECDEF	secretary of defense
SIGINT	signals intelligence
Sentry	Advanced Reconnaissance Satellite Project See WS-117L
Space Act	National Aeronautics and Space Act of 1958
SPI	Space Policy Institute (George Washington University, DC)
SSD	Space Systems Division
STG	Space Task Group
STS	Space Transportation System (space shuttle)
TCP	Technological Capabilities Panel
UN	United Nations
USAF	US Air Force
USSR	Union of Soviet Socialist Republics
VCSAF	vice-chief of staff of the Air Force
VFW	Veterans of Foreign Wars
WS	weapons system
WS-117L	USAF reconnaissance satellite program (renamed Sentry and finally Samos)
WPA	Works Progress Administration

Bibliography

Adair, Maj Robert E., USAF. Office of the Deputy Chief of Staff for Research and Development. *History of the Assistant for Manned Orbiting Laboratory: 1 January 1964 through 30 June 1964*, July 1964. IRIS 1002993, Air Force History Support Office.

Aeronautics and Astronautics Coordinating Board (AACB). *Intermediate-Class Launch Vehicles for Future DOD/NASA Manned Missions*. Report. November 1968. Folder: AACB Launch Vehicle Study, 1968, box: AACB no. 4, NASA Historical Data Reference Collection, NASA Headquarters, Washington, DC.

———. Launch Vehicle Panel. Minutes of the 9th Meeting of the Launch Vehicle Panel of the AACB, 5 January 1962. Folder: AACB Minutes & Reports, box: Arnold Levine, Selected Sources from the Author, NASA Historical Data Reference Collection, NASA Headquarters, Washington, DC.

———. Manned Space Flight Panel. A Survey of Information Exchange between NASA and DOD Relative to Manned Space Flight Activities, 16 November 1967, folder: Manned Space Flight Panel, AACB, box: AACB no. 1, NASA Historical Data Reference Collection, NASA Headquarters, Washington, DC.

———. Minutes of the 9th Meeting of the AACB, 19 July 1961. Box: AACB DOD/NASA, no. 4, NASA Historical Data Reference Collection, NASA Headquarters, Washington, DC.

———. *Study of DOD and NASA Launch Vehicle Requirements*. Box: AACB no. 3, NASA Historical Data Reference Collection, NASA Headquarters, Washington, DC.

———. Supporting Space Research and Technology Panel, Ad Hoc Subpanel on Reusable Launch Vehicle Technology. Final Report, 22 September 1966. IRIS 1003002, Air Force History Support Office.

"Air Force Secretary Brown: Tactical Air Power, A Vital Element in the Application of Military Forces," interview. *Armed Forces Management*, October 1966, 66–69.

Allen, Gen Lew, Jr., USAF. Series of oral history interviews, 30 September 1985–10 January 1986. K239.0512-1694, Air Force Historical Research Agency, Maxwell AFB, AL.

Allnutt, Robert F., associate administrator for Administrative Affairs, NASA. To Rep. James G. Fulton. Letter, 1 April 1969. Folder: MOL Correspondence, DOD subseries, Federal Agencies series, NASA Historical Data Reference Collection, NASA Headquarters, Washington, DC.

Alsop, Stewart. "Outer Space: The Next Battlefield." *Saturday Evening Post* 235 (28 July–4 August 1962): 15–19.

Ambrose, Mary Stone. *The National Space Program Phase II: Implementation of the National Aeronautics and Space Act of 1958: A Study of NASA's First Two Years of Operations with Emphasis on the Programming and Budgeting Aspects*, August 1960. Folder: Implementation of the Space Act of 1958, box: White House, Presidents, Eisenhower, Space Act Testimony, NASA Historical Data Reference Collection, NASA Headquarters, Washington, DC.

Ambrose, Stephen E. *Eisenhower: The President*. Vol. 2. New York: Simon and Schuster, 1984.

American Rocket Society. Space Flight Technical Committee. "Space Flight Program." Report, 10 October 1957. Folder: Eisenhower Administration, Space Correspondence, box: White House, Presidents, Eisenhower, Space Correspondence (1955–1960).

Anderson, Clinton P. To Kermit Gordon. NASA Historical Data Reference Collection, NASA Headquarters, Washington, DC. Letter, 9 November 1964. Folder: Space Committee, MOL, box 915, Clinton Anderson Papers, Library of Congress.

———. Office of Senator. Press Release, 20 December 1964. Folder: Space Committee, MOL, box 915, Clinton Anderson Papers, Library of Congress.

Anderson, Col John J., USAF, Office of the Deputy Chief of Staff for R&D. Memorandum for record, 10 November 1964. IRIS 1002995, Air Force History Support Office.

———. To Colonel Schultz. Memorandum, 18 December 1964. Air Force People in NASA. IRIS 1002995, Air Force History Support Office.

———. To General Electric. Memorandum for record, 20 August 1964. IRIS 1002994, Air Force History Support Office.

Armacost, Michael H. *The Politics of Weapons Innovation: The Thor-Jupiter Controversy.* New York: Columbia University Press, 1969.

Art, Robert J. *The TFX Decision: McNamara and the Military.* Boston: Little, Brown and Company, 1968.

Augenstein, Bruno W. Appendix 1, "Evolution of the U.S. Military Space Program, 1945–1960: Some Events in Study, Planning, and Program Development." In *International Security Dimensions of Space.* Edited by Yuri Ra'anan and Robert L. Pfaltzgraff Jr. Hamden, CT: Archon Books, 1984.

Barber and Associates, Richard J. *The Advanced Research Projects Agency, 1958–1974.* DOD contract no. MDA 903-74-C-0096. Springfield, VA: Defense Technical Information Center (DTIC), 1975.

Barree, Lt Col Alfred C., deputy chief of staff for R&D, USAF. Memorandum for record, 17 April 1967. IRIS 1003006, Air Force History Support Office.

Beard, Edmund. *Developing the ICBM: A Study in Bureaucratic Politics.* New York: Columbia University Press, 1976.

Beckler, David Z., Office of Defense Mobilization. To Arthur S. Flemming, Office of Defense Mobilization. Memorandum, 30 April 1956. Folder: Eisenhower Administration, Space Correspondence, box: White House, Presidents, Eisenhower, Space Correspondence (1955–1960), NASA Historical Data Reference Collection, NASA Headquarters, Washington, DC.

———. To William Y. Elliott. Memorandum, 18 April 1956. Folder: Eisenhower Administration, Space Correspondence, box: White House, Presidents, Eisenhower, Space Correspondence (1955–1960), NASA Historical Data Reference Collection, NASA Headquarters, Washington, DC.

BeLieu, Kenneth. To Lyndon Johnson. Memorandum, 17 December 1960. Folder: NASC 1960–1961, box: White House, National Aeronautics and Space Council, NASA Historical Data Reference Collection, NASA Headquarters, Washington, DC.

———. To Lyndon Johnson. Memorandum, 22 December 1960. Folder: NASC 1960–1961, box: White House, National Aeronautics and Space Council, NASA Historical Data Reference Collection, NASA Headquarters, Washington, DC.

Benson, Charles D., and William B. Faherty. *Moonport: A History of Apollo Launch Facilities and Operations.* NASA SP-4204. Washington, DC: Government Printing Office, 1978.

Berger, Carl. *The Air Force in Space: Fiscal Year 1961.* Washington, DC: USAF Historical Division Liaison Office, 1966.

———. *The Air Force in Space: Fiscal Year 1962.* Washington, DC: USAF Historical Division Liaison Office, June 1966.

Berlin Crisis, 1958–1962, The. National Security Archive microfiche document collection. Alexandria, VA: Chadwyck-Healey, Inc., 1991.

Beschloss, Michael R. *The Crisis Years: Kennedy and Khrushchev, 1960–1963.* New York: HarperCollins, 1991.

———. *Mayday: Eisenhower, Khrushchev and the U-2 Affair.* New York: Harper & Row, 1986.

Beukel, Erik. *American Perceptions of the Soviet Union as a Nuclear Adversary.* London: Pinter Publishers, 1989.

B. H. L. (only these initials survive as author). To Lyndon Johnson. Cover letter, 9 July 1962. Space Policy Institute (SPI) Archives document 1544, George Washington University, Washington, DC.

Bilstein, Roger. *Orders of Magnitude: A History of the NACA and NASA, 1915–1990.* NASA SP-4406. Washington, DC: Government Printing Office, 1989.

———. *Stages to Saturn: A Technological History of the Apollo-Saturn Launch Vehicles.* NASA SP-4206. Washington, DC: Government Printing Office, 1980.

Bisplinghoff, Raymond L. To Seamans. Memorandum, 22 November 1963. In *History of the Aeronautical Systems Division.* K243.011, Air Force Historical Research Agency, Maxwell AFB, AL. Document is now declassified.

Boggs, Marion W., deputy executive secretary, Executive Office of the President. NSC 6021, *Missiles and Military Satellite Programs,* 14 December 1960. Space Policy Institute Archives document 722.

Booda, Larry. "Air Force Outlines Broad Space Plans." *Aviation Week,* 5 December 1960, 26–28.

———. "Kennedy Asks $51.6 Billion for Defense." *Aviation Week and Space Technology*, 22 January 1962, 26–28.

Boone, Adm W. Fred, retired, deputy associate administrator for Defense Affairs. Report, 12 July 1963. Reprinted in *Exploring the Unknown: Selected Documents in the History of the U.S. Civil Space Program*. Vol. 2, *Relations with Other Organizations*. NASA SP-4407. Edited by John M. Logsdon, Dwayne A. Day, and Roger D. Launius. Washington, DC: Government Printing Office, 1996.

———. *NASA Office of Defense Affairs: The First Five Years, December 1, 1962 to January 1, 1968*. NASA HHR-32. Washington, DC: Government Printing Office, 1970.

———. To Earl Hilburn, AAD, Special Support to Military Services. Memorandum, 7 December 1965. Folder: DOD/USAF/NASA-Vietnam cooperation, DOD subseries, Federal Agencies series, NASA Historical Data Reference Collection, NASA Headquarters, Washington, DC.

———. To Webb. Memorandum, 9 January 1963. Folder: Webb, declassified papers, 1961–1968, Webb subseries, Administrators series, NASA Historical Data Reference Collection, NASA Headquarters, Washington, DC.

Bornet, Vaughn Davis. *The Presidency of Lyndon B. Johnson*. Lawrence, KS: The University Press of Kansas, 1983.

Bottome, Edgar M. *The Missile Gap: A Study of the Formulation of Military and Political Policy*. Rutherford, NJ: Farleigh Dickenson University Press, 1971.

Boushey, Homer, director of Advanced Technology and deputy chief of staff for Development. To CSAF. Memorandum, 30 October 1959. Folder: X-20 Dyna-Soar Documentation, DOD subseries, Federal Agencies series, NASA Historical Data Reference Collection, NASA Headquarters, Washington, DC.

———. To director of Requirements. Memorandum, 8 January 1959. K140.11-3, Air Force History Support Office. In Boushey, Brig Gen Homer A., deputy chief of staff for Research and Development, USAF. "Who Controls the Moon Controls the Earth." *US News and World Report*. 7 February 1958, 54.

————. "Blueprints for Space." In *Man in Space: The United States Air Force Program for Developing the Spacecraft Crew.* Edited by Lt Col Kenneth F. Gantz. New York: Duell, Sloan, and Pearce, 1959.

Bowen, Lee. *An Air Force History of Space Activities, 1945–1959.* Washington, DC: USAF Historical Division Liaison Office, 1964, SHO-C-64/50.

————. *Threshold of Space: The Air Force in the National Space Program, 1945–1959.* USAF Historical Division Liaison Office, 1960. NSA-MUS document 314.

Bradlee, Benjamin. *Conversations with Kennedy.* New York: Norton, 1975.

Branyan, Robert L., and Lawrence H. Larsen, eds. and comps. *The Eisenhower Administration, 1953–1961: A Documentary History.* 2 volumes. New York: Random House, Inc., 1971.

Brooks, Courtney G., James M. Grimwood, and Loyd S. Swenson Jr. *Chariots for Apollo: A History of Manned Lunar Spacecraft.* NASA SP-4205. Washington, DC: Government Printing Office, 1979.

Brooks, Harvey. "Motivations for the Space Program: Past and Future." In Allan A. Needell, *The First 25 Years in Space.* Washington, DC: Smithsonian Institution Press, 1983.

Brooks, Overton. To T. Keith Glennan. Letter, 14 February 1961. IRIS 1002992, Air Force History Support Office.

————. To John Kennedy. Letter, 9 March 1961. Reprinted in *Exploring the Unknown: Selected Documents in the History of the U.S. Civil Space Program.* Vol. 2, *Relations with Other Organizations.* NASA SP-4407. Edited by John M. Logsdon, Dwayne A. Day, and Roger D. Launius. Washington, DC: Government Printing Office, 1996.

Brown, Col George S., USAF, executive to the CSAF. Memorandum for record, 25 July 1958. Folder: Man in Space, DOD subseries, Federal Agencies series, NASA Historical Data Reference Collection, NASA Headquarters, Washington, DC.

Brown, Harold. To the SECDEF. Memorandum, 14 November 1963. In *History of the Aeronautical Systems Division.* K243.011, Air Force Historical Research Agency, Maxwell AFB, AL. Document is now declassified.

———. To the SECDEF. Memorandum, 30 November 1963. In *History of the Aeronautical Systems Division.* K243.011, Air Force Historical Research Agency, Maxwell AFB, AL. Document is now declassified.

———, director, Defense Research and Engineering. To the Senate Committee on Aeronautical and Space Sciences. Statement, 14 June 1962. Reprinted in USAF, *Air Force Information Policy Letter, Supplement for Commanders, Special Issue: Military Mission in Space, 1957–1962.* Director of Information, Office of the Secretary of the Air Force, 1962. Folder: DOD Space Policy, DOD subseries, Federal Agencies series, NASA Historical Data Reference Collection, NASA Headquarters, Washington, DC.

Brucker, Wilber. Army Position Paper, 15 October 1958. Folder: NASA/AOMC/JPL Transfer, box: Administrative History no. 6, shelf VI-C-6, NASA Historical Data Reference Collection, NASA Headquarters, Washington, DC.

Brundage, Percival. To the president. Memorandum, 30 April 1957. Box 6: Department of Defense subseries, Subject series, Office of the Staff Secretary: Records, White House Office, Dwight D. Eisenhower Library, Abilene, KS.

Bryant, F. B., co-chairman, NASA Management Working Group. To W. Fred Boone. Memorandum, 9 November 1966. Folder: USAF Satellites (General), DOD subseries, Federal Agencies series, NASA Historical Data Reference Collection, NASA Headquarters, Washington, DC.

Buckley, Edmond C., director of NASA Tracking and Data Acquisition. To Boone. Memorandum, 8 January 1963. Folder: DOD/USAF "Blue Gemini," DOD subseries, Federal Agencies series, NASA Historical Data Reference Collection, NASA Headquarters, Washington, DC.

Bulkeley, Rip. *The Sputniks Crisis and Early United States Space Policy: A Critique of the Historiography of Space.* Bloomington, IN: Indiana University Press, 1991.

Bundy, McGeorge. To Kennedy. Memorandum, 18 September 1963. Space Policy Institute Archives document 975.

———. To Webb. Memorandum, 23 February 1962. Folder: 3, National Security Council, box: White House, National Security Council, NASA Historical Data Reference Collection, NASA Headquarters, Washington, DC.

Burrows, William E. *Deep Black: Space Espionage and National Security*. New York: Berkley Books, 1986.

———. "A Study of Space Reconnaissance: Methodology for Researching a Classified System." In *A Spacefaring Nation: Perspectives on American Space History*. Edited by Martin J. Collins and Sylvia D. Fries. Washington, DC: Smithsonian Institution Press, 1991.

Bush, Vannevar. To James Webb. Letter, 11 April 1963. Space Policy Institute Archives document 978.

———. "New Vistas in Reconnaissance from Space." *Air Force Magazine*, March 1968, 46–56.

Butz, J. S. Jr. "Under the Spaceborne Eye: No Place to Hide." *Air Force and Space Digest*, May 1967, 93–98.

Cabell, Gen C. P., USAF, retired, consultant to the NASA administrator. Presentation, 21 December 1966. Folder: Apollo-AAP Planning, Skylab series, NASA Historical Data Reference Collection, NASA Headquarters, Washington, DC.

———. To Leonard Jaffee, director, Space Applications Programs. Memorandum, 21 June 1968. Folder: Resolution from Space, DOD subseries, Federal Agencies series, NASA Historical Data Reference Collection, NASA Headquarters, Washington, DC.

Callaghan, Richard L, special assistant to the administrator, NASA. To William Moyers, assistant to the president. Memorandum, 12 May 1964. Folder: Daily Reports to the White House, box: White House, NASA Reports to the White House, NASA Historical Data Reference Collection, NASA Headquarters, Washington, DC.

———, NASA associate administrator for Legislative Affairs. To Rep. Lester L. Wolff. Letter, 29 March 1967. Folder: USAF Manned Orbiting Laboratory, DOD subseries, Federal Agencies series, NASA Historical Data Reference Collection, NASA Headquarters, Washington, DC.

Cantwell, Gerald T. *The Air Force in Space: Fiscal Year 1963*. Washington, DC: USAF Historical Division Liaison Office, December 1966. Document is now declassified.

———. *The Air Force in Space: Fiscal Year 1964.* Washington, DC: USAF Historical Division Liaison Office, June 1967. NSA-MUS document 330.

———. *The Air Force in Space: Fiscal Year 1965.* Washington, DC: USAF Historical Division Liaison Office, April 1968. NSA-MUS document 331.

———. *The Air Force in Space: Fiscal Year 1966.* Washington, DC: Office of Air Force History, 1968.

———. *The Air Force in Space: Fiscal Year 1967, Part I.* Washington, DC: Office of Air Force History, 1969.

———. *The Air Force in Space: Fiscal Year 1968, Part I.* Washington, DC: Office of Air Force History, October 1970. NSA-MUS document 336.

———. *The Air Force–NASA Relationship in Space, 1958–1968.* Washington, DC: Office of Air Force History, October 1971, reprinted November 1990.

Carter, Jimmy. *Public Papers of the Presidents of the United States: Jimmy Carter, 1978.* Vol. 2, *Kennedy Space Center, Florida Remarks at the Congressional Space Medal of Honor Awards Ceremony.* Washington, DC: Government Printing Office, 1980.

Carter, Launor F., chief scientist of the Air Force. "An Interpretive Study of the Formulation of the Air Force Space Program," 4 February 1963. Space Policy Institute Archives unnumbered document.

Central Intelligence Agency (CIA). Memorandum, 30 May 1964. Folder: LBJ Library/Declassified Space Documents, box: White House, Presidents, Johnson, Correspondence, Declassified items, NASA Historical Data Reference Collection, NASA Headquarters, Washington, DC.

———. National Intelligence Estimate (NIE) 11-12-55, *Soviet Guided Missile Capabilities and Probable Programs.* 20 December 1955. Folder: CIA, box: Federal Agencies, CIA, NIE, shelf 11-B-3, NASA Historical Data Reference Collection, NASA Headquarters, Washington, DC.

———. National Intelligence Estimate 11-5-58, *Soviet Capabilities in Guided Missiles and Space Vehicles.* 19 August 1958. Folder 13, box 1, RG 263, records of the Central Intelligence Agency, National Archives and Records Administration.

————. National Intelligence Estimate 11-1-62, *The Soviet Space Program*. 5 December 1962. Folder: CIA NIE, box: Federal Agencies, CIA, NIE, shelf: 11-B-3, NASA Historical Data Reference Collection, NASA Headquarters, Washington, DC.

————. National Intelligence Estimate 11-4-65, *Main Trends in Soviet Military Policy*. 14 April 1965, reprinted in Donald P. Steury, comp., *Estimates on Soviet Military Power: 1954 to 1984*, Center for the Study of Intelligence. Washington, DC: CIA, December 1994.

————. National Intelligence Estimate 11-1-67, *The Soviet Space Program*. 2 March 1967. Folder: CIA National Intelligence Estimates, shelf 11-B-3, box: Federal Agencies, CIA, National Intelligence Estimates, NASA Historical Data Reference Collection, NASA Headquarters, Washington, DC.

Civilian Military Liaison Committee (CMLC). Minutes of CMLC Meeting, 13 January 1959. Folder: CMLC Minutes, January 1959. Box: Civilian Military Liaison Committee, NASA Historical Data Reference Collection, NASA Headquarters, Washington, DC.

————. Minutes of CMLC Meeting, 10 March 1959. Folder: CMLC Minutes, January 1959, box: Civilian Military Liaison Committee, NASA Historical Data Reference Collection, NASA Headquarters, Washington, DC.

————. Terms of Reference, 22 October 1958. Folder: CMLC Organization and Membership, box: Civilian Military Liaison Committee, NASA Historical Data Reference Collection, NASA Headquarters, Washington, DC.

Clark, Rear Adm John E., deputy director, ARPA. To James R. Killian Jr. Memorandum, 23 July 1958. Folder: Space Notebook, Piland, 1958–59 (4), box 16, OSAST, White House Office, Dwight D. Eisenhower Library, Abilene, KS.

Clements, Col Henry E., USAF. *The Coordination of Manned Spaceflight Operations between DOD and NASA*. Research Report No. 31. Washington, DC: Fort Lesley J. McNair, Industrial College of the Armed Forces, April 1969.

Clowse, Barbara Barksdale. *Brainpower for the Cold War: The Sputnik Crisis and the National Defense Education Act of 1958*. Westport, CT: Greenwood Press, 1981.

Colchagoff, Maj George, ARDC. To Lt Col R. C. Anderson, New Research Systems. Memorandum, 16 February 1956. Space Policy Institute Archives document 350.

Compton, W. David. *Where No Man Has Gone Before: A History of Apollo Lunar Exploration Missions.* NASA SP-4214. Washington, DC: Government Printing Office, 1989.

———, and Charles D. Benson. *Living and Working in Space: A History of Skylab.* NASA SP-4208. Washington, DC: Government Printing Office, 1983.

Costigliola, Frank. "Lyndon B. Johnson, Germany, and 'The End of the Cold War.'" In *Lyndon Johnson Confronts the World: American Foreign Policy, 1963–1968.* Edited by Warren I. Cohen and Nancy Bernkopf Tucker. Cambridge: Cambridge University Press, 1994.

Coughlin, William J. "Eulogy to a Dyna-Soar," editorial. *Missiles and Rockets*, 11 March 1963, 50.

———. "The Gleeful Conspiracy," editorial. *Missiles and Rockets*, 23 April 1962, 46.

———. "Speak Up, Mr. Secretary," editorial. *Missiles and Rockets*, 18 June 1962, 46.

———. "The Wall of Space," editorial. *Missiles and Rockets*, 2 December 1963, 48.

Coulman, Robert F. *Illusions of Choice: The F-111 and the Problem of Weapons Acquisition.* Princeton, NJ: Princeton University Press, 1977.

Coulter, Col John M., and Maj Benjamin L. Loret, USAF. "Manned Orbiting Space Stations." In *The U.S. Air Force in Space.* Edited by Eldon W. Downs. New York: Frederick A. Praeger, 1966.

"Council [NASC] Compiles List of Space 'Firsts.'" *Aviation Week and Space Technology*, 16 May 1966, 100.

Cronkite, Walter. Oral history interview, Lyndon Johnson, 5 July 1969. Folder: LBJ Interviews, box: White House, Presidents, Johnson, Pre-White House Interests, NASA Historical Data Reference Collection, NASA Headquarters, Washington, DC.

Davies, Merton E., and William R. Harris. *RAND's Role in the Evolution of Balloon and Satellite Observation Systems and*

Related US Space Technology. Santa Monica, CA: The RAND Corporation, 1988.

Davis, Lt Col Howard S., AFRMO, USAF. Memorandum for record, 8 October 1964. IRIS 1002995, Air Force History Support Office.

Davis, Maj Gen Leighton I., DOD representative for Project Mercury Support Operations. To the SECDEF. Report, 11 September 1963. Summary Report: DOD support of Project Mercury, July 1959–June 1963. Folder: DOD Support of Mercury, Mercury series, NASA Historical Data Reference Collection, NASA Headquarters, Washington, DC.

Dawson, Virginia P. "The Push from Within: Lewis Research Center's Transition to Space." In *A Spacefaring Nation: Perspectives on American Space History.* Edited by Martin J. Collins and Sylvia D. Fries. Washington, DC: Smithsonian Institution Press, 1991.

Day, Dwayne A. "CORONA: America's First Spy Satellite Program (Part 1)." *Quest: The History of Spaceflight Quarterly* 4, no. 2 (Summer 1995): 4–21.

———. "Invitation to Struggle: The History of Civilian-Military Relations in Space." In *Exploring the Unknown: Selected Documents in the History of the U.S. Civil Space Program.* Vol. 2, *Relations with Other Organizations.* NASA SP-4407. Edited by John M. Logsdon, Dwayne A. Day, and Roger D. Launius. Washington, DC: Government Printing Office, 1996.

———. "Space Policy-Making in the White House: The Early Years of the National Aeronautics and Space Council." In *Organizing for the Use of Space: Historical Perspectives on a Persistent Issue.* Edited by Roger Launius. AAS History Series. Vol. 18. San Diego, CA: Univelt, Inc., 1995

———, John M. Logsdon, and Brian Latell, eds. *Eye in the Sky: The Story of the Corona Spy Satellite.* Washington, DC: Smithsonian Institution Press, 1998.

Dean, Alan L. "Mounting a National Space Program." In *Science and Resources: Prospects and Implications of Technological Advance.* Edited by Henry Jarrett. Baltimore: The Johns Hopkins University Press, 1959.

Depoe, Stephen P. "Space and the 1960 Presidential Campaign: Kennedy, Nixon, and 'Public Time.'" *Western Journal of Speech Communication* 55 (Spring 1991): 215–33.

DeVorkin, David H. *Science with a Vengeance: How the Military Created the US Space Sciences after World War II.* New York: Springer-Verlag, 1992.

Diamond, Edwin. *The Rise and Fall of the Space Age.* Garden City, NY: Doubleday & Company, Inc., 1964.

Divine, Robert A. "Lyndon B. Johnson and the Politics of Space." Chapter in *The Johnson Years: Vietnam, the Environment, and Science.* Vol. 2. Lawrence, KS: The University of Kansas Press, 1987.

———. *The Sputnik Challenge.* Oxford: Oxford University Press, 1993.

"DOD/NASA Support Costs Revealed." *Armed Forces Management,* June 1969.

"DOD Space Position Defended." *Missiles and Rockets,* 4 February 1963.

Donovan, John C. *The Cold Warriors: A Policy-Making Elite.* Lexington, MA: D. C. Heath and Company, 1974.

Doolittle, James H. To Hugh Dryden. Letter, 28 March 1958. Folder: Testimony on Space Act, box: White House, Presidents, Eisenhower, National Aeronautics and Space Act, NASA Historical Data Reference Collection, NASA Headquarters, Washington, DC.

———. Oral history interview, 21 April 1969. K239.0512-625, Air Force Historical Research Agency, Maxwell AFB, AL.

Douglas Aircraft Corporation, RAND Corporation. "Preliminary Design of an Experimental World-Circling Spaceship." Report No. SM-11827, 2 May 1946. In *Exploring the Unknown: Selected Documents in the History of the U.S. Civil Space Program.* Vol. 1, *Organizing for Exploration.* NASA SP-4407. Edited by John M. Logsdon, Linda J. Lear, Jannelle Warren-Findley, Ray A. Williamson, and Dwayne A. Day. Washington, DC: Government Printing Office, 1995.

———. Missile and Space Systems Division. DAC-58060. *Evaluation of the Usefulness of the MOL to Accomplish Early NASA Mission Objectives.* Vol. 1, *Summary.* October 1967. Multiple folders labeled: NASA-MOL, DOD subseries, Federal Agencies series, NASA Historical Data Reference Collection, NASA Headquarters, Washington, DC.

———. *NASA Use of MOL for Extended Orbital Missions.* July 1968. Multiple folders labeled: NASA-MOL, DOD subseries, Federal Agencies series, NASA Historical Data Reference Collection, NASA Headquarters, Washington, DC.

Downs, Anthony. *Inside Bureaucracy.* Boston: Little, Brown and Company, 1967.

Draper, C. S. Oral history interview, 2 June 1974. Folder: Kennedy-Pre-White House, box: White House, Presidents, Kennedy, Kennedy Library materials, interviews through 1960. NASA Historical Data Reference Collection, NASA Headquarters, Washington, DC.

Dryden, Hugh. Address to the Institute for Aeronautical Sciences, 27 January 1958. Folder: NACA to NASA Transition, box: White House, Presidents, Eisenhower, Space Documentation (1957–1960). NASA Historical Data Reference Collection, NASA Headquarters, Washington, DC.

———. To NASA historian Eugene Emme. Letter, 8 September 1965. Space Policy Institute Archives document 484.

———. To James Killian. Letter, 18 July 1958. Folder: Space Notebook, Piland, 1958–59 (4), box 16, OSAST, White House Office, Dwight D. Eisenhower Library, Abilene, KS.

———. To Eugene M. Zuckert. Letter, 1 April 1964. Folder: 1964, Manned Lunar Landing Program, Personnel, Military, box 43, Samuel Phillips Papers, Library of Congress.

———. To James S. Lay, executive secretary, NSC. Memorandum, 16 September 1959. Folder: National Security Council, 1955–1980, box: White House, National Security Council, NASA Historical Data Reference Collection, NASA Headquarters, Washington, DC.

———. Oral history interview, 26 March 1964. Folder: Dryden, Mercury Tape, Dryden subseries, Deputy Administrators series, NASA Historical Data Reference Collection, NASA Headquarters, Washington, DC.

———. Oral history interview, 1 September 1965. Folder: Dryden, Mercury Tape, Dryden subseries, Deputy Administrators series, NASA Historical Data Reference Collection, NASA Headquarters, Washington, DC.

———. To Johnson. Report, March 1965. Folder: Eisenhower Library-Space Race, box: Presidents, Eisenhower, Photos, Presidential Library [document may be misfiled], NASA Historical Data Reference Collection, NASA Headquarters, Washington, DC.

———. Speech. American Aeronautics and Astronautics Society, 30 December 1961. Folder: NASC 1962–1972, box: White House, National Aeronautics and Space Council, NASA Historical Data Reference Collection, NASA Headquarters, Washington, DC.

———. Statement to the House Committee on Science and Astronautics, 17 May 1962. Folder: Aeronautics and Astronautics Coordinating Board, box: Aeronautics and Astronautics Coordinating Board, NASA Historical Data Reference Collection, NASA Headquarters, Washington, DC.

Dulles, Allen. To the president. Memorandum, 3 August 1960. Folder: Allen Dulles (1), box 13, Administration series, Ann Whitman file, Dwight D. Eisenhower Library, Abilene, KS.

Eisenhower, Dwight D. Speech. The Naval War College, 3 October 1961. In "Eisenhower at the Naval War College." U.S. Naval Institute *Proceedings*, June 1971, 18–24.

———. To Robert Altschul, National Planning Association. Letter, 25 October 1957. Folder: October 1957 DDE Dictation, box 27, DDE Diary Series, Ann Whitman file, Dwight D. Eisenhower Library, Abilene, KS.

———. To James Killian. Letter, 16 July 1959. *President Dwight D. Eisenhower's Office Files, 1953–1961, Part 1: Eisenhower Administration Series*, microfilmed from the Dwight D. Eisenhower Library, Project Coordinator Robert E. Lester, part of the series *Research Collections in American Politics: Microforms from Major Archival and Manuscript Collections*, William E. Leuchtenburg, general editor (Bethesda, MD: University Publications of America, 1990), reel 19.

———. To Thomas Gates. Letter, 10 June 1960. Folder: Reconnaissance Satellites 1960, box 15, Executive Secretary Subject file subseries, NSC Staff Papers Series, White House Office, Dwight D. Eisenhower Library, Abilene, KS.

———. To T. Keith Glennan. Letter, 14 January 1960. Folder: Glennan, Dr. (First name) Keith-NASA, box 15, Administration series, Ann Whitman file, Dwight D. Eisenhower Library, Abilene, KS.

————. To Professor Loyd Swenson. Letter, 5 August 1965. Primary author of *This New Ocean*, the history of Project Mercury. NASA Historical Data Reference Collection, NASA Headquarters, Washington, DC.

————. To Senator Stuart Symington. Letter, 29 October 1957. Folder: Eisenhower Administration-Space Correspondence, box: White House, Presidents, Eisenhower, Space Correspondence (1955–1960), NASA Historical Data Reference Collection, NASA Headquarters, Washington, DC.

————. To SECDEF. Memorandum, 24 March 1958. K140.11-11, Air Force History Support Office.

————. To SECDEF and the NACA chairman. Memorandum, 2 April 1958. Folder: National Aeronautics and Space Administration (1), box 44, Confidential File, White House Central Files, Dwight D. Eisenhower Library, Abilene, KS.

————. To the secretaries of state and defense, the attorney general, the chairman of the AEC, and the director of CIA. Memorandum, 26 August 1960. In *CORONA: America's First Satellite Program.* Edited by Kevin C. Ruffner. Washington, DC: Center for the Study of Intelligence, 1995.

————. "Year One of the Space Age," first annual report of the president to Congress, February 1959. *United States Aeronautics and Space Activities, 1 January–31 December 1958.*

————. *Public Papers of the Presidents of the United States: Dwight D. Eisenhower, 1957.* "The President's News Conference," no. 210, 9 October 1957. Washington, DC: Government Printing Office, 1958.

————. *Public Papers of the Presidents of the United States: Dwight D. Eisenhower, 1957.* Radio and Television Address to the American People on "Science in National Security," no. 230, 7 November 1957. Washington, DC: Government Printing Office, 1958.

————. *Public Papers of the Presidents of the United States: Dwight D. Eisenhower, 1957.* Radio and Television Address to the American People on "Our Future Security," no. 234, 13 November 1957. Washington, DC: Government Printing Office, 1958.

———. *Public Papers of the Presidents of the United States: Dwight D. Eisenhower, 1958.* "The President's News Conference," no. 28, 5 February 1958. Washington, DC: Government Printing Office, 1959.

———. *Public Papers of the Presidents of the United States: Dwight D. Eisenhower, 1958.* "The President's News Conference," no. 56, 26 March 1958. Washington, DC: Government Printing Office, 1959.

———. *Public Papers of the Presidents of the United States: Dwight D. Eisenhower, 1958.* "President's Message Relayed from the Atlas Satellite," no. 322, 19 December 1958. Washington, DC: Government Printing Office, 1959.

———. *Public Papers of the Presidents of the United States: Dwight D. Eisenhower, 1959.* "The President's News Conference," no. 172, 29 July 1959. Washington, DC: Government Printing Office, 1960.

———. *Public Papers of the Presidents of the United States: Dwight D. Eisenhower, 1960–61.* "Special Message to the Congress Recommending Amendments to the National Aeronautics and Space Act," no. 11, 4 January 1960. Washington, DC: Government Printing Office, 1961.

———. *Public Papers of the Presidents of the United States: Dwight D. Eisenhower, 1960–61.* "The President's News Conference," no. 21, 26 January 1960. Washington, DC: Government Printing Office, 1961.

———. *Public Papers of the Presidents of the United States: Dwight D. Eisenhower, 1960–61.* "The President's News Conference," no. 24, 3 February 1960. Washington, DC: Government Printing Office, 1961.

———. *Public Papers of the Presidents of the United States: Dwight D. Eisenhower, 1960–61.* "Statement by the president on US Achievements in Space," no. 264, 7 August 1960. Washington, DC: Government Printing Office, 1961.

———. *Public Papers of the Presidents of the United States: Dwight D. Eisenhower, 1960–61.* "Remarks at a Luncheon for Latin American Delegates to the UN General Assembly, New York City," no. 303, 22 September 1960. Washington, DC: Government Printing Office, 1961.

———. *Public Papers of the Presidents of the United States: Dwight D. Eisenhower, 1960–61.* "Letter in Reply to a Proposal for a Meeting of the President and Chairman Khrushchev," no. 313, 2 October 1960. Washington, DC: Government Printing Office, 1961.

———. *Public Papers of the Presidents of the United States: Dwight D. Eisenhower, 1960–61.* "Annual Message to the Congress on the State of the Union," no. 410, 12 January 1961. Washington, DC: Government Printing Office, 1961.

———. *Public Papers of the Presidents of the United States: Dwight D. Eisenhower, 1960–61.* "Annual Budget Message to the Congress: Fiscal Year 1962," no. 414, 16 January 1961. Washington, DC: Government Printing Office, 1961.

———. US Aeronautics and Space Activities, 1 January to 31 December 1959. Message from the President of the United States Transmitting the Second Annual Report of the Nation's Activities in the Fields of Aeronautics and Space, 25 February 1960.

———. US Aeronautics and Space Activities: 1 January–31 December 1960, Report to Congress from the President of the United States, 18 January 1961. NSA-MUS document 324.

———. *Waging Peace: 1956–1961.* New York: Doubleday & Company, Inc., 1965.

———. "Are We Headed in the Wrong Direction?" *Saturday Evening Post* 235 (11–18 April 1962): 19–25.

———. "Spending Into Trouble." *Saturday Evening Post* 236 (18–25 May 1963): 15–19.

———. "Why I Am a Republican." *Saturday Evening Post* 237 (11–18 April 1964): 17–19.

Eisenhower, John S. D. Memorandum of conference, 13 October 1960. Folder: National Aeronautics and Space Administration (8), box 18, Alphabetical subseries, Subject series, Office of the Staff Secretary, Dwight D. Eisenhower Library, Abilene, KS.

———. Oral history interview. 28 February 1967. Dwight D. Eisenhower Library, Abilene, KS.

Elliott, Derek W. *Finding an Appropriate Commitment: Space Policy Development Under Eisenhower and Kennedy, 1954–1963.* PhD diss., George Washington University, 1992.

Emme, Eugene M. *A History of Space Flight.* New York: Holt, Rinehart and Winston, Inc., 1965.

————. Chronology of Man-In-Space R&D Program, August 1962. Folder: USAF Man-in-Space Chronology, DOD subseries, Federal Agencies series, NASA Historical Data Reference Collection, NASA Headquarters, Washington, DC.

————. "Historical Origins of NASA." *Airpower Historian* 10 (January 1963): 18–23.

————. "Historical Perspectives on Apollo." *Journal of Spacecraft and Rockets* 5 (April 1968): 369–81.

————. *The Impact of Air Power: National Security and World Politics*. Princeton, NJ: D. Van Nostrand Co., 1959.

————. "Presidents and Space." In *Between Sputnik and the Shuttle: New Perspectives on American Astronautics*, American Astronautical Society History Series, Vol. 3. Edited by Frederick C. Durant III. San Diego, CA: Univelt, Inc., 1981.

Enthoven, Alain, and K. Wayne Smith. *How Much is Enough: Shaping the Defense Program*. New York, 1971.

Etzioni, Amitai. *The Moon-Doggle: Domestic and International Implications of the Space Race*. New York: Doubleday & Company, 1964.

Evans, Brig Gen Harry L., vice-director of the MOL Program. To the Air Force personnel director, Personnel Support for NASA. Memorandum, 10 September 1966. IRIS 1003002, Air Force History Support Office.

————. To the deputy chief of staff for R&D, USAF. Memorandum, 6 December 1965. IRIS 1003002, Air Force History Support Office.

Executive Order 10783. Transferring Certain Functions from the Department of Defense to the National Aeronautics and Space Administration, 1 October 1958. Space Policy Institute Archives document 1124.

Ezell, Edward Clinton. "The Apollo Program: History Must Judge." In *Ten Years Since Tranquillity Base*. Edited by Richard P. Hallion. Washington, DC: Smithsonian Institution Press, 1979.

————, and Linda Neuman Ezell. *The Partnership: A History of the Apollo-Soyuz Test Project*. NASA SP-4209. Washington, DC: Government Printing Office, 1978.

Ezell, Linda Neuman. *NASA Historical Data Book.* Vol. 2, *Programs and Projects, 1958–1968.* NASA SP-4012. Washington, DC: Government Printing Office, 1988.

———. *NASA Historical Data Book.* Vol. 3, *Programs and Projects, 1969–1978.* NASA SP-4012. Washington, DC: Government Printing Office, 1988.

Faget, Maxim, aeronautical research engineer. To Hugh Dryden. Letter, 5 June 1958. Space Policy Institute Archives document 1490.

———. To NACA associate director. Memorandum, 5 March 1958. Folder: USAF MIS/MISS, DOD subseries, Federal Agencies series, NASA Historical Data Reference Collection, NASA Headquarters, Washington, DC.

Ferer, Lt Col Benjamin H., office of the assistant for advanced technology. Memorandum for record, 14 March 1960. IRIS 1003000. Air Force History Support Office.

Ferguson, Gen James, AFSC/CC. Speech, 20 May 1968. James Ferguson file, Biographical series, NASA Historical Data Reference Collection, NASA Headquarters, Washington, DC.

———. Speech, 25 June 1968. James Ferguson file, Biographical series, NASA Historical Data Reference Collection, NASA Headquarters, Washington, DC.

———. Testimony on *DOD Appropriations for 1963: Hearings before the Subcommittee of the Committee on Appropriations.* 87th Cong., 2d sess., 27 February 1962.

Fink, Donald. "CIA Control Bid Slowed Decision on MOL." *Aviation Week and Space Technology,* 25 September 1961, 21.

———. "Senate Space Head Pushes MOL Merger." *Aviation Week and Space Technology,* 7 December 1964, 16.

Firestone, Bernard J. *The Quest for Nuclear Stability: John F. Kennedy and the Soviet Union.* Westport, CT: Greenwood Press, 1982.

Flax, Alexander H., assistant secretary of the Air Force for R&D. To George Mueller. Letter, 28 August 1964. IRIS 1002994, Air Force History Support Office.

———. To Harold Brown. Memorandum, 18 January 1964. IRIS 1002995, Air Force History Support Office.

———. To the director, Defense Research and Engineering. Memorandum, 29 March 1965. IRIS 1002996, Air Force History Support Office.

———. To Eugene Zuckert. Memorandum, 17 August 1964. IRIS 1002994, Air Force History Support Office.

"For $1.5 Billion: A New Air Force Eye in the Sky." *Newsweek*, 6 September 1965, 46–47.

Forrestal, James, SECDEF. *First Report of the SECDEF, 1948*. Washington, DC: Government Printing Office, 1948.

Foster, John S., director, Defense Research and Engineering. To the SECDEF. Memorandum, 6 October 1966. Box: Arnold Levine, selected sources from the author, NASA Historical Data Reference Collection, NASA Headquarters, Washington, DC.

———. To the Senate Committee on Aeronautical and Space Sciences. Statement, 26 March 1968. John Foster file, Biographical Series, NASA Historical Data Reference Collection, NASA Headquarters, Washington, DC.

———. To the Senate Space Committee. Statement, 26 March 1968. Department of Defense Activities in Space, 1967. Folder: Space 1968, box 917, Clinton Anderson Papers, Library of Congress.

———, and Homer E. Newell, NASA associate administrator. Terms of Reference, 6 August 1968. Folder: NASA-DOD Agreements, box 10, RG 220, Records of the National Aeronautics and Space Council, National Archives and Records Administration.

———, and Robert Seamans. To Charles Schultze. Joint letter, 4 April 1967. Folder: Air Force Eastern Test Range, DOD subseries, Federal Agencies series, NASA Historical Data Reference Collection, NASA Headquarters, Washington, DC.

Fries, Sylvia Doughty. "Introduction." In *A Spacefaring Nation: Perspectives on American Space History*. Edited by Martin J. Collins and Sylvia D. Fries. Washington, DC: Smithsonian Institution Press, 1991.

Frutkin, Arnold. *International Cooperation in Space*. Englewood Cliffs, NJ: Prentice-Hall, 1965.

Futrell, Robert Frank. *Ideas, Concepts, Doctrine: Basic Thinking in the United States Air Force*. Vol. 1, *Basic Thinking in the United States Air Force, 1907–1960.* Maxwell AFB, AL: Air University Press, 1989.

———. *Ideas, Concepts, Doctrine: Basic Thinking in the United States Air Force.* Vol. 2, *Basic Thinking in the United States Air Force, 1961–1984.* Maxwell AFB, AL: Air University Press, 1989.

Gaddis, John Lewis. *Russia, the Soviet Union, and the United States: An Interpretive History.* New York: John Wiley and Sons, Inc., 1978.

———. *Strategies of Containment: A Critical Appraisal of Postwar American National Security Policy.* Oxford: Oxford University Press, 1982.

Gale, Oliver M. "Post-Sputnik Washington from an Inside Office." *Cincinnati Historical Society Bulletin* 31 (1973): 224–52.

———. "Problems of Congress in Formulating Outer Space Legislation," 7 March 1958. Reprinted in House Select Committee, *Astronautics and Space Exploration.* Hearings, 85th Cong., 2d sess., 1958.

———. "Reasons for Confusion about Space Law," 11 May 1958. Folder: National Aeronautics and Space Act of 1958, box: White House, Presidents, Eisenhower, Space Act, NASA Historical Data Reference Collection, NASA Headquarters, Washington, DC.

Gantz, Lt Col Kenneth F., ed. *Man in Space: The United States Air Force Program for Developing the Spacecraft Crew.* New York: Duell, Sloan and Pearce, 1959.

Gardner, Richard N., deputy assistant secretary of State for International Organization Affairs. "Cooperation in Outer Space." *Foreign Affairs* 41, no. 2 (January 1963), 344–60.

Gardner, Trevor, chairman. *Report of the Air Force Space Study Committee,* 21 March 1961. Space Policy Institute Archives document 1525.

Gark, Lyle S., acting secretary of the Air Force. To Herbert York, director, Defense Research and Engineering. Letter, 12 January 1961. Folder: USAF Docs/Correspondence (1957–61), DOD subseries, Federal Agencies series, NASA Historical Data Reference Collection, NASA Headquarters, Washington, DC.

Garthoff, Raymond L. *Assessing: Estimates by the Eisenhower Administration of Soviet Intentions and Capabilities.* Washington, DC: The Brookings Institution, 1991.

———. "Banning the Bomb in Outer Space." *International Security* 5 (1980/81): 25–40.

Gates, Thomas, deputy SECDEF. To service secretaries, director, Defense Research and Engineering, chairman of the JCS, assistant secretaries of Defense, and ARPA director. Memorandum, 10 August 1959. IRIS 1002999, Air Force History Support Office.

———, SECDEF. Address, 27 January 1960. In *Air Force Information Policy Letter, Supplement for Commanders, Special Issue: Military Mission in Space, 1957–1962.* Director of Information, Office of the Secretary of the Air Force, 1962. Folder: DOD Space Policy, DOD subseries, Federal Agencies series, NHDRC.

Gavin, Lt Gen James M., USA, retired. *War and Peace in the Space Age.* New York: Harper and Brothers, 1958.

Gawdiak, Ihor, and Helen Fodor. *NASA Historical Data Book.* Vol. 4, *NASA Resources, 1969–1978.* NASA SP-4012. Washington, DC: Government Printing Office, 1994.

Gemini Program and Planning Board. Minutes of Meeting, 1 February 1965. IRIS 1002996, Air Force History Support Office.

Gibbs, Col Asa, Air Force liaison officer for Project Vanguard. To Lt Gen Donald L. Putt, commander of the Air Research and Development Command. Memorandum, 17 January 1958. Folder: USAF Documents/Correspondence (1957–1961), DOD subseries, Federal Agencies series, NASA Historical Data Reference Collection, NASA Headquarters, Washington, DC.

Gibney, Frank. "The Missile Mess." *Harper's Magazine,* January 1960, 38–45.

Gilpatric, Roswell. Agreements for Support of Manned Lunar Landing Program, 2 October 1961. IRIS 1003003, Air Force History Support Office.

———. To Clinton P. Anderson. Letter, 27 September 1963. Folder: 3 Armed Services, box 584, Clinton Anderson Papers, Library of Congress.

———. "In Gilpatric Speech . . . Military Space Move Left to Russians." *Missiles and Rockets,* 10 September 1962.

———. Oral history interview, 30 June 1970. From The John F. Kennedy Presidential Oral History Collection, Part I: The White House and Executive Departments, microfilmed from the holdings of the John F. Kennedy Library. Frederick, MD: University Publications of America, 1988, reel 5.

———. Statement before the Senate Committee on Aeronautical and Space Sciences, 13 June 1962. Congressional Appearances, Roswell Gilpatric Papers, Kennedy Library. Cited by Derek W. Elliott, *Finding an Appropriate Commitment: Space Policy Development under Eisenhower and Kennedy, 1954–1963* (PhD diss., George Washington University, 1992).

———. To the Senate Space Committee. Testimony, November 1963. Inserted into the Congressional Record, 20 November 1963, 21350.

———. To Committee on Science and Astronautics. Testimony, March 1961. *Defense Space Interests.* Hearings. 87th Cong., 1st sess., March 1961.

Gilruth, Robert R., director, NASA Manned Spacecraft Center. To George E. Mueller, NASA associate administrator for Manned Spaceflight. Letter, 25 March 1966. Space Policy Institute Archives document 363.

———. Oral history interviews, 27 February 1987 and 2 May 1987. National Air and Space Museum.

Ginter, R. D., director, Office of Tracking and Data Acquisition, NASA. To deputy associate administrator for Defense Affairs. Memorandum, 10 December 1964. Folder: DOD NASA Support, DOD subseries, Federal Agencies series, NASA Historical Data Reference Collection, NASA Headquarters, Washington, DC.

———, director of Special Programs Office, NASA. To the deputy associate administrator for Advanced Research and Technology. Memorandum, 4 October 1968. Folder: DOD NASA Support, DOD subseries, Federal Agencies series, NASA Historical Data Reference Collection, NASA Headquarters, Washington, DC.

———. To the commanding general, Headquarters, Army Materiel Command. Memorandum, 24 April 1969. Folder:

DOD NASA Support, DOD subseries, Federal Agencies series, NASA Historical Data Reference Collection, NASA Headquarters, Washington, DC.

————. To NASA assistant administrator for International Affairs. Memorandum, 24 June 1969. Folder: DOD NASA Support, DOD subseries, Federal Agencies series, NASA Historical Data Reference Collection, NASA Headquarters, Washington, DC.

Glennon, John P., ed. *Foreign Relations of the United States, 1955–1957.* Vol. 11, *United Nations and General International Matters.* Washington, DC: Government Printing Office, 1988.

Glennan, T. Keith. To Overton Brooks. Letter, 27 January 1961. Folder: 7-4 FAA/NASA/JCS/CIA/CAP, box 39, Thomas White Papers, Library of Congress.

————. To Deputy Administrator Hugh Dryden. Letter, 28 August 1959. Folder: Department of Defense Liaison, box: White House, Presidents, Eisenhower, DOD/CIA Information, NASA Historical Data Reference Collection, NASA Headquarters, Washington, DC.

————. To Thomas Gates. Letter, 22 July 1959. Folder: DOD Support of Mercury (1959–1963), Mercury series, NASA Historical Data Reference Collection, NASA Headquarters, Washington, DC.

————. To Thomas Gates. Letter, 19 December 1960. Folder: Department of Defense, Glennan subseries, Administrators series, NASA Historical Data Reference Collection, NASA Headquarters, Washington, DC.

————. To Roy Johnson. Letter, 17 November 1958. Folder: ARPA (Documentation), DOD subseries, Federal Agencies series, NASA Historical Data Reference Collection, NASA Headquarters, Washington, DC.

————. To James R. Killian Jr. Letter, 22 June 1959. Folder: PSAC Correspondence 1959, box: White House, President's Science Advisory Committee, NASA Historical Data Reference Collection, NASA Headquarters, Washington, DC.

————. To Neil McElroy. Letter, 15 October 1958. Space Policy Institute Archives document 488.

————. To Neil McElroy. Letter, 1 December 1958. Space Policy Institute Archives document 486.

————. To all military service secretaries and Deputy SECDEF Gates. Letter, 31 October 1958. Folder: Organizational Developments, 1958 Miscellaneous, box: Administrative History no. 6, shelf VI-C-6, NASA Historical Data Reference Collection, NASA Headquarters, Washington, DC.

————. To Rep. James M. Quigley. Letter, 4 April 1960. Folder: April 1960, Chronological, Glennan subseries, Administrators series, NASA Historical Data Reference Collection, NASA Headquarters, Washington, DC.

————. To Thomas White. Letter, 27 January 1961. Folder: 7-4 FAA/NASA/JCS/CIA/CAP, box 39, Thomas White Papers, Library of Congress.

————. To NASA leaders. Memorandum, 2 January 1960. Folder: Chronological January 1960, Glennan subseries, Administrators series, NASA Historical Data Reference Collection, NASA Headquarters, Washington, DC.

————. Notes on discussion. "How Important in the Current Scheme of Things is the Matter of Competing in the Space Field Aggressively and Ultimately Successfully with the Soviet Union," 23 September 1959. Folder: Glennan Speeches and Congressional Statements, Glennan subseries, Administrators series, NASA Historical Data Reference Collection, NASA Headquarters, Washington, DC.

————. To the Industrial College of the Armed Forces. Speech, 20 November 1959. Folder: Glennan Speeches and Congressional statements, Glennan subseries, Administrators series, NASA Historical Data Reference Collection, NASA Headquarters, Washington, DC.

————. To the NASA Authorization Subcommittee of the Senate Committee on Aeronautical and Space Sciences. Statement, 28 March 1960. Folder: Glennan Speeches and Congressional Statements, Glennan subseries, Administrators series, NASA Historical Data Reference Collection, NASA Headquarters, Washington, DC.

————. To the Subcommittee on Governmental Organization for Space Activities of the Senate Committee on Aeronautical and Space Sciences. Statement, 24 March 1959. Folder:

Glennan Speeches and Congressional statements, Glennan subseries, Administrators series, NASA Historical Data Reference Collection, NASA Headquarters, Washington, DC.

———. To the Subcommittee on Independent Offices of the Senate Appropriations Committee. Statement, 19 May 1960. Folder: Glennan Speeches and Congressional Statements, Glennan subseries, Administrators series, NASA Historical Data Reference Collection, NASA Headquarters, Washington, DC.

———. Oral history interview, 5 April 1974. Folder: Glennan interview, 5 April 1974, Glennan subseries, Administrators series, NASA Historical Data Reference Collection, NASA Headquarters, Washington, DC.

———. Oral history interview, 29 May 1987. National Air and Space Museum.

———. Interview by Martin Collins and Dr. Allan Needell, 29 May 1987. Transcript 5. National Air and Space Museum.

———. "Our Plans for Outer Space." *Saturday Evening Post* 231 (28 February 1959): 99–102.

Goddard, Robert H., and G. Edward Pendary, eds. *The Papers of Robert H. Goddard and Esther C. Goddard*. 3 vols. New York: McGraw-Hill Book Company, 1970.

———. *A Method of Reaching Extreme Altitudes*. Washington, DC: Smithsonian Institute. Smithsonian Miscellaneous Collections, Vol. 71, no. 2, publication 2540, 1919.

Goldman, Nathan C. *Space Policy: An Introduction*. Ames, IA: Iowa State University Press, 1992.

Golovin, Nicholas E. Chronological file entry, 30 October 1961. Folder: Chronological file, July–September 1961, box 6, Nicholas Golovin Papers, Library of Congress.

———. Chronological file entry, 22 November 1961. Folder: Chronological file, July–September 1961, box 6, Nicholas Golovin Papers, Library of Congress.

———. Chronological file entry, 1–5 December 1961. Folder: Chronological file, October–December 1961, box 6, Nicholas Golovin Papers, Library of Congress.

———. To Donald Hornig. Memorandum, 27 November 1964. Folder: Man in Space, 1964, box 401, RG 349, Records of the Office of Science and Technology, National Archives and Records Administration. Document is now declassified.

———. To Donald Hornig. Memorandum, 5 July 1967. Folder: NASA/OMSF, box 716, RG 359, Records of the Office of Science and Technology, National Archives and Records Administration. Document is now declassified.

———. To Jerome Wiesner. Memorandum, 21 December 1962. Folder: Withdrawn items, box 166, RG 359, Records of the Office of Science and Technology, National Archives and Records Administration. Document is now declassified.

——— et al. "Summary Report: NASA-DOD Large Launch Vehicle Planning Group." 3 vols. NASA-DOD LLVPG 105, 24 September 1962. Reprinted in *Exploring the Unknown: Selected Documents in the History of the U.S. Civil Space Program.* Vol. 2, *Relations with Other Organizations.* NASA SP-4407. Edited by John M. Logsdon, Dwayne A. Day, and Roger D. Launius. Washington, DC: Government Printing Office, 1996.

———, and L. Kavanau. To the Launch Vehicle Panel of the AACB. Memorandum, 31 August 1961. Folder: AACB Minutes & Reports, box: Arnold Levine, Selected Sources from the author, NASA Historical Data Reference Collection, NASA Headquarters, Washington, DC.

Goodpaster, Andrew. Memorandum of conference with the president, 30 March 1956. Folder: Staff Memos: March 1956, box: 15, DDE Diary Series, Ann Whitman file, Dwight D. Eisenhower Library, Abilene, KS.

———. Memorandum of conference, 5 April 1956. Folder: Staff Memos: April 1956, DDE Diary Series, Ann Whitman file, Dwight D. Eisenhower Library, Abilene, KS.

———. Memorandum of conference, 8 October 1957, 5:00 p.m. Folder: October 1957 Staff Notes (2) box 27, DDE Diary series, Ann Whitman file, Dwight D. Eisenhower Library, Abilene, KS.

———. Memorandum of conference, 8 October 1957, 8:30 p.m. Folder: October 1957 Staff Notes (2), box 27, DDE Diary Series, Ann Whitman file, Dwight D. Eisenhower Library, Abilene, KS.

————. Memorandum of conference, 11 October 1957. Folder: October 1957 Staff Notes (2), box 27, DDE Diary series, Ann Whitman file, Dwight D. Eisenhower Library, Abilene, KS.

————. Memorandum of conference, 30 October 1957, dated 31 October 1957. Folder: October 1957 Staff Notes (1), box 27, DDE Diary Series, Ann Whitman file, Dwight D. Eisenhower Library, Abilene, KS.

————. Memorandum of conference, 4 February 1958. Folder: Staff Notes February 1958, box 30, DDE Diary series, Ann Whitman file, Dwight D. Eisenhower Library, Abilene, KS.

————. Memorandum of conference, 11 March 1958. Folder: Missiles, January–March 1958 (2), box 12, OSAST, White House Office, Dwight D. Eisenhower Library, Abilene, KS.

————. Memorandum of conference, 17 July 1958, dated 18 July 1958. Folder: Staff Memos, July 1958 (1), box 35, DDE Diary series, Ann Whitman file, Dwight D. Eisenhower Library, Abilene, KS.

————. Memorandum of conference, 23 September 1958. From The Diaries of Dwight D. Eisenhower, 1953–1961. Microfilmed from the Holdings of the Dwight D. Eisenhower Library, Robert Lester, ed. Part of the Research Collections in American Politics: Microforms from Major Archival and Manuscript Collections, William Leuchtenburg, general editor. Bethesda, MD: University Publications of America, 1986. As deposited in the Library of Congress.

————. Memorandum of conference, 31 October 1958. Folder: Staff Notes, October 1958, box 36, DDE Diary series, Ann Whitman file, Dwight D. Eisenhower Library, Abilene, KS.

————. Memorandum of conference, 28 November 1958, dated 9 December 1958. Folder: Budget, Military FY 1960 (4), box 3, Department of Defense subseries, Subject series, Office of the Staff Secretary, White House Office, Dwight D. Eisenhower Library, Abilene, KS.

————. Memorandum of conference, 18 December 1958, dated 22 December 1958. Folder: Staff Notes, December 1958 (1), box 38, DDE Diary Series, Ann Whitman file, Dwight D. Eisenhower Library, Abilene, KS.

————. Memorandum of conference, 17 February 1959, dated 24 February 1959. Folder: Staff Notes, February 1959 (1), box 39, DDE Diary series, Ann Whitman file, Dwight D. Eisenhower Library, Abilene, KS.

———. Memorandum of conference, 16 September 1959. Folder: Department of Defense, Vol. 3 (8), box 2, Department of Defense subseries, Subject series, Office of the Staff Secretary, White House Office, Dwight D. Eisenhower Library, Abilene, KS.

———. Memorandum of conference, 21 September 1959, dated 23 September 1959. Folder: Staff Notes, September 1959 (1), box 44, DDE Diary series, Ann Whitman file, Dwight D. Eisenhower Library, Abilene, KS.

———. Memorandum of conference, 21 October 1959, dated 23 October 1959. Folder: Meetings with the President, box 12, OSAST, White House Office, Dwight D. Eisenhower Library, Abilene, KS.

———. Memorandum of conference, 26 October 1959, dated 30 October 1959. Folder: Staff Notes, October 1959 (1), box 45, DDE Diary series, Ann Whitman file, Dwight D. Eisenhower Library, Abilene, KS.

———. Memorandum of conference, 16 November 1959, dated 2 December 1959. Folder: Budget, Military FY 1961 (4), Department of Defense subseries, Subject series, Office of the Staff Secretary, White House Office, Dwight D. Eisenhower Library, Abilene, KS.

———. Memorandum of conference, 17 November 1959, dated 1 December 1959. Folder: Staff Notes, November 1959 (2), box 45, DDE Diary series, Ann Whitman file, Dwight D. Eisenhower Library, Abilene, KS.

———. Memorandum of conference, 18 November 1959, dated 20 January 1960. Folder: Department of Defense, Vol. 3 (8), box 2, Department of Defense subseries, Subject series, Office of the Staff Secretary: Records, White House Office, Dwight D. Eisenhower Library, Abilene, KS.

———. Memorandum of conference, 21 November 1959, dated 2 January 1960. Folder: Staff Notes, November 1959 (2), box 45, DDE Diary series, Ann Whitman file, Dwight D. Eisenhower Library, Abilene, KS.

———. Memorandum of conference, 11 January 1960, dated 14 January 1960. Folder: Staff Notes, January 1960 (2),

Eisenhower, Space Statements, NASA Historical Data Reference Collection, NASA Headquarters, Washington, DC.

———. Transcript of press conference, 5 October 1957. James C. Hagerty Papers, Dwight D. Eisenhower Library, Abilene, KS., on file with the personal papers of the Historian, NASA Historical Data Reference Collection, NASA Headquarters, Washington, DC.

Haines, Gerald. "The National Reconnaissance Office (NRO): Its Origins, Creation, and Early Years." In *Eye in the Sky: The Story of the Corona Spy Satellite*. Edited by Dwayne A. Day, John M. Logsdon, and Brian Latell. Washington, DC: Smithsonian Institution Press, 1998.

Hale, R. W., NASC staffer. To Edward C. Welsh. Memorandum, 19 November 1962. Folder: National Aeronautics and Space Administration, box 22, RG 220, Records of the National Aeronautics and Space Council, National Archives and Records Administration.

———. To Edward C. Welsh, NASC executive secretary. Memorandum, 7 March 1966. Folder: NASC Meeting, 3 March 1966, box 4, RG 220, Records of the National Aeronautics and Space Council, National Archives and Records Administration.

Hall, Albert G., deputy director, Defense Research and Engineering for Space. Speech, 5 February 1964. Albert G. Hall file, Biographical series, NASA Historical Data Reference Collection, NASA Headquarters, Washington, DC.

———. Speech, 29 July 1965. Reprinted and issued as DOD News Release no. 488-65. Albert Hall file, Biographical Series, NASA Historical Data Reference Collection, NASA Headquarters, Washington, DC.

Hall, R. Cargill. "Early U.S. Satellite Proposals." *Technology and Culture* 4, no. 4 (Fall, 1963): 410–34.

———. "The Eisenhower Administration and the Cold War: Framing American Astronautics to Serve National Security." *Prologue: Quarterly Journal of the National Archives and Record Administration* 27 (Spring 1995): 58–72.

———. "From Concept to National Policy: Strategic Reconnaissance in the Cold War." *Prologue: Quarterly Journal of the*

National Archives and Record Administration 28 (Summer 1996): 107–25.

———. "Instrumented Exploration and Utilization of Space: The American Experience." In *Two Hundred Years of Flight in America.* Edited by Eugene M. Emme. American Astronautical Society History Series, Vol. 1. San Diego, CA: Univelt, Inc., 1977.

———. "Civil-Military Relations in America's Early Space Program." In *The USAF in Space: 1945 to the Twenty-First Century.* Proceedings of Air Force Historical Foundation Symposium, Andrews AFB, MD, 21–22 September 1995. Washington, DC: Government Printing Office, 1998.

———. *Lunar Impact: A History of Project Ranger.* NASA SP-4201. Washington, DC: Government Printing Office, 1977.

———. "Origins of U.S. Space Policy: Eisenhower, Open Skies, and Freedom of Space." In *Exploring the Unknown: Selected Documents in the History of the U.S. Civil Space Program.* Vol. 1, *Organizing for Exploration.* Edited by John M. Logsdon, Linda J. Lear, Jannelle Warren-Findley, Ray A. Williamson, and Dwayne A. Day. NASA SP-4407. Washington, DC: Government Printing Office, 1995.

———. "Project Apollo in Retrospect." In *Blueprint for Space.* Washington, DC: Smithsonian Institution Press, 1992.

———. "Thirty Years into the Mission: NASA at the Crossroads." In *Reading Selections: Space Issues Symposium.* Maxwell AFB, AL: Air War College, 1988.

Hall, R. Cargill, and Jacob Neufeld, eds. *The USAF in Space: 1945 to the Twenty-First Century.* Proceedings of the Air Force Historical Foundation Symposium, Andrews AFB, MD, 21–22 September 1995. Washington, DC: Government Printing Office, 1998.

Hallion, Richard P., ed. *Apollo: Ten Years Since Tranquillity Base.* Washington, DC: Smithsonian Institution Press, 1979.

———. *On the Frontier: Flight Research at Dryden, 1946–1981.* NASA SP-4303. Washington, DC: Government Printing Office, 1984.

———. *The Hypersonic Revolution: Eight Case Studies in the History of Hypersonic Technology.* 2 vols. Wright-Patterson AFB,

OH: Special Staff Office, Aeronautical Systems Division, 1987.

Hammond, Paul. *Cold War and Détente: The American Foreign Policy Process Since 1945.* New York: Harcourt Brace Jovanovich, Inc., 1975.

Hansen, James R. *Enchanted Rendezvous: John C. Houbolt and the Genesis of the Lunar-Orbit Rendezvous Concept.* Monographs in Aerospace History Series, no. 4. Washington, DC: NASA, December 1995.

———. *Engineer in Charge: A History of Langley Aeronautical Laboratory, 1917–1958.* NASA SP-4305. Washington, DC: Government Printing Office, 1987.

———. *Spaceflight Revolution: NASA Langley Research Center from Sputnik to Apollo.* NASA SP-4308. Washington, DC: Government Printing Office, 1995.

Harlow, Bryce. Memorandum for record, 13 January 1960. Folder: Staff Notes, January 1960 (2), box 47, DDE Diary, Ann Whitman file, Dwight D. Eisenhower Library, Abilene, KS.

———. Memorandum for record, 14 January 1960. Folder: Staff Notes, January 1960 (2), box 47, DDE Diary, Ann Whitman file, Dwight D. Eisenhower Library, Abilene, KS.

———. Oral history of Deputy Assistant to the President for Congressional Affairs, 11 June 1974. Folder: Bryce Harlow interview, box: White House, Presidents, Eisenhower (cont.), DOD/CIA Information, Eisenhower, John S.D.-Lodge H.C., NASA Historical Data Reference Collection, NASA Headquarters, Washington, DC.

Harris, Elwyn D. *Standard Spacecraft Procurement Analysis: A Case Study in NASA-DOD Coordination in Space Programs.* Santa Monica, CA: RAND Corporation, R-2619-RC, May 1980.

Harvey, Dodd L., and Linda C. Ciccoritti. *U.S.-Soviet Cooperation in Space.* Miami, FL: Monographs in International Affairs, Center for Advanced International Studies, 1974.

Harvey, Mose. "Preeminence in Space: Still a Critical National Issue." *Orbis* 12 (Winter 1969): 959–83.

————. To Dr. Robert J. Seamans Jr. Memorandum, 2 July 1965. Safe no. 1, drawer 2, folder: MOL/AES, 1–3. Document is now declassified.

Hayes, Peter L. "Struggling towards Space Doctrine: U.S. Military Space Plans, Programs, and Perspectives during the Cold War." PhD diss., Boston, MA: Fletcher School of Law and Diplomacy, Tufts University, 1994.

Hechler, Ken. *The Endless Space Frontier: A History of the House Committee on Science and Astronautics, 1959–1978*, Vol. 4. America Astronautical Society History Series. San Diego, CA: Univelt, Inc., 1982.

Henry, Beverly Z., Jr., NASA aeronautical research engineer. To NASA Associate Director. Memorandum, 5 October 1959. Folder: Skylab/AAP Documentation 1959, Skylab series, NASA Historical Data Reference Collection, NASA Headquarters, Washington, DC.

Herring, George C., ed. *Lyndon B. Johnson National Security Files, Agency File, 1963–1969*. Microfilmed from the holdings of the Lyndon B. Johnson Library. Robert E. Lester, project coordinator. Bethesda, MD: University Publications of America, 1993.

————. "Introduction." In *Guide to Lyndon B. Johnson National Security Files, Agency File, 1963–1969*. A microfilm project. Bethesda, MD: University Publications of America, 1993.

Hilburn, Earl, NASA deputy associate administrator for Industry Affairs. To the Associate Administrator. Memorandum, 1 July 1965. Folder: MOL Correspondence, DOD subseries, Federal Agencies series, NASA Historical Data Reference Collection, NASA Headquarters, Washington, DC.

Hirsch, Richard, and Joseph Trento. *The National Aeronautics and Space Administration.* New York: Praeger Publishers, 1973.

Hitch, Charles. *Decision Making for Defense.* Berkeley, CA: University of California Press, 1965.

————, assistant SECDEF, comptroller. Testimony on *Defense Space Interests: Hearings before the Committee on Science and Astronautics.* 87th Cong., 1st sess., March 1961.

————. "Plans, Programs, and Budgets in the DOD." *Operations Research* 11 (January–February 1963): 1–17.

Holaday, William. Memorandum for record, 22 July 1959. Folder: CMLC, box: Civilian Military Liaison Committee, NASA Historical Data Reference Collection, NASA Headquarters, Washington, DC.

———. To the SECDEF and NASA Administrator. Memorandum, 30 September 1959. Folder: CMLC, box: Civilian Military Liaison Committee, NASA Historical Data Reference Collection, NASA Headquarters, Washington, DC.

———. Transcript of testimony before the House Committee on Science and Astronautics, 10 March 1960. Folder: CMLC, box: Civilian Military Liaison Committee, NASA Historical Data Reference Collection, NASA Headquarters, Washington, DC.

Holmes, Jay. *America on the Moon: The Enterprise of the Sixties.* New York: J. B. Lippincott Company, 1962.

Horelick, Arnold L., and Myron Rush. *Strategic Power and Soviet Foreign Policy.* Chicago: University of Chicago Press, 1966.

Horner, Richard. Oral history interview, 13 March 1974. Box: Emme/Roland interviews on early NASA history, shelf V-A-1, NASA Historical Data Reference Collection, NASA Headquarters, Washington, DC.

Hotz, Robert. "No Tears for MOL." *Aviation Week and Space Technology,* 30 June 1969, 11.

Houchin, Roy, II. "The Diplomatic Demise of Dyna-Soar: The Impact of International and Domestic Political Affairs on the Dyna-Soar X-20 Project, 1957–1963." *Aerospace Historian,* December 1988, 274–80.

———. "Interagency Rivalry?" *Quest: The History of Spaceflight Quarterly* 4 (Winter 1995): 36–39.

———. "Why the X-20 Program was Proposed." *Quest: The History of Spaceflight Quarterly* 3 (Winter 1994): 4–12.

———. "The Rise and Fall of Dyna-Soar: A History of Air Force Hypersonic R&D, 1945–1963." PhD diss., Auburn University, 1994.

Houston, Robert S., Richard P. Hallion, and Ronald G. Boston. "Transiting from Air to Space: The North American X-15," in *The Hypersonic Revolution: Eight Case Studies in the History of Hypersonic Technology,* 2 vols. Edited by Richard P. Hallion. Wright-Patterson AFB, OH: Special Staff Office, Aeronautical Systems Division, 1987.

Humphrey, Hubert. NASC Meeting, opening statement by the vice president, 3 March 1966. Folder: NASC Meeting 3 March 1966, box 4, RG 220, Records of the National Aeronautics and Space Council, National Archives and Records Administration.

———. NASC Meeting, introductory statement by the vice president, 15 November 1966. Folder: NASC Meeting, 15 November 1966, box 4, RG 220, Records of the National Aeronautics and Space Council, National Archives and Records Administration.

———. NASC Meeting, 22 June 1967. Folder: NASC Meeting, 22 June 1967, box 4, RG 220, Records of the National Aeronautics and Space Council, National Archives and Records Administration.

———. To James Webb. Memorandum, 21 August 1965. Folder: Space Surveillance, DOD subseries, Federal Agencies series, NASA Historical Data Reference Collection, NASA Headquarters, Washington, DC, 1.

"Humphrey: 'Space Program is Here to Stay,'" *Technology Week* 5 (September 1966): 13.

Hungerford, Maj John B., Jr., USAF. *Organization for Military Space—A Historical Perspective.* Air Command and Staff College, Report No. 92-1235. Air University, Maxwell AFB, AL, 1982.

Hunley, J. D., ed. *The Birth of NASA: The Diary of T. Keith Glennan.* NASA SP-4105. Washington, DC: Government Printing Office, 1993.

Hunsaker, Jerome C. (NACA chairman from 1941–1956). "Forty Years of Aeronautical Research." In *Forty-Fourth Annual Report of the National Advisory Committee for Aeronautics, 1958, Final Report.* Washington, DC: Government Printing Office, 1959.

Huntington, Samuel. *The Common Defense: Strategic Programs in National Politics.* New York: Columbia University Press, 1961.

"Is U.S. Running Alone in the Race to the Moon? Interview with Sir Bernard Lovell." *US News and World Report*, 12 August 1963, 70–71.

"Interview, Samuel Phillips." *Data*, 22 April 1968, 11–12.

Johnson, Lyndon B., executive office of the president. Memorandum, 31 August 1960. H. R. 12049, in *Congressional Record*.

———. "The Politics of the Space Age." *Saturday Evening Post*, 29 February 1964, 22–24.

———. *Public Papers of the Presidents of the United States: Lyndon B. Johnson, 1963–64*. "Remarks to Employees of the Department of State." Vol. 1, no. 26, 5 December 1963. Washington, DC: Government Printing Office, 1965.

———. *Public Papers of the Presidents of the United States: Lyndon B. Johnson, 1963–64*. "Annual Message to the Congress on the State of the Union." Vol. 1, no. 91, 8 January 1964. Washington, DC: Government Printing Office, 1965.

———. *Public Papers of the Presidents of the United States: Lyndon B. Johnson, 1963–64*. "Remarks on Foreign Affairs at the Associated Press Luncheon in New York City." Vol. 1, no. 272, 20 April 1964. Washington, DC: Government Printing Office, 1965.

———. *Public Papers of the Presidents of the United States: Lyndon B. Johnson, 1963–64*. "The President's News Conference." Vol. 1, no. 294, 25 April 1964. Washington, DC: Government Printing Office, 1965.

———. *Public Papers of the Presidents of the United States: Lyndon B. Johnson, 1963–64*. "Remarks Following a Briefing with Space Scientists on the Successful Flight to the Moon." Vol. 2, no. 493, 1 August 1964. Washington, DC: Government Printing Office, 1965.

———. *Public Papers of the Presidents of the United States: Lyndon B. Johnson, 1963–64*. "Remarks after Inspecting Space Facilities at Cape Kennedy." Vol. 2, no. 572, 15 September 1964. Washington, DC: Government Printing Office, 1965.

———. *Public Papers of the Presidents of the United States: Lyndon B. Johnson, 1963–64*. "Radio and Television Report to the American People on Recent Events in Russia, China, and Great Britain." Vol. 2, no. 686, 18 October 1964. Washington, DC: Government Printing Office, 1965.

———. *Public Papers of the Presidents of the United States: Lyndon B. Johnson, 1965.* "Remarks Following a Briefing at the National Aeronautics and Space Administration." Vol. 1, no. 81, 25 February 1965. Washington, DC: Government Printing Office, 1966.

———. *Public Papers of the Presidents of the United States: Lyndon B. Johnson, 1965.* "The President's News Conference at the LBJ Ranch." Vol. 1, no. 117, 20 March 1965. Washington, DC: Government Printing Office, 1966.

———. *Public Papers of the Presidents of the United States: Lyndon B. Johnson, 1965.* "Remarks at the Presentation of the National Aeronautics and Space Administration Awards." Vol. 2, no. 134, 26 March 1965. Washington, DC: Government Printing Office, 1966.

———. *Public Papers of the Presidents of the United States: Lyndon B. Johnson, 1965.* "Commencement Address at Catholic University." Vol. 2, no. 302, 6 June 1965. Washington, DC: Government Printing Office, 1966.

———. *Public Papers of the Presidents of the United States: Lyndon B. Johnson, 1966.* "Statement by the President on the Need for a Treaty Governing Exploration of Celestial Bodies." Vol. 1, no. 209, 7 May 1966. Washington, DC: Government Printing Office, 1967.

———. *Public Papers of the Presidents of the United States: Lyndon B. Johnson, 1967.* "Annual Message to the Congress on the State of the Union." Vol. 1, no. 3, 10 January 1967. Washington, DC: Government Printing Office, 1968.

———. *Public Papers of the Presidents of the United States: Lyndon B. Johnson, 1967.* "Annual Budget Message to the Congress, Fiscal Year 1968." Vol. 1, no. 13, 24 January 1967. Washington, DC: Government Printing Office, 1968.

———. *Public Papers of the Presidents of the United States: Lyndon B. Johnson, 1967.* "Message to the Congress Transmitting Annual Report of the United States Arms Control and Disarmament Agency." Vol. 1, no. 59, 17 February 1967. Washington, DC: Government Printing Office, 1968.

———. *Public Papers of the Presidents of the United States: Lyndon B. Johnson, 1967.* "Statement by the President upon Signing Appropriations Bill for the National Aeronautics and Space Administration." Vol. 2, no. 362, 21 August 1967. Washington, DC: Government Printing Office, 1968.

———. *Public Papers of the Presidents of the United States: Lyndon B. Johnson, 1967.* "Remarks at Ceremony Marking the Entry Into Force of the Outer Space Treaty." Vol. 2, no. 425, 10 October 1967. Washington, DC: Government Printing Office, 1968.

———. *Public Papers of the Presidents of the United States: Lyndon B. Johnson, 1967.* "Remarks Following an Inspection of NASA's Michoud Assembly Facility Near New Orleans." Vol. 2, no. 533, 12 December 1967. Washington, DC: Government Printing Office, 1968.

———. *Public Papers of the Presidents of the United States: Lyndon B. Johnson.* "A Conversation with the President," Joint Interview for Use by Television Networks. Vol. 2, no. 554, 19 December 1967. Washington, DC: Government Printing Office, 1968.

———. *Public Papers of the Presidents of the United States: Lyndon B. Johnson, 1968–69.* "Remarks before the U.N. General Assembly Following Its Endorsement of the Nuclear Nonproliferation Treaty." Vol. 1, no. 308, 12 June 1968. Washington, DC: Government Printing Office, 1970.

———. *Public Papers of the Presidents of the United States: Lyndon B. Johnson, 1968–69.* "Remarks at the Signing of the Nuclear Nonproliferation Treaty." Vol. 2, no. 349, 1 July 1968. Washington, DC: Government Printing Office, 1970.

———. *Public Papers of the Presidents of the United States: Lyndon B. Johnson, 1968–69.* "Special Message to the Senate on the Astronaut Assistance and Return Agreement." Vol. 2, no. 387, 15 July 1968. Washington, DC: Government Printing Office, 1970.

———. *Public Papers of the Presidents of the United States: Lyndon B. Johnson, 1968–69.* "Remarks in New Orleans Before the 50th Annual National Convention of the American Legion." Vol. 2, no. 472, 10 September 1968. Washington, DC: Government Printing Office, 1970.

————. *Public Papers of the Presidents of the United States: Lyndon B. Johnson, 1968–69.* "The President's Toast and Responses at a Dinner Honoring Members of the Space Program." Vol. 2, no. 616, 9 December 1968. Washington, DC: Government Printing Office, 1970.

————. *Public Papers of the Presidents of the United States: Lyndon B. Johnson, 1968–69.* "Annual Budget Message to the Congress, Fiscal Year 1970," Vol. 2, no. 678, 15 January 1969. Washington, DC: Government Printing Office, 1970.

————. To the Heads of Departments and Agencies. Memorandum, 30 November 1963. Folder: Johnson Correspondence, NASA, box: White House, Presidents, Johnson, Correspondence, Declassified items, NASA Historical Data Reference Collection, NASA Headquarters, Washington, DC.

————. To James Webb. Memorandum, 29 September 1967. Folder: Johnson Correspondence, NASA, box: White House, Presidents, Johnson, Correspondence, Declassified items, NASA Historical Data Reference Collection, NASA Headquarters, Washington, DC.

————. To John F. Kennedy. Memorandum, 28 April 1961. In *Exploring the Unknown: Selected Documents in the History of the U.S. Civil Space Program.* Vol. 1, *Organizing for Exploration.* NASA SP-4407. Edited by John M. Logsdon, Linda J. Lear, Jannelle Warren-Findley, Ray A. Williamson, and Dwayne A. Day. Washington, DC: Government Printing Office, 1995.

————. To John F. Kennedy, 13 May 1963. Memorandum. In *Exploring the Unknown: Selected Documents in the History of the U.S. Civil Space Program.* Vol. 1, *Organizing for Exploration.* NASA SP-4407. Edited by John M. Logsdon, Linda J. Lear, Jannelle Warren-Findley, Ray A. Williamson, and Dwayne A. Day. Washington, DC: Government Printing Office, 1995.

————. To Robert McNamara and James Webb. Letters, 22 July 1963. Folder: Johnson Correspondence, NASA, box: White House, Presidents, Johnson, Correspondence, Declassified Items, NASA Historical Data Reference Collection, NASA Headquarters, Washington, DC.

————. Statement of Democratic leader Lyndon B. Johnson to the meeting of the Democratic Conference on 7 January 1958. Folder: Armed Services, ICBM-Sputnik, box 584, Clinton P. Anderson Papers, Library of Congress.

————. Statement opening the vice president's ad hoc meeting on space, 3 May 1961. Space Policy Institute Archives document 1121. In *Exploring the Unknown: Selected Documents in the History of the U.S. Civil Space Program*. Vol. 1, *Organizing for Exploration*. NASA SP-4407. Edited by John M. Logsdon, Linda J. Lear, Jannelle Warren-Findley, Ray A. Williamson, and Dwayne A. Day. Washington, DC: Government Printing Office, 1995. Reprints the transcript of the meeting but not Johnson's opening statement.

————. *United State Aeronautics and Space Activities, 1963*. Report to Congress from the president, 27 January 1964. NSA-MUS document 329.

————. *US Aeronautics and Space Activities, 1964*. Report to Congress from the president, 27 January 1965. NASA Historical Data Reference Collection, NASA Headquarters, Washington, DC.

————. *US Aeronautics and Space Activities, 1965*. Report to Congress from the president, 31 January 1966. NSA-MUS document 332.

————. *US Aeronautics and Space Activities, 1966*. Report to the Congress from the president, 1 January 1967. NSA-MUS document 333.

————. *US Aeronautics and Space Activities, 1967*. Report to the Congress from the president, January 1968. NSA-MUS document 334.

————. *US Aeronautics and Space Activities, 1968*. Report to the Congress from the president, January 1969. NSA-MUS document 335.

————. *US Aeronautics and Space Report of the President, 1969*. Report to the Congress from the president, January 1970. NASA Historical Data Reference Collection, NASA Headquarters, Washington, DC.

————. *The Vantage Point: Perspectives on the Presidency, 1963–1969*. New York: Holt, Rinehart and Winston, 1971.

———. "The Vision of a Greater America." *The General Electric Forum for National Security and Free World Progress* V (July–September 1962).

Johnson, Roy. To all ARPA staff. Memorandum, 14 October 1958. Space Policy Institute Archives document 1439.

Johnson, Robert H. *Improbable Dangers: US Conceptions of Threat in the Cold War and After*. New York: St. Martin's Press, 1994.

Johnson, S. Paul. To James R. Killian Jr. Memorandum, 21 February 1958. "Preliminary Observations on the Organization for the Exploitation of Outer Space." In *Exploring the Unknown: Selected Documents in the History of the U.S. Civil Space Program*. Vol. 1, *Organizing for Exploration*. NASA SP-4407. Edited by John M. Logsdon, Linda J. Lear, Jannelle Warren-Findley, Ray A. Williamson, and Dwayne A. Day. Washington, DC: Government Printing Office, 1995.

Johnson, U. Alexis. Representative of the Department of State. Concurred to by representatives of the DOD, CIA, ACDA, NRO, NASA, and OST. Report of the National Aeronautics and Space Administration 156 Committee, 2 July 1962. Folder: NSAM 136–56, box 3, RG 59, General Records of the Department of State, National Archives and Records Administration. Declassification date: 31 December 1996.

Kaplan, Fred. *Wizards of Armageddon*. New York: Simon and Schuster, 1983.

Kaplan, Joseph. To Alan Waterman. Letter, 14 March 1955. Folder: OCB 000.91 Natural & Physical Sciences (2), box: 11, OCB Central files subseries, NSC: Staff Papers series, White House Office, Dwight D. Eisenhower Library, Abilene, KS.

Kash, Don E. *The Politics of Space Cooperation*. West Lafayette, IN: Purdue University Studies, 1967.

Kaufmann, William W. *The McNamara Strategy*. New York: Harper & Row, 1964.

Kearns, Doris. *Lyndon Johnson and the American Dream*. New York: Harper & Row Publishers, 1976.

Kecskemeti, Paul. *The Satellite Rocket Vehicle: Political and Psychological Problems*. Santa Monica, CA: Project RAND,

4 October 1950. RAND Report RM-567. Space Policy Institute Archives document 1284.

Keefer, Edward C., and David W. Mabon, eds. *Foreign Relations of the United States, 1958–1960*. Vol. *3, National Security Policy: Arms Control and Disarmament.* Washington, DC: Government Printing Office, 1996.

Kelly, C. Brian. "Ten Years in the Outer Realm." *Data,* June 1968, 22–24.

Kennan, Erlend A., and Edmund H. Harvey. *Mission to the Moon: A Critical Examination of NASA and the Space Program.* New York: William Morrow & Company, 1969.

Kennan, George. "The Sources of Soviet Conduct." *Foreign Affairs* 25 (July 1947): 566–82.

Kennedy, John F., executive office of the president. *U.S. Aeronautics and Space Activities, 1961.* Message to the Congress from the President of the United States, 31 January 1962. NSA-MUS document 326.

———. "If the Soviets Control Space, They Can Control Earth." *Missiles and Rockets,* 10 October 1960, 12–13.

———. *Public Papers of the Presidents of the United States: John F. Kennedy, 1961.* "Inaugural Address," no. 1, 20 January 1961. Washington, DC: Government Printing Office, 1962.

———. *Public Papers of the Presidents of the United States: John F. Kennedy, 1961.* "The President's News Conference," no. 8, 25 January 1961. Washington, DC: Government Printing Office, 1964.

———. *Public Papers of the Presidents of the United States: John F. Kennedy, 1961.* "Annual Message to the Congress on the State of the Union," no. 11, 30 January 1961. Washington, DC: Government Printing Office, 1962.

———. *Public Papers of the Presidents of the United States: John F. Kennedy, 1961.* "Special Message to the Congress on the Defense Budget," no. 99, 28 March 1961. Washington, DC: Government Printing Office, 1962.

———. *Public Papers of the Presidents of the United States: John F. Kennedy, 1961.* "The President's News Conference," no. 119, 12 April 1961. Washington, DC: Government Printing Office, 1962.

———. *Public Papers of the Presidents of the United States: John F. Kennedy, 1961.* "The President's News Conference," no. 139, 21 April 1961. Washington, DC: Government Printing Office, 1962.

———. *Public Papers of the Presidents of the United States: John F. Kennedy, 1961.* "Address 'The President and the Press' Before the American Newspaper Publishers Association, New York City," no. 153, 27 April 1961. Washington, DC: Government Printing Office, 1962.

———. *Public Papers of the Presidents of the United States: John F. Kennedy, 1961.* "Special Message to the Congress on Urgent National Needs," no. 205, 25 May 1961. Washington, DC: Government Printing Office, 1962.

———. *Public Papers of the Presidents of the United States: John F. Kennedy, 1961.* "Address in New York City before the General Assembly of the United Nations," no. 387, 25 September 1961. Washington, DC: Government Printing Office, 1962.

———. *Public Papers of the Presidents of the United States: John F. Kennedy, 1961.* "The Presidents News Conference," no. 415, 11 October 1961. Washington, DC: Government Printing Office, 1962.

———. *Public Papers of the Presidents of the United States: John F. Kennedy, 1961.* "Address in Los Angeles at a Dinner of the Democratic Party of California," no. 477, 18 November 1961. Washington, DC: Government Printing Office, 1962.

———. *Public Papers of the Presidents of the United States: John F. Kennedy, 1962.* "The President's News Conference," no. 245, 14 June 1962. Washington, DC: Government Printing Office, 1963.

———. *Public Papers of the Presidents of the United States: John F. Kennedy, 1962.* "Address at Rice University in Houston on the Nation's Space Effort," no. 373, 12 September 1962. Washington, DC: Government Printing Office, 1963.

———. *Public Papers of the Presidents of the United States: John F. Kennedy, 1962.* "The President's News Conference," no. 515, 20 November 1962. Washington, DC: Government Printing Office, 1963.

———. *Public Papers of the Presidents of the United States: John F. Kennedy, 1963.* "Annual Message to the Congress on the State of the Union," no. 12, 14 January 1963. Washington, DC: Government Printing Office, 1964.

———. *Public Papers of the Presidents of the United States: John F. Kennedy, 1963.* "Commencement Address at American University in Washington," no. 232, 10 June 1963. Washington, DC: Government Printing Office, 1964.

———. *Public Papers of the Presidents of the United States: John F. Kennedy, 1963.* "Remarks in the Rudolph Wilde Platz, Berlin," no. 269, 26 June 1963. Washington, DC: Government Printing Office, 1964.

———. *Public Papers of the Presidents of the United States: John F. Kennedy, 1963.* "Remarks in Naples at NATO Headquarters," no. 291, 2 July 1963. Washington, DC: Government Printing Office, 1964.

———. *Public Papers of the Presidents of the United States: John F. Kennedy, 1963.* "The President's News Conference," no. 305, 17 July 1963. Washington, DC: Government Printing Office, 1964.

———. *Public Papers of the Presidents of the United States: John F. Kennedy, 1963.* "Address before the 18th General Assembly of the United Nations," no. 366, 20 September 1963. Washington, DC: Government Printing Office, 1964.

———. *Public Papers of the Presidents of the United States: John F. Kennedy, 1963.* "Address at the University of Maine," no. 426, 19 October 1963. Washington, DC: Government Printing Office, 1964.

———. *Public Papers of the Presidents of the United States: John F. Kennedy, 1963.* "The President's News Conference," no. 448, 31 October 1963. Washington, DC: Government Printing Office, 1964.

———. *Public Papers of the Presidents of the United States: John F. Kennedy, 1963.* "The President's News Conference," no. 459, 14 November 1963. Washington, DC: Government Printing Office, 1964.

———. *Public Papers of the Presidents of the United States: John F. Kennedy, 1963.* "Remarks in San Antonio at the Dedication of the Aerospace Medical Health Center," no.

472, 21 November 1963. Washington, DC: Government Printing Office, 1964.

———. *Public Papers of the Presidents of the United States: John F. Kennedy, 1963.* "Remarks Intended for Delivery to the Texas Democratic State Committee in the Municipal Auditorium in Austin," no. 478, 22 November 1963. Washington, DC: Government Printing Office, 1964.

———. To Rep. Albert Thomas. Letter, 23 September 1963. Reprinted in House, Committee on Appropriations, *Independent Offices Appropriations Bill, 1964*, Report No. 824, 88th Cong., 1st sess., 7 October 1963.

———. To Overton Brooks. Letter, 23 March 1961. Reprinted in *Exploring the Unknown: Selected Documents in the History of the U.S. Civil Space Program*. Vol. 2, *Relations with Other Organizations*. NASA SP-4407. Edited by John M. Logsdon, Dwayne A. Day, and Roger D. Launius. Washington, DC: Government Printing Office, 1996.

———. To Nikita Khrushchev. Letter, 22 February 1962. Folder: Kennedy Correspondence (NASA), White House, Presidents, Kennedy, 25 May 1961 speech through JFK/NASA Correspondence, NASA Historical Data Reference Collection, NASA Headquarters, Washington, DC.

———. To Vice Pres. Lyndon B. Johnson. Memorandum, 20 April 1961. In *Exploring the Unknown: Selected Documents in the History of the U.S. Civil Space Program*. Vol. 1, *Organizing for Exploration*. NASA SP-4407. Edited by John M. Logsdon, Linda J. Lear, Jannelle Warren-Findley, Ray A. Williamson, and Dwayne A. Day. Washington, DC: Government Printing Office, 1995.

———. To Vice Pres. Lyndon B. Johnson. Memorandum, 9 April 1963. In *Exploring the Unknown: Selected Documents in the History of the U.S. Civil Space Program*. Vol. 1, *Organizing for Exploration*. NASA SP-4407. Edited by John M. Logsdon, Linda J. Lear, Jannelle Warren-Findley, Ray A. Williamson, and Dwayne A. Day. Washington, DC: Government Printing Office, 1995.

Kesaris, Paul L., ed. *Presidential Campaigns: The John F. Kennedy 1960 Campaign, Part I: Polls, Issues, and Strategy.* A microfilm collection from the holdings of the John F.

Kennedy Library. Part of the series edited by William Leuchtenburg, *Research Collections in American Politics: Microforms from Major Archival and Manuscript Collections.* Frederick, MD: University Publications of America, Inc., 1986.

―――. *Presidential Campaigns: The John F. Kennedy Campaign, Part II: Speeches, Press Conferences, and Debates.* A microfilm collection from the holdings of the John F. Kennedy Library. Part of the series edited by William Leuchtenburg, *Research Collections in American Politics: Microforms from Major Archival and Manuscript Collections.* Frederick, MD: University Publications of America, Inc., 1986.

―――, and Robert Lester. Project Coordinators. *President John F. Kennedy's Office Files, 1961–1963, Part 1: Special Correspondence, Speeches, Legislative, and Press Conference Files.* In *Research Collections in American Politics: Microforms from Major Archival and Manuscript Collections.* Edited by William Leuchtenberg. Frederick, MD: University Publications of America, Inc., 1989. Also including *Part V: Countries.*

Khrushchev, Nikita. *Khrushchev Remembers: The Last Testament.* Boston: Little, Brown and Company, 1974.

Killian, James R., Jr. A Brief Summary prepared for the first NASC meeting, 5 August 1958. Tab 1-2, box 1, Record Group (RG) 200, National Archives and Record Administration, 14. Document is now declassified.

―――. Memorandum, 30 December 1957. In *Exploring the Unknown: Selected Documents in the History of the U.S. Civil Space Program.* Vol. 1, *Organizing for Exploration.* NASA SP-4407. Edited by John M. Logsdon, Linda J. Lear, Jannelle Warren-Findley, Ray A. Williamson, and Dwayne A. Day. Washington, DC: Government Printing Office, 1995.

―――. To the Dallas Council on World Affairs. Speech, 23 September 1960. Folder: Ad Hoc Man-in-Space Panel, box 65, Subject files, 1957-62, RG 359, Office of Science and Technology, National Archives and Records Administration.

———. *The Education of a College President*. Cambridge, MA: The MIT Press, 1985.

———. *Sputnik, Scientists, and Eisenhower: A Memoir of the First Special Assistant to the President for Science and Technology*. Cambridge, MA: The MIT Press, 1977.

———. Oral history interviews, 9 November 1969 through 16 July 1970. Dwight D. Eisenhower Library, Abilene, KS.

———. Oral history interview, 23 July 1974. Biographical series, Killian file, NASA Historical Data Reference Collection, NASA Headquarters, Washington, DC.

———, Percival Brundage, and Nelson Rockefeller. To President Eisenhower. Memorandum, 5 March 1958. Folder: No. 174, Space Program, 1958–60, box 169, President's Advisory Committee on Government Organization, Dwight D. Eisenhower Library, Abilene, KS.

Killebrew, Maj Timothy, USAF. *Military Man in Space: A History of the Air Force Efforts to Find a Manned Space Mission*, Air Command and Staff College Report No. 87-1425. Maxwell, AFB, AL: Air University, May 1987.

Kinnard, Douglas. *President Eisenhower and Strategy Management: A Study in Defense Politics*. Lexington, KY: University Press of Kentucky, 1977.

———. "President Eisenhower and the Defense Budget." *Journal of Politics* 39 (August 1977): 596–623.

Kistiakowsky, George B. To Staff Secretary Goodpaster. Memorandum, September 1959. Folder: Kistiakowsky (2), box 23, Administration series, Ann Whitman file, Dwight D. Eisenhower Library, Abilene, KS.

———. *A Scientist at the White House: The Private Diary of President Eisenhower's Special Assistant for Science and Technology*. Cambridge, MA: Harvard University Press, 1976.

———. Transcript of oral history interview, 22 May 1974. File: Kistiakowsky, box: Emme/Roland interviews on early NASA history, shelf V-A-1, NASA Historical Data Reference Collection, NASA Headquarters, Washington, DC.

Klass, Philip J. "Big Bird Nears Full Operational Status." *Aviation Week and Space Technology*, 25 September 1972, 17.

———. "Military Satellites Gain Vital Data." *Aviation Week and Space Technology*, 15 September 1970, 55–61.

————. *Secret Sentries in Space.* New York: Random House, 1971.

Kohler, Foy D. "An Overview of US-Soviet Space Relations." Foreword in *U.S.-Soviet Cooperation in Space* by Dodd L. Harvey and Linda C. Ciccoritti. Miami, FL: Monographs in International Affairs, Center for Advanced International Studies, 1974.

Kolcum, Edward H. "Defense May Ease Impact of X-20 Loss." *Aviation Week and Space Technology,* 18 March 1963, 31.

Konecci, Eugene B., NASC staffer. To Edward Welsh, NASC executive secretary. Memorandum, 17 June 1965. Folder: Space Projects-MOL, box 21, RG 220, Records of the National Aeronautics and Space Council, National Archives and Records Administration.

Koppes, Clayton R. "The Militarization of the American Space Program." *Virginia Quarterly Review* 60 (1984): 1–20.

Kraemer, Sylvia K. "NASA and the Challenge of Organizing for Exploration." In *Organizing for the Use of Space: Historical Perspectives on a Persistent Issue.* Edited by Roger Launius. Vol. 18, AAS History Series. San Diego, CA: Univelt, Inc., 1995.

Kunz, Diane B., ed. "Introduction: The Crucial Decade." In *The Diplomacy of the Crucial Decade: American Foreign Relations during the 1960s.* New York: Columbia University Press, 1994.

Lambright, W. Henry. "James Webb and the Uses of Administrative Power." In *Leadership and Innovation: A Biographical Perspective on Entrepreneurs in Government.* Edited by Jameson W. Doig and Erwin C. Hargrove. Baltimore: The Johns Hopkins University Press, 1987.

————. *Powering Apollo: James E. Webb of NASA.* Baltimore: The Johns Hopkins University Press, 1995.

————. *Presidential Management of Science and Technology: The Johnson Presidency.* Austin, TX: University of Texas Press, 1985.

Lapidus, Robert. "Sputnik and Its Repercussions: A Historical Catalyst." *Aerospace Historian* 17 (Summer–Fall 1970): 88–93.

Lapp, Ralph. Advisory Committee on Science and Technology of the Democratic Advisory Council. Position Paper, prepared for Senator John F. Kennedy, 7 September 1960. Papers of the historian, NASA Historical Data Reference Collection, NASA Headquarters, Washington, DC.

Lapp, R. E., Frank McClure, and Trevor Gardner. Position Paper, 31 August 1960. In Paul L. Kesaris, *Presidential Campaigns: The John F. Kennedy 1960 Campaign, Part I: Polls, Issues, and Strategy.* A microfilm collection from the holdings of the John F. Kennedy Library. Part of the series edited by William Leuchtenburg. *Research Collections in American Politics: Microforms from Major Archival and Manuscript Collections.* Frederick, MD: University Publications of America, Inc., 1986, reel 4.

Large Launch Vehicle Planning Group. Summary Report: NASA-DOD Large Launch Vehicle Planning Group, 24 September 1962. Reprinted in *Exploring the Unknown: Selected Documents in the History of the U.S. Civil Space Program.* Vol. 2, *Relations with Other Organizations.* NASA SP-4407. Edited by John M. Logsdon, Dwayne A. Day, and Roger D. Launius. Washington, DC: Government Printing Office, 1996.

Launius, Roger D. *NASA: A History of the U.S. Civil Space Program.* Malabar, FL: Krieger Publishing Company, 1994.

———, ed. "Early U.S. Civil Space Policy, NASA, and the Aspiration of Space Exploration." In *Organizing for the Use of Space: Historical Perspectives on a Persistent Issue.* American Astronautical Society History Series, Vol. 18. San Diego, CA: Univelt, Inc., 1996.

———. "Eisenhower, Sputnik, and the Creation of NASA." *Prologue: The Quarterly Journal of the National Archives and Record Administration,* Summer 1996, 127–40.

———. " 'Never was Life More Interesting': The National Advisory Committee for Aeronautics, 1936–1945." *Prologue: Quarterly Journal of the National Archives,* Winter 1992, 361–73.

———, and Dennis R. Jenkins. *To Reach the High Frontier: A History of U.S. Launch Vehicles.* Lexington, KY: University Press of Kentucky, 2002.

Lehrer, Max. To Lyndon Johnson. Memorandum, 31 October 1960. Space Policy Institute Archives document 498.

LeMay, Curtis E. Oral history interview, January 1965. K239-0512-714, Air Force Historical Research Agency, Maxwell AFB, AL.

———. To CSAF White. Status Report, 15 December 1959. Project Dynasoar. Folder: 4–5 Missiles/Space/Nuclear, box 36, Thomas White Papers, Library of Congress.

———, vice-CSAF. To Zuckert. Letter, 19 October 1962. Serving as a cover for the revised Air Force Five-Year Space Plan. Folder: 6-1962, Box B128, Curtis LeMay Papers, Library of Congress.

Lester, Robert, ed. *The Diaries of Dwight D. Eisenhower, 1953–1961.* Microfilmed from the Holdings of the Dwight D. Eisenhower Library. Part of the *Research Collections in American Politics: Microforms from Major Archival and Manuscript Collections,* William Leuchtenburg, general editor. Bethesda, MD: University Publications of America, 1986.

Levine, Alan J. *The Missile and Space Race.* Westport, CT: Praeger, 1994.

Levine, Arnold S. "Management of Large-Scale Technology." In *A Spacefaring People: Perspectives on Early Spaceflight.* Edited by Alex Roland. NASA SP-4405. Washington, DC: Government Printing Office, 1985.

———. *Managing NASA in the Apollo Era.* NASA SP-4102. Washington, DC: Government Printing Office, 1982.

Levine, Arthur L. *The Future of the U.S. Space Program.* New York: Praeger Publishers, 1975.

Levy, Lillian. "Conflict in the Race for Space." In *Space: Its Impact on Man and Society.* Edited by Lillian Levy. New York: W.W. Norton & Company, 1965.

Link, Mae Mills. *Space Medicine in Project Mercury.* NASA SP-4003. Washington, DC: Government Printing Office, 1965.

Lockheed Corporation. Press Release, May 1995. Archives unnumbered document. Space Policy Institute, Center for International Science and Technology Policy, George Washington University.

Lloyd, O. B., Jr., director, NASA Office of Public Services and Information. "Letter to the Editor." *Missiles and Rockets*, 14 May 1962, 7.

Logsdon, John M. "The Apollo Decision in Historical Perspective." In *Apollo: Ten Years Since Tranquillity Base*. Edited by Richard P. Hallion. Washington, DC: Smithsonian Institution Press, 1979.

———. *The Decision to Go to the Moon: Project Apollo and the National Interest*. Cambridge, MA: The MIT Press, 1970.

———. "Evaluating Apollo." Remarks at a symposium, 19 July 1989, *Apollo in its Historical Context*. Organized by the Space Policy Institute, Center for International Science and Technology Policy, George Washington University.

———. "The Evolution of Civilian Space Exploitation." *Futures, The Journal of Forecasting and Planning* 14 (October 1982): 393–404.

———. "The Evolution of U.S. Space Policy and Plans." In *Exploring the Unknown: Selected Documents in the History of the U.S. Civil Space Program*. Vol. 1, *Organizing for Exploration*. NASA SP-4407. Edited by John M. Logsdon, Linda J. Lear, Jannelle Warren-Findley, Ray A. Williamson, and Dwayne A. Day. Washington, DC: Government Printing Office, 1995.

———. *NASA's Implementation of the Lunar Landing Decision*. Washington, DC: NASA HHN-81, August 1969.

———. "National Leadership and Presidential Power." Chapter 7 in *The Myth of Presidential Leadership*. Edited by Roger D. Launius. Urbana, IL: University of Illinois Press, 1997.

———. "Opportunities for Policy Historians: The Evolution of the U.S. Civilian Space Program." In *A Spacefaring People: Perspectives on Early Spaceflight*. NASA SP-4405. Edited by Alex Roland. Washington, DC: Government Printing Office, 1985.

———. Oral history interview, Robert C. Seamans Jr., 5 December 1967. Folder: Seamans/Logsdon interview, Seamans subseries, Deputy Administrators series, NASA Historical Data Reference Collection, NASA Headquarters, Washington, DC.

————, Linda J. Lear, Jannelle Warren-Findley, Ray A. Williamson, and Dwayne A. Day, eds. *Exploring the Unknown: Selected Documents in the History of the U.S. Civil Space Program.* Vol. 1, *Organizing for Exploration.* NASA SP-4407. Washington, DC: Government Printing Office, 1995.

————, Dwayne A. Day, and Roger D. Launius, eds. *Exploring the Unknown: Selected Documents in the History of the U.S. Civil Space Program.* Vol. 2, *Relations with Other Organizations.* NASA SP-4407. Washington, DC: Government Printing Office, 1996.

————, and Alain Dupas. "Was the Race to the Moon Real?" *Scientific American,* June 1994, 16–23.

Low, George M., director, Spacecraft and Flight Missions, Office of Manned Space Flight, NASA. To director of the Manned Spacecraft Center, NASA-DOD Operational and Management Plan for Gemini. Memorandum with attachment, 7 February 1962. Space Policy Institute Archives document 450.

Mabon, David, and David Patterson, eds. *Foreign Relations of the United States, 1961–1963.* Vol. 7, *Arms Control and Disarmament.* Washington, DC: Government Printing Office, 1995.

Macdonald, Duncan E. To Brian O'Brien, chairman, NAS Advisory Committee to USAF Systems Command, for transmission to General Schriever. Report, 27 May 1966. 168.7171-158, Air Force Historical Research Agency, Maxwell AFB, AL.

MacIntyre, Malcolm. To ARPA director Roy Johnson, Advanced Program Areas for Military Space Systems. Memorandum, 14 April 1959. Space Policy Institute Archives document 274.

Maier, Charles S. "Science, Politics, and Defense in the Eisenhower Era." In *A Scientist at the White House: The Private Diary of President Eisenhower's Special Assistant for Science and Technology,* by George B. Kistiakowsky, xxix. Cambridge, MA: Harvard University Press, 1976.

Malloy, James A. "The Dryden-Blagonravov Era of Space Cooperation, 1962–1965." *Aerospace Historian,* March 1977, 40–45.

Mandelbau, Leonard. "Apollo: How the United States Decided to Go to the Moon." *Science* 163 (February 1969): 649–54.

Martin, Col John L., deputy director of advanced technology, deputy chief of staff for development. To secretary of the Air Force, Legislative Liaison. Report, 10 November 1959. K140.11-7, Air Force History Support Office.

Mayhall, Gene, and D. O. Appleton. "Military Applications of Space Are Inevitable: Interview with Maj Gen Sam Phillips." *Data*, June 1968, 17–19.

McDougall, Walter A. *The Heavens and the Earth: A Political History of the Space Age*. New York: Basic Books, Inc., Publishers, 1985.

———. "Technocracy and Statecraft in the Space Age—Toward the History of a Saltation." *American Historical Review* 87 (October 1982): 1010–40.

McElroy, Neil. To chairman of the Joint Chiefs of Staff. Memorandum, 18 September 1959. In Briefing Book for Air Force witnesses before the House Committee on Science and Astronautics on the subject of DOD Space Directive 5160.32. K160.8636-4, Air Force History Support Office, 1961.

———. Transcript of remarks at SECDEF luncheon, 1 November 1957. NSA-MUS document 270.

McMillan, Brockway, assistant SAF for Research and Development. To Zuckert. Memorandum, 15 March 1963. Documents in the possession of Maj Roy Houchin, Air Command and Staff College, Maxwell AFB AL.

———, undersecretary of the Air Force. To Colonel Schultz, assistant for Manned Orbiting Laboratory, AFRMO. Memorandum, 2 July 1964. IRIS 1002993, Air Force History Support Office.

———, undersecretary of the Air Force, and Robert C. Seamans, NASA associate administrator. DOD/NASA Agreement on the NASA Manned Lunar Mapping and Survey Program, 20 April 1964. Space Policy Institute Archives document 228.

McNamara, Robert S. *The Essence of Security: Reflections in Office*. New York: Harper & Row, 1968.

———. To Eugene Zuckert. Memorandum, undated but probably 14 or 15 October 1961. 168.7050-54, Air Force Historical Research Agency, Maxwell AFB, AL. Document is now declassified.

———. To DOD Directive 5160.32. Cover letter, 6 March 1961. Folder: Defense, 1961, box 17, RG 220, Records of the National Aeronautics and Space Council, National Archives and Records Administration.

———. To Harold Brown. Two memoranda, 18 and 19 January 1963. Folder: 6-1963, box B129, Curtis LeMay Papers. Library of Congress.

———. Interview in *Missiles and Rockets*. 22 October 1962. In USAF booklet, *The Military Mission in Space: A Selection of Published Views, August 1962–June 1963*, 7 June 1963. Folder: Military Mission in Space, DOD subseries, Federal Agencies series, NASA Historical Data Reference Collection, NASA Headquarters, Washington, DC.

———. To James Webb. Letter, 16 September 1963. In *Exploring the Unknown*, Vol. 2, *Relations with Other Organizations*. NASA SP-4407. Edited by John M. Logsdon, Dwayne A. Day, and Roger D. Launius. Washington, DC: Government Printing Office, 1996.

———. To James Webb. Letter, 25 September 1964. IRIS 1003002, Air Force History Support Office.

———. To James Webb. Letter, 2 November 1966. Folder: NASA-DOD Correspondence, box: Arnold Levine, Selected Sources from the author, NASA Historical Data Reference Collection, NASA Headquarters, Washington, DC.

———. To Lyndon B. Johnson. Memorandum, 21 April 1961. In *Exploring the Unknown*. Vol. 1, *Organizing for Exploration*, 424–26.

———. Proposed Agreement between the Department of Defense and the National Aeronautics and Space Administration, Gemini Program Management, 12 January 1963. Space Policy Institute Archives document 888.

———. To the president. Memorandum, 26 October 1966. In *Lyndon B. Johnson National Security Files, Agency File, 1963–1969*. Edited by George C. Herring. Microfilmed from the holdings of the Lyndon B. Johnson Library, Robert E. Lester, Project Coordinator, Bethesda, MD: University Publications of America, 1993, reel 9, 1.

―――. To the secretary of the Air Force [Zuckert] Memorandum, 22 February 1962. Folder: Reading file, January–May 1962, box 114, RG 200, Robert McNamara Papers, National Archives and Records Administration. Document is now declassified.

―――. To the secretary of the Air Force. Memorandum, 15 March 1963. Folder: Reading file, February–May 1963, box 117, RG 200, Robert McNamara Papers, National Archives and Records Administration. Document is now declassified.

―――. To the secretary of the Air Force. Memorandum, 25 May 1963. Folder: 31 May–18 June 1963, Reading file, box 117, RG 200, Robert McNamara Papers, National Archives and Records Administration. Document is now declassified.

―――. To the secretary of the Air Force. Memorandum, 20 June 1963. Folder: Reading file, 20–29 June 1963, box 118, RG 200, Robert McNamara Papers, National Archives and Records Administration.

―――. To the secretary of the Air Force [Harold Brown]. Memorandum, 9 December 1967. Folder: Reading file: October 1967–February 1968, box 131, RG 200, Robert McNamara Papers, National Archives and Records Administration. Document is now declassified.

―――. To the vice president. Memorandum, 9 August 1963. Folder: NASC, 1962–1972, box: National Aeronautics and Space Council, NASA Historical Data Reference Collection, NASA Headquarters, Washington, DC.

―――. Note to author, 15 October 1996, after several requests for oral history interview.

―――. Press Conference, 23 June 1961. Folder: News Conferences and Press Briefings, 1961, box 182, RG 200, Robert McNamara Papers, National Archives and Records Administration.

―――. To Lyndon Johnson. Report, 3 May 1963. Reprinted in *Exploring the Unknown*, Vol. 2, *Relations with Other Organizations*. NASA SP-4407. Edited by John M. Logsdon, Dwayne A. Day, and Roger D. Launius. Washington, DC: Government Printing Office, 1996.

―――. Statement before the House Armed Services Committee on the 1965 Defense Budget, 27 January 1964.

Folder: Unclassified Statement, FY 65, box 32, RG 200, Robert McNamara Papers, National Archives and Records Administration.

———. Statement before the House Armed Services Committee on the 1966 Defense Budget, 18 February 1965. Folder: Unclassified Statement, 1966 Defense Budget, box 44, RG 200, Robert McNamara Papers, National Archives and Records Administration.

———. Statement before a Joint Session of the Senate Armed Services Committee and the Senate Subcommittee on Department of Defense Appropriations on the FY 1968 Defense Budget, 23 January 1967. Folder: Unclassified Statement, box 69, RG 200, Robert McNamara Papers, National Archives and Records Administration. Document is now declassified.

———. Statement before the Senate Armed Services Committee, 4 April 1961. Folder: Statements to Congressional Committees, FY 62, box 11, RG 200, Robert S. McNamara Papers, National Archives and Records Administration.

———. Statement before the Senate Armed Services Committee, 19 January 1962. Folder: Miscellaneous Budget, box 114, RG 200, Robert S. McNamara Papers, National Archives and Records Administration. Document is now declassified.

———. Statement on *Military Posture: Hearings before the Senate Armed Services Committee on the FY 1964–68 Defense Program*. 88th Congress, 1st sess., 21 January 1963. Folder: Statement to Congress 1964 Budget, box 22, RG 200, Robert McNamara Papers, National Archives and Records Administration.

———. Transcript of News Briefing. 10 December 1963. Folder: Dyna-Soar, DOD subseries, Federal Agencies series, NASA Historical Data Reference Collection, NASA Headquarters, Washington, DC, 1–2.

———. Transcript of News Conference. 12 December 1964. Folder: Space Projects-MOL, box 21, RG 220, Records of the National Aeronautics and Space Council, National Archives and Records Administration.

————, and James Webb. Joint Statement, 25 January 1965. Folder: Space Projects-MOL, box 21, RG 220, Records of the National Aeronautics and Space Council, National Archives and Records Administration.

Means, Paul. "Vega-Agena-B Mix-Up Cost Millions." Missiles and Rockets, 20 June 1960, 19–20.

Medaris, Maj Gen John B., USA, retired. *Countdown for Decision.* With Arthur Gordon. New York: G. P. Putnam's Sons, 1960.

Meier, Kenneth J. *Politics and the Bureaucracy: Policymaking in the Fourth Branch of Government.* North Scituate, MA: Duxbury Press, 1979.

Meilinger, Col Phillip S., USAF. *10 Propositions Regarding Air Power.* Washington, DC: Government Printing Office, 1995.

Miller, Rep. George. Speech. *Congressional Record,* 18670–18674, September 1962.

Miller, Col James E., director, Program Support, AFSC. To the directorate of NASA Program Support, deputy commander for Space, AFSC, USAF. Report, 9 December 1965. IRIS 1003002, Air Force History Support Office.

Miller, Susan, comp. *Statements of John F. Kennedy on Space Exploration, 1952–1963.* NASA HHN-26, 1964. NASA Historical Data Reference Collection, NASA Headquarters, Washington, DC.

Minnich, L. A. Legislative Leadership Meeting—Supplementary Notes, 7 January 1958. Folder: January 1958 Staff Notes, box 30, DDE Diary Series, Ann Whitman file, Dwight D. Eisenhower Library, Abilene, KS.

————. Legislative Leadership Meeting—Supplementary Notes, 4 February 1958. Folder: Staff Notes, February 1958, box 30, DDE Diary series, Ann Whitman file, Dwight D. Eisenhower Library, Abilene, KS.

————. Legislative Leadership Meeting—Supplementary Notes, 10 March 1959. Folder: Staff Notes, March 1–15 (1), box 39, DDE Diary series, Ann Whitman file, Dwight D. Eisenhower Library, Abilene, KS.

Minutes of Cabinet Meeting, 18 October 1957. Folder: Cabinet Meeting of 18 October 1957, box 9, Cabinet series, Ann Whitman file, Dwight D. Eisenhower Library, Abilene, KS.

———. Fourth meeting of the Joint Research and Development Board of the War and Navy Departments, 6 March 1947. Folder: Navy/BuAer: Earth Satellite Vehicle, DOD subseries, Federal Agencies series, NASA Historical Data Reference Collection, NASA Headquarters, Washington, DC.

———. 3 January 1958. Folder: January 1958 Staff Notes, box 30, DDE Diary Series, Ann Whitman file, Dwight D. Eisenhower Library, Abilene, KS.

"Missiles Gain, Space Suffers at DOD." *Missiles and Rockets*, 31 January 1966, 15–18.

"MOL Delayed by Funding Cut." *Aviation Week and Space Technology*, 21 April 1969, 17.

Montgomery, Maj Gen R. M., assistant vice-chief of staff. To multiple USAF recipients. Memorandum, 23 April 1962. IRIS 1003001, Air Force History Support Office.

Moore, Col W. L., Wright Air Development Division. Dyna-Soar Program Status. Remarks. USAF-NASA Conference on Lifting Manned Hypervelocity and Reentry Vehicles, Part 2, 11–14 April 1960. IRIS 1003000, Air Force History Support Office.

Morgan, Iwan. "Eisenhower and the Balanced Budget." In *Reexamining the Eisenhower Presidency*. Edited by Shirley Anne Warshaw. Westport, CT: Greenwood Press, 1993.

Mrozinski, R. V., NASC. To William Moore, director of Technical Staff, NASA. Memorandum, 1 April 1968. Folder: MOL 1968 and 1969, DOD subseries, Federal Agencies series, NASA Historical Data Reference Collection, NASA Headquarters, Washington, DC.

Mueller, George E., deputy associate administrator for Manned Space Flight. To James Webb. Letter, 26 September 1963. Folder: 1964, Manned Lunar Landing Program, box 43, Samuel Phillips Papers, Library of Congress.

———. To W. Fred Boone. Memorandum, 3 May 1966. Folder: USAF Manned Orbiting Laboratory, DOD subseries, Federal Agencies series, NASA Historical Data Reference Collection, NASA Headquarters, Washington, DC.

———. To the NASA associate administrator. Memorandum, 15 January 1964. Folder: NASA-DOD, box: Arnold S. Levine,

Selected Sources from the author, NASA Historical Data Reference Collection, NASA Headquarters, Washington, DC.

———. Oral history interview, 8 November 1988. National Air and Space Museum.

Murray, Bruce. "Can Space Exploration Survive the End of the Cold War?" *Space Policy*, February 1991, 23–34.

Murray, Charles, and Catherine Bly Cox. *Apollo: The Race to the Moon.* New York: Simon and Schuster, 1989.

National Advisory Committee for Aeronautics (NACA), chairman and SECDEF. To the president. Joint memorandum, April 1958. Folder: Department of Defense Liaison, box: White House, Presidents, Eisenhower, DOD/CIA information, NASA Historical Data Reference Collection, NASA Headquarters, Washington, DC.

———. Agreement for Air Force Support for NACA Research Activities, 8 July 1957. Folder: Minutes of CMLC Meeting–13 January 1959, box: Civilian Military Liaison Committee, NASA Historical Data Reference Collection, NASA Headquarters, Washington, DC.

———. Minutes of Meeting, 21 February 1957. Folder: Dyna-Soar, DOD subseries, Federal Agencies series, NASA Historical Data Reference Collection, NASA Headquarters, Washington, DC.

———. NACA Research on Missiles, 1958. Folder: Testimony on Space Act, box: White House, Presidents, Eisenhower, National Aeronautics and Space Act, Space Act Testimony, NASA Historical Data Reference Collection, NASA Headquarters, Washington, DC.

———. A National Research Program for Space Technology, 14 January 1958. Folder: NACA documents, box: Administrative History, Pre-NASA Documents, NASA/DOD, shelf VI-C-6, NASA Historical Data Reference Collection, NASA Headquarters, Washington, DC.

———. On Assignment to the National Advisory Committee for Aeronautics Certain Officers of the United States Army for Reserve for Extended Active Duty, 27 July 1956, and similar agreements for the Air Force and Navy, same date. Box: Civilian-Military Liaison Committee, NASA Historical Data Reference Collection, NASA Headquarters, Washington, DC.

———. Wartime Role of NACA in Support of Department of Defense, 21 March 1957. Folder: Copies of Agreements, DOD subseries, Federal Agencies series, NASA Historical Data Reference Collection, NASA Headquarters, Washington, DC.

National Aeronautics and Space Administration (NASA). *A National Space Vehicle Program: A Report to the President*, 27 January 1959. Folder: NASC Papers (1), box 5, OCB Secretariat Series, NSC Staff series, White House Office, Dwight D. Eisenhower Library, Abilene, KS.

———. *Aeronautical and Astronautical Events of 1961*. Washington, DC: Government Printing Office, 1962.

———. *Aeronautics and Astronautics, 1965: Chronology on Science, Technology, and Policy*. Washington, DC: Government Printing Office, 1966.

———. *Astronautics and Aeronautics, 1966: Chronology on Science, Technology, and Policy*. NASA SP-4007. Washington, DC: Government Printing Office, 1967.

———. *Aeronautics and Space Report of the President, Fiscal Year 1995 Activities*. Washington, DC: Government Printing Office, 1996.

———. Agreement between the Department of the Air Force and the NASA Concerning Air Force Assistance to NASA in the Procurement of Research and Development and/or the Performance of Field Service Functions, 15 October 1959. Folder: Copies of Agreements, DOD subseries, Federal Agencies series, NASA Historical Data Reference Collection, NASA Headquarters, Washington, DC.

———. Agreement with the Air Force, GEMINI B/MOL Control Center and Network Support Procedures, 28 January 1965. Folder: MOL 1965–67, DOD subseries, Federal Agencies series, NASA Historical Data Reference Collection, NASA Headquarters, Washington, DC.

———. Answers to Vice Pres. Hubert Humphrey's 21 Questions on MOL, 29 July 1965. Safe no.1, drawer 2, folder: MOL/AES. Document is now declassified.

———. Chronology, NASA Participation in X-20 Project, 13 March 1963. Folder: Dyna Soar Proposals & Evaluation, DOD subseries, Federal Agencies series, NASA Historical Data Reference Collection, NASA Headquarters, Washington, DC.

———. Considerations Preparatory to Establishing a NASA Position on ABMA, 20 August 1959. Folder: NASA/AOMC/JPL Transfer, box: Administrative History no. 6, shelf VI-C-6, NASA Historical Data Reference Collection, NASA Headquarters, Washington, DC.

———. FY1969 Budget Briefing, 29 January 1968. Folder: Webb Budget Briefing, Webb subseries, Administrators series, NASA Historical Data Reference Collection, NASA Headquarters, Washington, DC.

———. Internal position paper on Project Gemini, marked "Confidential," 7 January 1963. Folder: Webb, declassified papers, 1961–1968, Webb subseries, Administrators series, NASA Historical Data Reference Collection, NASA Headquarters, Washington, DC.

———. "Inventory of NASA Interagency Relationships." 13 October 1967. Folder: Copies of Agreements, DOD subseries, Federal Agencies series, NASA Historical Data Reference Collection, NASA Headquarters, Washington, DC.

———. Minutes of NASA Dyna Soar Coordinating Committee, 30 March 1962. As an attachment to a memorandum from a NASA official, Thomas F. Dixon, to Deputy Administrator Hugh L. Dryden. Folder: X-20, Correspondence, DOD subseries, Federal Agencies series, NASA Historical Data Reference Collection, NASA Headquarters, Washington, DC.

———. Minutes of Williamsburg Conference, 21 October 1960. Folder: Aeronautics and Astronautics Coordinating Board, box: Aeronautics and Astronautics Coordinating Board, NASA Historical Data Reference Collection, NASA Headquarters, Washington, DC.

———. News Release no. 61-115, 25 May 1961. Folder: JFK, Miscellaneous Clippings, box: White House, Presidents, Kennedy, Biography materials, NASA Historical Data Reference Collection, NASA Headquarters, Washington, DC.

———. News Release no. 62-249, 21 November 1962. Space Policy Institute Archives document 1580.

———. News Release no. 64-78, 15 April 1964. Folder: DOD/USAF "Blue Gemini," DOD subseries, Federal Agencies series, NASA Historical Data Reference Collection, NASA Headquarters, Washington, DC.

———. News Release, 25 January 1965. Folder: USAF Manned Orbiting Laboratory, DOD subseries, Federal Agencies series, NASA Historical Data Reference Collection, NASA Headquarters, Washington, DC.

———. Office of Defense Affairs. Report, 3 July 1964. In Boone, *NASA Office of Defense Affairs.*

———. Office of Defense Affairs. Report, 8 November 1965. Folder: NASA's Contributions to National Security, DOD subseries, Federal Agencies series, NASA Historical Data Reference Collection, NASA Headquarters, Washington, DC.

———. Position Summary Paper, 6 January 1967. Folder: NASA-MOL, DOD subseries, Federal Agencies series, NASA Historical Data Reference Collection, NASA Headquarters, Washington, DC.

———. Position Paper, 26 January 1968. Folder: PSAC Correspondence, 1968, box: White House, President's Science Advisory Committee, NASA Historical Data Reference Collection, NASA Headquarters, Washington, DC.

———. *Preliminary History of NASA: 1963–1969.* Final Edition. Administrative Histories Project, 15 January 1969.

———. *Selected Statements of President Kennedy on Defense Topics, December 1957–August 1, 1962,* 15 August 1962. Folder: Kennedy, Statements on Defense, box: White House, Presidents, Kennedy, Defense Statements, NASA Historical Data Reference Collection, NASA Headquarters, Washington, DC.

———. Manned Spacecraft Center. *Gemini Program Mission Support: Gemini V,* August 1965. Box: Gemini Program, GT4 & GT5, shelf VII-A-4, NASA Historical Data Reference Collection, NASA Headquarters, Washington, DC.

———. Office of Manned Space Flight. Report on Complementary Nature of AAP and MOL Programs, 10 June 1968. Folder: Comparison of AAP and MOL, box 114, James Webb Papers, Harry S. Truman Library, Independence, MO.

———. Office of Programming. Program Review Document, *Manned Space Science and Advanced Manned Missions,* 7 October 1965. Shelf 10-A-4, box: Program Reviews (23 November 1963–15 February 1966), NASA Historical Data Reference Collection, NASA Headquarters, Washington, DC.

———. Office of Program Planning and Evaluation. NASA Long Range Plan, 12 January 1961. Folder: NASA Long Range Plan, box 1, National Aeronautics and Space Administration: Documents Relating to the space program, 1953–62, Dwight D. Eisenhower Library, Abilene, KS.

———. Long Range Plan, January 1962. Folder: NASA Long Range Plan, box 1, National Aeronautics and Space Administration: Documents relating to the space program, 1953–1962, Dwight D. Eisenhower Library, Abilene, KS.

———. *NASA Pocket Statistics, 1996 Edition.* Washington, DC: Government Printing Office, 1996.

National Aeronautics and Space Council (NASC). Minutes of Meeting, 3 December 1958. In "Pentagon Shows Caution on Space," 10 October 1962, 25. Folder: National Aeronautics and Space Council (Unclassified) 1958," box: White House, National Aeronautics and Space Council, 1958–59, NASA Historical Data Reference Collection, NASA Headquarters, Washington, DC.

———. Minutes of Meeting, 2 March 1959. Folder: National Aeronautics and Space Council 1959, box: White House, National Aeronautics and Space Council, 1958–59, NASA Historical Data Reference Collection, NASA Headquarters, Washington, DC.

———. Minutes of Meeting, 27 April 1959. Folder: Summary of National Aeronautics and Space Council Meetings, 1958–1960, box: White House, National Aeronautics and Space Council, NASA Historical Data Reference Collection, NASA Headquarters, Washington, DC.

———. Minutes of Meeting no. 7, 26 October 1959. Folder: 7th Meeting, box 2, RG 220, National Archives and Records Administration.

———. Selected Congressional Testimony, 1968. Folder: Space Projects, MOL, box 21, RG 220, Records of the National Aeronautics and Space Council, National Archives and Records Administration.

———. Summary Minutes, 18 August 1961. Folder: NASC Meeting, 18 August 1961, box: 1, RG 220, Records of the National Aeronautics and Space Council, National Archives and Records Administration.

———. Summary Minutes, 17 July 1963. Folder: NASC meeting, 17 July 1963, box 3, RG 220, Records of the National Aeronautics and Space Council, National Archives and Records Administration.

———. Summary Minutes, 31 July 1963. Folder: NASC Meeting, 31 July 1963, box 3, RG 220, Records of the National Aeronautics and Space Council, National Archives and Records Administration.

———. Summary Minutes, 9 July 1965. Folder: Official Record Copy, NASC Meeting, 9 July 1965. box 4, RG 220, Records of the National Aeronautics and Space Council, National Archives and Records Administration. Document is now declassified.

———. Summary Minutes, 15 June 1966. Folder: NASC Meeting, 15 June 1966, box 4, RG 220, Records of the National Aeronautics and Space Council, National Archives and Records Administration.

———. Summary Minutes, 22 June 1967. Folder: NASC Meeting, 22 June 1967, box 4, RG 220, Records of the National Aeronautics and Space Council, National Archives and Records Administration.

———. Summary Minutes, 14 November 1967. Folder: NASC Meeting, 14 November 1967, box 4, RG 220, Records of the National Aeronautics and Space Council, National Archives and Records Administration.

———. US Policy on Outer Space, 26 January 1960. Space Policy Institute Archives document 92. Also referred to as NSC 5918.

National Security Council (NSC). "Comments on the Report to the President by the Technological Capability Panel of the Science Advisory Committee," 8 June 1955. Folder: NSC 5522, Technological Capabilities Panel (2), box 16, Policy Papers subseries, NSC series, Office of the Special Assistant for National Security Affairs, Records: 1952–1961, White House Office. Dwight D. Eisenhower Library (DDEL), Abilene, KS.

————. Conference in the president's office, 8:30 a.m., 8 October 1957. Folder: Earth Satellites (1), box 7, Briefing Notes subseries, NSC series, OSANSA: Records, White House Office, Dwight D. Eisenhower Library, Abilene, KS.

————. 250th Meeting of the NSC. Memorandum of discussion, 26 May 1955. Folder: 250th Meeting of the NSC, box 6, NSC Series, Ann Whitman file, Dwight D. Eisenhower Library, Abilene, KS.

————. 310th Meeting of the NSC. Memorandum of discussion, 24 January 1957. Folder: 310th Meeting of the NSC, box 8, NSC series, Ann Whitman file, Dwight D. Eisenhower Library, Abilene, KS.

————. 339th Meeting of the NSC. Memorandum of discussion, 11 October 1957. Folder: 339th Meeting of the NSC, box 9, NSC series, Ann Whitman file, Dwight D. Eisenhower Library, Abilene, KS.

————. 376th Meeting of the NSC. Memorandum of discussion, 14 August 1958, dated 15 August 1958. Folder: 376th Meeting of NSC, box 10, NSC series, Ann Whitman file, Dwight D. Eisenhower Library, Abilene, KS.

————. 406th Meeting of the NSC. Memorandum of discussion, 13 May 1959. Folder: 406th Meeting of the NSC, box 11, NSC series, Ann Whitman file, Dwight D. Eisenhower Library, Abilene, KS.

————. 415th meeting of the NSC. Memorandum of discussion, 30 July 1959. Folder: 415th Meeting of the NSC, box 11, NSC series, Ann Whitman file, Dwight D. Eisenhower Library, Abilene, KS.

————. 466th Meeting of the NSC. Memorandum of discussion, 7 November 1960, dated 8 November 1960. Folder: 466th Meeting of the NSC, box 13, NSC series, Ann Whitman file, Dwight D. Eisenhower Library, Abilene, KS.

————. 469th Meeting of the NSC. Memorandum of discussion, 8 December 1960, dated 9 December 1960. Folder: 469th Meeting of the NSC, box 13, NSC series, Ann Whitman file, Dwight D. Eisenhower Library, Abilene, KS.

————. 470th Meeting of the NSC. Memorandum of discussion, 20 December 1960, dated 21 December 1960. Folder: 470th

Meeting of the NSC, box: 13, NSC series, Ann Whitman file, Dwight D. Eisenhower Library, Abilene, KS.

———. National Security Action Memorandum (NSAM) 129, *US-USSR Cooperation in the Exploration of Space*, 23 February 1962. NSA PD document 803.

———. NSAM 144, *Assignment of Highest National Priority to the Apollo Manned Lunar Landing Program*, 11 April 1962. NSA PD document 824.

———. NSAM 156, no title, 26 May 1962. Folder: NSAM 136–156, box 3, RG 59, General Records of the Department of State, National Archives and Records Administration. Document is now declassified.

———. NSAM 183, *Explanation and Defense of US Space Program*, 27 August 1962, signed by McGeorge Bundy. Folder: NSAM 136–156, box 3, RG 59, General Records of the Department of State, National Archives and Records Administration. Document is now declassified.

———. NSAM 271, *Cooperation with the USSR on Outer Space Matters*, 12 November 1963. NSA PD document 1026.

———. NSAM 352, *Bridge Building*, 8 July 1966. NSA PD document 1147.

———. NSC 5520, *Draft Statement of Policy on U.S. Scientific Satellite Program*, 20 May 1955. Space Policy Institute Archives document 86. Available in *Exploring the Unknown: Selected Documents in the History of the U.S. Civil Space Program*. Vol. 1, *Organizing for Exploration*. NASA SP-4407. Edited by John M. Logsdon, Linda J. Lear, Jannelle Warren-Findley, Ray A. Williamson, and Dwayne A. Day. Washington, DC: Government Printing Office, 1995.

———. NSC 5810/1, *Basic National Security Policy*, 5 May 1958.

———. NSC 5814/1, *Preliminary U.S. Policy on Outer Space*, 18 August 1958. Space Policy Institute Archives document 87. Document is now declassified.

———. NSC 5918, *U.S. Policy on Outer Space*, 17 December 1959. Document is now declassified.

———. NSC 6021, *Missiles and Military Satellite Programs*, 14 December 1960. Space Policy Institute Archives document 722.

———. NSC 6108, *Certain Aspects of Missile and Space Programs*, 18 January 1961. Space Policy Institute Archives document 278.

———. Operations Coordinating Board (OCB). Memorandum of meeting, 13 June 1957. Folder: OCB Working Group on Earth Satellites, box 1, National Aeronautics and Space Administration series, Dwight D. Eisenhower Library, Abilene, KS.

———. OCB. Memorandum of meeting, 8 October 1957. Folder: OCB Working Group on Earth Satellites, box: White House, Presidents, Eisenhower, Space Correspondence (1955–1960), NASA Historical Data Reference Collection, NASA Headquarters, Washington, DC.

———. OCB. Pentagon briefing memorandum, 12 October 1955. Folder: OCB 000.91 Natural and Physical Sciences (3), box 11, OCB Central file Subseries, NSC Staff Papers series, White House Office, Dwight D. Eisenhower Library, Abilene, KS.

———. OCB. "Public Information Program with Respect to the Implementation of NSC 5520," July 1955. Folder: OCB 000.91 Natural & Physical Sciences (2), box 11, OCB Central file subseries, NSC Staff Papers series, White House Office, Dwight D. Eisenhower Library, Abilene, KS.

———. Planning Board. Report, 9 November 1956. Folder: NSC 5520, Satellite Program (1), box 16, Policy Papers subseries, NSC Series, OSANSA, White House Office, Dwight D. Eisenhower Library, Abilene, KS.

Neu, Charles E. "The Rise of the National Security Bureaucracy." In *The New American State: Bureaucracies and Politics since World War II*. Edited by Louis Galambos. Baltimore: The Johns Hopkins Press, 1987.

Neufeld, Jacob. *The Air Force in Space, Fiscal Years 1969–1970*. Washington, DC: Office of Air Force History, July 1972.

———. *The Development of Ballistic Missiles in the United States Air Force 1945–1960*. Washington, DC: Office of Air Force History, 1990.

———, ed. *Research and Development in the United States Air Force*. Washington, DC: Center for Air Force History, 1993.

Neufeld, Michael J. *The Rocket and the Reich: Peenemunde and the Coming of the Ballistic Missile Era.* New York: Free Press, 1995.

Neustadt, Richard. *Presidential Power and the Modern Presidents: The Politics of Leadership from Roosevelt to Reagan.* New York: The Free Press, 1990.

———. To Senator Kennedy. Memorandum, 20 December 1960. Space Policy Institute Archives document 1178.

Newell, Homer E. *Beyond the Atmosphere: Early Years of Space Science.* NASA SP-4211. Washington, DC: Government Printing Office, 1980.

Newkirk, Roland W., Ivan D. Ertel, and Courtney G. Brooks. *Skylab: A Chronology.* NASA SP-4011. Washington, DC: Government Printing Office, 1977.

Nieburg, H. L. *In the Name of Science.* Chicago: Quadrangle Books, 1966.

Nixon, Richard M. To heads of executive departments and agencies. Memorandum, 25 January 1969. Folder: NASA-DOD Agreements, box 10, RG 220, Records of the National Aeronautics and Space Council, National Archives and Records Administration.

———. To Robert P. Mayo, director of the BOB. Memorandum, 24 March 1969. Folder: NASA-DOD Agreements, box 10, RG 220, Records of the National Aeronautics and Space Council, National Archives and Records Administration.

Oberth, Hermann. *Ways to Spaceflight,* trans. Agence Tunisienne de Public-Relations. Washington, DC: NASA TT F-622, 1972. Biographical series, Hermann Oberth file, NASA Historical Data Reference Collection, NASA Headquarters, Washington, DC.

O'Neill, Col J. W., ARDC/BMD, deputy chief of staff for plans and operations. To Col Donald Heaton, NASA technical assistant to the director, Launch Vehicle Programs. Memorandum, 30 July 1960. Folder: USAF Documents/Correspondence, 1957–1961, DOD subseries, Federal Agencies series, NASA Historical Data Reference Collection, NASA Headquarters, Washington, DC.

Packard, Robert F., Office of International Scientific Affairs, State Department. To the executive secretary, National Aeronautics and Space Council. Memorandums, 9 and 24 April 1963. Space Policy Institute Archives document 972.

Paine, Thomas. To Clinton Anderson. Letter, 22 July 1969. Folder: Space Committee, General, 91st Congress, box 921, Clinton Anderson Papers, Library of Congress.

———. To Melvin Laird. Letter, 11 August 1969. Folder: NASA/ DOD Cooperation/Space Merger, DOD subseries, Federal Agencies series, NASA Historical Data Reference Collection, NASA Headquarters, Washington, DC.

Payne, Seth. "After Apollo Blasts Off, What Next in Space?" *Product Engineering*, 7 October 1968, 9–12.

Pealer, Donald. "Manned Orbiting Laboratory, Part I." *Quest: The History of Spaceflight Quarterly* 4 (Fall 1995): 4–16.

———. "Manned Orbiting Laboratory, Part II." *Quest: The History of Spaceflight Quarterly* 4 (Winter 1995): 28–35.

———. "Manned Orbiting Laboratory, Part III." *Quest: The History of Spaceflight Quarterly* 5 (Summer 1996): 16–23.

Peebles, Curtis. *The Corona Project: America's First Spy Satellites.* Annapolis, MD: Naval Institute Press, 1997.

———. "The Guardians." *Spaceflight*, 20 November 1978, 381– 400.

———. "The Origins of the U.S. Space Shuttle." *Spaceflight* 21 (November 1979): 435–42.

Phillips, Samuel. Oral history interview, 22 July 1970. Phillips file, Biographical series, NASA Historical Data Reference Collection, NASA Headquarters, Washington, DC.

———. Report of the Joint Air Force–NASA Military Requirements Review Group, September 1964. Folder: 1964, Manned Lunar Landing Program, Personnel, Military, box 43, Samuel Phillips Papers, Library of Congress.

Piland, Robert O. To James Killian. Memorandum, 28 February 1958. Folder: Missiles (2), box 12, OSAST, White House Office, Dwight D. Eisenhower Library, Abilene, KS.

———. To James Killian. Memorandum, 14 April 1958. Folder: National Aeronautics and Space Administration (1), box 44, Confidential file, White House Central files, Dwight D. Eisenhower Library, Abilene, KS.

———. To James Killian. Memorandum, 15 June 1958. Space Policy Institute Archives document 1120.

———. To James Killian. Memorandum, 2 July 1958, dated 8 July 1958. Space Policy Institute Archives document 1127.

———. To James Killian. Memorandum, Attachment 1, 3 November 1958. Folder: Missiles April–December 1958 (3), box 12, OSAST, Dwight D. Eisenhower Library, Abilene, KS.

———. To James Killian. Memorandum. Folder: Space Notebook, Piland, 1958–59 (4), box 16, OSAST, White House Office, Dwight D. Eisenhower Library, Abilene, KS.

Piper, Robert F. *The Space Systems Division: Background, 1957–1962*, AFSC Historical Publications Series 62-27, SSEH-40. Air Force Systems Command, 1963.

"Policy Split Over Boosters Reported." *Missiles and Rockets*, 18 September 1961, 94–95.

Polmar, Norman, and Timothy Laur. *Strategic Air Command: People, Aircraft and Missiles*. 2d ed. Baltimore: Nautical & Aviation Publishing Company, 1990.

Portree, David S. F. *Thirty Years Together: A Chronology of U.S.-Soviet Space Cooperation*. NASA Contractor Report 185707. Washington, DC: NASA, Johnson Space Center, 1993.

Power, Thomas. To Chief of Staff Thomas D. White. Letter, 18 August 1958. Folder: Command, SAC, box 16, Thomas D. White Papers, Library of Congress.

Prados, John. *The Soviet Estimate: U.S. Intelligence Analysis and Russian Military Strength*. New York: The Dial Press, 1982.

President's Science Advisory Committee (PSAC). *Introduction to Outer Space*, 26 March 1958. Space Policy Institute Archives document 2.

———. Report, 16 December 1960. In *Exploring the Unknown: Selected Documents in the History of the U.S. Civil Space Program*. Vol. 1, *Organizing for Exploration*. NASA SP-4407. Edited by John M. Logsdon, Linda J. Lear, Jannelle Warren-Findley, Ray A. Williamson, and Dwayne A. Day. Washington, DC: Government Printing Office, 1995.

———. To George Kistiakowsky, Strategic Systems Panel. Report, 17 September 1960. Box 12, OSAST, White House Office, Dwight D. Eisenhower Library, Abilene, KS.

———. To internal. Memorandum, 2 February 1959. Folder: PSAC Correspondence 1959, box: White House, President's Science Advisory Committee, Correspondence 1959–1964,

NASA Historical Data Reference Collection, NASA Headquarters, Washington, DC.

Public Law (PL) 85-568, "National Aeronautics and Space Act of 1958." 72 Statute 426. 29 July 1958. Record Group 255, National Archives and Records Administration, Washington, DC: NASA Historical Data Reference Collection, NASA Headquarters, Washington, DC.

Putt, Donald. To CSAF. Memorandum, 22 November 1957. K168-8636-25, Air Force Historical Research Agency, Maxwell AFB, AL.

———. To commander, Air Research and Development Command, Hypersonic Research Aircraft. Memorandum, 31 January 1958. Reprinted in *Exploring the Unknown: Selected Documents in the History of the U.S. Civil Space Program.* Vol. 2, *Relations with Other Organizations*, NASA SP-4407. Edited by John M. Logsdon, Dwayne A. Day, and Roger D. Launius. Washington, DC: Government Printing Office, 1996.

———. To Hugh Dryden. Letter, 31 January 1958. Reprinted in *Exploring the Unknown: Selected Documents in the History of the U.S. Civil Space Program.* Vol. 2, *Relations with Other Organizations*, NASA SP-4407. Edited by John M. Logsdon, Dwayne A. Day, and Roger D. Launius. Washington, DC: Government Printing Office, 1996.

Putt, Lt Gen Donald E, USAF. Oral history interview, 30 April 1974. File: Donald E. Putt, Biographical series, NASA Historical Data Reference Collection, NASA Headquarters, Washington, DC.

Quarles, Donald. To acting secretary of the NASC. Memorandum, 15 April 1959. Folder: NASC 1958–1959, box: White House, National Aeronautics and Space Council, NASA Historical Data Reference Collection, NASA Headquarters, Washington, DC.

———. To the president. Memorandum, 7 October 1957. Box 7, Briefing Notes subseries, NSC Series, Office of the Special Assistant for National Security Affairs: Records, 1952–61, White House Office, Dwight D. Eisenhower Library, Abilene, KS.

———. To Maurice Stans, BOB director. Letter, 1 April 1958. K168.8636-23, Air Force Historical Research Agency, Maxwell AFB, AL.

"A Quiet Retirement." *Time*, 9 September 1966, 24–25.

Rabi, I. I. To Arthur Flemming, Office of Defense Mobilization. Letter, 10 October 1956. Folder: Eisenhower Administration, Space Correspondence, box: White House, Presidents, Eisenhower, Space Correspondence (1955–1960), NASA Historical Data Reference Collection, NASA Headquarters, Washington, DC.

RAND Corporation. Conference on Methods for Studying the Psychological Effects of Unconventional Weapons, 26–28 January 1949. Space Policy Institute Archives document 1297.

Rathjens, George W., member of PSAC Strategic Systems Panel. To other members of the PSAC Strategic Systems Panel. Memorandum, 23 September 1960. Folder: Missiles July–September 1960 (6), box 12, OSAST, White House Office, Dwight D. Eisenhower Library, Abilene, KS.

Redford, Emmette S., and Orion F. White. *What Manned Space Program after Reaching the Moon?: Government Attempts to Decide, 1962–1968*. Syracuse, NY: The Inter-University Case Program, January 1971.

Reeves, Richard. *President Kennedy: Profile of Power*. New York: Simon & Schuster, 1993.

Richelson, Jeffrey. *America's Secret Eyes in Space: The U.S. Keyhole Spy Satellite Program*. New York: HarperCollins, 1990.

———, comp. *Military Uses of Space, 1945–1991*. Alexandria, VA: The National Security Archives and Chadwyck-Healey, Inc., 1991.

———. *Presidential Directives on National Security from Truman to Clinton*. Alexandria, VA: The National Security Archives and Chadwyck-Healey, Inc., 1994.

———. "The Keyhole Satellite Program." *The Journal of Strategic Studies* 7 (June 1984): 121–53.

Roland, Alex, ed. *A Spacefaring People: Perspectives on Early Spaceflight*. NASA SP-4405. Washington, DC: Government Printing Office, 1985.

———. Minutes of PSAC meeting, 10 December 1957. In *The Papers of the President's Science and Advisory Committee, 1957–1961*. In *Science and Technology: Research Collections in U.S. Public Policy*. Bethesda, Md.: University Publications of America, 1986. Microfilmed from the holdings of the Dwight D. Eisenhower Library, Abilene, KS.

Roland, Alex. *Model Research: The National Advisory Committee for Aeronautics, 1915–1958*. 2 Vols. NASA SP-4103. Washington, DC: Government Printing Office, 1985.

———. "Barnstorming in Space." In *Space Policy Reconsidered*. Edited by Radford Byerly Jr. Boulder, CO: Westview Press, 1989.

———. "The Lonely Race to Mars: The Future of Manned Spaceflight." In *Space Policy Alternatives*. Edited by Radford Byerly Jr. Boulder, CO: Westview Press, 1992.

Rosen, Milton, OMSF director of Launch Vehicles and Propulsion. To D. Brainerd Holmes, director, OMSF, Large Launch Vehicle Programs. Memorandum, 6 November 1961. Space Policy Institute Archives document 1597.

———, senior scientist, NASA Office of Defense Affairs. To Admiral Boone. Memorandum, 8 October 1965. Folder: USAF Gemini Role Documentation, Gemini series, NASA Historical Data Reference Collection, NASA Headquarters, Washington, DC.

Rosenberg, Max. *The Air Force in Space, 1959–1960*. SHO-S-62/112. Washington, DC: USAF Historical Division Liaison Office, 1962.

Rosholt, Robert L. *An Administrative History of NASA: 1958–1963*. NASA SP-4101. Washington, DC: Government Printing Office, 1966.

Rostow, Walt W. To John F. Kennedy. Memorandum, 7 November 1960. In *President John F. Kennedy's Office Files, 1961–1963, Part 1: Special Correspondence, Speeches, Legislative, and Press Conference Files*. Paul Kesaris and Robert Lester, Project Coordinators. In Research Collections in American Politics, Microforms from Major Archival and Manuscript Collections. William Leuchtenberg, general editor. Frederick, MD: University Publications of America, 1989. Reel 5.

Rotunda, Donald T. *The Legislative History of the National Aeronautics and Space Act of 1958*. NASA HHN-125. NASA Historical Office, 1972.

Rubel, John H., deputy director, Defense Research and Engineering. To Edward Welsh. Letter, 10 April 1962. Folder: NASC meeting, 21 March 1962, box 1, RG 220, Records of the National Aeronautics and Space Council, National Archives and Records Administration.

———, signing for Harold Brown. To the SECDEF. Memorandum, 20 February 1962. Folder: Reading file, January–May 1962, box 114, RG 200, Robert McNamara Papers, National Archives and Records Administration. Document is now declassified.

———, acting director, Defense Research and Engineering. To Office of the SECDEF. Report on Military Space Projects, 20 October 1960. NSA-MUS document 322.

———, deputy director, Defense Research and Engineering. Speech, 9 October 1962. Released as DOD News Release no. 1642-62. DOD, Office of Public Affairs. John Rubel file, Biographical series, NASA Historical Data Reference Collection, NASA Headquarters, Washington, DC.

———, and Robert C. Seamans. To Robert McNamara and James Webb. Memorandum, 7 December 1961. Folder: DOD/USAF "Blue Gemini," DOD subseries, Federal Agencies series, NASA Historical Data Reference Collection, NASA Headquarters, Washington, DC.

Salter, R. M., and J. E. Lipp. *Utility of a Satellite Vehicle for Reconnaissance*. RAND Report R-217. Santa Monica, CA: April 1951. Space Policy Institute Archives document 1296.

Schichtle, Cass. *The National Space Program from the Eighties to the Fifties*, National Security Affairs Monograph Series 83-6. Washington, DC: National Defense University, 1983.

Schilling, Warren R. "Scientists, Foreign Policy, and Politics." In *Components of Defense Policy*. Edited by Davis Dobrow. Chicago: Rand McNally & Company, 1965.

Schlesinger, Arthur M., Jr. *A Thousand Days: John F. Kennedy in the White House*. Boston: Houghton Mifflin Company, 1965.

Schoettle, Enid Curtis Bok. "The Establishment of NASA." In *Knowledge and Power: Essays on Science and Government*. Sanford A. Lakoff. New York: Free Press, 1966.

Schriever, Lt Gen Bernard A., USAF, commander, ARDC. "Comments." In *The First 25 Years in Space*. Edited by Allan A. Needell. Washington, DC: Smithsonian Institution Press, 1983.

———. "Does the Military Have a Role in Space?" In *Space: Its Impact on Man and Society*. Edited by Lillian Levy. New York: W. W. Norton & Company, 1965.

———. Oral history interview, 20 June 1973. K239.0512-676, Air Force Historical Research Agency, Maxwell AFB, AL.

———. Oral history interview, 29 June 1977. K239.0512-1492. Air Force Historical Research Agency, Maxwell AFB, AL.

———. To the Allegheny Conference on Community Development, Pittsburgh, PA. Speech, 21 November 1960. Inserted into the *Congressional Record*, 6 January 1961, Appendix, A-93–A-94, by Rep. James Fulton.

———. To the American Institute of Aeronautics and Astronautics. Speech, 12 October 1965. IRIS 1013465, Air Force Historical Research Agency, Maxwell AFB, AL.

———. To the Aviation Writers Association. Speech, May 1966. Reprinted in the *Congressional Record*, 22 June 1966, A3272-74.

———. To CSAF Curtis E. LeMay. Letter, n.d. but sometime shortly after Schriever's testimony to the Senate Armed Services Committee on 20 July 1961. Folder: USAF Documents/Correspondence, DOD subseries, Federal Agencies, NASA Historical Data Reference Collection, NASA Headquarters, Washington, DC.

———. To CSAF Thomas White. Letter, 15 September 1959. Folder: Command, SAC, box 16, Thomas White Papers, Library of Congress.

———. To T. Keith Glennan. Letter, 11 January 1961. Folder: Glennan (Select Correspondence), Glennan subseries, Administrators series, NASA Historical Data Reference Collection, NASA Headquarters, Washington, DC.

———. To General Curtin, Office of the Secretary of the Air Force. Memorandum, 13 February 1961. Space Policy Institute Archives document 32.

———. To Lyndon Johnson. Memorandum, 30 April 1961. 168.7171-151, Air Force Historical Research Agency, Maxwell AFB, AL.

———. To Thomas White. Letter, 20 October 1960. Folder: 2-6, ARDC, box 33, Thomas White Papers, Library of Congress.

Schultz, Col Kenneth W., USAF, assistant for Manned Orbiting Laboratory, Office of the Deputy Chief of Staff for Research and Development. Congressional Preparation Instruction Letter no. 1, Chief of Staff Policy Book, 1965, September 1964. IRIS 1002994, Air Force History Support Office.

———. Congressional Preparation Instruction no. 7, 27 November 1964. IRIS 1002995, Air Force History Support Office.

———. Inputs on MOL to the Chief of Staff's Posture Statement, 6 November 1964, IRIS 1002995, Air Force History Support Office.

———. Letter of Nonconcurrence, 29 September 1964. IRIS 1002994, Air Force History Support Office.

Schultze, Charles L., BOB director. To James Webb. Letter, 13 August 1966. Folder: Space-NASA-1966, box 611, RG 359, Office of Science and Technology, National Archives and Records Administration. Document is now declassified.

———. To Lyndon Johnson. Memorandum, 20 September 1966. Space Policy Institute Archives document 859.

Schwartz, Leonard. "Manned Orbiting Laboratory—For War or Peace?" *International Affairs* 43 (January 1967): 51–64.

Schwiebert, Ernest G. "USAF Ballistic Missiles: 1954–1964." *Air Force/Space Digest*, May 1964, 51–166.

Science Advisory Committee. Notes, 15 October 1957. Folder: Eisenhower Administration, Space Correspondence, box: White House, Presidents, Eisenhower, Space Correspondence (1955–1960), NASA Historical Data Reference Collection, NASA Headquarters, Washington, DC.

Schwiebert, Ernest G. *A History of U.S. Air Force Ballistic Missiles*. New York: Praeger, 1965.

Seamans, Robert C., Jr. *Aiming at Targets*. NASA SP-4106. Washington, DC: Government Printing Office, 1996.

———. Memorandum for record, 19 December 1963. Folder: DOD-NASA Coordination, box 17, RG 220, Records of the National Aeronautics and Space Council, National Archives and Records Administration.

———. Statement on *FY 67 Budget: Hearings before the Subcommittee on Manned Space Flight of the House Committee on Science and Astronautics*, 18 February 1966. Folder: Seamans, House of Representatives, Seamans subseries, Deputy Administrators series, NASA Historical Data Reference Collection, NASA Headquarters, Washington, DC.

———. To Aerospace Corporation. Speech, 29 August 1961. Defense Technical Information Center AD-B185 903.

———. To Bernard Schriever. Letter, 12 April 1965. Folder: NASA/DOD Cooperation/Documentation, DOD subseries, Federal Agencies series, NASA Historical Data Reference Collection, NASA Headquarters, Washington, DC.

———. To Dr. Albert C. Hall, deputy director for Space; director, Defense Research and Engineering. Memorandum, 5 August 1965. Safe no. 1, drawer 2, Folder: MOL/AES, NASA Historical Data Reference Collection, NASA Headquarters, Washington, DC, 1. Document is now declassified.

———. To Harold Brown, Letter. 6 January 1964. Folder: Defense 1964–1965, box 17, RG 220, Records of the National Aeronautics and Space Council, National Archives and Records Administration.

———. To Harold Brown, Letter. 27 September 1965. Folder: MOL Correspondence, DOD subseries, Federal Agencies series, NASA Historical Data Reference Collection, NASA Headquarters, Washington, DC.

———. To Harold Brown. Letter, 22 November 1966. Folder: USAF Gemini Role Documentation, Gemini series, NASA Historical Data Reference Collection, NASA Headquarters, Washington, DC.

———. To Hugh Dryden and James E. Webb. Memorandum, 11 September 1963. Space Policy Institute Archives document 1456.

———. To James E. Webb. Memorandum, 7 July 1961. Folder: AACB Minutes & Reports, box: Arnold Levine, Selected

Sources from the author, NASA Historical Data Reference Collection, NASA Headquarters, Washington, DC.

———. To James E. Webb. Memorandum, 7 July 1965. Folder: MOL Correspondence II, DOD subseries, Federal Agencies series, NASA Historical Data Reference Collection, NASA Headquarters, Washington, DC.

———. To John S. Foster. Letter, 22 November 1966. Folder: DOD/USAF/NASA, Vietnam cooperation, DOD subseries, Federal Agencies series, NASA Historical Data Reference Collection, NASA Headquarters, Washington, DC.

———. To John S. Foster. Letter, 17 July 1967. Folder: DOD/USAF/NASA, Vietnam cooperation, DOD subseries, Federal Agencies series, NASA Historical Data Reference Collection, NASA Headquarters, Washington, DC.

———. To Lt Gen H. M. Wade, deputy chief of staff for personnel, USAF. Letter, 4 January 1967. Folder: Military Personnel Detailed to NASA, DOD subseries, Federal Agencies series, NASA Historical Data Reference Collection, NASA Headquarters, Washington, DC.

———. To Thomas White. Letter, 28 February 1961. Folder: 7–4 FAA/NASA/JCS/CIA/CAP, box 39, Thomas White Papers, Library of Congress.

———. Oral history interview, 27 March 1964. Folder: JFK Library Interview, Seamans subseries, Deputy Administrators series, NASA Historical Data Reference Collection, NASA Headquarters, Washington, DC.

———. Oral history interview, 26 May 1966. Folder: Gemini interview, Seamans subseries, Deputy Administrators series, NASA Historical Data Reference Collection, NASA Headquarters, Washington, DC.

———. Oral history interview series, 8 May 1968 and 3 June 1968. Folder: Exit Interview, Seamans subseries, Deputy Administrator series, NASA Historical Data Reference Collection, NASA Headquarters, Washington, DC.

———. Oral history interviews of September 1973–March 1974. K239.0512-687A, Air Force Historical Research Agency, Maxwell AFB, AL., 539. Document is now declassified.

————. Oral history interviews, 2 November 1987; 19 January 1988; 15 December 1988. National Air and Space Museum, Washington, DC.

————, and John S. Foster, director, Defense Research and Engineering. Memorandum of agreement, 21 March 1966. Folder: MSF Panel AACB, box: Aeronautics and Astronautics Coordinating Board no. 2, NASA Historical Data Reference Collection, NASA Headquarters, Washington, DC.

"Secretary McNamara Seeks NASA-AF Cooperation on Manned Orbiting Laboratory." *Journal of the Armed Forces*, 26 December 1964, 21.

Shapley, Deborah. *Promise and Power: The Life and Times of Robert McNamara*. Boston: Little, Brown and Company, 1993.

————. "Robert McNamara: Success and Failure." In *Leadership and Innovation: A Biographical Perspective on Entrepreneurs in Government*. Edited by Jameson W. Doig and Erwin C. Hargrove. Baltimore: The Johns Hopkins University Press, 1987.

Shapley, Willis, Military Division, BOB. To BOB director. Memorandum, 28 October 1958. Folder: National Aeronautics and Space Council 1958, box: White House, National Aeronautics and Space Council, 1958–59, NASA Historical Data Reference Collection, NASA Headquarters, Washington, DC.

————. To BOB director. Memorandum, 6 December 1963. File: Space Projects, MOL, Space Stations, box 21, RG 220, Records of the National Aeronautics and Space Council, National Archives and Records Administration.

————. Oral history interview by John Logsdon, 14 December 1967. File: Willis Shapley, Biographical series, NASA Historical Data Reference Collection, NASA Headquarters, Washington, DC.

Sharp, Dudley. To the CSAF. Memorandum, 13 September 1960. Folder: 4–5 Missiles/Space/Nuclear, box 36, Thomas White Papers, Library of Congress.

Shaughnessy, Col John J., chief, Plans Group, Legislative liaison, USAF. Statement on MOL Cancellation. Reprinted in *Congressional Record*, 10 June 1969.

Sheldon, Charles S. *A Comparison of the United States and Soviet Space Programs*. Paper no. 10, Washington, DC: George Washington University, Program of Policy Studies in Science and Technology, June 1965. Space Policy Institute Archives unnumbered document.

———. To NASC Executive Secretary Edward Welsh. Memorandum, 24 November 1964. Tab 2, box 23, RG 220, Records of the National Aeronautics and Space Council, National Archives and Records Administration. Document is now declassified.

Sickman, Philip. "The Fantastic Weaponry," *Fortune*, June 1962.

Sidders, Carl, and Robert Bickett. *Air Force Support of Army, Navy and NASA Space Programs.* Office of Information, Western Contract Management Region, AFSC, 29 August 1961. Folder: NASA/DOD Cooperation/Space Merger?, DOD subseries, Federal Agencies series, NASA Historical Data Reference Collection, NASA Headquarters, Washington, DC.

Siddiqi, Asif A. *Challenge to Apollo: The Soviet Union and the Space Race, 1945–1974.* NASA SP-2000-4408. Washington, DC: Government Printing Office, 2000.

Sidey, Hugh. *John F. Kennedy, President.* New York: Atheneum, 1963.

Siegel, Gerald. Oral history interview, 8 June 1977. Box: Emme/Roland interviews on early NASA history, shelf V-A-1, NASA Historical Data Reference Collection, NASA Headquarters, Washington, DC.

Siepert, Albert F. To James Webb. Memorandum, 8 February 1963. In *Exploring the Unknown: Selected Documents in the History of the U.S. Civil Space Program.* Vol. 1, *Organizing for Exploration.* NASA SP-4407. Edited by John M. Logsdon, Linda J. Lear, Jannelle Warren-Findley, Ray A. Williamson, and Dwayne A. Day. Washington, DC: Government Printing Office, 1995.

Smith, Col James G., assistant for public affairs, Office of the DOD Manager for Manned Space Flight Support Operations, USAF. *Department of Defense Support: Apollo 11,* July 1969. Folder: 1969, Manned Lunar Landing Program,

Apollo 11, Department of Defense, box 108, Samuel Phillips Papers, Library of Congress.

Smith, Richard Austin. "Canaveral, Industry's Trial by Fire." *Fortune*, June 1962, 135–39, 200, 204, 206, 211–12.

Sorensen, Theodore C. *Kennedy*. New York: Harper & Row, Publishers, 1965.

Soule, Hartley A., NACA research airplane projects leader. To NACA Headquarters. Memorandum, 10 February 1958. Folder: Dyna Soar Proposals & Evaluation, DOD subseries, Federal Agencies series, NASA Historical Data Reference Collection, NASA Headquarters, Washington, DC.

Space Policy Institute. *The Legislative Origins of the Space Act: Proceedings of a Videotaped Workshop*. Washington, DC: George Washington University, 3 April 1992.

Spires, David. *Beyond Horizons: A Half-Century of Air Force Space Leadership*. Peterson AFB, CO: Air Force Space Command, USAF; and US Government Printing Office, 1997.

Spriggs, James O., NASA Ad Hoc Subpanel on Security Practices of the Supporting Space Research and Technology Panel of the AACB. Memorandum for record, 16 February 1967. Folder: Resolution from Space, DOD subseries, Federal Agencies series, NASA Historical Data Reference Collection, NASA Headquarters, Washington, DC.

Staats, Elmer. Oral history interview, 13 July 1964. Folder: Kennedy Library, box: White House, Presidents, Kennedy, Photographs, Presidential Library, NASA Historical Data Reference Collection, NASA Headquarters, Washington, DC.

Staff Secretary, Office of the. Memorandum of conference with the president, 9 October 1957. Folder: Missiles and Satellites, Vol. 1 (3), box 6, Department of Defense subseries, Subject Series, Office of the Staff Secretary: Records, Dwight D. Eisenhower Library, Abilene, KS.

Stans, Maurice, BOB director. To President Eisenhower. Memorandum, 29 July 1958. Folder: Staff Memos, July 1958 (1), box 35, DDE Diary series, Ann Whitman file, Dwight D. Eisenhower Library, Abilene, KS.

Stares, Paul B. *The Militarization of Space: U.S. Policy, 1945–1984*. Ithaca, NY: Cornell University Press, 1985.

Steinberg, Gerald M. *Satellite Reconnaissance: The Role of Informal Bargaining.* New York: Praeger Publishers, 1983.

Steury, Donald P., comp. *Estimates on Soviet Military Power: 1954 to 1984.* Washington, DC: CIA, Center for the Study of Intelligence, December 1994.

———, ed. *Intentions and Capabilities: Estimates on Soviet Strategic Forces, 1950–1983.* Washington, DC: CIA, Center for the Study of Intelligence, 1996.

Sylvester, Arthur, Office of the assistant SECDEF for public affairs. To Kennedy. Memorandum, 26 January 1961. NSA-MUS document 639.

Swenson, Loyd S., Jr., James M. Grimwood, and Charles C. Alexander. *This New Ocean: A History of Project Mercury.* NASA SP-4201. Washington, DC: Government Printing Office, 1966.

Technological Capabilities Panel (TCP). *The Report to the President by the Technological Capabilities Panel of the Science Advisory Committee.* Vol. 2, *Meeting the Threat of Surprise Attack.* Washington, DC, February 1955. Space Policy Institute Archives document 1410.

Teller, Edward. Report of the Teller Ad Hoc Committee, 28 October 1957. K140.11-3, Washington, DC: Air Force History Support Office.

Tsiolkovsky, K. E. *Works on Rocket Technology.* NASA Translation TT F-243, November 1965, from the Publishing House of the Defense Industry, Moscow, 1947. Box: Works on Rocket Technology by K. E. Tsiolkovsky, shelf: 3-B-7, NASA Historical Data Reference Collection, NASA Headquarters, Washington, DC.

Tucker, Samuel, ed. *A Modern Design for Defense Decision: A McNamara-Hitch-Enthoven Anthology.* Washington, DC, 1966.

Twining, Nathan. *Neither Liberty Nor Safety.* New York: Holt, Rinehart and Winston, 1966.

———. To the SECDEF. Memorandum, 11 August 1958. In *Exploring the Unknown: Selected Documents in the History of the U.S. Civil Space Program.* Vol. 1, *Organizing for Exploration.* NASA SP-4407. Edited by John M. Logsdon, Linda J. Lear, Jannelle Warren-Findley, Ray A. Williamson, and

Dwayne A. Day. Washington, DC: Government Printing Office, 1995.

Ulam, Adam. *The Rivals: America & Russia Since World War II*. New York: Penguin Books, 1971.

"USAF Pushes Pied Piper Space Vehicle." *Aviation Week*, 14 October 1957, 26.

US Air Force. Advanced Systems: DYNASOAR, Internal Air Force Document. In *Briefing Book for Air Force Witnesses before the House Committee on Science and Astronautics on the Subject of DOD Space Directive 5160.32*. K160.8636-4, Air Force History Support Office, 13 March 1961.

———. "Air Force Competency in Space Operations." *Air Force Information Policy Letter for Commanders*, 1 December 1960. Vol. 14, no. 17. Folder: NASA-USAF Policy Relations, Other Agency Agreements subseries, Federal Agencies series, NASA Historical Data Reference Collection, NASA Headquarters, Washington, DC .

———. *Air Force Information Policy Letter, Supplement for Commanders, Special Issue: Military Mission in Space, 1957–1962*. Director of Information, Office of the Secretary of the Air Force, dated only 1962. Folder: DOD Space Policy, DOD subseries, Federal Agencies series, NASA Historical Data Reference Collection, NASA Headquarters, Washington, DC.

———. *Air Force Space Plan*, September 1961. Space Policy Institute Archives unnumbered document.

———. "The Air Force Space Study Program," n.d. K140.11-13, Air Force History Support Office.

———. *Air Force Support of Project Gemini: Inputs from the Major Commands*, 1967. K110.8-50, Air Force History Support Office.

———. Air Force Systems Command (AFSC). *Chronology of Early Air Force Man-in-Space Activity*. AFSC Historical Publications Series 65-21-1, 1965. NSA-MUS document 446.

———. Air Force Systems Command. *The Genesis of the Air Force Systems Command*. Historical Publication 62-102260, 1962.

———. Air Force Systems Command. *History of the Aeronautical Systems Division, July–December 1963*. Vol. 4, *Termination of the X-20A Dyna-Soar (Documents)*. AFSC Historical Publications Series 64-51-4. K243.011, Air Force Historical Research Agency, Maxwell AFB, AL. Document is now declassified.

———. Briefing. Armed Forces Policy Council, 5 November 1957. K140.11-3, Air Force History Support Office.

———. Briefings. Secretary of the Air Force, 28–29 January 1959. K140.11-13, Air Force History Support Office.

———. *Commander's Congressional Policy Book.* Vol. 2, Tab C-1, *Dynasoar Program (X-20).* 168.7171-52, Air Force Historical Research Agency, Maxwell AFB, AL. Document is now declassified.

———. Current Status Report, Strategic Mission Area, 620 A-DYNA SOAR, February 1960. IRIS 1003000, Air Force History Support Office.

———. deputy chief of staff for development. Memorandum for record, 6 February 1958. Folder: USAF Documents/Correspondence (1957–1961), DOD subseries, Federal Agencies series, NASA Historical Data Reference Collection, NASA Headquarters, Washington, DC.

———. deputy chief of staff for research and development, Office of the Assistant for Manned Orbiting Laboratory (AFRMO). Congressional Preparation Instruction no. 14, 26 January 1965. IRIS 1002996, Air Force History Support Office.

———. Detachment 2, at NASA Manned Spacecraft Center. Report on Preliminary Evaluation, 13 September 1966. IRIS 1002997, Air Force History Support Office.

———. director of information, Office of the SAF. *Air Force Information Policy Letter, Supplement for Commanders, Special Issue: Military Mission in Space, 1957–1962.* Folder: DOD Space Policy, DOD subseries, Federal Agencies series, NASA Historical Data Reference Collection, NASA Headquarters, Washington, DC, 1962.

———. Headquarters. Development Directive for System 620A, DYNA SOAR (Step I), Hypersonic Glider System, 12 October 1960. IRIS 100300, Air Force History Support Office.

———. *History of the Aeronautical Systems Division, January–June 1962.* Vol. 1, *Narrative.* AFSC Historical Publication Series 62-52-1. K243.011, Air Force Historical Research

Agency, Maxwell AFB, AL, 1962. Document is now declassified.

———. *History of the Directorate of Space, Deputy Chief of Staff for Research and Development: January–June 1967.* K140.01-1, Air Force History Support Office, July 1967.

———. *History of the Directorate of Space, Deputy Chief of Staff for Research and Development: January–June 1968.* K140.01-1, Air Force History Support Office, July 1968.

———. History Office. *Space and Missile Systems Organization: A Chronology, 1954–1979.* 1985.

———. Insert for the Record to the House DOD Subcommittee on Appropriations, 1 April 1965. IRIS 1002996, Air Force History Support Office.

———. Memorandum for record, 5 April 1955. K140.11-11, Air Force Historical Research Agency, Maxwell AFB, AL.

———. *The Military Mission in Space: A Selection of Published Views, August 1961–June 1963.* Folder: Military Mission in Space, DOD subseries, Federal Agencies series, NASA Historical Data Reference Collection, NASA Headquarters, Washington, DC: 7 June 1963.

———. "NASA-USAF Cooperation," *Air Force Information Policy Letter for Commanders,* 1 February 1961. Vol. 15, no. 3. Folder: USAF Space, AFCHO, DOD subseries, Federal Agencies series, NASA Historical Data Reference Collection, NASA Headquarters, Washington, DC.

———. Office of the CSAF. To deputy CSAF for development and deputy CSAF for material. Record of Decision, 17 November 1959. Folder: Air Force Council Decisions 1959, box 25, Thomas White Papers, Library of Congress.

———. Office of the secretary of the Air Force. *Air Force Information Fact Sheet, X-20 Dyna-Soar,* January 1963. Folder: X-20 Dyna-Soar Documentation, DOD subseries, Federal Agencies series, NASA Historical Data Reference Collection, NASA Headquarters, Washington, DC.

———. Outline of History of USAF Man-in-Space R&D Program, August 1962. K140.11-7, Air Force History Support Office.

———. *Primary Experiments Data for the Manned Orbiting Laboratory System (MOL) Program,* March 1965. SSMM-67. Space Policy Institute Archives unnumbered document.

————. *Project Forecast: Policy and Military Considerations Report*, January 1964. K168.154-12, Air Force Historical Research Agency, Maxwell AFB, AL. Document is now declassified.

————. Research and Development Command, Ballistic Missiles Division. Air Force Ballistic Missile Division's Responsibility for "Man-in-Space" Program. K140.11–7, Air Force History Support Office.

————. Research and Development Command, Ballistic Missiles Division. Military Lunar Base Program or S. R. 183 Lunar Observatory Study, April 1960. Space Policy Institute Archives document 1212.

————. Secretary of the Air Force. Message. To all subordinate commanders and offices, 13 August 1965. IRIS 1003002, Air Force History Support Office.

————. Signal Detection Experiment for MOL, 28 May 1964. IRIS 1002993, Air Force History Support Office.

————. Space Systems Division, Headquarters. Partial Systems Package Plan for Military Orbital Development System (MODS) System No. 648C, 4 June 1962. K243.8636-9, Air Force Historical Research Agency, Maxwell AFB, AL. Document is now declassified.

————. *Study Pertaining to the National Lunar Program.* Vol. 1, *Summary Report*, April 1964. IRIS 880570, Air Force History Support Office as well as K140.22-2, Air Force Historical Research Agency, Maxwell AFB, AL. Document is now declassified.

————. Report, 22 December 1964. IRIS 1002995, Air Force History Support Office.

————. To multiple recipients. Memorandum, 20 April 1964. Folder: 208, box B208, LeMay Papers, Library of Congress, cover letter and 1–2. Document is now declassified.

————. To the secretary of the Air Force. Memorandum. IRIS 1002993, Air Force History Support Office.

————. Unclassified Supporting Witness Statement. April 1965. IRIS 1002996, Air Force History Support Office.

————. Unclassified Supporting Witness Statement. Manned Space Programs, 9 March 1965. IRIS 1002996, Air Force History Support Office.

————. *USAF Manned Military Space Development Plan.* Vol. 2. IRIS 1002991, Air Force History Support Office.

————. *USAF/NASA Coordination in Space Problems.* Internal Document. In *Briefing Book for Air Force Witnesses before the House Committee on Science and Astronautics on the Subject of DOD Space Directive 5160.32.* K160.8636-4, Air Force History Support Office, 16 March 1961.

US Army. Ballistic Missile Agency. ABMA Report no. D-TR-1-58, *Development Proposal for Project Adam,* 17 April 1958. Folder: ABMA (Project Adam), DOD subseries, Federal Agencies series, NASA Historical Data Reference Collection, NASA Headquarters, Washington, DC.

————. Briefing. Satellite Reconnaissance System, 19 November 1957. Folder: Space, November 1957 (2), box 15, Office of the Special Assistant for Science and Technology (OSAST), White House Office, Dwight D. Eisenhower Library, Abilene, KS.

————. Redstone Arsenal, Scientific Information Center. *Project Horizon.* Vol. 1, *Summary, A U.S. Army Study for the Establishment of a Lunar Military Outpost.* 8 June 1959. Folder: Project Horizon, DOD subseries, Federal Agencies series, NASA Historical Data Reference Collection, NASA Headquarters, Washington, DC.

US Bureau of the Budget (BOB), Bureau of the Military Division. Discussion Paper, FY 1966 Budget Policy Considerations on Manned Orbiting Laboratory (MOL) and Extended Apollo Systems (EAS), 19 November 1964. Folder: NASA/MOL, DOD subseries, Federal Agencies series, NASA Historical Data Reference Collection, NASA Headquarters, Washington, DC.

————. Draft Staff Report on Special Space Review, August 1962. Folder: DOD and NASA Space Programs 1962, box 20: National Archives and Records Administration RG 200, Robert S. McNamara Papers.

————. Military Division. To Dr. Henry J.E. Reid, director, Langley Research Center, Hampton, VA. Memorandum, 23 November 1959. Folder: National Aeronautics and Space Council 1959, box: White House, National Aeronautics and

Space Council, 1958–59, NASA Historical Data Reference Collection, NASA Headquarters, Washington, DC.

———. Report, 30 October 1968. Reprinted in *Exploring the Unknown: Selected Documents in the History of the U.S. Civil Space Program*. 2 Vols. NASA SP-4407. Edited by John M. Logsdon, Dwayne A. Day, and Roger D. Launius. Washington, DC: Government Printing Office, 1996.

———. To the BOB director, Military Division of the BOB. Memorandum, 10 June 1958. Folder: Pre-NASA Documents, NACA-DOD Talks, box: Administrative History, Pre-NASA Documents, NASA/DOD, NASA Historical Data Reference Collection, NASA Headquarters, Washington, DC.

US Department of Defense (DOD). Chronology of Significant Events and Decisions Relating to the US Missile and Earth Satellite Development Programs, Supplement 1, October 1957–October 1958. Joint Chiefs of Staff. Historical Division, Joint Secretariat, 15 December 1958.

———. Agreement between the DOD, Army, Navy, Air Force, and NASA Concerning the Detailing of Military Personnel for Service with NASA, 13 April 1959. Space Policy Institute Archives document 1537.

———. Agreement between the DOD and NASA Concerning the Reimbursement of Costs, 12 November 1959. In *Exploring the Unknown: Selected Documents in the History of the U.S. Civil Space Program*. Vol. 2, *Relations with Other Organizations*, NASA SP-4407. Edited by John M. Logsdon, Dwayne A. Day, and Roger D. Launius. Washington, DC: Government Printing Office, 1996.

———. Army-Navy. General Proposal for Organization for Command and Control of Military Operations in Space, 1959. Box 5, OSAST, White House Office, Dwight D. Eisenhower Library, Abilene, KS.

———. Background Briefing on the Manned Orbiting Laboratory, 25 August 1965. NSA-MUS document 452.

———. Directive 5030.18, *Department of Defense Support of the National Aeronautics and Space Administration*, 24 February 1962. Folder: DOD Space Policy, DOD subseries, Federal Agencies series, NASA Historical Data Reference Collection, NASA Headquarters, Washington, DC.

————. Directive 5105.23, *National Reconnaissance Office*, 27 March 1964. In *Exploring the Unknown: Selected Documents in the History of the U.S. Civil Space Program.* Vol. 1, *Organizing for Exploration*. NASA SP-4407. Edited by John M. Logsdon, Linda J. Lear, Jannelle Warren-Findley, Ray A. Williamson, and Dwayne A. Day. Washington, DC: Government Printing Office, 1995.

————. Directive 5160.32, *Development of Space Systems*, 6 March 1961. Reprinted in *Exploring the Unknown: Selected Documents in the History of the U.S. Civil Space Program.* Vol. 2, *Relations with Other Organizations*, NASA SP-4407. Edited by John M. Logsdon, Dwayne A. Day, and Roger D. Launius. Washington, DC: Government Printing Office, 1996.

————. Directive S-5200.13, *Security Policy for Military Space Programs*, 23 March 1962. Folder: Defense 1962, box 17, RG 220, Records of the National Aeronautics and Space Council, National Archives and Records Administration.

————, manager for Manned Space Flight Support Operations. Overall Plan, Revised, Department of Defense Support for Project Gemini Operations, 21 October 1964. K243.04-34, Air Force Historical Research Agency, Maxwell AFB, AL.

————. News Release No. 491-69, 10 June 1969. Folder: USAF MOL, 1968, DOD subseries, Federal Agencies series, NASA Historical Data Reference Collection, NASA Headquarters, Washington, DC.

————. Office of the Director, Defense Research and Engineering. To the assistant secretary of the Air Force for R&D. Memorandum, 11 December 1963, attachment 1. Space Policy Institute Archives document 1655.

————. Press Release no. 1556-63, 10 December 1963. Folder: Dyna-Soar, DOD subseries, Federal Agencies series, NASA Historical Data Reference Collection, NASA Headquarters, Washington, DC.

————, director, Defense Research and Engineering. Report on DYNA SOAR, 20 February 1962. Folder: Reading file, January–May 1962, box 114, RG 200, Robert S. McNamara Papers, National Archives and Records Administration. Document is now declassified.

———. To Robert McNamara. Report, 20 February 1962. Folder: Reading file, January–May 1962, box 114, RG 200, Robert McNamara Papers, National Archives and Records Administration. Document is now declassified.

———. Space Program Budget for Fiscal Years 1968–1970, 31 December 1968. Tab 5-7, Box 13, RG 220, Records of the National Aeronautics and Space Council. Document is now declassified.

———. Space Program Data Sheet on Manned Orbiting Laboratory, 23 July 1964. IRIS 1002993, Air Force History Support Office.

US Department of State (DOS), director of Intelligence and Research. Note, 5 November 1963. Folder: Space Projects, Manned Lunar Landing, box 21, RG 220, Records of the National Aeronautics and Space Council, National Archives and Records Administration.

———. *Documents on Disarmament, 1963*. Washington, DC: Government Printing Office, 1964.

———. Internal memorandum, 14 September 1960. Folder: 12 Satellite and Missile Programs, box 6, RG 59, General Records of the Department of State, Bureau of European Affairs, Office of Soviet Union Affairs, Subject files 1957–1963, National Archives and Records Administration.

———. Memorandum of conference, 3 June 1961, Vienna, 12:45 a.m. In the document collection, National Security Archives, *The Berlin Crisis, 1958–1962*. Alexandria, VA: Chadwyck-Healey, Inc., 1991, document 2074.

———. Memorandum of conference, 4 June 1961, Vienna, 3:15 p.m. In the document collection, National Security Archives, *The Berlin Crisis, 1958–1962*. Alexandria, VA: Chadwyck-Healey, Inc., 1991, document 2077.

———. Memorandum of conference, 4 June 1961, Vienna. In the document collection, National Security Archives, *The Berlin Crisis, 1958–1962*. Alexandria, VA: Chadwyck-Healey, Inc., 1991, document 2079.

———. Paper on SAMOS Satellite, 18 July 1960. Folder: 12 Satellite and Missile Programs, box 6, RG 59, General Records of the Department of State, Bureau of European

Affairs, Office of Soviet Union Affairs, Subject files 1957–1963, National Archives and Records Administration.

———. Policy Planning Council, special assistant for Soviet Library of Congress Politico-Military Affairs. Planning Implications for National Security of Outer Space in the 1970s, 30 January 1964. Space Policy Institute Archives document 1538.

———. Report on Space Goals after the Lunar Landing, October 1966. Space Policy Institute Archives document 849.

———. Secretary of State Rusk. To McGeorge Bundy, president's special assistant for National Security Affairs. Memorandum of conference, 5 September 1961. Reprinted in *Foreign Relations of the United States, 1961–1963*. Vol. 2, *Arms Control and Disarmament*. Edited by David Mabon and David Patterson. Washington, DC: Government Printing Office, 1995.

———. To McGeorge Bundy. Report, 5 June 1962. Space Policy Institute Archives document 1539.

———. To American Embassy Moscow. Airgram, 22 March 1963. Folder: SP Space and Astronautics, USSR, box 4186, RG 59, General Records of the Department of State, Central Foreign Policy file, National Archives and Records Administration.

———. *United States Treaties and Other International Agreements*. Vol. 18, pt. 3. Washington, DC: Government Printing Office, 1967.

US House. Committee on Appropriations. *Department of Defense Appropriations for 1960*. Hearings. Part 6: "Research, Development, Test, and Evaluation." 86th Cong., 1st sess., 1959.

———. Committee on Appropriations, *Department of Defense Appropriations for 1965*. Hearings. Part 5. 88th Cong., 2d sess., 1964.

———. Committee on Appropriations, *Department of Defense Appropriations for 1966*. Hearings. Parts 3 (March) and 5 (April). 89th Cong., 1st sess., April 1965.

———. Committee on Appropriations, Subcommittee on Independent Offices, *Independent Offices Appropriations Bill, 1964*, 88th Cong., 1st sess., H. R. 8747. To accompany H. R. 8747, "Minority views of Rep. Louis C. Wyman," H. R. 824, 7 October 1963.

————. Committee on Appropriations, Subcommittee on Independent Offices, *Independent Offices Appropriations for 1966.* Hearings. Part 2. 89th Cong., 1st sess., April, 1965.

————. Committee on Appropriations, Subcommittee on Independent Offices, *Independent Offices Appropriations for 1967.* Hearings. Part 2. 89th Cong., 2d sess., 1966.

————. Committee on Armed Services. *DOD Authorizations for Fiscal Year 1967.* Hearings. 89th Cong., 2d sess., March 1966.

————. Committee on Government Operations. *Government Operations in Space (Analysis of Civil-Military Roles and Relationships).* 13th Report. 89th Cong, 1st sess., H. R. 445, 4 June 1965.

————. Committee on Naval Affairs. *National Advisory Committee for Aeronautics.* H. R. 1423. 63d Cong., 3d sess., 27 February 1915.

————. Committee on Science and Astronautics. *Amending the National Aeronautics and Space Act of 1958.* H. R. 1633. 86th Cong., 2d sess., March 1960.

————. Committee on Science and Astronautics. *Defense Space Interests.* Hearings. 87th Cong., 1st sess., March 1961.

————. Committee on Science and Astronautics. *Military Astronautics (Preliminary Report).* H. R. 360. 87th Cong., 1st sess., 5 May 1961.

————. Committee on Science and Astronautics. *Space, Missiles, and the Nation.* 86th Cong., 2d sess., 5 July 1960.

————. *Comparison of H. R. 12575 as Passed by the House and as Passed by the Senate,* 18 June 1958. Committee Print. 85th Cong., 2d sess., 1958.

————. *Establishment of the National Space Program.* 85th Cong., 2d sess., 24 May 1958. H. R. 1770.

————. *Legislative History of the Space Act of 1958 Establishing the World's First Civilian Space Agency: A Detailed Documentation of the Enactment by Congress of a New Law.* H. R. Rep. 86th Cong., 2d sess., August 1960.

————. *Message from the President of the United States to the Congress Transmitting Second Annual Report on U. S. Aeronautics and Space Activities.* House Document No. 349, 86th Cong., 2d sess., 24 February 1960.

————. *Military Posture.* Hearings. 88th Cong., 1st sess., 1963.

————. *Missile and Space Ground Support Operations.* H. R. 1340. 89th Cong., 2d sess., 21 March 1966.

————. *NASA Authorization, 1969.* Hearings. Part 1. 90th Cong., 2d sess., February 1968.

————. *National Aeronautics and Space Act of 1958, Conference Report.* H. R. 2166. 85th Cong., 2d sess., 15 July 1958.

————. Public Law 85-825, Advanced Research Projects Agency, 12 February 1958. Space Policy Institute Archives document 1541.

————. Select Committee on Astronautics and Space Exploration. *Astronautics and Space Exploration.* Hearings, 85th Cong., 2d sess., 1958.

————. Select Committee on Astronautics and Space Exploration. *Establishment of the National Space Program.* H. R. 1770. 85th Cong., 2d sess., 24 May 1958.

————. Select Committee on Astronautics and Space Exploration. *The National Space Program.* 85th Cong., 2d sess., 21 May 1958. H. R. 1758.

————. Select Committee on Astronautics and Space Exploration. Statement of Brig Gen Homer A. Boushey, USAF. 85th Cong. 2d sess. H. R. 11881, 23 April 1958, 522.

————. Subcommittee on NASA Oversight. *The NASA-DOD Relationship.* 88th Cong., 2d sess., 1964.

————. *Transfer of the Development Operations Division of the Army Ballistic Missile Agency to National Aeronautics and Space Administration.* Hearings. 86th Cong., 2d sess., February 1960.

US Navy, Bureau of Aeronautics. "Investigation of the Possibility of Establishing a Space Ship in an Orbit Above the Surface of the Earth," ADR Report R-48, November 1945. Folder: Navy/BuAer: Earth Satellite Vehicle, DOD subseries, Federal Agencies series, NASA Historical Data Reference Collection, NASA Headquarters, Washington, DC.

US Senate. Committee on Aeronautical and Space Sciences. *Khrushchev's Statement to American Businessmen.* Staff report, 7 November 1963. Folder: Senate Committee on Space and Astronautics, 5 of 5, box 908, Clinton P. Anderson Papers, Library of Congress.

———. Committee on Aeronautical and Space Sciences. *Manned Orbiting Laboratory*. Hearings. 89th Cong., 2d sess., 24 February 1966.

———. Committee on Aeronautical and Space Sciences. *Soviet Space Programs: 1962–1965*. Report prepared by the Legislative Reference Service. 89th Cong., 2d sess., December 1966.

———. Committee on Appropriations. *Independent Offices Appropriations, Fiscal Year 1969*. Hearings. 90th Cong., 2d sess., March 1968.

———. Committee on Armed Services. Preparedness Investigating Subcommittee. *Inquiry into Satellite and Missile Programs*. 85th Cong., 1st and 2d sess., 1958.

———. *Investigation of Governmental Organization for Space Activities*. Hearings. 86th Cong., 1st sess., 1959.

———. *NASA Authorization for Fiscal Year 1963*. Hearings. 87th Cong., 2d sess., June 1962.

———. *National Aeronautics and Space Act of 1958*. S. Rep.1701. 85th Cong., 2d sess., 11 June 1958.

———. Special Committee on Space and Astronautics. *A Bill Establishing a National Aeronautics and Space Agency, with a Section-by-Section Analysis and Staff Explanation and Comments*. 85th Cong., 2d sess., 1958, Confidential Committee Print. S 3609.

———. Special Committee on Space and Astronautics. *Final Report*. S. Rep. 100. 86th Cong., 1st sess., 11 March 1959.

———. Special Committee on Space and Astronautics. *National Aeronautics and Space Act*. Hearings. 85th Cong., 2d sess., 1958.

———. Subcommittee on Government Organization for Space Activities. *Government Organization for Space Activities*. S. Rep. 806. 86th Cong., 1st sess., 25 August 1959.

———. Subcommittee on NASA Authorizations. *Transfer of the Von Braun Team to NASA*, H. J. Res. 567. 86th Cong., 2d sess., 18 February 1960.

Vance, Cyrus R., deputy SECDEF. To the president. Memorandum, 12 February 1966. NSA-MUS document 454.

Vandenberg, Hoyt S., vice-CSAF. Policy statement, 12 January 1948. Reprinted in *Exploring the Unknown: Selected Docu-*

ments in the History of the U.S. Civil Space Program. Vol. 2,
Relations with Other Organizations, NASA SP-4407. Edited
by John M. Logsdon, Dwayne A. Day, and Roger D. Launius.
Washington, DC: Government Printing Office, 1996.

Van Dyke, Vernon. Pride and Power: The Rationale of the Space
Program. Urbana, IL: University of Illinois Press, 1964.

Van Nimmen, Jane, Leonard C. Bruno, and Robert Rosholt.
NASA Historical Data Book. Vol. 1, NASA Resources 1958–
1968. NASA SP-4012. Washington, DC: Government Printing Office, 1988.

"Washington Roundup." Aviation Week and Space Technology,
4 November 1968. In Erlend A. Kennan and Edmund H.
Harvey, Mission to the Moon: A Critical Examination of NASA
and the Space Program. New York: William Morrow & Company, 1969, 228.

"Washington Roundup, Space Lab Merger." Aviation Week and
Space Technology, 30 November 1964, 17.

Waterman, Alan. To Donald Quarles. Confidential letter from
Waterman, 13 May 1955. Space Policy Institute Archives
unnumbered document.

———. To Robert Murphy. Confidential Memorandum, 18
March 1955. Folder: OCB 000.91 Natural & Physical Sci-
ences (2), box: 11, OCB Central files subseries, NSC: Staff
Papers series, White House Office, Dwight D. Eisenhower
Library, Abilene, KS.

Watson, George M., Jr. Office of Air Force History. Series of Oral
History Interviews of Eugene Zuckert, 3–5, 9 December
1986. K239.0512-1763, Air Force Historical Research
Agency, Maxwell AFB, AL.

———. The Office of the Secretary of the Air Force, 1947–1965.
Washington, DC: Center for Air Force History, 1993.

Webb, James E. Agreement Covering a Possible New Manned
Earth Orbital Research and Development Project, with attachment, 17 August 1963 for Webb's signature, 14 September 1963 for McNamara's signature. Reprinted in Exploring the Unknown: Selected Documents in the History of
the U.S. Civil Space Program. Vol. 2, Relations with Other
Organizations, NASA SP-4407. Edited by John M. Logs-

don, Dwayne A. Day, and Roger D. Launius. Washington, DC: Government Printing Office, 1996.

———. Memorandum for record, 24 February 1961. Space Policy Institute Archives document 984.

———. Memorandum of understanding, 14 January 1966. IRIS 1003002, Air Force History Support Office.

———. "NASA and the USAF: A Space Age Partnership." *The Airman*, August 1964, 6–11. National Aeronautics and Space Council, NASA Historical Data Reference Collection, NASA Headquarters, Washington, DC.

———. Press Conference. Transcript, 16 September 1968. Folder: Webb, Press Conferences, Webb subseries, Administrators series, NASA Historical Data Reference Collection, NASA Headquarters, Washington, DC.

———. Report, 30 January 1967. Folder: Administrator's Reference and Backup Book, box 133, James Webb Papers, Harry S. Truman Library, Independence, MO.

———. Speech, 11–12 February 1963. NASA-Industry Program Plans Conference. Extracted from *The Military Mission in Space: A Selection of Published Views, August 1962–June 1963*. Folder: Military Mission in Space, DOD subseries, Federal Agencies series, NASA Historical Data Reference Collection, NASA Headquarters, Washington, DC.

———. Speech, 8 April 1965. US Naval Academy. Folder: Speech file I, 23 February–8 April 1965, box 220, James Webb Papers, Harry S. Truman Library, Independence, MO.

———. Statement to the Senate Appropriations Committee, Subcommittee on Independent Offices, 17 May 1968. Folder: Subcommittee on Independent Offices, May 1968, box 155, James Webb Papers.

———. Statement to the Senate Committee on Aeronautical and Space Sciences, 4 March 1964. Folder: Senate Committee on Aeronautical and Space Sciences, box: 153, James Webb Papers, Harry S. Truman Library, Independence, MO.

———. Testimony to the House Appropriations Committee, 5 February 1968. Excerpted in NASC document, Selected Congressional Testimony, 1968. Folder: Space Projects, MOL, box 21, RG 220, Records of the National Aeronautics and Space Council, National Archives and Records Administration.

———. To David Bell. Letter, 17 March 1961. Folder: Apollo 1961 Decision Documentation, box: White House, Presidents, Kennedy, Correspondence, Apollo Decision Documentation, NASA Historical Data Reference Collection, NASA Headquarters, Washington, DC.

———. To Dr. Arthur E. Raymond, NASA Consultant. Letter, 13 February 1963. Folder: Webb Correspondence, January–June 1963, Webb subseries, Administrators series, NASA Historical Data Reference Collection, NASA Headquarters, Washington, DC.

———. To Edward C. Welsh. Letter, 7 May 1963. Folder: NASC 1962–1972, box: White House.

———. To Eugene Zuckert. Letter, 28 May 1962. Folder: Webb, declassified papers, 1961–1968, Webb subseries, Administrators series, NASA Historical Data Reference Collection, NASA Headquarters, Washington, DC.

———. To Eugene Zuckert. Letter, 11 December 1963. Folder: 1964, Manned Lunar Landing Program, box 43, Samuel Phillips Papers, Library of Congress, 1–3.

———. To Gen Maxwell Taylor. Letter, 16 January 1964. Folder: DOD Joint Chiefs of Staff, DOD subseries, Federal Agencies series, NASA Historical Data Reference Collection, NASA Headquarters, Washington, DC.

———. To Hugh Dryden and Robert Seamans. Letter, 4 May 1962. Folder: Webb, Correspondence, Jan–Jun 1962, Webb subseries, Administrators series, NASA Historical Data Reference Collection, NASA Headquarters, Washington, DC.

———. To John F. Kennedy. Report, 30 November 1962. In *Exploring the Unknown: Selected Documents in the History of the U.S. Civil Space Program*. Vol. 1, *Organizing for Exploration*. NASA SP-4407. Edited by John M. Logsdon, Linda J. Lear, Jannelle Warren-Findley, Ray A. Williamson, and Dwayne A. Day. Washington, DC: Government Printing Office, 1995.

———. To John F. McConnell, CSAF. Letter, 29 December 1966. Folder: DOD/USAF/NASA, Vietnam cooperation, DOD subseries, Federal Agencies series, NASA Historical

Data Reference Collection, NASA Headquarters, Washington, DC.

———. To Lyndon Johnson. Letter, 4 May 1961. Folder: NASC 1960–1961, box: White House, National Aeronautics & Space Council, NASA Historical Data Reference Collection, NASA Headquarters, Washington, DC.

———. To Lyndon Johnson, Letter, 3 May 1963. Folder: NASC meeting 7 May 1963, box 3, RG 22, National Aeronautics and Space Council, National Archives and Records Administration.

———. To Lyndon Johnson. Letter, 28 January 1964. Tab 4, box 23, RG 220, Records of the National Aeronautics and Space Council, National Archives and Records Administration. Document is now declassified.

———. To Lyndon Johnson. Letter, 16 May 1966. Folder: Johnson Correspondence, NASA, box: White House, Presidents, Johnson, Correspondence, Declassified items, NASA Historical Data Reference Collection, NASA Headquarters, Washington, DC.

———. To Lyndon Johnson. Letter, 26 August 1966, Space Policy Institute Archives document 860.

———. To Lyndon Johnson. Memorandum, 18 December 1964. NSA PD document 1045.

———. To Robert McNamara. Letter, 7 July 1961. Folder: Webb correspondence, 1961, Webb subseries, Administrators series, NASA Historical Data Reference Collection, NASA Headquarters, Washington, DC.

———. To Robert McNamara. Letter, 16 January 1963. Reprinted in *Exploring the Unknown: Selected Documents in the History of the U.S. Civil Space Program*. Vol. 2, *Relations with Other Organizations*, NASA SP-4407. Edited by John M. Logsdon, Dwayne A. Day, and Roger D. Launius. Washington, DC: Government Printing Office, 1996.

———. To Robert McNamara. Letter, 24 April 1963. Space Policy Institute Archives document 1457.

———. To Robert McNamara. Letter, 14 October 1964. IRIS 1002995, Air Force History Support Office.

———. To Robert Seamans. Letter, 18 January 1963. Reprinted in *Exploring the Unknown: Selected Documents in the History of the U.S. Civil Space Program.* Vol. 2, *Relations with Other Organizations*, NASA SP-4407. Edited by John M. Logsdon, Dwayne A. Day, and Roger D. Launius. Washington, DC: Government Printing Office, 1996.

———. To Robert C. Seamans. Memorandum, 13 December 1963. Folder: December 1963, box 35, James Webb Papers, Harry S. Truman Library, Independence, MO.

———. To Senator Everett Dirksen. Letter, 9 August 1966. In *Exploring the Unknown: Selected Documents in the History of the U.S. Civil Space Program.* Vol. 1, *Organizing for Exploration.* NASA SP-4407. Edited by John M. Logsdon, Linda J. Lear, Jannelle Warren-Findley, Ray A. Williamson, and Dwayne A. Day. Washington, DC: Government Printing Office, 1995.

———. To the vice president. Memorandum, 10 May 1963. Folder: Eisenhower, Defense Policy, box: White House, Presidents, Eisenhower, Space Documentation, NASA Historical Data Reference Collection, NASA Headquarters, Washington, DC.

———. To the vice president. Memorandum, 30 July 1963. Folder: Johnson, Declassified Space Correspondence, box: White House, Presidents, Johnson, Correspondence, NASA Historical Data Reference Collection, NASA Headquarters, Washington, DC.

———. To the vice president. Memorandum, 9 August 1963. Folder: Johnson Correspondence, NASA, box: White House, Presidents, Johnson, Correspondence, Declassified Items, NASA Historical Data Reference Collection, NASA Headquarters, Washington, DC.

———. To W. Fred Boone. Letter, 13 February 1963. Folder: X-20 Correspondence, DOD subseries, Federal Agencies series, NASA Historical Data Reference Collection, NASA Headquarters, Washington, DC.

———. Transcript of oral history interview, 11 April 1974. File: James Webb, box: Emme/Roland interviews on early NASA history, shelf: V-A-1, NASA Historical Data Reference Collection, NASA Headquarters, Washington, DC.

———. Transcript of oral history interviews, 15 March 1985. National Air and Space Museum, Washington, DC.

———. Transcript of oral history interviews, 15 October 1985. National Air and Space Museum, Washington, DC.

———. "Why Spend $20 Billion to Go to the Moon?" *U.S. News and World Report*, 3 July 1961.

———, and Robert McNamara. To Vice Pres. Lyndon B. Johnson. Memorandum, 8 May 1961. Space Policy Institute Archives document 300.

———, and Eugene Zuckert. Memorandum of understanding, 7 August 1962. Folder: Dyna-Soar, DOD subseries, Federal Agencies series, NASA Historical Data Reference Collection, NASA Headquarters, Washington, DC.

———, and Robert McNamara. Agreement between the National Aeronautics and Space Administration and the Department of Defense Concerning the Gemini Program, 21 January 1963. Reprinted in *Exploring the Unknown: Selected Documents in the History of the U.S. Civil Space Program.* Vol. 2, *Relations with Other Organizations*, NASA SP-4407. Edited by John M. Logsdon, Dwayne A. Day, and Roger D. Launius. Washington, DC: Government Printing Office, 1996.

———, and Roswell Gilpatric. Joint Memorandum, 14 February 1961. Space Policy Institute Archives document 26.

Weitzen, William, deputy for operations, research, and development. To Richard Horner, assistant secretary of the Air Force for R&D. Memorandum, 18 December 1957. K168. 8636-9, Air Force Historical Research Agency, Maxwell AFB, AL. Document is now declassified.

Welsh, Edward C., NASC executive secretary. "Khrushchev, Address to Third World Meeting of Journalists." Report Attachment, 25 October 1963. Folder: Space Projects, Manned Lunar Landing, box 21, RG 220, Records of the National Aeronautics and Space Council, National Archives and Records Administration.

———. Oral history interview, 20 February 1969. Folder: LBJ Speeches, Press Conferences (1968), box: White House, Presidents, Johnson, Chronological, Press Conferences, NASA Historical Data Reference Collection, NASA Headquarters, Washington, DC.

———. Testimony to the House Appropriations Committee. 5 February 1968. Excerpted in NASC document, Selected

Congressional Testimony, 1968, folder: Space Projects, MOL, box 21, RG 220, Records of the National Aeronautics and Space Council, National Archives and Records Administration.

———. To the vice president. Memorandum, 19 January 1963. Folder: Defense 1963, box 17, RG 220, Records of the National Aeronautics and Space Council, National Archives and Records Administration.

———. To the vice president. Memorandum, 4 March 1968. Folder: National Aeronautics and Space Administration, 1967 & 1968, box 23, RG 220, Records of the National Aeronautics and Space Council, National Archives and Records Administration.

———. To the American Legion. Message, 6 October 1962. Folder: McNamara-Webb Report/Logsdon Interviews, box: White House, Presidents, Kennedy, Correspondence, Apollo Decision Documentation, NASA Historical Data Reference Collection, NASA Headquarters, Washington, DC.

Wheelon, Albert D. "Lifting the Veil on CORONA." *Space Policy* 11 (November 1995): 249–60.

Whisenand, Maj Gen James, assistant deputy CSAF for research and technology. To CSAF LeMay. Memorandum, 30 January 1963. Folder: 208, box B208, Curtis LeMay Papers, Library of Congress, 1. Document is now declassified.

White House/BOB. Memorandum, 8 April 1958. Box: White House, Presidents, Eisenhower, National Aeronautics and Space Act, W/H and BOB Space Act documents from National Archives (Record Group 51), NASA Historical Data Reference Collection, NASA Headquarters, Washington, DC.

———, Office of the Press Secretary. Statement of the President, 25 August 1965. NSA-MUS document 453.

White, Thomas, CSAF. To Gens Truman H. Landon and Wilson, Air Force deputy chiefs of staff for Personnel and Development, with copies to Bernard Schriever and Curtis LeMay, among others. Letter, 14 April 1960. Folder: Civilian vs. Military Role in Space, DOD subseries, Federal Agencies series, NASA Historical Data Reference Collection, NASA Headquarters, Washington, DC.

———. To James E. Webb. Letter, 31 January 1961. Folder: Webb nomination, Webb subseries, Administrators series, NASA Historical Data Reference Collection, NASA Headquarters, Washington, DC.

———. To Joseph Charyk. Memorandum, 27 October 1959. Folder: Chief of Staff Signed Memos, box 26, Thomas White Papers, Library of Congress.

———. To Overton Brooks. Letter, 19 January 1961. IRIS 1002992, Air Force History Support Office.

———. To T. Keith Glennan. Letter, 21 March 1961. Reprinted in *Briefing Book for Air Force witnesses before the House Committee on Science and Astronautics on the Subject of DOD Space Directive 5160.32.* K160.8636-4, Air Force History Support Office, 1961.

———. "Space Control and National Security." *Air Force Magazine, Space Weapons: A Handbook of Military Astronautics.* New York: Frederick A. Praeger, 1959.

———, and Hugh Dryden. Memorandum of understanding, 20 May 1958. Reprinted in *Exploring the Unknown: Selected Documents in the History of the U.S. Civil Space Program.* Vol. 2, *Relations with Other Organizations*, NASA SP-4407. Edited by John M. Logsdon, Dwayne A. Day, and Roger D. Launius. Washington, DC: Government Printing Office, 1996.

———, and T. Keith Glennan. Memorandum of understanding, 14 November 1958. Folder: X-20 Correspondence, DOD subseries, Federal Agencies series, NASA Historical Data Reference Collection, NASA Headquarters, Washington, DC.

Whitman, Ann. Diary entry, 22 November 1957. Folder: November '57 A.C.W. DIARY (1), box 9, Administration series, Ann Whitman file, Dwight D. Eisenhower Library, Abilene, KS.

Wiesner, Jerome, Kenneth BeLieu, Trevor Gardner, Donald F. Hornig, Edwin H. Land, Maxwell Lehrer, Edward M. Purcell, Bruno B. Rossi, and Harry J. Watters. Oral history interview, 24 July 1974. Wiesner file, Biographical series, NASA Historical Data Reference Collection, NASA Headquarters, Washington, DC.

———. To the President-Elect of the Ad Hoc Committee on Space. Report, 10 January 1961. Space Policy Institute

Archives document 1238. In addition, Section 5 of the Wiesner Report is the military space section. Folder: Space, Man-in-Space, 1962, box 167, RG 359, Office of Science and Technology, National Archives and Records Administration. Document is now declassified.

Wilford, John Noble. "A Spacefaring People: Keynote Address." In *A Spacefaring People: Perspectives on Early Spaceflight.* Edited by Alex Roland. NASA SP-4405. Washington, DC: Government Printing Office, 1985.

Wilson, Charles. To the secretaries of the Army, Navy, and Air Force. Memorandum, 28 March 1955. K140.11-11, Air Force Historical Research Agency, Maxwell AFB, AL.

Wilson, George C. "Defense Denies Bid for NASA Programs." *Aviation Week and Space Technology,* 25 June 1962, 34–35.

———. "President Will Outline U.S. Space Goals." *Aviation Week and Space Technology,* 11 January 1965, 12.

Wilson, Glen P. "Lyndon Johnson and the Legislative Origins of NASA." *Prologue: Quarterly Journal of the National Archives* 25 (Winter 1993): 362–71.

Wilson, Lt Gen Roscoe C., deputy CSAF for Development. Internal memorandum, dated only March 1961. Response to Mr. Brooks. IRIS 1002992, Air Force History Support Office.

———. To Gen Thomas White. Letter, 7 April 1960. Folder: 7-4 FAA/NASA/JCS/CIA/CAP, box 39, Thomas D. White Papers, Library of Congress.

Witkin, Richard. *The Challenge of the Sputniks.* New York: Doubleday, 1958.

Witze, Claude. "How Our Space Policy Evolved." *Air Force/Space Digest,* April 1962, 83–92.

Wolf, Richard I. *The United States Air Force Basic Documents on Roles and Missions.* Washington, DC: Office of Air Force History, 1987.

Wolfe, Tom. *The Right Stuff.* New York: Sloan, Duell, and Pearce, 1979.

Wood, Clotaire, assistant to the director for research management, NACA. Transmittal of copies of proposed memorandum of understanding between Air Force and NACA for joint NACA-Air Force project for a recoverable manned satellite test vehicle, 11 April 1958. Space Policy Institute Archives document 117.

———. Tabling of Proposed memorandum of understanding between Air Force and NACA, dated 20 May 1958. Project for a Recoverable Manned Satellite Test Vehicle, Memorandum for files, 11 April 1958. Space Policy Institute Archives document 288.

———. To T. Keith Glennan, Memorandum, 29 October 1958. Folder: Air Force Space, AFCHO, DOD subseries, Federal Agencies series, NASA Historical Data Reference Collection, NASA Headquarters, Washington, DC.

Worthman, Col Paul E., USAF. "The Promise of Space." *Air University Review* 20 (January–February 1969): 120–27.

Yarymovych, Michael. Oral history interview, 2 February 1976. Michael Yarymovych file, Biographical series, NASA Historical Data Reference Collection, NASA Headquarters, Washington, DC.

York, Herbert F., and G. Allen Greb. "Strategic Reconnaissance." *Bulletin of the Atomic Scientists,* April 1977.

York, Herbert F. Oral history interview, 16 June 1964. Folder: Kennedy Library, box: White House, Presidents, Kennedy, Photographs—Presidential Library, NASA Historical Data Reference Collection, NASA Headquarters, Washington, DC.

———. Transcript of oral history interview, 12 June 1973. Washington, DC: NASA Headquarters, NASA Historical Data Reference Collection, NASA Headquarters, Washington, DC.

———. Transcript of oral history interview, 24 January 1989. Washington, DC: National Air and Space Museum.

———. *Race to Oblivion: A Participant's View of the Arms Race.* New York: Simon and Schuster, 1970.

Young, R. P., NASA executive officer. To Lacklen, director of personnel. Letter, 10 September 1963. Folder: Military Personnel Detailed to NASA, DOD subseries, Federal Agencies series, NASA Historical Data Reference Collection, NASA Headquarters, Washington, DC.

Zuckert, Eugene. Oral history interview, 25 July 1964. *The John F. Kennedy Presidential Oral History Collection, Part I: The White House and Executive Departments.* Microfilmed from the holdings of the John F. Kennedy Library (Frederick, MD: University Publications of America, 1988).

———. "The Secretary of the Air Force Speaks on Space Programs." *Air Force Information Policy Letter for Commanders* 1, no. 2, 15 January 1962. Folder: USAF Space, DOD subseries, Federal Agencies series, NASA Historical Data Reference Collection, NASA Headquarters, Washington, DC.

———. Speech, 6 March 1962. In *Air Force Information Policy Letter for Commanders*, Vol. 16, 15 March 1962. Folder: USAF Space, DOD subseries, Federal Agencies series, NASA Historical Data Reference Collection, NASA Headquarters, Washington, DC.

———. Statement in the *GE Forum*, 10 January 1962. Reprinted in the collection from the Air Force Office of Information. *Policy Statements on Military Space*, 19 September 1962. 168.7171-65, Air Force Historical Research Agency, Maxwell AFB, AL.

———. To Webb. Letter, 25 August 1961. IRIS 1003003, Air Force History Support Office.

———. Zuckert. To McNamara. Memorandum, 26 February 1963. Folder: 6-1963, Air Force, box B129, Curtis LeMay Papers, Library of Congress.

Index